R

IN A NUTSHELL

R

IN A NUTSHELL

Joseph Adler

O'REILLY®

Beijing · Cambridge · Farnham · Köln · Sebastopol · Tokyo

R in a Nutshell
by Joseph Adler

Published by O'Reilly Media, Inc., 1005 Gravenstein Highway North, Sebastopol, CA 95472.

O'Reilly books may be purchased for educational, business, or sales promotional use. Online editions are also available for most titles (*http://my.safaribooksonline.com*). For more information, contact our corporate/institutional sales department: (800) 998-9938 or *corporate@oreilly.com*.

Editor: Mike Loukides		**Cover Designer:** Karen Montgomery	
Production Editor: Sumita Mukherji		**Interior Designer:** David Futato	
Production Services: Newgen North America, Inc.		**Illustrator:** Robert Romano	

Printing History:

December 2009: First Edition.

 This book uses RepKover™, a durable and flexible lay-flat binding.

ISBN: 978-0-596-80170-0

[M] [12/10]

1290128814

Table of Contents

Part II.   The R Language

Part III. Working with Data

Preface

It's been 10 years since I was first introduced to R. Back then, I was a young product development manager at DoubleClick, a company that sells advertising software for managing online ad sales. I was working on inventory prediction: estimating the number of ad impressions that could be sold for a given search term, web page, or demographic characteristic. I wanted to play with the data myself, but we couldn't afford a piece of expensive software like SAS or MATLAB. I looked around for a little while, trying to find an open source statistics package, and stumbled on R. Back then, R was a bit rough around the edges, and was missing a lot of the features it has today (like fancy graphics and statistics functions). But R was intuitive and easy to use; I was hooked. Since that time, I've used R to do many different things: estimate credit risk, analyze baseball statistics, and look for Internet security threats. I've learned a lot about data, and matured a lot as a data analyst.

R, too, has matured a great deal over the past 10 years. R is used at the world's largest technology companies (including Google, Microsoft, and Facebook), the largest pharmaceutical companies (including Johnson & Johnson, Merck, and Pfizer), and at hundreds of other companies. It's used in statistics classes at universities around the world and by statistics researchers to try new techniques and algorithms.

Why I Wrote This Book

This book is designed to be a concise guide to R. It's not intended to be a book about statistics or an exhaustive guide to R. In this book, I tried to show all the things that R can do and to give examples showing how to do them. This book is designed to be a good desktop reference.

I wrote this book because I like R. R is fun and intuitive in ways that other solutions are not. You can do things in a few lines of R that could take hours of struggling in a spreadsheet. Similarly, you can do things in a few lines of R that could take pages of Java code (and hours of Java coding). There are some excellent books on R, but

I couldn't find an inexpensive book that gave an overview of everything you could do in R. I hope this book helps you use R.

When Should You Use R?

I think R is a great piece of software, but it isn't the right tool for every problem. Clearly, it would be ridiculous to write a video game in R, but it's not even the best tool for all data problems.

R is very good at plotting graphics, analyzing data, and fitting statistical models using data that fits in the computer's memory. It's not as good at storing data in complicated structures, efficiently querying data, or working with data that doesn't fit in the computer's memory.

Typically, I use a tool like Perl to preprocess large files before using them in R. It's technically possible to use R for these problems (by reading files one line at a time and using R's regular expression support), but it's pretty awkward. To hold large data files, I usually use a database like MySQL, PostgreSQL, SQLite, or Oracle (when someone else is paying the license fee).

R License Terms

R is an open source software package, licensed under the GNU General Public License (GPL).* This means that you can install R for free on most desktop and server machines. (Comparable commercial software packages sell for hundreds or thousands of dollars.) If R were a poor substitute for the commercial software packages, this might have limited appeal. However, I think R is *better* than its commercial counterparts in many respects.

Capability
> You can find implementations for hundreds (maybe thousands) of statistical and data analysis algorithms in R. No commercial package offers anywhere near the scope of functionality available through the Comprehensive R Archive Network (CRAN).

Community
> There are now hundreds of thousands (if not millions) of R users worldwide. By using R, you can be sure that you're using the same software that your colleagues are using.

---

* There is some controversy about GPL licensed software, and what it means to you as a corporate user. Some users are afraid that any code that they write in R will be bound by the GPL. If you are not writing extensions to R, you do not need to worry about this issue. R is an interpreter, and the GPL does not apply to a program just because it is executed on a GPL licensed interpreter.

If you are writing extensions to R, they might be bound by the GPL. For more information, see the GNU foundation's FAQ on the GPL: *http://www.gnu.org/licenses/gplfaq*. However, for a definite answer, see an attorney. If you are worried about a specific application, see an attorney.

Performance

R's performance is comparable, or superior, to most commercial analysis packages. R requires you to load data sets into memory before processing. If you have enough memory to hold the data, R can run very quickly. Luckily, memory is cheap. You can buy 32 GB of server RAM for less than the cost of a single desktop license of a comparable piece of commercial statistical software.

Examples

I have tried to provide many unique examples in this book, illustrating how to use different functions in R. I deliberately decided to use new and original examples, and not to rely on the data sets included with R. When I'm trying to solve a problem, I try to find examples of similar solutions. There are already good examples for many functions in the R help files. I tried to provide new examples to help users figure out how to solve their problems quickly. The examples are available by from O'Reilly Media at *http://oreilly.com/catalog/9780596801700*.

Additionally, the example data is also available through CRAN as an R package. To install the `nutshell` package, type the following command on the R console:

```
install.packages("nutshell")
```

How This Book Is Organized

I've broken this book into five parts:

- Part I, *R Basics*, covers the basics of getting and running R. It's designed to help get you up and running if you're a new user, including a short tour of the many things you can do with R.
- Part II, *The R Language*, discusses the R language in detail. This section picks up where the first section leaves off, describing the R language in detail.
- Part III, *Working with Data*, covers data processing in R: loading data into R, transforming data, summarizing data, and plotting data. Summary statistics and charts are an important part of statistical analysis, but many laypeople don't think of these things as statistical analysis. So, I cover these topics without using too much math in order to keep them accessible.
- Part IV, *Statistics with R*, covers statistical tests and models in R.
- Finally, I included an Appendix describing functions and data sets included with the base distribution of R.

If you are new to R, install R and start with Chapter 3. Next, take a look at Chapter 5 to learn some of the rules of the R language. If you plan to use R for plotting, statistical tests, or statistical models, take a look at the appropriate chapter. Make sure you look at the first few sections of the chapter, because these provide an overview of how all the related functions work. (For example, don't skip straight to "Random forests for regression" on page 416 without reading "Example: A Simple Linear Model" on page 373.)

Conventions Used in This Book

The following typographical conventions are used in this book:

Italic
> Indicates new terms, URLs, email addresses, filenames, and file extensions.

`Constant width`
> Used for program listings, as well as within paragraphs to refer to program elements such as variable or function names, databases, data types, environment variables, statements, and keywords.

`Constant width bold`
> Shows commands or other text that should be typed literally by the user.

`Constant width italic`
> Shows text that should be replaced with user-supplied values or by values determined by context.

 This icon indicates a warning or caution.

Using Code Examples

This book is here to help you get your job done. In general, you may use the code in this book in your programs and documentation. You do not need to contact us for permission unless you're reproducing a significant portion of the code. For example, writing a program that uses several chunks of code from this book does not require permission. Selling or distributing a CD-ROM of examples from O'Reilly books does require permission. Answering a question by citing this book and quoting example code does not require permission. Incorporating a significant amount of example code from this book into your product's documentation does require permission.

We appreciate, but do not require, attribution. An attribution usually includes the title, author, publisher, and ISBN. For example: "*R in a Nutshell* by Joseph Adler. Copyright 2010 O'Reilly Media, Inc., 978-0-596-80170-0."

If you feel your use of code examples falls outside fair use or the permission given above, feel free to contact us at *permissions@oreilly.com*.

How to Contact Us

Please address comments and questions concerning this book to the publisher:

> O'Reilly Media, Inc.
> 1005 Gravenstein Highway North
> Sebastopol, CA 95472
> 800-998-9938 (in the United States or Canada)

707-829-0515 (international or local)
707 829-0104 (fax)

We have a web page for this book, where we list errata, examples, and any additional information. You can access this page at:

http://oreilly.com/catalog/9780596801700

To comment or ask technical questions about this book, send email to:

bookquestions@oreilly.com

For more information about our books, conferences, Resource Centers, and the O'Reilly Network, see our website at:

http://www.oreilly.com

Safari® Books Online

 Safari® Books Online is an on-demand digital library that lets you easily search over 7,500 technology and creative reference books and videos to find the answers you need quickly.

With a subscription, you can read any page and watch any video from our library online. Read books on your cell phone and mobile devices. Access new titles before they are available for print, and get exclusive access to manuscripts in development and post feedback for the authors. Copy and paste code samples, organize your favorites, download chapters, bookmark key sections, create notes, print out pages, and benefit from tons of other time-saving features.

O'Reilly Media has uploaded this book to the Safari® Books Online service. To have full digital access to this book and others on similar topics from O'Reilly and other publishers, sign up for free at *http://my.safaribooksonline.com*.

Acknowledgments

Many people helped support the writing of this book. First, I'd like to thank all of my technical reviewers. These folks check to make sure the examples work, look for technical and mathematical errors, and make many suggestions on writing quality. It's not possible to write a quality technical book without quality technical reviewers: Peter Goldstein, Aaron Mandel, and David Hoaglin are the reason that this book reads as well as it does.

I'd like to thank Randall Munroe, author of the xkcd comic. He kindly allowed us to reprint two of his (excellent) comics in this book. You can find his comics (and assorted merchandise) at *http://www.xkcd.com*.

Additionally, I'd like to thank everyone who provided or suggested example data. Aaron Schatz of Football Outsiders (*http://www.footballoutsiders.com*) provided me with play-by-play data from the 2005 NFL season (the field goal data is from its database). Sandor Szalma of Johnson & Johnson suggested GSE2034 as an example of gene expression data.

Finally, I'd like to thank my wife, Sarah, and my daughter, Zoe. Writing a book takes a lot of time, and they were very understanding when I needed to work. They were also very understanding when I dragged them to the San Diego Zoo to look at the harpy eagles.

R Basics

This part of the book covers the basics of R: how to get R, how to install it, and how to use packages in R. It also includes a quick tutorial on R and an overview of the features of R.

1

Getting and Installing R

This chapter explains how to get R and how to install it on your computer.

R Versions

Today, R is maintained by a team of developers around the world. Usually, there is an official release of R twice a year, in April and in October. I used version 2.9.2 in this book. (Actually, it was 2.8.1 when I started writing the book and was updated three times while I was writing. I installed the updates, but they didn't change very much content.)

R hasn't changed that much in the past few years: usually there are some bug fixes, some optimizations, and a few new functions in each release. There have been some changes to the language, but most of these are related to somewhat obscure features that won't affect most users. (For example, the type of NA values in incompletely initialized arrays was changed in R 2.5.) Don't worry about using the exact version of R that I used in this book; any results you get should be very similar to the results shown in this book. If there are any changes to R that affect the examples in this book, I'll try to add them to the official errata online.

Additionally, I've given some example filenames below for the current release. The filenames usually have the release number in them. So, don't worry if you're reading this book and don't see a link for *R-2.9.1-win32.exe*, but see a link for *R-3.0.1-win32.exe* instead; just use the latest version, and you'll be fine.

Getting and Installing Interactive R Binaries

R has been ported to every major desktop computing platform. Because R is open source, developers have ported R to many different platforms. Additionally, R is available with no license fee.

If you're using a Mac or Windows machine, you'll probably want to download the files yourself and then run the installers. (If you're using Linux, I recommend using a port management system like Yum to simplify the installation and updating process; see "Linux and Unix Systems" on page 5.) Here's how to find the binaries.

1. Visit the official R website (*http://www.r-project.org/*). On the site, you should see a link to "Download."
2. The download link (*http://cran.r-project.org/mirrors.html*) actually takes you to a list of mirror sites. The list is organized by country. You'll probably want to pick a site that is geographically close, because it's likely to also be close on the Internet, and thus fast. I usually use the link for the University of California, Los Angeles (*http://cran.stat.ucla.edu/*), because I live in California.
3. Find the right binary for your platform and run the installer.

There are a few things to keep in mind, depending on what system you're using.

Building R from Source

It's standard practice to build R from source on Linux and Unix systems, but not on Mac OS X or Windows platforms. It's pretty tricky to build your own binaries on Mac OS X or Windows, and it doesn't yield a lot of benefits for most users. Building R from source won't save you space (you'll probably have to download a lot of other stuff, like LaTeX), and it won't save you time (unless you already have all the tools you need and have a really, really slow Internet connection). The best reason to build your own binaries is to get better performance out of R, but I've never found R's performance to be a problem, even on very large data sets. If you're interested in how to build your own R, see "Building Your Own" on page 141.

Windows

Installing R on Windows is just like installing any other piece of software on Windows, which means that it's easy if you have the right permissions, difficult if you don't. If you're installing R on your personal computer, this shouldn't be a problem. However, if you're working in a corporate environment, you might run into some trouble.

If you're an "Administrator" or "Power User" on Windows XP, installation is straightforward: double-click the installer and follow the on-screen instructions.

There are some known issues with installing R on Microsoft Windows Vista. In particular, some users have problems with file permissions. Here are two approaches for avoiding these issues:

- Install R as a standard user in your own file space. This is the simplest approach.
- Install R as the default Administrator account (if it is enabled and you have access to it). Note that you will also need to install packages as the Administrator user.

For a full explanation, see *http://cran.r-project.org/bin/windows/base/rw-FAQ.html #Does-R-run-under-Windows-Vista_003f*.

Currently, CRAN only releases 32-bit builds of R for Microsoft Windows. These are tested on 64-bit versions of Windows and should run correctly.

Mac OS X

The current version of R runs on both PowerPC- and Intel-based Mac systems running Mac OS X 10.4.4 (Tiger) and higher. If you're using an older operating system, or an older computer, you can find older versions on the website that may work better with your system.

You'll find three different R installers for Mac OS X: a three-way universal binary for Mac OS X 10.5 (Leopard) and higher, a legacy universal binary for Mac OS X 10.4.4 and higher *with* supplemental tools, and a legacy universal binary for Mac OS X 10.4.4 and higher *without* supplemental tools. See the CRAN download site for more details on the differences between these versions.

As with most applications, you'll need to have the appropriate permissions on your computer to install R. If you're using your personal computer, you're probably OK: you just need to remember your password. If you're using a computer managed by someone else, you may need that person's help to install R.

The universal binary of R is made available as an installer package; simply download the file and double-click the package to install the application. The legacy R installers are packaged on a disk image file (like most Mac OS X applications). After you download the disk image, double-click it to open it in the finder (if it does not automatically open). Open the volume and double-click the R.mpkg icon to launch the installer. Follow the directions in the installer, and you should have a working copy of R on your computer.

Linux and Unix Systems

Before you start, make sure that you know the system's root password or have sudo privileges on the system you're using. If you don't, you'll need to get help from the system administrator to install R.

Installation using package management systems

On a Linux system, the easiest way to install R is to use a package management system. These systems automate the installation process: they fetch the R binaries (or sources), get any other software that's needed to run R, and even make upgrading to the latest version easy.

For example, on Red Hat (or Fedora), you can use Yum (which stands for "Yellow Dog Updater, Modified") to automate the installation. On an x86 Linux platform, open a terminal window and type:

```
sudo yum install R.i386
```

You'll be prompted for your password, and if you have sudo privileges, R should be installed on your system. Later, you can update R by typing:

```
sudo yum update R.i386
```

And, if there is new version available, your R installation will be upgraded to the latest version.

If you're using another Unix system, you may also be able to install R. (For example, R is available through the FreeBSD Ports system at *http://www.freebsd.org/cgi/ cvsweb.cgi/ports/math/R/*.) I haven't tried these versions but have no reason to think they don't work correctly. See the documentation for your system for more information about how to install software.

Installing R from downloaded files

If you'd like, you can manually download R and install it later. Currently, there are precompiled R packages for several flavors of Linux, including Red Hat, Debian, Ubuntu, and SUSE. Precompiled binaries are also available for Solaris.

On Red Hat–style systems, you can install these packages through the Red Hat Package Manager (RPM). For example, suppose that you downloaded the file *R-2.8.0-2.fc10.i386.rpm* to the directory *~/Downloads*. Then you could install it with a command like:

```
rpm -i ~/Downloads/R-2.8.0-2.fc10.i386.rpm
```

For more information on using RPM, or other package management systems, see your user documentation.

2

The R User Interface

If you're reading this book, you probably have a problem that you would like to solve in R. You might want to:

- Check the statistical significance of experimental results
- Plot some data to help understand it better
- Analyze some genome data

The R system is a software environment for statistical computing and graphics. It includes many different components. In this book, I'll use the term "R" to refer to a few different things:

- A computer language
- The interpreter that executes code written in R
- A system for plotting computer graphics described using the R language
- The Windows, Mac OS, or Linux application that includes the interpreter, graphics system, standard packages, and user interface

This chapter contains a short description of the R user interface and the R console, and describes how R varies on different platforms. If you've never used an interactive language, this chapter will explain some basic things you will need to know in order to work with R. We'll take a quick look at the R graphical user interface (GUI) on each platform and then talk about the most important part: the R console.

The R Graphical User Interface

Let's get started by launching R and taking a look at R's graphical user interface on different platforms. When you open the R application on Windows or Max OS X, you'll see a command window and some menu bars. On most Linux systems, R will simply start on the command line.

Windows

By default, R is installed into *%ProgramFiles%R* (which is usually *C:\Program Files \R*) and installed into the Start menu under the group R. When you launch R in Windows, you'll see something like the user interface shown in Figure 2-1. Inside the R GUI window, there is a menu bar, a toolbar, and the R console.

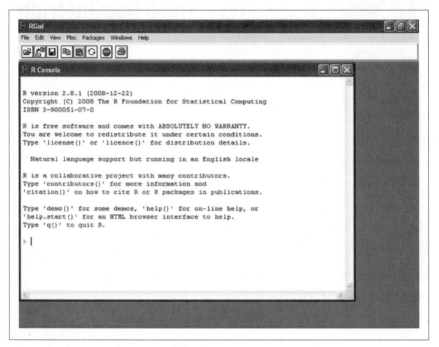

Figure 2-1. R user interface on Windows XP

Mac OS X

The default R installer will add an application called *R* to your Applications folder that you can run like any other application on your Mac. When you launch the R application on Mac OS X systems, you'll see something like the screen shown in Figure 2-2. Like the Windows system, there is a menu bar, a toolbar with common functions, and an R console window.

On a Mac OS system, you can also run R from the terminal without using the GUI. To do this, first open a terminal window. (The terminal program is located in the Utilities folder inside the Applications folder.) Then enter the command "R" on the command line to start R.

Linux and Unix

On Linux systems, you can start R from the command line by typing:

```
R
```

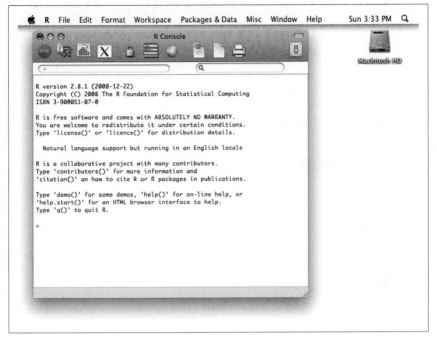

Figure 2-2. R user interface on Mac OS X

Notice that it's a capital "R"; filenames on Linux are case sensitive.

Unlike the default applications for Mac OS and Windows, this will start an interactive R session on the command line itself. If you prefer, you can launch R in an application window similar to the user interface on other platforms. To do this, use the following command:

```
R -g Tk &
```

This will launch R in the background running in its own window, as shown in Figure 2-3. Like the other platforms, there is a menu bar with some common function, but unlike the other platforms, there is no toolbar. The main window acts as the R console.

Additional R GUIs

If you're a typical desktop computer user, you might find it surprising to discover how little functionality is implemented in the standard R GUI. The standard R GUI only implements very rudimentary functionality through menus: reading help, managing multiple graphics windows, editing some source and data files, and some other basic functionality. There are no menu items, buttons, or palettes for loading data, transforming data, plotting data, building models, or doing any interesting work with data. Commercial applications like SAS, SPSS, and S-PLUS include UIs with much more functionality.

Several projects are aiming to build an easier to use GUI for R:

Rcmdr
> The Rcmdr project is an R package that provides an alternative GUI for R. You can install it as an R package. It provides some buttons for loading data, and menu items for many common R functions.

Rkward
> Rkward is a slick GUI frontend for R. It provides a palette and menu-driven UI for analysis, data editing tools, and an IDE for R code development. It's still a young project, and currently works best on Linux platforms (though Windows builds are available). It is available from *http://sourceforge.net/apps/mediawiki/rkward/.*

R Productivity Environment
> Revolution Computing recently introduced a new IDE called the R Productivity Environment. This IDE provides many features for analyzing data: a script editor, object browser, visual debugger, and more. The R Productivity Environment is currently available only for Windows, as part of REvolution R Enterprise.

You can find a list of additional projects at *http://www.sciviews.org/_rgui/.* This book does not cover any of these projects in detail. However, you should still be able to use this book as a reference for all of these packages because they all use (and expose) R functions.

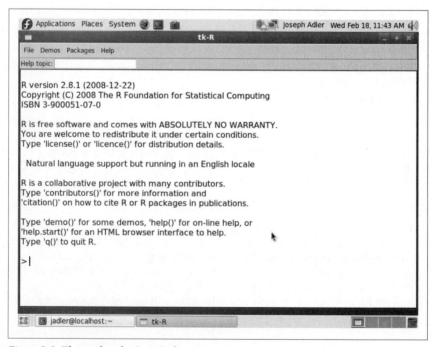

Figure 2-3. Tk interface for R on Fedora

The R Console

The R console is the most important tool for using R. The R console is a tool that allows you to type commands into R and see how the R system responds. The commands that you type into the console are called *expressions*. A part of the R system called the *interpreter* will read the expressions and respond with a result or an error message. Sometimes, you can also enter an expression into R through the menus.

If you've used a command line before (for example, the *cmd.exe* program on Windows) or a language with an interactive interpreter such as LISP, this should look familiar.* If not: don't worry. Command-line interfaces aren't as scary as they look. R provides a few tools to save you extra typing, to help you find the tools you're looking for, and to spot common mistakes. Besides, you have a whole reference book on R that will help you figure out how to do what you want.

Personally, I think that a command-line interface is the best way to analyze data. After I finish working on a problem, I want a record of every step that I took. (I want to know how I loaded the data, if I took a random sample, how I took the sample, whether I created any new variables, what parameters I used in my models, etc.) A command-line interface makes it very easy to keep a record of everything I do and then re-create it later if I need to.

When you launch R, you will see a window with the R console. Inside the console, you will see a message like this:

```
R version 2.9.2 (2009-08-24)
Copyright (C) 2009 The R Foundation for Statistical Computing
ISBN 3-900051-07-0

R is free software and comes with ABSOLUTELY NO WARRANTY.
You are welcome to redistribute it under certain conditions.
Type 'license()' or 'licence()' for distribution details.

  Natural language support but running in an English locale

R is a collaborative project with many contributors.
Type 'contributors()' for more information and
'citation()' on how to cite R or R packages in publications.

Type 'demo()' for some demos, 'help()' for on-line help, or
'help.start()' for an HTML browser interface to help.
Type 'q()' to quit R.

[R.app GUI 1.29 (5464) i386-apple-darwin8.11.1]

[Workspace restored from /Users/josephadler/.RData]

>
```

---

* Incidentally, R has quite a bit in common with LISP: both languages allow you to compute expressions on the language itself, both languages use similar internal structures to hold data, and both languages use lots of parentheses.

This window displays some basic information about R: the version of R you're running, some license information, quick reminders about how to get help, and a *command prompt*.

By default, R will display a greater-than sign (">") in the console (at the beginning of a line, when nothing else is shown) when R is waiting for you to enter a command into the console. R is prompting you to type something, so this is called a *prompt*. This book includes many examples of expressions that I entered into R (and that you can enter into R) and the responses from the R system. In each of these cases, I have shown the prompt from R as a way to differentiate between the commands I entered into R and the responses from the R system.

What this means is that you should *not* type a command prompt (">") if you see one at the beginning of a line. If you want to duplicate my results, type whatever appears after the prompt. For example, I might include a snippet that looks like this:

```
> 17 + 3
[1] 20
```

This means:

- I entered "17 + 3" into the R command prompt.
- The computer responded by writing "[1] 20" (I'll explain what that means in Chapter 3).

If you would like to try this yourself, then type "17 + 3" at the command prompt and press the Enter key. You should see a response like the one shown above.

Sometimes, an R command doesn't fit on a single line. If you enter an incomplete command on one line, the R prompt will change to a plus sign ("+"). Here's a simple example:

```
> 1 * 2 * 3 * 4 * 5 *
+ 6 * 7 * 8 * 9 * 10
[1] 3628800
```

This could cause confusion in some cases (such as in long expressions that contain sums or inequalities). On most platforms, command prompts, user-entered text, and R responses are displayed in different colors to help clarify the differences. Table 2-1 presents a summary of the default colors.

Table 2-1. Text colors in R interactive mode

Platform	Command prompt	User input	R output
Mac OS X	Purple	Blue	Black
Microsoft Windows	Red	Red	Blue
Linux	Black	Black	Black

Command-Line Editing

On most platforms, R provides tools for looking through previous commands.[†] You will probably find the most important line edit commands are the up and down arrow keys. By placing the cursor at the end of the line, you can scroll through previous commands by pressing the up arrow or the down arrow. The up arrow lets you look at earlier commands, and the down arrow lets you look at later commands. If you would like to repeat a previous command with a minor change (such as a different parameter), or if you need to correct a mistake (such as a missing parenthesis), you can do this easily.

You can also type `history()` to get a list of previously typed commands.[‡]

R also includes automatic completions for function names and filenames. Type the "tab" key to see a list of possible completions for a function or filenames.

Batch Mode

R's interactive mode is convenient for most ad hoc analyses, but typing in every command can be inconvenient for some tasks. Suppose that you wanted to do the same thing with R multiple times. (For example, you may want to load data from an experiment, transform it, generate three plots as Portable Document Format [PDF] files, and then quit.) R provides a way to run a large set of commands in sequence and save the results to a file. This is called *batch mode*.

One way to run R in batch mode is from the system command line (not the R console). By running R from the system command line, it's possible to run a set of commands without starting R. This makes it easier to automate analyses, as you can change a couple of variables and rerun an analysis. For example, to load a set of commands from the file *generate_graphs.R*, you would use a command like this:

```
% R CMD BATCH generate_graphs.R
```

R would run the commands in the input file *generate_graphs.R*, generating an output file called *generate_graphs.Rout* with the results. You can also specify the name of the output file. For example, to put the output in a file labeled with today's date (on a Mac or Unix system), you could use a command like this:

```
% R CMD BATCH generate_graphs.R generate_graphs_`date "+%y%m%d"`.log
```

If you're generating graphics in batch mode, remember to specify the output device and filenames. For more information about running R from the command line, including a list of the available options, run R from the command line with the `--help` option:

```
% R --help
```

[†] On Linux and Mac OS X systems, the command line uses the GNU `readline` library and includes a large set of editing commands. On Windows platforms, a smaller number of editing commands are available.

[‡] As of this writing, the `history` command does not work completely correctly on Mac OS X. The `history` command will display the last saved history, not the history for the current session.

You can also run commands in batch mode from inside R. To do this, you can use the source command; see the help file for source for more information.

Using R Inside Microsoft Excel

If you're familiar with Microsoft Excel, or if you work with a lot of data files in Excel format, you might want to run R directly from inside Excel. The RExcel software lets you do just that (on Microsoft Windows systems). You can find information about this software at *http://rcom.univie.ac.at/*. This site also includes a single installer that will install R plus all the other software you need to use RExcel.

If you already have R installed, you can install RExcel as a package from CRAN. The following set of commands will download RExcel, configure the RCOM server, install RDCOM, and launch the RExcel installer:

```
> install.packages("RExcelInstaller", "rcom", "rsproxy")
> # configure rcom
> library(rcom)
> comRegisterRegistry()
> library(RExcelInstaller)
> # excecute the following command in R to start the installer for RDCOM
> installstatconnDCOM()
> # excecute the following command in R to start the installer for REXCEL
> installRExcel()
```

Follow the prompts within the installer to install RExcel.

After you have installed RExcel, you will be able to access RExcel from a menu item. If you are using Excel 2007, you will need to select the "Add-Ins" ribbon to find this menu as shown in Figure 2-4. To use RExcel, first select the R Start menu item. As a simple test, try doing the following:

1. Enter a set of numeric values into a column in Excel (for example, B1:B5).
2. Select the values you entered.
3. On the RExcel menu, go to the item "Put R Var" > "Array."
4. A dialog box will open, asking you to name the object that you are creating in Excel. Enter "v" and press the Enter key. This will create an array (in this case, just a vector) in R with the values that you entered with the name v.
5. Now, select a blank cell in Excel.
6. On the RExcel menu, go to the item "Get R Value" > "Array."
7. A dialog box will open, prompting you to enter an R expression. As an example, try entering (v - mean(v)) / sd(v). This will rescale the contents of v, changing the mean to 0 and the standard deviation to 1.
8. Inspect the results that have been returned within Excel.

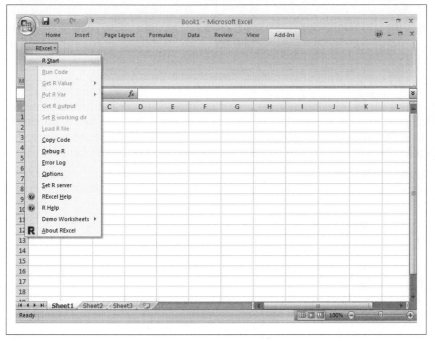

Figure 2-4. Accessing RExcel in Microsoft Excel 2007

For some more interesting examples of how to use RExcel, take a look at the Demo Worksheets under this menu. You can use Excel functions to evaluate R expressions, use R expressions in macros, and even plot R graphics within Excel.

Other Ways to Run R

There are several open source projects that allow you to combine R with other applications:

As a web application
> The Rapache software allows you to incorporate analyses from R into a web application. (For example, you might want to build a server that shows sophisticated reports using R lattice graphics.) For information about this project, see *http://biostat.mc.vanderbilt.edu/rapache/*.

As a server
> The Rserve software allows you to access R from within other applications. For example, you can produce a Java program that uses R to perform some calculations. As the name implies, Rserve is implemented as a network server, so a single Rserve instance can handle calculations from multiple users on different machines. One way to use Rserve is to install it on a heavy-duty server with lots of CPU power and memory, so that users can perform calculations that they couldn't easily perform on their own desktops. For more about this project, see *http://www.rforge.net/Rserve/index.html*.

Inside Emacs

The ESS (Emacs Speaks Statistics) package is an add-on for Emacs that allows you to run R directly within Emacs. For more on this project, see *http://ess.r-project.org/*.

A Short R Tutorial

This chapter contains a short tutorial of R with a lot of examples.

If you've never used R before, this is a great time to start it up and try playing with it. There's no better way to learn something than by trying it yourself. You can follow along by typing in the same text that's shown in the book. Or, try changing it a little bit to see what happens. (For example, if the sample code says 3 + 4, try typing 3 - 4 instead.)

If you've never used an interactive language before, take a look at Chapter 2 before you start. That chapter contains an overview of the R environment, including the console. Otherwise, you might find the presentation of the examples—and the terminology—confusing.

Basic Operations in R

Let's get started using R. When you enter an expression into the R console and press the Enter key, R will evaluate that expression and display the results (if there are any). If the statement results in a value, R will print that value. For example, you can use R to do simple math:

```
> 1 + 2 + 3
[1] 6
> 1 + 2 * 3
[1] 7
> (1 + 2) * 3
[1] 9
```

The interactive R interpreter will automatically print an object returned by an expression entered into the R console. Notice the funny "[1]" that accompanies each returned value. In R, any number that you enter in the console is interpreted as a *vector*. A vector is an ordered collection of numbers. The "[1]" means that the index of the first item displayed in the row is 1. In each of these cases, there is also only one element in the vector.

You can construct longer vectors using the c(...) function. (c stands for "combine.") For example:

```
> c(0, 1, 1, 2, 3, 5, 8)
[1] 0 1 1 2 3 5 8
```

is a vector that contains the first seven elements of the Fibonacci sequence. As an example of a vector that spans multiple lines, let's use the sequence operator to produce a vector with every integer between 1 and 50:

```
> 1:50
 [1]  1  2  3  4  5  6  7  8  9 10 11 12 13 14 15 16 17 18 19 20 21 22
[23] 23 24 25 26 27 28 29 30 31 32 33 34 35 36 37 38 39 40 41 42 43 44
[45] 45 46 47 48 49 50
```

Notice the numbers in the brackets on the lefthand side of the results. These indicate the index of the first element shown in each row.

When you perform an operation on two vectors, R will match the elements of the two vectors pairwise and return a vector. For example:

```
> c(1, 2, 3, 4) + c(10, 20, 30, 40)
[1] 11 22 33 44
> c(1, 2, 3, 4) * c(10, 20, 30, 40)
[1]  10  40  90 160
> c(1, 2, 3, 4) - c(1, 1, 1, 1)
[1] 0 1 2 3
```

If the two vectors aren't the same size, R will repeat the smaller sequence multiple times:

```
> c(1, 2, 3, 4) + 1
[1] 2 3 4 5
> 1 / c(1, 2, 3, 4, 5)
[1] 1.0000000 0.5000000 0.3333333 0.2500000 0.2000000
> c(1, 2, 3, 4) + c(10, 100)
[1]  11 102  13 104
> c(1, 2, 3, 4, 5) + c(10, 100)
[1]  11 102  13 104  15
Warning message:
In c(1, 2, 3, 4, 5) + c(10, 100) :
  longer object length is not a multiple of shorter object length
```

Note the warning if the second sequence isn't a multiple of the first.

In R, you can also enter expressions with characters:

```
> "Hello world."
[1] "Hello world."
```

This is called a *character vector* in R. This example is actually a character vector of length 1. Here is an example of a character vector of length 2:

```
> c("Hello world", "Hello R interpreter")
[1] "Hello world"        "Hello R interpreter"
```

(In other languages, like C, "character" refers to a single character, and an ordered set of characters is called a *string*. A string in C is equivalent to a character value in R.)

You can add comments to R code. Anything after a pound sign ("#") on a line is ignored:

```
> # Here is an example of a comment at the beginning of a line
> 1 + 2 + # and here is an example in the middle
+ 3
[1] 6
```

Functions

In R, the operations that do all of the work are called *functions*. We've already used a few functions above (you can't do anything interesting in R without them). Functions are just like what you remember from math class. Most functions are in the following form:

```
f(argument1, argument2, ...)
```

Where f is the name of the function, and argument1, argument2, . . . are the arguments to the function. Here are a few more examples:

```
> exp(1)
[1] 2.718282
> cos(3.141593)
[1] -1
> log2(1)
[1] 0
```

In each of these examples, the functions only took one argument. Many functions require more than one argument. You can specify the arguments by name:

```
> log(x=64, base=4)
[1] 3
```

Or, if you give the arguments in the default order, you can omit the names:

```
> log(64,4)
[1] 3
```

Not all functions are of the form f(...). Some of them are in the form of operators.[*] For example, we used the addition operator ("+") above. Here are a few examples of operators:

```
> 17 + 2
[1] 19
> 2 ^ 10
[1] 1024
> 3 == 4
[1] FALSE
```

---

[*] When you enter a binary or unary operator into R, the R interpreter will actually translate the operator into a function; there is a function equivalent for each operator. We'll talk about this more in Chapter 5.

We've seen the first one already: it's just addition. The second operator is the exponentiation operator, which is interesting because it's not a commutative operator. The third operator is the equality operator. (Notice that the result returned is FALSE; R has a Boolean data type.)

Variables

Like most other languages, R lets you assign values to variables and refer to them by name. In R, the assignment operator is <-. Usually, this is pronounced as "gets." For example, the statement:

```
x <- 1
```

is usually read as "x gets 1." (If you've ever done any work with theoretical computer science, you'll probably like this notation: it looks just like algorithm pseudocode.)

After you assign a value to a variable, the R interpreter will substitute that value in place of the variable name when it evaluates an expression. Here's a simple example:

```
> x <- 1
> y <- 2
> z <- c(x,y)
> # evaluate z to see what's stored as z
> z
[1] 1 2
```

Notice that the substitution is done at the time that the value is assigned to z, not at the time that z is evaluated. Suppose that you were to type in the preceding three expressions and then change the value of y. The value of z would not change:

```
> y <- 4
> z
[1] 1 2
```

I'll talk more about the subtleties of variables and how they're evaluated in Chapter 8.

R provides several different ways to refer to a member (or set of members) of a vector. You can refer to elements by location in a vector:

```
> b <- c(1,2,3,4,5,6,7,8,9,10,11,12)
> b
 [1]  1  2  3  4  5  6  7  8  9 10 11 12
> # let's fetch the 7th item in vector b
> b[7]
[1] 7
> # fetch items 1 through 6
> b[1:6]
[1] 1 2 3 4 5 6
> # fetch only members of b that are congruent to zero (mod 3)
> # (in non-math speak, members that are multiples of 3)
> b[b %% 3 == 0]
[1]  3  6  9 12
```

You can fetch multiple items in a vector by specifying the indices of each item as an integer vector:

```
> # fetch items 1 through 6
> b[1:6]
[1] 1 2 3 4 5 6
> # fetch 1, 6, 11
> b[c(1,6,11)]
[1]  1  6 11
```

You can fetch items out of order. Items are returned in the order that they are referenced:

```
> b[c(8,4,9)]
[1] 8 4 9
```

You can also specify which items to fetch through a logical vector. As an example, let's fetch only multiples of 3 (by selecting items that are congruent to 0 mod 3):

```
> b %% 3 == 0
 [1] FALSE FALSE  TRUE FALSE FALSE  TRUE FALSE FALSE  TRUE FALSE FALSE
[12]  TRUE
> b[b %% 3 == 0]
[1]  3  6  9 12
```

In R, there are two additional operators that can be used for assigning values to symbols. First, you can use a single equals sign ("=") for assignment.† This operator assigns the symbol on the left to the object on the right. In many other languages, all assignment statements use equals signs. If you are more comfortable with this notation, you are free to use it. However, I will be using only the <- assignment operator in this book because I think it is easier to read. Whichever notation you prefer, be careful because the = operator does not mean "equals." For that, you need to use the == operator:

```
> one <- 1
> two <- 2
> # This means: assign the value of "two" to the variable "one"
> one = two
> one
[1] 2
> two
[1] 2
> # let's start again
> one <- 1
> two <- 2
> # This means: does the value of "one" equal the value of "two"
> one == two
[1] FALSE
```

In R, you can also assign an object on the left to a symbol on the right:

```
> 3 -> three
> three
[1] 3
```

† Note that you cannot use the <- operator when passing arguments to a function; you need to map values to argument names using the "=" symbol. Using the <- operator in a function will assign the value to the variable in the current environment and then pass the value returned to the function. This might be what you want, but it probably isn't.

In some programming contexts, this notation might help you write clearer code. (It may also be convenient if you type in a long expression and then realize that you have forgotten to assign the result to a symbol.)

A function in R is just another object that is assigned to a symbol. You can define your own functions in R, assign them a name, and then call them just like the built-in functions:

```
> f <- function(x,y) {c(x+1, y+1)}
> f(1,2)
[1] 2 3
```

This leads to a very useful trick. You can often type the name of a function to see the code for it. Here's an example:

```
> f
function(x,y) {c(x+1, y+1)}
```

Introduction to Data Structures

In R, you can construct more complicated data structures than just vectors. An *array* is a multidimensional vector. Vectors and arrays are stored the same way internally, but an array may be displayed differently and accessed differently. An array object is just a vector that's associated with a dimension attribute. Here's a simple example.

First, let's define an array explicitly:

```
> a <- array(c(1,2,3,4,5,6,7,8,9,10,11,12),dim=c(3,4))
```

Here is what the array looks like:

```
> a
     [,1] [,2] [,3] [,4]
[1,]    1    4    7   10
[2,]    2    5    8   11
[3,]    3    6    9   12
```

And here is how you reference one cell:

```
> a[2,2]
[1] 5
```

Now, let's define a vector with the same contents:

```
> v <- c(1,2,3,4,5,6,7,8,9,10,11,12)
> v
 [1]  1  2  3  4  5  6  7  8  9 10 11 12
```

A matrix is just a two-dimensional array:

```
> m <- matrix(data=c(1,2,3,4,5,6,7,8,9,10,11,12),nrow=3,ncol=4)
> m
     [,1] [,2] [,3] [,4]
[1,]    1    4    7   10
[2,]    2    5    8   11
[3,]    3    6    9   12
```

Arrays can have more than two dimensions. For example:

```
> w <- array(c(1,2,3,4,5,6,7,8,9,10,11,12,13,14,15,16,17,18),dim=c(3,3,2))
> w
, , 1

     [,1] [,2] [,3]
[1,]    1    4    7
[2,]    2    5    8
[3,]    3    6    9

, , 2

     [,1] [,2] [,3]
[1,]   10   13   16
[2,]   11   14   17
[3,]   12   15   18

> w[1,1,1]
[1] 1
```

R uses very clean syntax for referring to part of an array. You specify separate indices for each dimension, separated by commas:

```
> a[1,2]
[1] 4
> a[1:2,1:2]
     [,1] [,2]
[1,]    1    4
[2,]    2    5
```

To get all rows (or columns) from a dimension, simply omit the indices:

```
> # first row only
> a[1,]
[1]  1  4  7 10
> # first column only
> a[,1]
[1] 1 2 3
> # you can also refer to a range of rows
> a[1:2,]
     [,1] [,2] [,3] [,4]
[1,]    1    4    7   10
[2,]    2    5    8   11
> # you can even refer to a noncontiguous set of rows
> a[c(1,3),]
     [,1] [,2] [,3] [,4]
[1,]    1    4    7   10
[2,]    3    6    9   12
```

In all the examples above, we've just looked at data structures based on a single underlying data type. In R, it's possible to construct more complicated structures with multiple data types. R has a built-in data type for mixing objects of different types, called *lists*. Lists in R are subtly different from lists in many other languages. Lists in R may contain a heterogeneous selection of objects. You can name each component in a list. Items in a list may be referred to by either location or name.

Here is an example of a list with two named components:

```
> # a list containing a number and string
> e <- list(thing="hat", size="8.25")
> e
$thing
[1] "hat"

$size
[1] "8.25"
```

You may access an item in the list in multiple ways:

```
> e$thing
[1] "hat"
> e[1]
$thing

[1] "hat"
> e[[1]]
[1] "hat"
```

A list can even contain other lists:

```
> g <- list("this list references another list", e)
> g
[[1]]
[1] "this list references another list"

[[2]]
[[2]]$thing
[1] "hat"

[[2]]$size
[1] "8.25"
```

A *data frame* is a list that contains multiple named vectors that are the same length. A data frame is a lot like a spreadsheet or a database table. Data frames are particularly good for representing experimental data. As an example, I'm going to use some baseball data. Let's construct a data frame with the win/loss results in the National League (NL) East in 2008:

```
> teams <- c("PHI","NYM","FLA","ATL","WSN")
> w <- c(92, 89, 94, 72, 59)
> l <- c(70, 73, 77, 90, 102)
> nleast <- data.frame(teams,w,l)
> nleast
  teams  w   l
1   PHI 92  70
2   NYM 89  73
3   FLA 94  77
4   ATL 72  90
5   WSN 59 102
```

You can refer to the components of a data frame (or items in a list) by name using the $ operator:

```
> nleast$w
[1] 92 89 94 72 59
```

Here's one way to find a specific value in a data frame. Suppose that you wanted to find the number of losses by the Florida Marlins (FLA). One way to select a member of an array is by using a vector of Boolean values to specify which item to return from a list. You can calculate an appropriate vector like this:

```
> nleast$teams=="FLA"
[1] FALSE FALSE  TRUE FALSE FALSE
```

Then you can use this vector to refer to the right element in the losses vector:

```
> nleast$l[nleast$teams=="FLA"]
[1] 77
```

You can import data into R from another file or from a database. See Chapter 12 for more information on how to do this.

In addition to lists, R has other types of data structures for holding a heterogeneous collection of objects, including formal class definitions through S4 objects.

Objects and Classes

R is an object-oriented language. Every object in R has a type. Additionally, every object in R is a member of a *class*. We have already encountered several different classes: character vectors, numeric vectors, data frames, lists, and arrays.

You can use the `class` function to determine the class of an object. For example:

```
> class(teams)
[1] "character"
> class(w)
[1] "numeric"
> class(nleast)
[1] "data.frame"
> class(class)
[1] "function"
```

Notice the last example: a function is an object in R with the class `function`.

Some functions are associated with a specific class. These are called *methods*. (Not all functions are tied closely to a particular class; the class system in R is much less formal than that in a language like Java.)

In R, methods for different classes can share the same name. These are called *generic functions*. Generic functions serve two purposes. First, they make it easy to guess the right function name for an unfamiliar class. Second, generic functions make it possible to use the same code for objects of different types.

For example, + is a generic function for adding objects. You can add numbers together with the + operator:

```
> 17 + 6
[1] 23
```

You might guess that the addition operator would work similarly with other types of objects. For example, you can also use the + operator with a date object and a number:

```
> as.Date("2009-09-08") + 7
[1] "2009-09-15"
```

By the way, the R interpreter calls the generic function `print` on any object returned on the R console. Suppose that you define x as:

```
> x <- 1 + 2 + 3 + 4
```

When you type:

```
> x
[1] 10
```

the interpreter actually calls the function `print(x)` to print the results. This means that if you define a new class, you can define a print method to specify how objects from that new class are printed on the console. Some functions take advantage of this functionality to do other things when you enter an expression on the console.[‡]

I'll talk about objects in more depth in Chapter 7, and classes in Chapter 10.

Models and Formulas

To statisticians, a *model* is a concise way to describe a set of data, usually with a mathematical formula. Sometimes, the goal is to build a *predictive* model with *training* data to predict values based on other data. Other times, the goal is to build a *descriptive* model that helps you understand the data better.

R has a special notation for describing relationships between variables. Suppose that you are assuming a linear model for a variable y, predicted from the variables x1, x2, ..., xn. (Statisticians usually refer to y as the *dependent* variable, and x1, x2, ..., xn as the *independent* variables.) In equation form, this implies a relationship like:

$$y = c_0 + c_1 x_1 + c_2 x_2 + \cdots + c_n x_n + \varepsilon$$

In R, you would write the relationship as y ~ x1 + x2 + ... + xn, which is a formula object.

So, let's try to use a linear regression to estimate the relationship. The formula is `dist~speed`. We'll use the `lm` function to estimate the parameters of a linear model. The `lm` function returns an object of class `lm`, which we will assign to a variable called `cars.lm`:

```
> cars.lm <- lm(formula=dist~speed,data=cars)
```

[‡] A very important example of this is lattice graphics. Plotting functions in the lattice library return lattice objects but don't plot results. If you call a lattice function on the R console, the console will print the object, thus plotting the results. However, if you call a lattice function within another function, or in a script, R will not plot the results unless you explicitly print the lattice object.

Now, let's take a quick look at the results returned:

```
> cars.lm

Call:
lm(formula = dist ~ speed, data = cars)

Coefficients:
(Intercept)        speed
    -17.579        3.932
```

As you can see, printing an lm object shows you the original function call (and thus the data set and formula) and the estimated coefficients. For some more information, we can use the summary function:

```
> summary(cars.lm)

Call:
lm(formula = dist ~ speed, data = cars)

Residuals:
    Min      1Q  Median      3Q     Max
-29.069  -9.525  -2.272   9.215  43.201

Coefficients:
             Estimate Std. Error t value Pr(>|t|)
(Intercept) -17.5791     6.7584  -2.601   0.0123 *
speed         3.9324     0.4155   9.464 1.49e-12 ***
---
Signif. codes:  0 '***' 0.001 '**' 0.01 '*' 0.05 '.' 0.1 ' ' 1

Residual standard error: 15.38 on 48 degrees of freedom
Multiple R-squared: 0.6511,     Adjusted R-squared: 0.6438
F-statistic: 89.57 on 1 and 48 DF,  p-value: 1.490e-12
```

As you can see, the summary option shows you the function call, the distribution of the residuals from the fit, the coefficients, and information about the fit. By the way, it is possible to simply call the lm function or to call summary(lm(...)) and not assign a name to the model object:

```
> lm(dist~speed,data=cars)

Call:
lm(formula = dist ~ speed, data = cars)

Coefficients:
(Intercept)        speed
    -17.579        3.932

> summary(lm(dist~speed,data=cars))

Call:
lm(formula = dist ~ speed, data = cars)

Residuals:
    Min      1Q  Median      3Q     Max
```

```
   -29.069  -9.525  -2.272   9.215  43.201

Coefficients:
            Estimate Std. Error t value Pr(>|t|)
(Intercept) -17.5791     6.7584  -2.601   0.0123 *
speed         3.9324     0.4155   9.464 1.49e-12 ***
---
Signif. codes:  0 '***' 0.001 '**' 0.01 '*' 0.05 '.' 0.1 ' ' 1

Residual standard error: 15.38 on 48 degrees of freedom
Multiple R-squared: 0.6511,    Adjusted R-squared: 0.6438
F-statistic: 89.57 on 1 and 48 DF,  p-value: 1.490e-12
```

In some cases, this can be more convenient. However, you often want to perform additional analyses, such as plotting residuals, calculating additional statistics, or updating a model to add or subtract variables. By assigning a name to the model, you can make your code easier to understand and modify. Additionally, refitting a model can be very time consuming for complex models and large data sets. By assigning the model to a variable name, you can avoid these problems.

Charts and Graphics

R includes several packages for visualizing data: `graphics`, `grid`, and `lattice`. Usually, you'll find that functions within the `graphics` and `lattice` packages are the most useful.[§] If you're familiar with Microsoft Excel, you'll find that R can generate all of the charts that you're familiar with: column charts, bar charts, line plots, pie charts, and scatter plots. Even if that's all you need, R makes it much easier than Excel to automate the creation of charts and customize them. However, there are many, many more types of charts available in R, many of them quite intuitive and elegant.

To make this a little more interesting, let's work with some real data. We're going to look at all field goal attempts in the National Football League (NFL) in 2005.[||] For those of you who aren't familiar with American football, here's a quick explanation. A team can attempt to kick a football between a set of goalposts to receive 3 points. If it misses the field goal, possession of the ball reverts to the other team (at the spot on the field where the kick was attempted). We're going to take a look at kick attempts in the NFL in 2005.

First, let's take a quick look at the distribution of distances. R provides a function, `hist`, that can do this quickly for us. Let's start by loading the appropriate data set. (The data set is included in the `nutshell` package; see the Preface for information on how to obtain this package.)

```
> library(nutshell)
> data(field.goals)
```

[§] Other packages are available for visualizing data. For example, the RGobi package provides tools for creating interactive graphics.

[||] The data was provided by Aaron Schatz of *Pro Football Prospectus*. For more information, see the Football Outsiders website at *http://www.footballoutsiders.com/*, or you can find its annual books at most bookstores—both online and "brick and mortar."

Let's take a quick look at the names of the columns in the `field.goals` data frame:

```
> names(field.goals)
[1] "home.team"    "week"         "qtr"          "away.team"
[5] "offense"      "defense"      "play.type"    "player"
[9] "yards"        "stadium.type"
```

Now, let's just try the `hist` command:

```
> hist(field.goals$yards)
```

This produces a chart like the one shown in Figure 3-1. (Depending on your system, if you try this yourself, you may see a differently colored and formatted chart. I tweaked a few graphical parameters so the charts would look good in print.) I wanted to see more detail about the number of field goals at different distances, so I modified the `breaks` argument to add more bins to the histogram:

```
> hist(field.goals$yards, breaks=35)
```

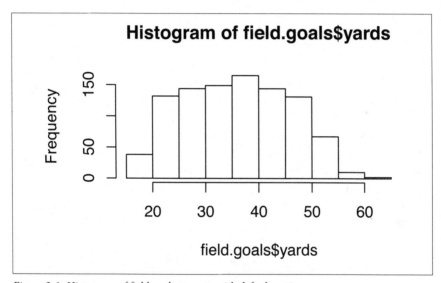

Figure 3-1. Histogram of field goal attempts with default settings

You can see the results of this command in Figure 3-2. R also features many other ways to visualize data. A great example is a strip chart. This chart just plots one point on the *x*-axis for every point in a vector. As an example, let's look at the distance of blocked field goals. We can distinguish blocked field goals with the `play.type` variable in the `field.goals` data frame. Let's take a quick look at how many blocked field goals there were in 2005. We'll use the `table` function to tabulate the results:

```
> table(field.goals$play.type)

FG aborted FG blocked   FG good    FG no
         8         24       787      163
```

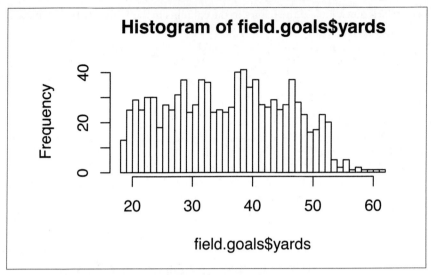

Figure 3-2. Histogram of field goal distances, showing more bins

Now, we'll select only observations with blocked field goals. We'll add a little jitter so we can see individual points. Finally, we will also change the appearance of the points using the pch argument:

```
> stripchart(field.goals[field.goals$play.type=="FG blocked",]$yards,
+            pch=19, method="jitter")
```

The results are shown in Figure 3-3.

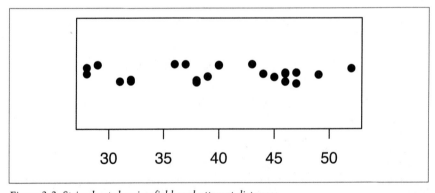

Figure 3-3. Strip chart showing field goal attempt distances

As a second example, let's use the **cars** data set, which is included in the **base** package. The **cars** data set consists of a set of 50 observations:

```
> data(cars)
> dim(cars)
[1] 50  2
```

```
> names(cars)
[1] "speed" "dist"
```

Each observation contains the speed of the car and the distance required to stop.
Let's take a quick look at the contents of this data set:

```
> summary(cars)
     speed             dist
 Min.    : 4.0    Min.    :   2.00
 1st Qu.:12.0    1st Qu.:  26.00
 Median :15.0    Median :  36.00
 Mean    :15.4    Mean    :  42.98
 3rd Qu.:19.0    3rd Qu.:  56.00
 Max.    :25.0    Max.    : 120.00
```

Let's plot the relationship between vehicle speed and stopping distance:

```
> plot(cars, xlab = "Speed (mph)", ylab = "Stopping distance (ft)",
       las = 1, xlim = c(0, 25))
```

The plot is shown in Figure 3-4. At a quick glance, we see that stopping distance is
roughly proportional to speed.

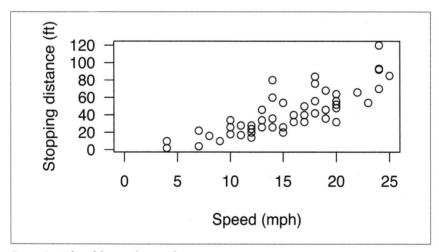

Figure 3-4. Plot of data in the cars data set

Let's try one more example, this time using lattice graphics. Lattice graphics provide
some great tools for drawing pretty charts, particularly charts that compare different
groups of points. By default, the **lattice** package is not loaded; you will get an error
if you try calling a lattice function without loading the library. To load the library,
use the following command:

```
> library(lattice)
```

We will talk more about R packages in Chapter 4.

For example data, we'll look at how American eating habits changed between 1980 and 2005.#

The consumption data set is available in the `nutshell` package. It contains 48 observations, each showing the amount of a commodity consumed (or produced) in a specific year. Data is available only for years that are multiples of 5 (so there are six unique years between 1980 and 2005). The amount of food consumed is given by `Amount`, the type of food is given by `Food`, and the year is given by `Year`.

Two of the variables are numeric vectors: `Amount` and `Year`. However, two of them are an important data type that we haven't seen yet: factors. A *factor* is an R object type that is used to compactly represent a vector of categorical values. Factors are used in many modeling functions. You can create a factor from another vector (typically a character vector) using the `factor` function. In this data frame, the values `Food` and `Units` are factors. (We'll discuss vectors in more detail in "Vectors" on page 82.)

To help reveal trends in the data, I decided to use the `dotplot` function. (This function resembles line charts in Excel.) Specifically, we'd like to look at how the `Amount` varies by `Year`. We'd like to separately plot the trend for each value of the `Food` variable. For lattice graphics, we specify the data that we want to plot through a formula, in this case, `Amount ~ Year | Food`. A *formula* is an R object that is used to express a relationship between a set of variables.

If you'd like, you can try plotting the relationship using the default settings:

```
> library(lattice)
> dotplot(Amount~Year|Food, consumption)
```

I found the default plot hard to read: the axis labels were too big, the scale for each plot was the same, and the stacking didn't look right to me. So, I tuned the presentation a little bit. Here is the version that produced Figure 3-5:

```
> dotplot(Amount ~ Year | Food,data=consumption,
  aspect="xy",scales=list(relation="sliced", cex=.4))
```

The `aspect` option changes the aspect ratios of each plot to try to show changes from 45° angles (making changes easier to see). The `scales` option changes how the axes are drawn. I'll discuss lattice plots in more detail in Chapter 15, explaining how to use different options to tune the look of your charts.

Getting Help

R includes a help system to help you get information about installed packages. To get help on a function, for example, `glm` you would type:

```
> help(glm)
```

#I obtained the data from the *2009 Statistical Abstract of the United States*, a terrific book of data about the United States that is published by the Census Bureau. I took a subset of the data, only keeping consumption for the largest categories. You can find this data at *http://www.census.gov/compendia/statab/cats/health_nutrition/food_consumption_and_nutrition.html.*

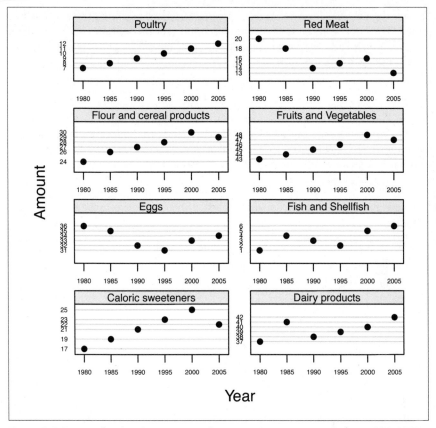

Figure 3-5. Lattice plot showing American changes in American eating habits, 1980–2005

or, equivalently:

```
> ?glm
```

To search for help on an operator, you need to place the operator in backquotes:

```
> ?`+`
```

If you'd like to try the examples in a help file, you can use the `example` function to automatically try them. For example, to see the example for `glm`, type:

```
> example(glm)
```

You can search for help on a topic, for example, "regression," using the `help.search` function:

```
> help.search("regression")
```

This can be very helpful if you can't remember the name of a function; R will return a list of relevant topics. There is a shorthand for this command as well:

```
> ??regression
```

To get the help file for a package, you can sometimes use one of the commands above. However, you can also use the help option for the library command to get more complete information. For example, to get help on the grDevices library, you would use the following function:

```
> library(help="grDevices")
```

Some packages (especially packages from Bioconductor) include at least one vignette. A *vignette* is a short document that describes how to use the package, complete with examples. You can view a vignette using the vignette command. For example, to view the vignette for the affy package (assuming that you have installed this package), you would use the following command:

```
> vignette("affy")
```

To view available vignettes for all attached packages, you can use the following command:

```
> vignette(all=FALSE)
```

To view vignettes for all installed packages, try this command:

```
> vignette(all=TRUE)
```

4

R Packages

A *package* is a related set of functions, help files, and data files that have been bundled together. Packages in R are similar to modules in Perl, libraries in C/C++, and classes in Java.

Typically, all of the functions in the package are related: for example, the `stats` package contains functions for doing statistical analysis. To use a package, you need to load it into R (see "Loading Packages" on page 38 for directions on loading package).

R offers an enormous number of packages: packages that display graphics, packages for performing statistical tests, and packages for trying the latest machine learning techniques. There are also packages designed for a wide variety of industries and applications: packages for analyzing microarray data, packages for modeling credit risks, and packages for social sciences.

Some of these packages are included with R: you just have to tell R that you want to use them. Other packages are available from public package repositories. You can even make your own packages. This chapter explains how to use packages.

An Overview of Packages

To use a package in R, you first need to make sure that it has been installed into a local *library*.* By default, packages are read from one system-level library, but you can add additional libraries.

---

* If you're a C/C++ programmer, don't get confused; library means something different in R.

Next, you need to load the packages into your curent session. You might be wondering why you need to load packages into R in order to use them. First, R's help system slows down significantly when you add more packages to search. (I know this from personal experience: I loaded dozens of packages into R while writing this book, and the help system slowed to a crawl.) Second, it's possible that two packages have objects with the same name. If every package were loaded into R by default, you might think you were using one function but really be using another. Even worse, it's possible for there to be internal conflicts: two packages may both use functions with names like "fit" that work very differently, resulting in strange and unexpected results. By only loading packages that you need, you can minimize the chance of these conflicts.

Listing Packages in Local Libraries

To get the list of packages loaded by default, you can use the getOption command to check the value of the defaultPackages value:

```
> getOption("defaultPackages")
[1] "datasets"  "utils"      "grDevices" "graphics"  "stats"
[6] "methods"
```

This command omits the base package; the base package implements many key features of the R language and is always loaded.

If you would like to see the list of currently loaded packages, you can use the .packages command (note the parentheses around the outside):

```
> (.packages())
[1] "stats"      "graphics" "grDevices" "utils"      "datasets" "methods"
[7] "base"
```

To show all packages available, you can use the all.available option with the packages command:

```
> (.packages(all.available=TRUE))
 [1] "KernSmooth" "MASS"       "base"       "bitops"     "boot"
 [6] "class"      "cluster"    "codetools"  "datasets"   "foreign"
[11] "grDevices"  "graphics"   "grid"       "hexbin"     "lattice"
[16] "maps"       "methods"    "mgcv"       "nlme"       "nnet"
[21] "rpart"      "spatial"    "splines"    "stats"      "stats4"
[26] "survival"   "tcltk"      "tools"      "utils"
```

You can also enter the library() command with no arguments and a new window will pop up showing you the set of available packages.

Included Packages

R comes with a number of different packages (see Table 4-1 for a list). Some of these packages (like base, graphics, grDevices, methods, and utils) implement basic features of the R language or R environment. Other packages provide commonly used statistical modeling tools (like cluster, nnet, and stats). Other packages implement sophisticated graphics (grid and lattice), contain examples (datasets), or contain other frequently used functions. In many cases, you won't need to get any other packages.

Table 4-1. Packages included with R

Package name	Loaded by default	Description
base	✓	Basic functions of the R language, including arithmetic, I/O, programming support
boot		Bootstrap resampling
class		Classification algorithms, including nearest neighbors, self-organizing maps, and learning vector quantization
cluster		Clustering algorithms
codetools		Tools for analyzing R code
datasets	✓	Some famous data sets
foreign		Tools for reading data from other formats, including Stata, SAS, and SPSS files
graphics	✓	Functions for base graphics
grDevices	✓	Device support for base and grid graphics, including system-specific functions
grid		Tools for building more sophisticated graphics than the base graphics
KernSmooth		Functions for kernel smoothing
lattice		An implementation of Trellis graphics for R: prettier graphics than the default graphics
MASS		Functions and data used in the book *Modern Applied Statistics with S* by Venables and Ripley; contains a lot of useful statistics functions
methods	✓	Implementation of formal methods and classes introduced in S version 4 (called S4 methods and classes)
mgcv		Functions for generalized additive modeling and generalized additive mixed modeling
nlme		Linear and nonlinear mixed-effects models
nnet		Feed-forward neural networks and multinomial log linear models
rpart		Tools for building recursive partitioning and regression tree models
spatial		Functions for Kriging and point pattern analysis
splines		Regression spline functions and classes

Package name	Loaded by default	Description
stats	✓	Functions for statistics calculations and random number generation; includes many common statistical tests, probability distributions, and modeling tools
stats4		Statistics functions as S4 methods and classes
survival		Survival analysis functions
tcltk		Interface to Tcl/Tk; used to create platform-independent UI tools
tools		Tools for developing packages
utils	✓	A variety of utility functions for R, including package management, file reading and writing, and editing

Loading Packages

By default, not all packages are loaded into R. If you try to use a function from a package that hasn't been loaded, you'll get an error:

```
> # try to use rpart before loading it
> fit <- rpart(Kyphosis ~ Age + Number + Start, data=kyphosis)
Error: could not find function "rpart"
```

To load a package in R, you can use the `library()` command. For example, to load the package `rpart` (which contains functions for building recursive partition trees), you would use the following command:

```
> library(rpart)
```

(There is a similar command, `require()`, that takes slightly different arguments. For more about `require`, see the R help files.)

If you're more comfortable using a GUI, you can browse for packages and load them using the GUI. If you choose to use this interface to find packages, make sure that you include the appropriate `library` command with your scripts to prevent errors later.

Loading Packages on Windows and Linux

On Microsoft Windows, you can use the `library` function to load packages. Alternatively, you can select "Load package. . ." from the Packages menu in the GUI. This will bring up a window showing a list of packages that you can choose to load.

Loading Packages on Mac OS X

The Mac OS X R environment is a little fancier than the other versions. Like the other versions, you can use the `library()` function. Otherwise, you can select "Package Manager" from the Packages and Data menu. The Package Manager UI, as shown in Figure 4-1, lets you see which packages are loaded, load packages, and even browse the help file for a package.

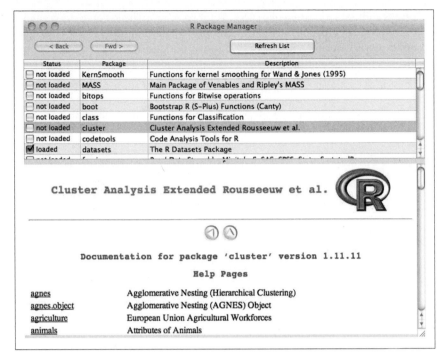

Figure 4-1. Mac OS X Package Manager

Exploring Package Repositories

The packages included with R are very useful; many users will never need to use any other features. However, you can find thousands of additional packages online.

The two biggest sources of packages are CRAN (Comprehensive R Archive Network) and Bioconductor, but some packages are available elsewhere. (If you know Perl, you'll notice that CRAN is very similar to CPAN: the Comprehensive Perl Archive Network.) CRAN is hosted by the R Foundation (the same nonprofit organization that oversees R development). The archive contains a very large number of packages (there were 1,698 packages on February 24, 2009), covering a wide number of different applications. CRAN is hosted on a set of mirror sites around the world. Try to pick an archive site near you: you'll minimize download times *and* help reduce the server load on the R Foundation.

Bioconductor is an open isource project for building tools to analyze genomic data. Bioconductor tools are built using R and are distributed as R packages. The Bioconductor packages are distributed separately from R, and most are not available on CRAN. There are dozens of different packages available directly through the Bioconductor project.

R-Forge is another interesting place to look for packages. The R-Forge site contains projects that are in progress, and it provides tools for developers to collaborate. You may find some interesting packages on this site, but please be sure to read the disclaimers and documentation because many of these packages are works in progress.

R includes the ability to download and install packages from other repositories. However, I don't know of other public repositories for R packages. Most R projects simply use CRAN to host their packages. (I've even seen some books that use CRAN to distribute sample code and sample data.)

Exploring Packages on the Web

R provides good tools for installing packages within the GUI but doesn't provide a good way to find a specific package. Luckily, it's pretty easy to find a package on the Web.

You can browse through the set of available packages with your web browser. Here are some places to look for packages.

Repository	URL
CRAN	See *http://cran.r-project.org/web/packages/* for an authoritative list, but you should try to find your local mirror and use that site instead
Bioconductor	*http://www.bioconductor.org/packages/release/Software.html*
R-Forge	*http://r-forge.r-project.org/*

However, you can also try to find packages with a search engine. I've had good luck finding packages by using Google to search for "R package" plus the name of the application. For example, searching for "R package multivariate additive regression splines" can help you find the mda package, which contains the mars function. (Of course, I discovered later that the earth package is a better choice for this algorithm, but we'll get to that later.)

Finding and Installing Packages Inside R

Once you figure out what package you want to install, the easiest way to do it is inside R.

Windows and Linux GUIs

Installing packages through the Windows GUI is pretty straightforward.

1. (Optional). By default, R is set to fetch packages from the "CRAN" and "CRAN (extra)" categories. To pick additional sets of packages, choose "Select repositories. . ." from the Packages menu. You can choose multiple repositories.

2. From the Packages menu, select "Install package(s). . . ."

3. If this is the first time that you are installing a package during this session, R will ask you to pick a mirror. (You'll probably want to pick a site that is

geographically close, because it's likely to also be close on the Internet, and thus fast.)

4. Click the name of the package that you want to install and press the "OK" button.

R will download and install the packages that you have selected.

Note that you may run into issues installing packages, depending on the permissions assigned to your user account. If you are using Windows XP, and your account is a member of the Administrators group, you should have no problems. If you are using Windows Vista, and you installed R in your own directory, you should have no issues. Otherwise, you may need to run R as an Administrator in order to install supplementary packages.

Mac OS X GUI

On Mac OS X, there is a slightly different user interface for package installation. It shows a little more information than the Windows version, but it's a little more confusing to use.

1. From the Package and Data menu, select "Package Installer." (See Figure 4-1 for a picture of the installer window.)
2. (Optional) In the top lefthand corner of the window is a menu that allows you to select the category of packages that you would like to download. Initially, this is set to "CRAN (binaries)."
3. Click the "Get List" button to display the available set of packages.
4. You can use the search box to filter the list to show only packages that match the name you are looking for. (Note: you have to click the "Get List" button before the search will return results.)
5. Select the set of packages that you want to install and press the "Install Selected" button.

By default, R will install packages at the system level, making them available to all users. If you do not have the appropriate permissions to install packages globally, or if you would like to install them elsewhere, then select an alternative location. Additionally, R will not install the additional packages on which your packages depend. You will get an error if you try to load a package and have not installed other packages on which it is dependent.

R console

You can also install R packages directly from the R console. Table 4-2 shows the set of commands for installing packages from the console. As a simple example, suppose that you wanted to install the packages tree and maptree. You could accomplish this with the following command:

```
> install.packages(c("tree","maptree"))
trying URL 'http://cran.cnr.Berkeley.edu/bin/macosx/universal/contrib/
2.9/tree_1.0-26.tgz'
Content type 'application/x-gzip' length 103712 bytes (101 Kb)
```

```
opened URL
==================================================
downloaded 101 Kb

trying URL 'http://cran.cnr.Berkeley.edu/bin/macosx/universal/contrib/
2.9/maptree_1.4-5.tgz'
Content type 'application/x-gzip' length 101577 bytes (99 Kb)
opened URL
==================================================
downloaded 99 Kb


The downloaded packages are in
        /var/folders/gj/gj60srEiEVq4hTWB5lvMak+++TM/-Tmp-//RtmpIXUWDu/
downloaded_packages
```

This will install the packages to the default library specified by the variable `.Library`. If you'd like to remove these packages after you're done, you can use **remove.packages**. You need to specify the library where the packages were installed:

```
> remove.packages(c("tree", "maptree"),.Library)
```

Table 4-2. Common package installation commands

Command	Description
installed.packages	Returns a matrix with information about all currently installed packages.
available.packages	Returns a matrix of all packages available on the repository.
old.packages	Returns a matrix of all currently installed packages for which newer versions are available.
new.packages	Returns a matrix showing all currently uninstalled packages available from the package repositories.
download.packages	Downloads a set of packages to a local directory.
install.packages	Installs a set of packages from the repository.
remove.packages	Removes a set of installed packages.
update.packages	Updates installed packages to the latest versions.
setRepositories	Sets the current list of package repositories.

Installing from the command line

You can also install downloaded packages from the command line. (There is actually a set of different commands that you can issue to R directly from the command line, without launching the full R shell.) To do this, you run R with the "CMD INSTALL" option. For example, suppose that you had downloaded the package **aplpack** ("Another Plotting PACKage"). For Mac OS X, the binary file is called *aplpack_1.1.1.tgz*. To install this package, change to the directory where the package is located and issue the following command:

```
R CMD INSTALL aplpack_1.1.1.tgz
```

If successful, you'll see a message like the following:

```
* Installing to library '/Library/Frameworks/R.framework/Resources/library'
* Installing *binary* package 'aplpack' ...
* DONE (aplpack)
```

Custom Packages

Building your own packages is a good idea if you want to share code or data with other people, or if you just want to pack it up in a form that's easy to reuse, you should consider building your own package. This section explains the easy way to create your own packages.

Creating a Package Directory

To build a package, you need to place all of the package files (code, data, documentation, etc.) inside a single directory. You can create an appropriate directory structure using the R function `package.skeleton`:

```
package.skeleton(name = "anRpackage", list,
                 environment = .GlobalEnv,
                 path = ".", force = FALSE, namespace = FALSE,
                 code_files = character())
```

This function can also copy a set of R objects into that directory. Here's a description of the arguments to `package.skeleton`.

Argument	Description	Default
name	A character value specifying a name for the new package	"anRpackage" (as a side note, this may be the least useful default value for any R function)
list	A character vector containing names of R objects to add to the package	
environment	The environment in which to evaluate `list`	.GlobalEnv
path	A character vector specifying the path in the file system	"."
force	A Boolean value specifying whether to overwrite files, if a directory name already exists at `path`	FALSE
namespace	A Boolean value specifying whether to add a namespace to the package	FALSE
code_files	A character vector specifying the paths of files containing R code	character()

For this book, I created a package called `nutshell` containing most of the data sets used in this book:

```
> package.skeleton(name="nutshell",path="~/Documents/book/current/")
Creating directories ...
Creating DESCRIPTION ...
Creating Read-and-delete-me ...
Saving functions and data ...
```

```
Making help files ...
Done.
Further steps are described in
'~/Documents/book/current//nutshell/Read-and-delete-me'.
```

The package.skeleton function creates a number of files. There are directories named man (for help files), R (for R source files), and data (for data files). One of the most imporant is the DESCRIPTION file, at the root of the created directory. Here is the file that was generated by the package.skeleton function:

```
Package: nutshell
Type: Package
Title: What the package does (short line)
Version: 1.0
Date: 2009-08-18
Author: Who wrote it
Maintainer: Who to complain to <yourfault@somewhere.net>
Description: More about what it does (maybe more than one line)
License: What license is it under?
LazyLoad: yes
```

Many of these items are self-explanatory, although a couple of items require more explanation. Additionally, there are a few useful optional items:

LazyLoad

LazyLoad controls how objects (including data) are loaded into R. If you set LazyLoad to yes (the default), then data files in the packages are not loaded into memory. Instead, promise objects are loaded for each data package. You can still access the objects, but they take up (almost) no space.

LazyData

LazyData works like LazyLoad but specifies what to do (specifically) with data files.

Depends

If your package depends on other packages to be installed (or on certain versions of R), you can specify them with this line. For example, to specify that your package requires R 2.8 or later and the earth package, you would add the line:

```
Depends: R(>= 2.8), nnet
```

R includes a set of functions that help automate the creation of help files for packages: prompt (for generic documentation), promptData (for documenting data files), promptMethods (for documenting methods of a generic function), and promptClass (for documenting a class). See the help files for these functions for additional information.

You can add data files to the data directory in several different forms: as R data files (created by the save function and named with either a *.rda* or a *.Rdata* suffix), as comma-separated value files (with a *.csv* suffix), or as an R source file containing R code (with a *.R* suffix).

Building the Package

After you've added all the materials to the package, you can build it from the command line on your computer (not the R shell). It's usually a good idea to start by using the check command to make sure that. For the previous example, we would use the following command:

```
% R CMD CHECK nutshell
```

You can get more information about the CMD check command by entering "R CMD CHECK --help" on the command line. To build the package, you would use the following command:

```
% R CMD build nutshell
```

As above, help is available through the --help option. If you're really interested in how to build R packages, see the manual *Writing R Extensions*, available at *http://cran.r-project.org/doc/manuals/R-exts.pdf*.

The R Language

This part gives an overview of the R programming language.

In keeping with the "Nutshell" theme, this isn't an exhaustive explanation of the inner workings of R. It is a more organized and thorough overview of R than that given in the tutorial chapter with some useful reference tables.

5

An Overview of the R Language

Learning a computer language is a lot like learning a spoken language (only much simpler). If you're just visiting a foreign country, you might learn enough phrases to get by without really understanding how the language is structured. Similarly, if you're just trying to do a couple simple things with R (like drawing some charts), you can probably learn enough from examples to get by.

However, if you want to learn a new spoken language really well, you have to learn about syntax and grammar: verb conjugation, proper articles, sentence structure, and so on. The same is true with R: if you want to learn how to program effectively in R, you'll have to learn more about the syntax and grammar.

This chapter gives an overview of the R language, designed to help you understand R code and write your own. I'll assume that you've spent a little time looking at R syntax (maybe from reading Chapter 3). Here's a quick overview of how R works.

Expressions

R code is composed of a series of *expressions*. Examples of expressions in R include assignment statements, conditional statements, and arithmetic expressions. Here are a few examples of expressions:

```
> x <- 1
> if (1 > 2) "yes" else "no"
[1] "no"
> 127 %% 10
[1] 7
```

Expressions are composed of objects and functions. You may separate expressions with new lines or with semicolons. For example, here is a series of expressions separated by semicolons:

```
> "this expression will be printed"; 7 + 13; exp(0+1i*pi)
[1] "this expression will be printed"
[1] 20
[1] -1+0i
```

Objects

All R code manipulates *objects*. The simplest way to think about an object is as a "thing" that is represented by the computer. Examples of objects in R include numeric vectors, character vectors, lists, and functions. Here are some examples of objects:

```
> # a numerical vector (with five elements)
> c(1,2,3,4,5)
[1] 1 2 3 4 5

> # a character vector (with one element)
> "This is an object too"
[1] "This is an object too"

> # a list
> list(c(1,2,3,4,5),"This is an object too", " this is a list")
[[1]]
[1] 1 2 3 4 5

[[2]]
[1] "This is an object too"

[[3]]
[1] " this is a list"

> # a function
> function(x,y) {x + y}
function(x,y) {x + y}
```

Symbols

Formally, variable names in R are called *symbols*. When you assign an object to a variable name, you are actually assigning the object to a symbol in the current environment. (Somewhat tautologically, an *environment* is defined as the set of symbols that are defined in a certain context.) For example, the statement:

```
> x <- 1
```

assigns the symbol "*x*" to the object "1" in the current environment. For a more complete discussion of symbols and environments, see Chapter 8.

Functions

A *function* is an object in R that takes some input objects (called the *arguments* of the function) and returns an output object. All work in R is done by functions. Every statement in R—setting variables, doing arithmetic, repeating code in a loop—can be written as a function. For example, suppose that you had defined a variable animals pointing to a character vector with four elements: "cow," "chicken," "pig," and "tuba." Here is a statement that assigns this variable:

```
> animals <- c("cow", "chicken", "pig", "tuba")
```

Suppose that you wanted to change the fourth element to the word "duck." Normally, you would use a statement like this:

```
> animals[4] <- "duck"
```

This statement is parsed into a function call to the [<- function. So, you could actually use this equivalent expression:*

```
> `[<-`(animals,4,"duck")
```

In practice, you would probably never write this statement as a function call; the bracket notation is much more intuitive and much easier to read. However, it is helpful to know that every operation in R is a function. Because you know that this assignment is really a function call, it means that you can inspect the code of the underlying function, search for help on this function, or create methods with the same name for your own object classes.†

Here are a few more examples of R syntax and the corresponding function calls:

```
> # pretty assignment
> apples <- 3
> # functional form of assignment
> `<-`(apples,3)
> apples
[1] 3

> # another assignment statement, so that we can compare apples and oranges
> `<-`(oranges,4)
> oranges
[1] 4

> # pretty arithmetic expression
> apples + oranges
[1] 7
> # functional form of arithmetic expression
> `+`(apples,oranges)
[1] 7

> # pretty form of if-then statement
> if (apples > oranges) "apples are better" else "oranges are better"
[1] "oranges are better"
> # functional form of if-then statement
> `if`(apples > oranges,"apples are better","oranges are better")
[1] "oranges are better"
> x <- c("apple","orange","banana","pear")

> # pretty form of vector reference
> x[2]
[1] "orange"
```

---

* This expression acts slightly differently, because the result is not printed on the R console. However, the result is the same:

```
> animals
[1] "cow"     "chicken" "pig"     "duck"
```

† See Chapter 10 for more information on object-oriented programming using R.

```
> # functional form or vector reference
> `[`(x,2)
[1] "orange"
```

Objects Are Copied in Assignment Statements

In assignment statements, most objects are immutable. R will copy the object, not just the reference to the object. For example:

```
> u <- list(1)
> v <- u
> u[[1]] <- "hat"
> u
[[1]]
[1] "hat"

> v
[[1]]
[1] 1
```

This applies to vectors, lists, and most other primitive objects in R.

This is also true in function calls. Consider the following function, which takes two arguments: a vector x and an index i. The function sets the ith element of x to 4 and does nothing else:

```
> f <- function(x,i) {x[i] = 4}
```

Suppose that we define a vector w and call f with x = w and i = 1:

```
> w <- c(10, 11, 12, 13)
> f(w,1)
```

The vector w is copied when it is passed to the function, so it is not modified by the function:

```
> w
[1] 10 11 12 13
```

The value x is modified inside the context of the function. Technically, the R interpreter copies the object assigned to w and then assigns the symbol x to point at the copy. We will talk about how you can actually create mutable objects, or pass references to objects, when we talk about environments.

Everything in R Is an Object

In the last few sections, most examples of objects were objects that stored data: vectors, lists, and other data structures. However, everything in R is an object: functions, symbols, and even R expressions.

For example, function names in R are really symbol objects that point to function objects. (That relationship is, in turn, stored in an environment object.) You can assign a symbol to refer to a numeric object and then change the symbol to refer to a function:

```
> x <- 1
> x
[1] 1
> x(2)
Error: could not find function "x"
> x <- function(i) i^2
> x
function(i) i^2
> x(2)
[1] 4
```

You can even use R code to construct new functions. If you really wanted to, you could write a function that modifies its own definition.

Special Values

There are a few special values that are used in R.

NA

In R, the NA values are used to represent missing values. (NA stands for "not available.") You may encounter NA values in text loaded into R (to represent missing values) or in data loaded from databases (to replace NULL values).

If you expand the size of a vector (or matrix or array) beyond the size where values were defined, the new spaces will have the value NA (meaning "not available"):

```
> v <- c(1,2,3)
> v
[1] 1 2 3
> length(v) <- 4
> v
[1]  1  2  3 NA
```

Inf and -Inf

If a computation results in a number that is too big, R will return Inf for a positive number and -Inf for a negative number (meaning positive and negative infinity, respectively):

```
> 2 ^ 1024
[1] Inf
> - 2 ^ 1024
[1] -Inf
```

This is also the value returned when you divide by 0:

```
> 1 / 0
[1] Inf
```

NaN

Sometimes, a computation will produce a result that makes little sense. In these cases, R will often return NaN (meaning "not a number"):

```
> Inf - Inf
[1] NaN
> 0 / 0
[1] NaN
```

NULL

Additionally, there is a null object in R, represented by the symbol NULL. (The symbol NULL always points to the same object.) NULL is often used as an argument in functions to mean that no value was assigned to the argument. Additionally, some functions may return NULL. Note that NULL is not the same as NA, Inf, -Inf, or NaN.

Coercion

When you call a function with an argument of the wrong type, R will try to coerce values to a different type so that the function will work. There are two types of coercion that occur automatically in R: coercion with formal objects and coercion with built-in types.

With generic functions, R will look for a suitable method. If no exact match exists, R will search for a coercion method that converts the object to a type for which a suitable method does exist. (The method for creating coercion functions is described in "Creating Coercion Methods" on page 127.)

Additionally, R will automatically convert between built-in object types when appropriate. R will convert from more specific types to more general types. For example, suppose that you define a vector x as follows:

```
> x <- c(1, 2, 3, 4, 5)
> x
[1] 1 2 3 4 5
> typeof(x)
[1] "double"
> class(x)
[1] "numeric"
```

Let's change the second element of the vector to the word "hat." R will change the object class to character and change all the elements in the vector to char:

```
> x[2] <- "hat"
> x
[1] "1"    "hat" "3"    "4"    "5"
> typeof(x)
[1] "character"
> class(x)
[1] "character"
```

Here is an overview of the coercion rules:

- Logical values are converted to numbers: `TRUE` is converted to `1` and `FALSE` to `0`.
- Values are converted to the simplest type required to represent all information.
- The ordering is roughly logical < integer < numeric < complex < character < list.
- Objects of type `raw` are not converted to other types.
- Object attributes are dropped when an object is coerced from one type to another.

You can inhibit coercion when passing arguments to functions by using the `AsIs` function (or, equivalently, the `I` function). For more information, see the help file for `AsIs`.

Many newcomers to R find coercion nonintuitive. Strongly typed languages (like Java) will raise exceptions when the object passed to a function is the wrong type but will not try to convert the object to a compatible type. As John Chambers (who developed the S language) describes:

> In the early coding, there was a tendency to make as many cases "work" as possible. In the later, more formal, stages the conclusion was that converting richer types to simpler automatically in all situations would lead to confusing, and therefore untrustworthy, results.‡

In practice, I rarely encounter situations where values are coerced in undesirable ways. Usually, I use R with numeric vectors that are all the same type, so coercion simply doesn't apply.

The R Interpreter

R is an interpreted language. When you enter expressions into the R console (or run an R script in batch mode), a program within the R system, called the *interpreter*, executes the actual code that you wrote. Unlike C, C++, and Java, there is no need to compile your programs into an object language. Other examples of interpreted languages are LISP, Perl, and JavaScript.

All R programs are composed of a series of expressions. These expressions often take the form of function calls. The R interpreter begins by parsing each expression, translating syntactic sugar into functional form. Next, R substitutes objects for symbols (where appropriate). Finally, R evaluates each expression, returning an *object*. For complex expressions, this process may be recursive. In some special cases (such as conditional statements), R does not evaluate all arguments to a function. As an example, let's consider the following R expression:

```
> x <- 1
```

On an R console, you would typically type `x <- 1` and then press the Enter key. The R interpreter will first translate this expression into the following function call:

```
`<-`(x, 1)
```

---

‡ From [Chambers2008], p. 154.

Next, the interpreter evaluates this function. It assigns the constant value 1 to the symbol x in the current environment and then returns the value 1.

Let's consider another example. (We'll assume it's from the same session, so that the symbol x is mapped to the value 1.)

```
> if (x > 1) "orange" else "apple"
[1] "apple"
```

Here is how the R interpreter would evaluate this expression. I typed if (x > 1) "orange" else "apple" into the R console and pressed the Enter key. The entire line is the expression that was evaluated by the R interpreter. The R interpreter parsed this expression and identified it as a set of R expressions in an if-then-else control structure. To evaluate that expression, the R interpreter begins by evaluating the *condition* (x > 1). If the condition is true, then R would evaluate the next statement (in this example, "orange"). Otherwise, R would evaluate the statement after the else keyword (in this example, "apple"). We know that *x* is equal to 1. When R evaluates the condition statement, the result is false. So, R does not evaluate the statement after the condition. Instead, R will evaluate the expression after the else keyword. The result of this expression is the character vector "apple". As you can see, this is the value that is returned on the R console.

If you are entering R expressions into the R console, then the interpreter will pass objects returned to the console to the print function.

Some functionality is implemented internally within the R system. These calls are made using the .Internal function. Many functions use .Internal to call internal R system code. For example, the graphics function plot.xy is implemented using .Internal:

```
> plot.xy
function (xy, type, pch = par("pch"), lty = par("lty"), col = par("col"),
    bg = NA, cex = 1, lwd = par("lwd"), ...)
.Internal(plot.xy(xy, type, pch, lty, col, bg, cex, lwd, ...))
<environment: namespace:graphics>
```

In a few cases, the overhead for calling .Internal within an R function is too high. R includes a mechanism to define functions that are implemented completely internally.

You can identify these functions because the body of the function contains a call to the function .Primitive. For example, the assignment operator is implemented through a primitive function:

```
> `<-`
.Primitive("<-")
```

This mechanism is only used for a few basic functions where performance is critical. You can find a current list of these functions in [RInternals2009].

Seeing How R Works

To end this overview of the R language, I wanted to share a few functions that are convenient for seeing how R works. As you may recall, R expressions are R objects. This means that it is possible to parse expressions in R, or partially evaluate expressions in R, and see how R interprets them. This can be very useful for learning how R works or for debugging R code.

As noted above, the R interpreter goes through several steps when evaluating statements. The first step is to parse a statement, changing it into proper functional form. It is possible to view in the R interpreter to see how a given expression is evaluated. As an example, let's use the same R code fragment that we used in "The R Interpreter" on page 55:

```
> if (x > 1) "orange" else "apple"
[1] "apple"
```

To show how this expression is parsed, we can use the `quote()` function. This function will parse its argument but not evaluate it. By calling `quote`, an R expression returns a "language" object:

```
> typeof(quote(if (x > 1) "orange" else "apple"))
[1] "language"
```

Unfortunately, the `print` function for language objects is not very informative:

```
> quote(if (x > 1) "orange" else "apple")
if (x > 1) "orange" else "apple"
```

However, it is possible to convert a language object into a list. By displaying the language object as a list, it is possible to see how R evaluates an expression. This is the *parse tree* for the expression:

```
> as(quote(if (x > 1) "orange" else "apple"),"list")
[[1]]
`if`

[[2]]
x > 1

[[3]]
[1] "orange"

[[4]]
[1] "apple"
```

We can also apply the `typeof` function to every element in the list to see the type of each object in the parse tree:[§]

[§] As a convenient shorthand, you can omit the `as` function because R will automatically coerce the language object to a list. This means you can just use a command like:

```
> lapply(quote(if (x > 1) "orange" else "apple"),typeof)
```

Coercion is explained in "Coercion" on page 54.

```
> lapply(as(quote(if (x > 1) "orange" else "apple"), "list"),typeof)
[[1]]
[1] "symbol"

[[2]]
[1] "language"

[[3]]
[1] "character"

[[4]]
[1] "character"
```

In this case, we can see how this expression is interpreted. Notice that some parts of the if-then statement are not included in the parsed expression (in particular, the else keyword). Also, notice that the first item in the list is a *symbol*. In this case, the symbol refers to the if function. So, although the syntax for the if-then statement is different from a function call, the R parser translates the expression into a function call before evaluating the expression. The function name is the first item, and the arguments are the remaining items in the list.

For constants, there is only one item in the returned list:

```
> as.list(quote(1))
[[1]]
[1] 1
```

By using the quote function, you can see that many constructions in the R language are just *syntactic sugar* for function calls. For example, let's consider looking up the second item in a vector x. The standard way to do this is through R's bracket notation, so the expression would be x[2]. An alternative way to represent this expression is as a function: `[`(x,2). (Function names that contain special characters need to be encapsulated in backquotes.) Both of these expressions are interpreted the same way by R:

```
> as.list(quote(x[2]))
[[1]]
`[`

[[2]]
x

[[3]]
[1] 2

> as.list(quote(`[`(x,2)))
[[1]]
`[`

[[2]]
x

[[3]]
[1] 2
```

As you can see, R interprets both of these expressions identically. Clearly, the operation is not reversible (because both expressions are translated into the same parse tree). The deparse function can take the parse tree and turn it back into properly formatted R code. (The deparse function will use proper R syntax when translating a language object back into the original code.) Here's how it acts on these two bits of code:

```
> deparse(quote(x[2]))
[1] "x[2]"
> deparse(quote(`[`(x,2)))
[1] "x[2]"
```

As you read through this book, you might want to try using quote, substitute, typeof, class, and methods to see how the R interpreter parses expressions.

6

R Syntax

This chapter contains an overview of R syntax. It's not intended to be a formal or complete description of all valid syntax in R, but just a readable description of valid R expressions.

It is possible to write almost any R expression as a function call. However, it's confusing reading lots of embedded function calls, so R provides some special syntax to make code for common operations more readable.[*]

Constants

Let's start by looking at constants. Constants are the basic building blocks for data objects in R: numbers, character values, and symbols.

Numeric Vectors

Numbers are interpreted literally in R:

```
> 1.1
[1] 1.1
> 2
[1] 2
> 2^1023
[1] 8.988466e+307
```

You may specify values in hexadecimal notation by prefixing them with 0x:

```
> 0x1
[1] 1
> 0xFFFF
[1] 65535
```

---

[*] You could write R code as a series of function calls with lots of function calls. This would look a lot like LISP code, with all the parentheses. Incidentally, the S language was inspired by LISP and uses many of the same data structures and evaluation techniques that are used by LISP interpreters.

By default, numbers in R expressions are interpreted as double-precision floating-point numbers, even when you enter simple integers:

```
> typeof(1)
[1] "double"
```

If you really want an integer, you can use the sequence notation or the as function to obtain an integer:

```
> typeof(1:1)
[1] "integer"
> typeof(as(1,"integer"))
[1] "integer"
```

The sequence operator $a:b$ will return a vector of integers between a and b. To combine an arbitrary set of numbers into a vector, use the c function:

```
> v <- c(173,12,1.12312,-93)
```

R allows a lot of flexibility when entering numbers. However, there is a limit to the size and precision of numbers that R can represent:

```
# limits of precision
> (2^1023 + 1) == 2^1023
[1] TRUE
# limits of size
> 2^1024
[1] Inf
```

In practice, this is rarely a problem. Most R users will load data from other sources on a computer (like a database) that also can't represent very large numbers.

R also supports complex numbers. Complex values are written as *real_part* +*imaginary_part*i. For example:

```
> 0+1i ^ 2
[1] -1+0i
> sqrt(-1+0i)
[1] 0+1i
> exp(0+1i * pi)
[1] -1+0i
```

Note that the sqrt function returns a value of the same type as its input; it will return the value 0+1i when passed -1+0i but will return an NaN value when just passed the numeric value -1:

```
> sqrt(-1)
[1] NaN
Warning message:
In sqrt(-1) : NaNs produced
```

Character Vectors

A character object contains all of the text between a pair of quotes. Most commonly, character objects are denoted by double quotes:

```
> "hello"
[1] "hello"
```

A character string may also be enclosed by single quotes:

```
> 'hello'
[1] "hello"
```

This can be convenient if the enclosed text contains double quotes (or vice versa). Equivalently, you may also escape the quotes by placing a backslash in front of each quote:

```
> identical("\"hello\"",'"hello"')
[1] TRUE

> identical('\'hello\'',"'hello'")
[1] TRUE
```

These examples are all vectors with only one element. To stitch together longer vectors, use the c function:

```
[1] TRUE
> numbers <- c("one","two","three","four","five")
> numbers
[1] "one"    "two"    "three" "four"   "five"
```

Symbols

An important class of constants is symbols. A symbol is an object in R that refers to another object; a symbol is the name of a variable in R. For example, let's assign the numeric value 1 to the symbol x:

```
> x <- 1
```

In this expression, x is a symbol. The statement x <- 1 means "map the symbol x to the numeric value 1 in the current environment." (We'll discuss environments in Chapter 8.)

A symbol that begins with a character and contains other characters, numbers, periods, and underscores may be used directly in R statements. Here are a few examples of symbol names that can be typed without escape characters:

```
> x <- 1
> x1 <- 1
> X1 <- 2
> x1
[1] 1
> X1
[1] 2
> x1.1 <- 1
> x1.1_1 <- 1
```

Some symbols contain special syntax. In order to refer to these objects, you enclose them in backquotes. For example, to get help on the assignment operator (<-), you would use a command like this:

```
?`<-`
```

If you really wanted to, you could use backquotes to define a symbol that contains special characters or starts with a number:

```
> `1+2=3` <- "hello"
> `1+2=3`
[1] "hello"
```

Not all words are valid as symbols; some words are reserved in R. Specifically, you can't use if, else, repeat, while, function, for, in, next, break, TRUE, FALSE, NULL, Inf, NaN, NA, NA_integer_, NA_real_, NA_complex_, NA_character_, ..., ..1, ..2, ..3, ..4, ..5, ..6, ..7, ..8, or ..9.

You can redefine primitive functions that are not on this list. For example, when you start R, the symbol c normally refers to the primitive function c, which combines elements into vectors:

```
> c
function (..., recursive = FALSE)  .Primitive("c")
```

However, you can redefine the symbol c to point to something else:

```
> c <- 1
> c
[1] 1
```

Even after you redefine the symbol c, you can continue to use the "combine" function as before:

```
> v <- c(1,2,3)
> v
[1] 1 2 3
```

See Chapter 2 for more information on the combine function.

Operators

Many functions in R can be written as operators. An *operator* is a function that takes one or two arguments and can be written without parentheses.

One familiar set of operators is binary operators for arithmetic. R supports arithmetic operations:

```
> # addition
> 1 + 19
[1] 20

> # multiplication
> 5 * 4
[1] 20
```

R also includes notation for other mathematical operations, including modulus, exponents, and integer division:

```
> # modulus
> 41 %% 21
[1] 20?

> # exponents
> 20 ^ 1
[1] 20
```

```
> # integer division
> 21 %/% 2
[1] 10
```

You can define your own binary operators. User-defined binary operators consist of a string of characters between two "%" characters. To do this, create a function of two variables and assign it to an appropriate symbol. For example, let's define an operator %myop% that doubles each operand and then adds them together:

```
> `%myop%` <- function(a, b) {2*a + 2*b}
> 1 %myop% 1
[1] 4
> 1 %myop% 2
[1] 6
```

Some language constructs are also binary operators. For example, assignment, indexing, and function calls are binary operators:†

```
> # assignment is a binary operator
> # the left side is a symbol, the right is a value
> x <- c(1,2,3,4,5)

> # indexing is a binary operator too
> # the left side is a symbol, the right is an index
> x[3]
[1] 3

> # a function call is also a binary operator
> # the left side is a symbol pointing to the function argument
> # the right side are the arguments
> max(1,2)
[1] 2
```

There are also unary operators that take only one variable. Here are two familiar examples:

```
> # negation is a unary operator
> -7
[1] -7

> # ? (for help) is also a unary operator
> ?`?`
```

Order of Operations

You may remember from high school math that you always evaluate mathematical expressions in a certain order. For example, when you evaluate the expression 1 + 2 • 5, you first multiply 2 and 5 and then add 1. The same thing is true in computer

---

† Don't be confused by the closing bracket in an indexing operation or the closing parenthesis in a function call; although this syntax uses two symbols, both operations are still technically binary operators. For example, a function call has the form *f(arguments)*, where *f* is a function and *arguments* are the arguments for the function.

languages like R. When you enter an expression in R, the R interpreter will always evaluate some expressions first.

In order to resolve ambiguity, operators in R are always interpreted in the same order. Here is a summary of the precedence rules:

- Function calls and grouping expressions
- Index and lookup operators
- Arithmetic
- Comparison
- Formulas
- Assignment
- Help

Table 6-1 shows a complete list of operators in R and their precedence.

Table 6-1. Operator precedence[a]

Operators (in order of priority)	Description
({	Function calls and grouping expressions (respectively)
[[[	Indexing
:: :::	Access variables in a namespace
$ @	Component / slot extraction
^	Exponentiation (right to left)
- +	Unary minus and plus
:	Sequence operator
%any%	Special operators
* /	Multiply, divide
+ -	(Binary) add, subtract
< > <= >= == !=	Ordering and comparison
!	Negation
& &&	And
| ||	Or
~	As in formulas
-> ->>	Rightward assignment
=	Assignment (right to left)
<- <<-	Assignment (right to left)
?	Help (unary and binary)

[a] From the help(syntax) file

For a current list of built-in operators and their precedence, see the help file for syntax.

Assignments

Most assignments that we've seen so far simply assign an object to a symbol. For example:

```
> x <- 1

> y <- list(shoes="loafers", hat="Yankees cap", shirt="white")

> z <- function(a,b,c) {a ^ b / c}

> v <- c(1,2,3,4,5,6,7,8)
```

There is an alternative type of assignment statement in R that acts differently: assignments with a function on the lefthand side of the assignment operator. These statements replace an object with a new object that has slightly different properties. Here are a few examples:

```
> dim(v) <- c(2,4)

> v[2,2] <- 10

> formals(z) <- alist(a=1,b=2,c=3)
```

There is a little bit of magic going on behind the scenes. An assignment statement of the form:

> *fun*(*sym*) <- *val*

is really syntactic sugar for a function of the form:

> `*fun<-*`(*sym,val*)

Each of these functions replaces the object associated with **sym** in the current environment. By convention, *fun* refers to a property of the object represented by **sym**. If you write a method with the name *method_name<-*, then R will allow you to place *method_name* on the lefthand side of an assignment statement.

Expressions

R provides different constructs for grouping together expressions: semicolons, parentheses, and curly braces.

Separating Expressions

You can write a series of expressions on separate lines:

```
> x <- 1
> y <- 2
> z <- 3
```

Alternatively, you can place them on the same line, separated by semicolons:

```
> x <- 1; y <- 2; z <- 3
```

Parentheses

The parentheses notation returns the result of evaluating the expression inside the parentheses:

(*expression*)

The operator has the same precedence as a function call. In fact, grouping a set of expressions inside parentheses is equivalent to evaluating a function of one argument that just returns its argument:

```
> 2 * (5 + 1)
[1] 12
> # equivalent expression
> f <- function (x) x
> 2 * f(5 + 1)
[1] 12
```

Grouping expressions with parentheses can be used to override the default order of operations. For example:

```
> 2 * 5 + 1
[1] 11
> 2 * (5 + 1)
[1] 12
```

Curly Braces

Curly braces are used to evaluate a series of expressions (separated by new lines or semicolons) and return only the last expression:

{*expression_1*; *expression_2*; ... *expression_n*}

Often, curly braces are used to group a set of operations in the body of a function:

```
> f <- function() {x <- 1; y <- 2; x + y}
> f()
[1] 3
```

However, curly braces can also be used as expressions in other contexts:

```
> {x <- 1; y <- 2; x + y}
[1] 3
```

The contents of the curly braces are evaluated inside the current environment; a new environment is created by a function call but *not* by the use of curly braces:

```
> # when evaluated in a function, u and v are assigned
> # only inside the function environment
> f <- function() {u <- 1; v <- 2; u + v}
> u
Error: object "u" not found
> v
Error: object "v" not found
> # when evaluated outside the function, u and v are
> # assigned in the current environment
> {u <- 1; v <- 2; u + v}
[1] 3
```

```
> u
[1] 1
> v
[1] 2
```

For more information about variable scope and environments, see Chapter 8.

The curly brace notation is translated internally as a call to the `` `{` `` function. (Note, however, that the arguments are not evaluated the same way as in a standard function.)

Control Structures.

Nearly every operation in R can be written as a function, but it isn't always convenient to do so. Therefore, R provides special syntax that you can use in common program structures. We've already described two important sets of constructions: operators and grouping brackets. This section describes a few other key language structures and explains what they do.

Conditional Statements

Conditional statements take the form:

```
if (condition) true_expression else false_expression
```

or, alternatively:

```
if (condition) expression
```

Because the expressions *expression*, *true_expression*, and *false_expression* are not always evaluated, the function if has the type special:

```
> typeof(`if`)
[1] "special"
```

Here are a few examples of conditional statements:

```
> if (FALSE) "this will not be printed"
> if (FALSE) "this will not be printed" else "this will be printed"
[1] "this will be printed"
> if (is(x, "numeric")) x/2 else print("x is not numeric")
[1] 5
```

In R, conditional statements are not vector operations. If the *condition* statement is a vector of more than one logical value, only the first item will be used. For example:

```
> x <- 10
> y <- c(8, 10, 12, 3, 17)
> if (x < y) x else y
[1]  8 10 12  3 17
Warning message:
In if (x < y) x else y :
  the condition has length > 1 and only the first element will be used
```

If you would like a vector operation, use the `ifelse` function instead:

```
> a <- c("a","a","a","a","a")
> b <- c("b","b","b","b","b")
> ifelse(c(TRUE,FALSE,TRUE,FALSE,TRUE),a,b)
[1] "a" "b" "a" "b" "a"
```

Loops

There are three different looping constructs in R. Simplest is **repeat**, which just repeats the same expression:

```
repeat expression
```

To stop repeating the expression, you can use the keyword **break**. To skip to the next iteration in a loop, you can use the command **next**.

As an example, the following R code prints out multiples of 5 up to 25:

```
> i <- 5
> repeat {if (i > 25) break else {print(i); i <- i + 5;}}
[1] 5
[1] 10
[1] 15
[1] 20
[1] 25
```

If you do not include a **break** command, the R code will be an infinite loop. (This can be useful for creating an interactive application.)

Another useful construction is **while** loops, which repeat an expression while a condition is true:

```
while (condition) expression
```

As a simple example, let's rewrite the example above using a **while** loop:

```
> i <- 5
> while (i <= 25) {print(i); i <- i + 5}
[1] 5
[1] 10
[1] 15
[1] 20
[1] 25
```

You can also use **break** and **next** inside **while** loops. The **break** statement is used to stop iterating through a loop. The **next** statement skips to the next loop iteration without evaluating the remaining expressions in the loop body.

Finally, R provides **for** loops, which iterate through each item in a vector (or a list):

```
for (var in list) expression
```

Let's use the same example for a **for** loop:

```
> for (i in seq(from=5,to=25,by=5)) print(i)
[1] 5
[1] 10
[1] 15
```

```
[1] 20
[1] 25
```

You can also use break and next inside for loops.

There are two important properties of looping statements to remember. First, results are not printed inside a loop unless you explicitly call the print function. For example:

```
> for (i in seq(from=5,to=25,by=5)) i
```

Second, the variable *var* that is set in a for loop is changed in the calling environment:

```
> i <- 1
> for (i in seq(from=5,to=25,by=5)) i
> i
[1] 25
```

Like conditional statements, the looping functions `repeat`, `while`, and `for` have type special, because *expression* is not necessarily evaluated.

Looping Extensions

If you've used modern programming languages like Java, you might be disappointed that R doesn't provide iterators or foreach loops. Luckily, these mechanisms are available through add-on packages. (These packages were written by Revolution Computing and are available through CRAN.)

Iterators are an abstract object that return elements from another object. Using iterators can help make code easier to understand. Additionally, iterators can make code easier to parallelize. To use iterators, you'll need to install the iterators package. Iterators can return elements of a vector, array, data frame, or other object. You can even use an iterator to return values returned by a function (such as a function that returns random values). To create an iterator in R, you would use the iter function:

```
iter(obj, checkFunc=function(...) TRUE, recycle=FALSE,...)
```

The argument obj specifies the object, recycle specifies whether the iterator should reset when it runs out of elements, and checkFunc specifies a function that filters values returned by the iterator.

You fetch the next item with the function nextElem. This function will implicitly call checkFunc. If the next value matches checkFunc, it will be returned. If it doesn't match, the function will try another value. nextElem will continue checking values until it finds one that matches checkFunc, or it runs out of values. When there are no elements left, the iterator calls stop with the message "StopIteration."

For example, let's create an iterator that returns values between 1 and 5:

```
> library(iterators)
> onetofive <- iter(1:5)
> nextElem(onetofive)
[1] 1
> nextElem(onetofive)
[1] 2
> nextElem(onetofive)
```

```
[1] 3
> nextElem(onetofive)
[1] 4
> nextElem(onetofive)
[1] 5
> nextElem(onetofive)
Error: StopIteration
```

A second extension is the foreach loop, available through the foreach package. Foreach provides an elegant way to loop through multiple elements of another object (such as a vector, matrix, data frame, or iterator), evaluate an expression for each element, and return the results. Within the foreach function, you assign elements to a temporary value, just like in a for loop.

Here is the prototype for the foreach function:

```
foreach(..., .combine, .init, .final=NULL, .inorder=TRUE,
        .multicombine=FALSE,
        .maxcombine=if (.multicombine) 100 else 2,
        .errorhandling=c('stop', 'remove', 'pass'),
        .packages=NULL, .export=NULL, .noexport=NULL,
        .verbose=FALSE)
```

Technically, the foreach function returns a foreach object. To actually evaluate the loop, you need to apply the foreach loop to an R expression using the %do% or %dopar% operators. That sounds weird, but it's actually pretty easy to use in practice. For example, you can use a foreach loop to calculate the square roots of numbers between 1 and 5:

```
> sqrts.1to5 <- foreach(i=1:5) %do% sqrt(i)
> sqrts.1to5
[[1]]
[1] 1

[[2]]
[1] 1.414214

[[3]]
[1] 1.732051

[[4]]
[1] 2

[[5]]
[1] 2.236068
```

The %do% operator evaluates the expression in serial, while the %dopar% can be used to evaluate expressions in parallel. For more about parallel computing with R, see "Parallel Computation with R" on page 139.

Accessing Data Structures

R has some specialized syntax for accessing data structures. You can fetch a single item from a structure, or multiple items (possibly as a multidimensional array) using

R's index notation. You can fetch items by location within a data structure or by name.

Data Structure Operators

Table 6-2 shows the operators in R used for accessing objects in a data structure.

Table 6-2. Data structure access notation

Syntax	Objects	Description
x[i]	Vectors, lists	Returns objects from object x, described by i. i may be an integer vector, character vector (of object names), or logical vector. Does not allow partial matches. When used with lists, returns a list. When used with vectors, returns a vector.
x[[i]]	Vectors, lists	Returns a single element of x, matching i. i may be an integer or character vector of length 1. Allows partial matches (with exact=FALSE option).
x$n	Lists	Returns object with name n from object x.
x@n	S4 objects	Returns element stored in slot named n.

Although the single-bracket notation and double-bracket notation look very similar, there are three important differences. First, double brackets always return a single element, while single brackets may return multiple elements. Second, when elements are referred to by name (as opposed to by index), single brackets only match named objects exactly, while double brackets allow partial matches. Finally, when used with lists, the single-bracket notation returns a list, but the double-bracket notation returns a vector.

I'll explain how to use this notation below.

Indexing by Integer Vector

The most familiar way to look up an element in R is by numeric vector. As an example, let's create a very simple vector of 20 integers:

```
> v <- 100:119
```

You can look up individual elements by position in the vector using the bracket notation x[s], where x is the vector from which you want to return elements and s is a second vector representing the set of element indices that you would like to query. You can use an integer vector to look up a single element or multiple elements:

```
> v[5]
[1] 104
> v[1:5]
[1] 100 101 102 103 104
> v[c(1,6,11,16)]
[1] 100 105 110 115
```

As a special case, you can use the double-bracket notation to reference a single element:

```
> v[[3]]
[1] 102
```

The double-bracket notation works the same as the single-bracket notation in this case; see "Indexing by Name" on page 76 for an explanation of references that do not work with the single-bracket notation.

You can also use negative integers to return a vector consisting of all elements except the specified elements:

```
> # exclude elements 1:15 (by specifying indexes -1 to -15)
> v[-15:-1]
[1] 115 116 117 118 119
```

The same notation applies to lists:

```
> l <- list(a=1,b=2,c=3,d=4,e=5,f=6,g=7,h=8,i=9,j=10)
> l[1:3]
$a
[1] 1

$b
[1] 2

$c
[1] 3

> l[-7:-1]
$h
[1] 8

$i
[1] 9

$j
[1] 10
```

You can also use this notation to extract parts of multidimensional data structures:

```
> m <- matrix(data=c(101:112),nrow=3,ncol=4)
> m
     [,1] [,2] [,3] [,4]
[1,]  101  104  107  110
[2,]  102  105  108  111
[3,]  103  106  109  112
> m[3]
[1] 103
> m[3,4]
[1] 112
> m[1:2,1:2]
     [,1] [,2]
[1,]  101  104
[2,]  102  105
```

If you omit a vector specifying a set of indices for a dimension, then elements for all indices are returned:

```
> m[1:2,]
     [,1] [,2] [,3] [,4]
[1,]  101  104  107  110
[2,]  102  105  108  111
```

```
> m[3:4]
[1] 103 104
> m[,3:4]
     [,1] [,2]
[1,]  107  110
[2,]  108  111
[3,]  109  112
```

When selecting a subset, R will automatically coerce the result to the most appropriate number of dimensions. If you select a subset of elements that corresponds to a matrix, R will return a matrix object; if you select a subset that corresponds to only a vector, R will return a vector object. To disable this behavior, you can use the `drop=FALSE` option:

```
> a <- array(data=c(101:124),dim=c(2,3,4))
> class(a[1,1,])
[1] "integer"
> class(a[1,,])
[1] "matrix"
> class(a[1:2,1:2,1:2])
[1] "array"
> class(a[1,1,1,drop=FALSE])
[1] "array"
```

It is possible to create an array object with dimensions of length 1. However, when selecting subsets, R simplifies the returned objects.

It is also possible to replace elements in a vector, matrix, or array using the same notation:

```
> m[1] <- 1000
> m
     [,1] [,2] [,3] [,4]
[1,] 1000  104  107  110
[2,]  102  105  108  111
[3,]  103  106  109  112
> m[1:2,1:2] <- matrix(c(1001:1004),nrow=2,ncol=2)
> m
     [,1] [,2] [,3] [,4]
[1,] 1001 1003  107  110
[2,] 1002 1004  108  111
[3,]  103  106  109  112
```

It is even possible to extend a data structure using this notation. A special `NA` element is used to represent values that are not defined:

```
> v <- 1:12
> v[15] <- 15
> v
 [1]  1  2  3  4  5  6  7  8  9 10 11 12 NA NA 15
```

You can also index a data structure by a factor; the factor is interpreted as an integer vector.

R Syntax

Indexing by Logical Vector

As an alternative to indexing by an integer vector, you can also index through a logical vector. As a simple example, let's construct a vector of alternating true and false elements to apply to v:

```
> rep(c(TRUE,FALSE),10)
 [1]  TRUE FALSE  TRUE FALSE  TRUE FALSE  TRUE FALSE  TRUE FALSE  TRUE
[12] FALSE  TRUE FALSE  TRUE FALSE  TRUE FALSE  TRUE FALSE
> v[rep(c(TRUE,FALSE),10)]
 [1] 100 102 104 106 108 110 112 114 116 118
```

Often, it is useful to calculate a logical vector from the vector itself:

```
> # trivial example: return element that is equal to 103
> v[(v==103)]
> # more interesting example: multiples of three
> v[(v %% 3 == 0)]
[1] 102 105 108 111 114 117
```

The index vector does not need to be the same length as the vector itself. R will repeat the shorter vector, returning matching values:

```
> v[c(TRUE,FALSE,FALSE)]
[1] 100 103 106 109 112 115 118
```

As above, the same notation applies to lists:

```
> l[(l > 7)]
$h
[1] 8

$i
[1] 9

$j
[1] 10
```

Indexing by Name

With lists, each element may be assigned a name. You can index an element by name using the $ notation:

```
> l <- list(a=1,b=2,c=3,d=4,e=5,f=6,g=7,h=8,i=9,j=10)
> l$j
[1] 10
```

You can also use the single-bracket notation to index a set of elements by name:

```
> l[c("a","b","c")]
$a
[1] 1

$b
[1] 2

$c
[1] 3
```

You can also index by name using the double-bracket notation when selecting a single element. It is even possible to index by partial name using the exact=FALSE option:

```
> dairy <- list(milk="1 gallon", butter="1 pound", eggs=12)
> dairy$milk
[1] "1 gallon"
> dairy[["milk"]]
[1] "1 gallon"
> dairy[["mil"]]
NULL
> dairy[["mil",exact=FALSE]]
[1] "1 gallon"
```

Sometimes, an object is a list of lists. You can also use the double-bracket notation to reference an element in this type of data structure. To do this, use a vector as an argument. R will iterate through the elements in the vector, referencing sublists:

```
> fruit <- list(apples=6, oranges=3, bananas=10)
> shopping.list <- list (dairy, fruit)
> shopping.list
$dairy
$dairy$milk
[1] "1 gallon"

$dairy$butter
[1] "1 pound"

$dairy$eggs
[1] 12


$fruit
$fruit$apples
[1] 6

$fruit$oranges
[1] 3

$fruit$bananas
[1] 10


> shopping.list[[c("dairy", "milk")]]
[1] "1 gallon"
> shopping.list[[c(1,2)]]
[1] "1 pound"
```

R Code Style Standards

Standards for code style aren't the same as syntax, although they are sort of related. It is usually wise to be careful about code style to maximize the readability of your code, making it easier for you and others to maintain.

In this book, I've tried to stick to Google's R Style Guide, which is available at *http://google-styleguide.googlecode.com/svn/trunk/google-r-style.html*. Here is a summary of its suggestions:

Indentation
> Indent lines with two spaces, not tabs. If code is inside parentheses, indent to the innermost parentheses.

Spacing
> Use only single spaces. Add spaces between binary operators and operands. Do not add spaces between a function name and the argument list. Add a single space between items in a list, after each comma.

Blocks
> Don't place an opening brace ("{") on its own line. Do place a closing brace ("}") on its own line. Indent inner blocks (by two spaces).

Semicolons
> Omit semicolons at the end of lines when they are optional.

Naming
> Name objects with lowercase words, separated by periods. For function names, capitalize the name of each word that is joined together, with no periods. Try to make function names verbs.

Don't be confused by the object names. You don't have to name objects things like "field.goals" or "sanfrancisco.home.prices" or "top.bacon.searching.cities." It's just convention.

7

R Objects

All objects in R are built on top of a basic set of built-in objects. The *type* of an object defines how it is stored in R. Objects in R are also members of a *class*. Classes define what information objects contain, and how those objects may be used.

R provides some mechanisms for object-oriented programming (which doesn't just mean "programming with objects"). This chapter focuses on built-in objects and how to use them and not on the object-oriented programming system. We'll discuss object-oriented programming features like class definitions, inheritance, and methods in Chapter 10.

Primitive Object Types

Table 7-1 shows all of the built-in object types. I introduced these objects in Chapter 3, so they should seem familiar. I classified the object types into a few categories, to make them easier to understand.

Basic vectors
These are vectors containing a single type of value: integers, floating-point numbers, complex numbers, text, logical values, or raw data.

Compound objects
These objects are containers for the basic vectors: lists, pairlists, S4 objects, and environments. Each of these objects has unique properties (described below), but each of them contains a number of named objects.

Special objects
These objects serve a special purpose in R programming: any, NULL, and .... Each of these means something important in a specific context, but you would never create an object of these types.

R language
These are objects that represent R code; they can be evaluated to return other objects.

Functions

Functions are the workhorses of R; they take arguments as inputs and return objects as outputs. Sometimes, they may modify objects in the environment or cause side effects outside the R environment like plotting graphics, saving files, or sending data over the network.

Internal

These are object types that are formally defined by R, but which aren't normally accessible within the R language. In normal R programming, you will probably never encounter any of the objects.

We'll explore what each of these objects is used for in this chapter.

Table 7-1. Primitive object types in R

Category	Object type	Description	Example
Vectors	integer	Naturally produced from sequences. Can be coerced with the `integer()` function.	5:5 integer(5)
	double	Used to represent floating-point numbers (numbers with decimals and large numbers). On most modern platforms, this will be 8 bytes, or 64 bits. By default, most numerical values are represented as doubles. Can be coerced with the `double()` function.	1 -1 2 ** 50 double(5)
	complex	Complex numbers. To use, you must include both the real and the imaginary parts (even if the real part is 0).	2+3i 0+1i exp(0+1i * pi)
	character	A string of characters (just called a string in many other languages).	"Hello world."
	logical	Represents Boolean values.	TRUE FALSE
	raw	A vector containing raw bytes. Useful for encoding objects from outside the R environment.	raw(8) CharToRaw("Hello")
Compound	list	A (possibly heterogeneous) collection of other objects. Elements of a list may be named. Many other object types in R (such as data frames) are implemented as lists.	list(1, 2, "hat")
	pairlist	A data structure used to represent a set of name-value pairs. Pairlists are primarily used internally but can be created at the user level. Their use is deprecated in user-level programs, because standard list objects are just as efficient and more flexible.	.Options pairlist(apple=1, pear=2, banana=3)
	S4	An R object supporting modern object-oriented paradigms (inheritance, methods, etc.). See Chapter 10 for a full explanation.	

Category	Object type	Description	Example
	environment	An R environment describes the set of symbols available in a specific context. An environment contains a set of symbol-value pairs and a pointer to an enclosing environment. (For example, you could use any in the signature of a default generic function.)	.GlobalEnv new.env(parent = baseenv())
Special	any	An object used to mean that "any" type is OK. Used to prevent coercion from one type to another. Useful in defining slots in S4 objects or signatures for generic functions.	setClass("Something", representation(data="ANY"))
	NULL	An object that means "there is no object." Returned by functions and expressions whose value is not defined. The NULL object can have no attributes.	NULL
	...	Used in functions to implement variable-length argument lists, particularly arguments passed to other functions.	N/A
R language	symbol	A symbol is a language object that refers to other objects. Usually encountered when parsing R statements.	as.name(x) as.symbol(x) quote(x)
	promise	Promises are objects that are not evaluated when they are created but are instead evaluated when they are first used. They are used to implement delayed loading of objects in packages.	> x <- 1; > y <- 2; > z <- 3 > delayedAssign("v", c(x, y, z)) > # v is a promise
	language	R language objects are used when processing the R language itself.	quote(function(x) { x + 1})
	expression	An unevaluated R expression. Expression objects can be created with the `expression` function, and later evaluated with the `eval` function.	expression(1 + 2)
Functions	closure	An R function not implemented inside the R system. Most functions fall into this category. Includes user-defined functions, most functions included with R, and most functions in R packages.	f <- function(x) { x + 1} print
	special	An internal function whose arguments are not necessarily evaluated on call.	if [
	builtin	An internal function that evaluates its arguments.	+ ^
Internal	char	A scalar "string" object. A character vector is composed of `char`'s. (Users can't easily generate a `char` object but don't ever need to.)	N/A

Category	Object type	Description	Example
	bytecode	A data type reserved for a future byte-code compiler.	N/A
	externalptr	External pointer. Used in C code.	N/A
	weakref	Weak reference (internal only).	N/A

Vectors

When using R, you will frequently encounter the six basic vector types. R includes several different ways to create a new vector. The simplest one is the c function, which *combines* its arguments into a vector:

```
> # a vector of four numbers
> v <- c(.295, .300, .250, .287, .215)
> v
[1] 0.295 0.300 0.250 0.287 0.215
```

The c function also *coerces* all of its arguments into a single type:

```
> # creating a vector from four numbers and a char
> v <- c(.295, .300, .250, .287, "zilch")
> v
[1] "0.295" "0.3"   "0.25"  "0.287" "zilch"
```

You can use the c function to recursively assemble a vector from other data structures using the recursive=TRUE option:

```
> # creating a vector from four numbers and a list of
> # three more
> v <- c(.295, .300, .250, .287, list(.102, .200, .303), recursive=TRUE)
> v
[1] 0.295 0.300 0.250 0.287 0.102 0.200 0.303
```

But beware of using a list as an argument, as you will get back a list:

```
> v <- c(.295, .300, .250, .287, list(.102, .200, .303), recursive=TRUE)
> v
[1] 0.295 0.300 0.250 0.287 0.102 0.200 0.303
> typeof(v)
[1] "double"
> v <- c(.295, .300, .250, .287, list(1, 2, 3))
> typeof(v)
[1] "list"
> class(v)
[1] "list"
> v
[[1]]
[1] 0.295

[[2]]
[1] 0.3

[[3]]
[1] 0.25
```

```
[[4]]
[1] 0.287

[[5]]
[1] 1

[[6]]
[1] 2

[[7]]
[1] 3
```

Another useful tool for assembling a vector is the ":" operator. This operator creates a sequence of values from the first operand to the second operand:

```
> 1:10
 [1]  1  2  3  4  5  6  7  8  9 10
```

A more flexible function is the seq function:

```
> seq(from=5,to=25,by=5)
[1]  5 10 15 20 25
```

You can explicitly manipulate the length of a vector through the length attribute:

```
> w <- 1:10
> w
 [1]  1  2  3  4  5  6  7  8  9 10
> length(w) <- 5
> w
[1] 1 2 3 4 5
```

Note that when you expand the length of a vector, uninitialized values are given the NA value:

```
> length(w) <- 10
> w
 [1]  1  2  3  4  5 NA NA NA NA NA
```

Lists

An R *list* is an ordered collection of objects. Like vectors, you can refer to elements in a list by position:

```
> l <- list(1,2,3,4,5)
> l[1]
[[1]]
[1] 1

> l[[1]]
[1] 1
```

Additionally, each element in a list may be given a *name* and then be referred to by that name. For example, suppose that we wanted to represent a few properties of a parcel (a real, physical parcel, to be sent through the mail). Suppose the parcel is destined for New York, has dimensions of 2 inches deep by 6 inches wide by 9 inches long, and costs $12.95 to mail. The three properties are all different data types in R: a character, a numeric vector of length 3, and a vector of length 1. We could combine the information into an object like this:

```
> parcel <- list(destination="New York",dimensions=c(2,6,9),price=12.95)
```

It is then possible to refer to each component individually using the $ notation. For example, if we wanted to get the price, we would use the following expression:

```
> parcel$price
[1] 12.95
```

Lists are a very important building block in R, because they allow the construction of heterogeneous structures. For example, data frames are built on lists.

Other Objects

There are some other objects that you should know about if you're using R. Although most of these objects are not formally part of the R language, they are used in so many R packages, or get special treatment in R, that they're worth a closer look.

Matrices

A *matrix* is an extension of a vector to two dimensions. A matrix is used to represent two-dimensional data of a single type. A clean way to generate a new matrix is with the `matrix` function. As an example, let's create a matrix object with three columns and four rows. We'll give the rows the names "r1," "r2," "r3," and "r4," and the columns the names "c1," "c2," and "c3."

```
> m <- matrix(data=1:12,nrow=4,ncol=3,
+             dimnames=list(c("r1","r2","r3","r4"),
+                           c("c1","c2","c3")))
> m
   c1 c2 c3
r1  1  5  9
r2  2  6 10
r3  3  7 11
r4  4  8 12
```

It is also possible to transform another data structure into a matrix using the `as.matrix` function.

An important note: matrices are implemented as vectors, not as a vector of vectors (or as a list of vectors). Array subscripts are used for referencing elements and don't reflect the way the data is stored. (Unlike other classes, matrices don't have an explicit class attribute. We'll talk about attributes in "Attributes" on page 92.)

Arrays

An array is an extension of a vector to more than two dimensions. Vectors are used to represent multidimensional data of a single type. As above, you can generate an array with the `array` function:

```
> a <- array(data=1:24,dim=c(3,4,2))
> a
, , 1

     [,1] [,2] [,3] [,4]
[1,]    1    4    7   10
[2,]    2    5    8   11
[3,]    3    6    9   12

, , 2

     [,1] [,2] [,3] [,4]
[1,]   13   16   19   22
[2,]   14   17   20   23
[3,]   15   18   21   24
```

Like matrices, the underlying storage mechanism for an array is a vector. (Like matrices, and unlike most other classes, matrices don't have an explicit class attribute.)

Factors

When analyzing data, it's quite common to encounter categorical values. For example, suppose that you have a set of observations about people that includes eye color. You could represent the eye colors as a character array:

```
> eye.colors <- c("brown","blue","blue","green","brown","brown","brown")
```

This is a perfectly valid way to represent the information, but it can become inefficient if you are working with large names or a large number of observations. R provides a better way to represent categorical values, by using factors. A *factor* is an ordered collection of items. The different values that the factor can take are called *levels*.

Let's recode the eye colors as a factor:

```
> eye.colors <- factor(c("brown", "blue", "blue", "green",
+                         "brown", "brown", "brown"))
```

The `levels` function shows all the levels from a factor:

```
> levels(eye.colors)
[1] "blue"  "brown" "green"
```

Printing a factor shows slightly different information than printing a character vector. In particular, notice that the quotes are not shown and that the levels are explicitly printed:

```
> eye.colors
[1] brown blue  blue  green brown brown brown
Levels: blue brown green
```

In the eye color example, order did not matter. However, sometimes the order of the factors matters for a specific problem. For example, suppose that you had conducted a survey and asked respondents how they felt about the statement "melon is delicious with an omelet." Furthermore, suppose that you allowed respondents to give the following responses: Strongly Disagree, Disagree, Neutral, Agree, Strongly Agree.

There are multiple ways to represent this information in R. You could code these as integers (for example, on a scale of 1 to 5), although this approach has some drawbacks. This approach implies a specific quantitative relationship between values, which may or may not make sense. For example, is the difference between Strongly Disagree and Disagree the same as the difference between Disagree and Neutral? A numeric reponse also implies that you can calculate meaningful statistics based on the responses. Can you be sure that a Disagree response and an Agree response average out to Neutral?

To get around these problems, you can use an ordered factor to represent the response of this survey. Here is an example:

```
> survey.results <- factor(
+    c("Disagree", "Neutral", "Strongly Disagree",
+      "Neutral", "Agree", "Strongly Agree",
+      "Disagree", "Strongly Agree", "Neutral",
+      "Strongly Disagree", "Neutral", "Agree"),
+    levels=c("Strongly Disagree", "Disagree",
+      "Neutral", "Agree", "Strongly Agree"),
+    ordered=TRUE)
> survey.results
 [1] Disagree           Neutral         Strongly Disagree
 [4] Neutral            Agree           Strongly Agree
 [7] Disagree           Strongly Agree  Neutral
[10] Strongly Disagree Neutral          Agree
5 Levels: Strongly Disagree < Disagree < Neutral < ... < Strongly Agree
```

As you can see, R will display the order of the levels when you display an ordered factor.

Factors are implemented internally using integers. The levels attribute maps each integer to a factor level. Integers take up a small, fixed amount of storage space, so they can be more space efficient than character vectors. It's possible to take a factor and turn it into an integer array:

```
> # use the eye colors vector we used above
> eye.colors
[1] brown blue  blue  green brown brown brown
Levels: blue brown green
> class(eye.colors)
[1] "factor"
> # now create a vector by removing the class:
> eye.colors.integer.vector <- unclass(eye.colors)
> eye.colors.integer.vector
[1] 2 1 1 3 2 2 2
attr(,"levels")
[1] "blue"  "brown" "green"
> class(eye.colors.integer.vector)
[1] "integer"
```

It's possible to change this back to a factor by setting the class attribute:

```
> class(eye.colors.integer.vector) <- "factor"
> eye.colors.integer.vector
[1] brown blue  blue  green brown brown brown
Levels: blue brown green
> class(eye.colors.integer.vector)
[1] "factor"
```

Data Frames

Data frames are a useful way to represent tabular data. In scientific contexts, many experiments consist of individual observations, each of which involves several different measurements. Often, the measurements have different dimensions, and sometimes they are qualitative and not quantitative. In business contexts, data is often kept in database tables. Each table has many rows, which may consist of multiple "columns" representing different quantities and which may be kept in multiple formats. A data frame is a natural way to represent these data sets in R.

A data frame represents a table of data. Each column may be a different type, but each row in the data frame must have the same length:

```
> data.frame(a=c(1,2,3,4,5),b=c(1,2,3,4))
Error in data.frame(a = c(1, 2, 3, 4, 5), b = c(1, 2, 3, 4)) :
  arguments imply differing number of rows: 5, 4
```

Usually, each column is named, and sometimes rows are named as well. The columns in a data frame are often referred to as "variables."

Here is a simple example of a data frame, showing how frequently users search for the word "bacon" in different cities around the world.[*]

This data set is included in the `nutshell` package. Alternatively, you can create it manually with the following statement:

```
> top.bacon.searching.cities <- data.frame(
+      city = c("Seattle","Washington","Chicago",
+               "New York","Portland","St Louis",
+               "Denver","Boston","Minneapolis","Austin",
+               "Philadelphia","San Francisco","Atlanta",
+               "Los Angeles","Richardson"),
+      rank = c(100, 96, 94, 93, 93, 92, 90, 90, 89, 87,
+               85, 84, 82, 80, 80)
+  )
```

Here is what this data frame contains:

```
> top.bacon.searching.cities
            city rank
1        Seattle  100
```

---

[*] The data was taken from Google Insights, *http://www.google.com/insights/search/ #q=bacon&cmpt=q*. The query was run on September 5, 2009, for data from 2004 through 2009.

The fact that I could find this information is a sign that there is too much data in the world. It is probably good that you are learning to use R, or you would never be able to make sense of it all.

```
 2     Washington  96
 3        Chicago  94
 4       New York  93
 5       Portland  93
 6       St Louis  92
 7         Denver  90
 8         Boston  90
 9    Minneapolis  89
10         Austin  87
11   Philadelphia  85
12  San Francisco  84
13        Atlanta  82
14    Los Angeles  80
15     Richardson  80
```

Data frames are implemented as lists with class `data.frame`:

```
> typeof(top.bacon.searching.cities)
[1] "list"
> class(top.bacon.searching.cities)
[1] "data.frame"
```

This means that the same methods can be used to refer to items in lists and data frames. For example, to extract the rank column from this data frame, you could use the expression `top.bacon.searching.cities$rank`.

Formulas

Very often, you need to express a relationship between variables. Sometimes, you want to plot a chart showing the relationship between the two variables. Other times, you want to develop a mathematical model. R provides a *formula* class that lets you describe the relationship for both purposes.

Let's create a formula as an example:

```
> sample.formula <- as.formula(y~x1+x2+x3)
> class(sample.formula)
[1] "formula"
> typeof(sample.formula)
[1] "language"
```

This formula means "y is a function of x1, x2, and x3." Some R functions use more complicated formulas. For example, in "Charts and Graphics" on page 28, we plotted a formula of the form `Amount~Year|Food`, which means "Amount is a function of Year, conditioned on Food." Here is an explanation of the meaning of different items in formulas:

Variable names
> Represent variable names.

Tilde (~)
> Used to show the relationship between the response variables (to the left) and the stimulus variables (to the right).

Plus sign (+)
> Used to express a linear relationship between variables.

Zero (0)

　When added to a formula, indicates that no intercept term should be included. For example:

```
y~u+w+v+0
```

Vertical bar (|)

　Used to specify conditioning variables (in lattice formulas; see "Customizing Lattice Graphics" on page 308).

Identity function (I())

　Used to indicate that the enclosed expression should be interpreted by its arithmetic meaning. For example:

```
a+b
```

means that both *a* and *b* should be included in the formula. The formula:

```
I(a+b)
```

means that "*a* plus *b*" should be included in the formula.

Asterisk (*)

　Used to indicate interactions between variables. For example:

```
y~(u+v)*w
```

is equivalent to:

```
y~u+v+w+I(u*w)+I(v*w)
```

Caret (^)

　Used to indicate crossing to a specific degree. For example:

```
y~(u+w)^2
```

is equivalent to:

```
y~(u+w)*(u+w)
```

Function of variables

　Indicates that the function of the specified variables should be interpreted as a variable. For example:

```
y~log(u)+sin(v)+w
```

Some additional items have special meaning in formulas, for example, s() for smoothing splines in formulas passed to gam. We'll revisit formulas in Chapter 15 and Chapter 20.

Time Series

Many important problems look at how a variable changes over time, and R includes a class to represent this data: time series objects. Regression functions for time series (like ar or arima) use time series objects. Additionally, many plotting functions in R have special methods for time series.

To create a time series object (of class `"ts"`), use the `ts` function:

```
ts(data = NA, start = 1, end = numeric(0), frequency = 1,
   deltat = 1, ts.eps = getOption("ts.eps"), class = , names = )
```

The `data` argument specifies the series of observations; the other arguments specify when the observations were taken. Here is a description of the arguments to `ts`.

Argument	Description	Default
data	A vector or matrix representing a set of observations over time (usually numeric).	NA
start	A numeric vector with one or two elements representing the start of the time series. If one element is used, then it represents a "natural time unit." If two elements are used, then it represents a "natural time unit" and an offset.	1
end	A numeric vector with one or two elements representing the end of the time series. (Represented the same way as start.)	numeric(0)
frequency	The number of observations per unit of time.	1
deltat	The fraction of the sampling period between observations; frequency=1/deltat.	1
ts.eps	Time series comparison tolerance. The frequency of two time series objects is considered equal if the difference is less than this amount.	getOption("ts.eps")
class	The class to be assigned to the result.	"ts" for a single series, c("mts", "ts") for multiple series
names	A character vector specifying the name of each series in a multiple series object.	colnames(data) when not null, otherwise "Series1", "Series2", ...

The print method for time series objects can print pretty results when used with units of months or quarters (this is enabled by default and is controlled with the `calendar` argument to `print.ts`; see the help file for more details). As an example, let's create a time series representing eight consecutive quarters between Q2 2008 and Q1 2010:

```
> ts(1:8,start=c(2008,2),frequency=4)
     Qtr1 Qtr2 Qtr3 Qtr4
2008         1    2    3
2009    4    5    6    7
2010    8
```

As another example of a time series, we will look at the price of turkey. The U.S. Department of Agriculture has a program that collects data on the retail price of various meats. The data is taken from supermarkets representing approximately 20% of the U.S. market and then averaged by month and region. The turkey price data is included in the `nutshell` package as `turkey.price.ts`:

```
> library(nutshell)
> data(turkey.price.ts)
> turkey.price.ts
```

```
     Jan  Feb  Mar  Apr  May  Jun  Jul  Aug  Sep  Oct  Nov  Dec
2001 1.58 1.75 1.63 1.45 1.56 2.07 1.81 1.74 1.54 1.45 0.57 1.15
2002 1.50 1.66 1.34 1.67 1.81 1.60 1.70 1.87 1.47 1.59 0.74 0.82
2003 1.43 1.77 1.47 1.38 1.66 1.66 1.61 1.74 1.62 1.39 0.70 1.07
2004 1.48 1.48 1.50 1.27 1.56 1.61 1.55 1.69 1.49 1.32 0.53 1.03
2005 1.62 1.63 1.40 1.73 1.73 1.80 1.92 1.77 1.71 1.53 0.67 1.09
2006 1.71 1.90 1.68 1.46 1.86 1.85 1.88 1.86 1.62 1.45 0.67 1.18
2007 1.68 1.74 1.70 1.49 1.81 1.96 1.97 1.91 1.89 1.65 0.70 1.17
2008 1.76 1.78 1.53 1.90
```

R includes a variety of utility functions for looking at time series objects:

```
> start(turkey.price.ts)
[1] 2001    1
> end(turkey.price.ts)
[1] 2008    4
> frequency(turkey.price.ts)
[1] 12
> deltat(turkey.price.ts)
[1] 0.08333333
```

We'll revisit this time series later in the book.

Shingles

A shingle is a generalization of a factor to a continuous variable. A shingle consists of a numeric vector and a set of intervals. The intervals are allowed to overlap (much like roof shingles, hence the name shingles). Shingles are used extensively in the lattice package. Specifically, they allow you to easily use a continuous variable as a conditioning or grouping variable. See Chapter 15 for more information about the lattice package.

Dates and Times

R includes a set of classes for representing dates and times:

Date
> Represents dates but *not* times.

POSIXct
> Stores dates and times as seconds since January 1, 1970, 12:00 A.M.

POSIXlt
> Stores dates and times in separate vectors. The list includes sec (0–61)[†], min (0–59), hour (0–23), mday (day of month, 1–31), mon (month, 0–11), year (years since 1900), wday (day of week, 0–6), yday (day of year, 0–365), and isdst (flag for "is daylight savings time").

When possible, it's a good idea to store date and time values as date objects, not as strings or numbers. There are many good reasons for this. First, manipulating dates as strings is difficult. The date and time classes include functions for addition and subtraction. For example:

† This makes it possible to represent leap seconds.

```
> date.I.started.writing <- as.Date("2/13/2009","%m/%d/%Y")
> date.I.started.writing
[1] "2009-02-13"
> today <- Sys.Date()
> today
[1] "2009-08-03"
> today - date.I.started.writing
Time difference of 171 days
```

Additionally, R includes a number of other functions for manipulating time and date objects. Many plotting functions require dates and times.

Connections

R includes a special object type for receiving data from (or sending data to) applications or files outside the R environment. (Connections are like file pointers in C or filehandles in Perl.) You can create connections to files, URLs, zip compressed files, gzip compressed files, bzip compressed files, Unix pipes, network sockets, and FIFO (first in, first out) objects. You can even read from the system clipboard (to paste data into R).

To use connections, you create the connection, open the connection, use the connection, and close the connection. For example, suppose that you had saved some data objects into a file called *consumption.RData* and wanted to load the data. R saves files in a compressed format, so you would create a connection with the gzfile command. Here is how to load the file using a connection:

```
> consumption.connection <- gzfile(description="consumption.RData",open="r")
> load(consumption.connection)
> close(consumption.connection)
```

Most of the time, you don't have to explicitly open connections. Many functions for reading or writing files (such as save, load, or read.table) will implicitly open connections when you provide a filename or URL as argument. Connections can be useful for reading data from nonstandard file types (such as bz compressed files or network connections).

See the help file for connection for more information.

Attributes

Objects in R can have many properties associated with them, called *attributes*. These properties explain what an object represents and how it should be interpreted by R. Quite often, the only difference between two similar objects is that they have different attributes.[‡] Some important attributes are shown in Table 7-2. Many objects in R are used to represent numerical data, in particular, arrays, matrices, and data frames. So, many common attributes refer to properties of these objects.

---

‡ You might wonder why attributes exist; the same functionality could be implemented with lists or S4 objects. The reason is historical; attributes predate most of R's modern object mechanisms. See Chapter 10 for a full discussion of formal objects in R.

Table 7-2. Common attributes

Attribute	Description
class	The class of the object.
comment	A comment on the object; often a description of what the object means.
dim	Dimensions of the object.
dimnames	Names associated with each dimension of the object.
names	Returns the names attribute of an object. Results depend on object type; for example, returns the name of each data column in a data frame or each named object in an array.
row.names	The name of each row in an object (related to dimnames).
tsp	Start time for an object. Useful for time series data.
levels	Levels of a factor.

There is a standard way to query object attributes in R. For an object x and attribute a, you refer to the attribute through a(x). In most cases, there is a method to get the current value of the attribute and a method to set a new value of the attribute. (Changing attributes with these methods will alter the attributes in the current environment but will not affect the attributes in an enclosing environment.)

You can get a list of all attributes of an object using the `attributes` function. As an example, let's consider the matrix that we created in "Matrices" on page 84:

```
> m <- matrix(data=1:12,nrow=4,ncol=3,
+             dimnames=list(c("r1","r2","r3","r4"),
+                           c("c1","c2","c3")))
```

Now, let's take a look at the attributes of this object:

```
> attributes(m)
$dim
[1] 4 3

$dimnames
$dimnames[[1]]
[1] "r1" "r2" "r3" "r4"

$dimnames[[2]]
[1] "c1" "c2" "c3"
```

The `dim` attribute shows the dimensions of the object, in this case four rows by three columns. The `dimnames` attribute is a two-element list, consisting of the names for each respective dimension of the object (rows then columns). It is possible to access each of these attributes directly, using the `dim` and `dimnames` functions, respectively:

```
> dim(m)
[1] 4 3
> dimnames(m)
[[1]]
[1] "r1" "r2" "r3" "r4"

[[2]]
[1] "c1" "c2" "c3"
```

There are convenience functions for accessing the row and column names:

```
> colnames(m)
[1] "c1" "c2" "c3"
> rownames(m)
[1] "r1" "r2" "r3" "r4"
```

It is possible to transform this matrix into another object class simply by changing the attributes. Specifically, we can remove the dimension attribute (by setting it to NULL), and the object will be transformed into a vector:

```
> dim(m) <- NULL
> m
 [1]  1  2  3  4  5  6  7  8  9 10 11 12
> class(m)
[1] "integer"
> typeof(m)
[1] "integer"
```

Let's go back to an example that we used in "Introduction to Data Structures" on page 22. We'll construct an array a:

```
> a <- array(1:12,dim=c(3,4))
> a
     [,1] [,2] [,3] [,4]
[1,]    1    4    7   10
[2,]    2    5    8   11
[3,]    3    6    9   12
```

Now, let's define a vector with the same contents:

```
> b <- 1:12
> b
 [1]  1  2  3  4  5  6  7  8  9 10 11 12
```

You can use R's bracket notation to refer to elements in a as a two-dimensional array, but you can't refer to elements in b as a two-dimensional array, because b doesn't have any dimensions assigned:

```
> a[2,2]
[1] 5
> b[2,2]
Error in b[2, 2] : incorrect number of dimensions
```

At this point, you might wonder if R considers the two objects to be the same. Here's what happens when you compare them with the == operator:

```
> a == b
     [,1] [,2] [,3] [,4]
[1,] TRUE TRUE TRUE TRUE
[2,] TRUE TRUE TRUE TRUE
[3,] TRUE TRUE TRUE TRUE
```

Notice what is returned: an array with the dimensions of a, where each cell shows the results of the comparison. There is a function in R called all.equal that compares the data and attributes of two objects to show if they're "nearly" equal, and if they are not explains why:

```
> all.equal(a,b)
[1] "Attributes: < Modes: list, NULL >"
[2] "Attributes: < names for target but not for current >"
[3] "Attributes: < Length mismatch: comparison on first 0 components >"
[4] "target is matrix, current is numeric"
```

If you just want to check whether two objects are exactly the same, but don't care why, use the function identical:

```
> identical(a,b)
[1] FALSE
```

By assigning a dimension attribute to b, b is transformed into an array and the two-dimensional data access tools will work. The all.equal function will also show that the two objects are equivalent:

```
> dim(b) <- c(3,4)
> b[2,2]
[1] 5
> all.equal(a,b)
[1] TRUE
> identical(a,b)
[1] TRUE
```

Class

An object's class is implemented as an attribute. For simple objects, the class and type are often closely related. For compound objects, however, the two can be different.

Sometimes, the class of an object is listed with **attributes**. However, for certain classes (such as matrices and arrays), the class is implicit. To determine the class of an object, you can use the **class** function. You can determine the underlying type of object using the **typeof** function.

For example, here is the type and class for a simple numeric vector:

```
> x <- c(1, 2, 3)
> typeof(x)
[1] "double"
> class(x)
[1] "numeric"
```

It is possible to change the class of an object in R, just like changing any other attribute. For example, factors are implemented internally using integers and a map of the integers to the factor levels. (Integers take up a small, fixed amount of storage space, so they can be much more efficient than character vectors.) It's possible to take a factor and turn it into an integer array:

```
> eye.colors.integer.vector
[1] 2 1 1 3 2 2 2
attr(,"levels")
[1] "blue"  "brown" "green"
```

It is possible to create an integer array and turn it into a factor:

```
> v <- as.integer(c(1,1,1,2,1,2,2,3,1))
> levels(v) <- c("what","who","why")
> class(v) <- "factor"
> v
[1] what what what who  what who  who  why  what
Levels: what who why
```

Note that there is no guarantee that the implementation of factors won't change, so be careful using this trick in practice.

For some objects, you need to quote them to prevent them from being evaluated when the class or type function is called. For example, suppose that you wanted to determine the type of the symbol x and not the object to which it refers. You could do this like this:

```
> class(quote(x))
[1] "name"
> typeof(quote(x))
[1] "symbol"
```

Unfortunately, you can't actually use these functions on every type of object. Specifically, there is no way to isolate an any, ..., char, or promise object in R. (Checking the type of a promise object requires evaluating the promise object, converting it to an ordinary object.)

8

Symbols and Environments

So far, we've danced around the concept of environments without explicitly defining them. Every *symbol* in R is defined within a specific environment. An *environment* is an R object that contains the set of symbols available in a given context, the objects associated with those symbols, and a pointer to a parent environment. The symbols and associated objects are called a *frame*.

Every evaluation context in R is associated with an environment. When R attempts to resolve a symbol, it begins by looking through the current environment. If there is no match in the local environment, R will recursively search through parent environments looking for a match.

Symbols

When you define a variable in R, you are actually assigning a symbol to a value in an environment. For example, when you enter the statement:

```
> x <- 1
```

on the R console, it assigns the symbol x to a vector object of length 1 with the constant (double) value 1 in the global environment. When the R interpreter evaluates an expression, it evaluates all symbols. If you compose an object from a set of symbols, R will resolve the symbols at the time that the object is constructed:

```
> x <- 1
> y <- 2
> z <- 3
> v <- c(x, y, z)
> v
[1] 1 2 3
> # v has already been defined, so changing x does not change v
> x <- 10
> v
[1] 1 2 3
```

It is possible to delay evaluation of an expression so that symbols are not evaluated immediately:

```
> x <- 1
> y <- 2
> z <- 3
> v <- quote(c(x,y,z))
> eval(v)
[1] 1 2 3
> x <- 5
> eval(v)
[1] 5 2 3
```

It is also possible to create a promise object in R to delay evaluation of a variable until it is (first) needed. You can create a promise object through the delayedAssign function:

```
> x <- 1
> y <- 2
> z <- 3
> delayedAssign("v", c(x,y,z))
> x <- 5
> v
[1] 5 2 3
```

Promise objects are used within packages to make objects available to users without loading them into memory. Unfortunately, it is not possible to determine if an object is a promise object, nor is it possible to figure out the environment in which it was created.

Working with Environments

Like everything else in R, environments are objects. Table 8-1 shows the functions in R for manipulating environment objects.

Table 8-1. Manipulating environment objects

Function	Description
assign	Assigns the name x to the object value in the environment envir.
get	Gets the object associated with the name x in the environment envir.
exists	Checks that the name x is defined in the environment envir.
objects	Returns a vector of all names defined in the environment envir.
remove	Removes the list of objects in the argument list from the environment envir. (List is an unfortunate argument name, especially as the argument needs to be a vector.)
search	Returns a vector containing the names of attached packages. You can think of this as the search path in which R tries to resolve names. More precisely, it shows the list of chained parent environments.
searchpaths	Returns a vector containing the paths of attached packages.
attach	Adds the objects in the list, data frame, or data file what to the current search path.
detach	Removes the objects in the list, data frame, or data file what from the current search path.
emptyenv	Returns the empty environment object. All environments chain back to this object.

Function	Description
parent.env	Returns the parent of environment env.
baseenv	The environment of the base package.
globalenv or .GlobalEnv	Returns the environment for the user's workspace (called the "global environment"). See "The Global Environment" for an explanation of what this means.
environment	Returns the environment for function fun. When evaluated with no arguments (or fun=NULL), returns the current environment.
new.env	Returns a new environment object.

To show the set of objects available in the current environment (or, more precisely, the set of symbols in the current environment associated with objects), use the objects function:

```
> x <- 1
> y <- 2
> z <- 3
> objects()
[1] "x" "y" "z"
```

You can remove an object from the current environment with the rm function:

```
> rm(x)
> objects()
[1] "y" "z"
```

The Global Environment

When a user starts a new session in R, the R system creates a new environment for objects created during that session. This environment is called the *global environment*. The global environment is not actually the root of the tree of environments. It's actually the last environment in the chain of environments in the search path. Here's the list of parent environments for the global environment in my R installation:

```
> x <- .GlobalEnv
> while (environmentName(x) != environmentName(emptyenv())) {
+     print(environmentName(parent.env(x))); x <- parent.env(x)}
[1] "tools:RGUI"
[1] "package:stats"
[1] "package:graphics"
[1] "package:grDevices"
[1] "package:utils"
[1] "package:datasets"
[1] "package:methods"
[1] "Autoloads"
[1] "base"
[1] "R_EmptyEnv"
```

Every environment has a parent environment except for one: *the empty environment*. All environments chain back to the empty environment.

Environments and Functions

When a function is called in R, a new environment is created within the body of the function, and the arguments of the function are assigned to symbols in the local environment.*

As an example, let's create a function that takes four arguments and does nothing except print out the objects in the current environment:

```
> env.demo <- function(a, b, c, d) {print(objects())}
> env.demo(1, "truck", c(1,2,3,4,5), pi)
[1] "a" "b" "c" "d"
```

Notice that the `objects` function returns only the objects from the current environment, so the function `env.demo` only prints the arguments defined in that environment. All other objects exist in the parent environment, not in the local environment.

The parent environment of a function is the environment in which the function was created. If a function was created in the execution environment (for example, in the global environment), then the environment in which the function was called will be the same as the environment in which the function was created. However, if the function was created in another environment (such as a package), then the parent environment will *not* be the same as the calling environment.

Working with the Call Stack

Although the parent environment for a function is not always the environment in which the function was called, it is possible to access the environment in which a function was called.† Like many other languages, R maintains a stack of calling environments. (A stack is a data structure in which objects can only be added or subtracted from one end. Think about a stack of trays in a cafeteria; you can only add a tray to the top or take a tray off the top. Adding an object to a stack is called "pushing" the object onto the stack. Taking an object off of the stack is called "popping" the object off of the stack.) Each time a new function is called, a new environment is pushed onto the call stack. When R is done evaluating a function, the environment is popped off of the call stack.

Table 8-2 shows the functions for manipulating the call stack.

Table 8-2. Manipulating the call stack

Function	Description
`sys.call`	Returns a language object containing the current function call (including arguments).
`sys.frame`	Returns the calling environment.
`sys.nframe`	Returns the number of the current frame (the position on the call stack). Returns 0 if called on the R console.

---

* If you're familiar with other languages and language lingo, you could say that R is a *lexically scoped* language.

† This allows symbols to be accessed as though R were *dynamically scoped*.

Function	Description
`sys.function`	Returns the function currently being evaluated.
`sys.parent`	Returns the number of the parent frame.
`sys.calls`	Returns the calls for all frames on the stack.
`sys.frames`	Returns all environments on the stack.
`sys.parents`	Returns the parent for each frame on the stack.
`sys.on.exit`	Returns the expression used for on.exit for the current frame.
`sys.status`	Returns a list with the results of calls to `sys.calls`, `sys.parents`, and `sys.frames`.
`parent.frame`	Returns `sys.frame(sys.parent(n))`. In other words, returns the parent frame.

If you are writing a package where a function needs to know the meaning of a symbol in the calling context (and not in the context within the package), you can do so with these functions. Some common R functions, like modeling functions, use this trick to determine the meaning of symbols in the calling context. In specifying a model, you pass a formula object to a modeling function. The formula object is a language object; the symbol names are included in the formula but not in the data. You can specify a data object like a data frame, but you don't have to. When you don't specify the objects containing the variables, the model function will try to search through the calling environment to find the data.

Evaluating Functions in Different Environments

You can evaluate an expression within an arbitrary environment using the `eval` function:

```
eval(expr, envir = parent.frame(),
          enclos = if(is.list(envir) || is.pairlist(envir))
                      parent.frame() else baseenv())
```

The argument `expr` is the expression to be evaluated, and `envir` is an environment, data frame, or pairlist in which to evaluate `expr`. When `envir` is a data frame or pairlist, `enclos` is the enclosure in which to look for object definitions. As an example of how to use `eval`, let's create a function to time the execution of another expression. We'd like the function to record the starting time, evaluate its arguments (an arbitrary expression) in the parent environment, record the end time, and print the difference:

```
timethis <- function(...) {
    start.time <- Sys.time();
    eval(..., sys.frame(sys.parent(sys.parent())));
    end.time <- Sys.time();
    print(end.time - start.time);
}
```

As an example of how this works, we'll time an inefficient function that sets 10,000 elements in a vector to the value 1:

```
> create.vector.of.ones <- function(n) {
    return.vector <- NA;
```

```
        for (i in 1:n) {
            return.vector[i] <- 1;
        }
        return.vector;
    }
    # note that returned.vector is not defined
    > returned.vector
    Error: object 'returned.vector' not found
    # measure time to run function above with n=10000
    > timethis(returned.vector <- create.vector.of.ones(10000))
    Time difference of 1.485959 secs
    # notice that the function took about 1.5 seconds to run
    # also notice that returned.vector is now defined
    > length(returned.vector)
    [1] 10000
```

The timing part is neat, but the point of this function is to show that it is evaluating the expression in the calling environment. Most important, notice that the symbol returned.vector is now defined in that environment:

```
    > length(returned.vector)
    [1] 10000
```

This is a little off the subject, but here is a more efficient version of the same function:

```
    create.vector.of.ones.b <- function(n) {
        return.vector <- NA;
        length(return.vector) <- n;
        for (i in 1:n) {
            return.vector[i] <- 1;
        }
        return.vector;
    }
    > timethis(returned.vector <- create.vector.of.ones.b(10000))
    Time difference of 0.04076099 secs
```

Three useful shorthands are the functions evalq, eval.parent, and local. When you want to quote the expression, use evalq, which is equivalent to eval(quote(expr), ...). When you want to evaluate an expression within the parent environment, you can use the function eval.parent, which is equivalent to eval(expr, parent.frame(n)). When you want to evaluate an expression in a new environment, you can use the function local, which is equivalent to eval(quote(expr), envir=new.env()).

As an example of how to use eval.parent, we can shorten the timing function from the example above:

```
    timethis.b <- function(...) {
        start.time <- Sys.time();
        eval.parent(...);
        end.time <- Sys.time();
        print(end.time - start.time);
    }
```

Sometimes, it is convenient to treat a data frame or a list as an environment. This lets you refer to each item in the data frame or list by name as if you were using symbols. You can do this in R with the functions `with` and `within`:

```
with(data, expr, ...)
within(data, expr, ...)
```

The argument `data` is the data frame or list to treat as an environment, `expr` is the expression, and additional arguments in `...` are passed to other methods. The function `with` evaluates the expression and then returns the result, while the function `within` makes changes in the object `data` and then returns `data`.

Here are some examples of using `with` and `within`:

```
> example.list <- list(a=1, b=2, c=3)
> a+b+c
Error: object 'b' not found
> with(example.list, a+b+c)
[1] 6
> within(example.list, d<-a+b+c)
$a
[1] 1

$b
[1] 2

$c
[1] 3

$d
[1] 6
```

Adding Objects to an Environment

R provides a shorthand for adding objects to the current environment: `attach`. If you have saved a set of objects to a data file with `save`, you can load the objects into the current environment with `attach`.

Additionally, you can use `attach` to load all of the elements specified within a data frame or list into the current environment. Often, operators like `$` are convenient for accessing objects within a list or data frame, but sometimes it can be cumbersome to do so:

```
attach(what, pos = 2, name = deparse(substitute(what)),
       warn.conflicts = TRUE)
```

The argument `what` is the object to attach (called a *database*), `pos` specifies the position in the search path in which to attach the element within `what`, `name` is the name to use for the attached database (more on what this is used for below), and `warn.conflicts` specifies whether to warn the user if there are conflicts. The database can be a data frame, a list, or an R data file created with the `save` function.

When you're done, you can remove all the objects in a data frame from the current environment with the function detach:

```
detach(name, pos = 2, unload = FALSE)
```

In this function, the argument name specifies the name of the database to detach (which corresponds to the argument name from attach), pos is the position in the search path at which the database was attached, and unload specifies whether or not to unload the namespace and S4 methods when a database is detached.

Be careful using attach. Often, I find myself working with multiple data frames with identically named columns. Using attach can be confusing, because it is difficult to keep track of the data frame from which each object came. It is often better to use functions like transform to change values within a data frame or with to evaluate expressions using values in a data frame.

Exceptions

You may have noticed that R sometimes gives you an error when you enter an invalid expression. For example:

```
> 12 / "hat"
Error in 12/"hat" : non-numeric argument to binary operator
```

Other times, R may just give you a warning:

```
> if (c(TRUE,FALSE)) TRUE else FALSE
[1] TRUE
Warning message:
In if (c(TRUE, FALSE)) TRUE else FALSE :
  the condition has length > 1 and only the first element will be used
```

Like other modern programming languages, R includes the ability to signal exceptions when unusual events occur and catch exceptions when they occur. If you are writing your own R programs, it is usually a good idea to stop execution when an error occurs and alert the user (or calling function). Likewise, it is usually a good idea to catch exceptions from functions that are called within your programs.

It might seem strange to talk about exception handling in the context of environments, but exception handling and environments are closely linked. When an exception occurs, the R interpreter may need to abandon the current function and signal the exception in the calling environment.

This section explains how the error-handling system in R works.

Signaling Errors

If something occurs in your code that requires you to stop execution, you can use the stop function. For example, suppose that you had written a function called dowork(filename) to automatically generate some charts and save them to a file specified by the argument filename. Suppose that R couldn't write to the file, possibly because the directory didn't exist. To stop execution and print a helpful error message, you could structure your code like this:

```
> doWork <- function(filename) {
+    if(file.exists(filename)) {
+      read.delim(filename)
+    } else {
+      stop("Could not open the file: ", filename)
+    }
+ }
> doWork("file that doesn't exist")
Error in doWork("file that doesn't exist") :
  Could not open the file: file that doesn't exist
```

If something occurs in your code that you want to tell the user about, but which isn't severe enough to normally stop execution, you can use the `warning` function. Reusing the example above, if the file "filename" already exists, then the function will simply return the string `"la la la"`. If the file does not exist, then the function will warn the user that the file does not exist.

```
> doNoWork <- function(filename) {
+    if(file.exists(filename)) {
+      "la la la"
+    } else {
+      warning("File does not exist: ", filename)
+    }
+ }
> doNoWork("another file that doesn't exist")
Warning message:
In doNoWork("another file that doesn't exist") :
  File does not exist: another file that doesn't exist
```

If you just want to tell the user something, you can use the `message` function:

```
> doNothing("another input value")
This function does nothing.
```

Catching Errors

Suppose that you are writing a function in R called `foo` that calls another function called `bar`. Furthermore, suppose that `bar` sometimes generates an error, but you don't want `foo` to stop if the error is generated. For example, maybe `bar` tries to open a file, but signals an error when it can't open the file. If `bar` can't open the file, maybe you want `foo` to try doing something else instead.

A simple way to do this is to use the `try` function. This function hides some of the complexity of R's exception handling. Here's an example of how to use `try`:

```
> res <- try({x <- 1}, silent=TRUE)
> res
[1] 1
> res <- try({open("file that doesn't exist")}, silent=TRUE)
> res
[1] "Error in UseMethod(\"open\") : \n  no applicable method for 'open'
applied to an object of class \"character\"\n"
attr(,"class")
[1] "try-error"
```

The **try** function takes two arguments, expr and silent. The first argument, expr, is the R expression to be tried (often a function call). The second argument specifies whether the error message should be printed to the R console (or stderr); the default is to print errors. If the expression results in an error, then **try** returns an object of class "try-error".

A more capable function is **tryCatch**. The **tryCatch** function takes three sets of arguments: an expression to try, a set of handlers for different conditions, and a final expression to evaluate. For example, suppose that the following call was made to **tryCatch**:

```
> tryCatch(expression, handler1, handler2, ..., finally=finalexpr)
```

The R interpreter would first evaluate *expression*. If a condition occurs (an error or warning), R will pick the appropriate handler for the condition (matching the class of the condition to the arguments for the handler). After the expression has been evaluated, *finalexpr* will be evaluated. (The handlers will not be active when this expression is evaluated.)

## 9

Functions

Functions are the R objects that evaluate a set of input arguments and return an output value. This chapter explains how to create and use functions in R.

The Function Keyword

In R, function objects are defined with this syntax:

```
function(arguments) body
```

where *arguments* is a set of symbol names (and, optionally, default values) that will be defined within the body of the function, and *body* is an R expression. Typically, the body is enclosed in curly braces, but it does not have to be if the body is a single expression. For example, the following two definitions are equivalent:

```
> f <- function(x,y) x+y
```

```
> f <- function(x,y) {x+y}
```

Arguments

A function definition in R includes the names of arguments. Optionally, it may include default values. If you specify a default value for an argument, then the argument is considered optional:

```
> f <- function(x,y) {x+y}
> f(1,2)
[1] 3
> g <- function(x,y=10) {x+y}
> g(1)
[1] 11
```

If you do not specify a default value for an argument, and you do not specify a value when calling the function, you will get an error if the function attempts to use the argument:[*]

```
> f(1)
Error in f(1) :
  element 2 is empty;
  the part of the args list of '+' being evaluated was:
  (x, y)
```

In a function call, you may override the default value:

```
> g(1,2)
[1] 3
```

In R, it is often convenient to specify a variable-length argument list. You might want to pass extra arguments to another function, or you may want to write a function that accepts a variable number of arguments. To do this in R, you specify an ellipsis (...) in the arguments to the function.[†]

As an example, let's create a function that prints the first argument and then passes all the other arguments to the summary function. To do this, we will create a function that takes one argument: x. The arguments specification also includes an ellipsis to indicate that the function takes other arguments. We can then call the summary function with the ellipsis as its argument:

```
> v <- c(sqrt(1:100))
> f <- function(x,...) {print(x); summary(...)}
> f("Here is the summary for v.", v, digits=2)
[1] "Here is the summary for v."
   Min. 1st Qu.  Median    Mean 3rd Qu.    Max.
    1.0     5.1     7.1     6.7     8.7    10.0
```

Notice that all of the arguments after x were passed to summary.

---

[*] Note that you will only get an error if you try to use the uninitialized argument within the function; you could easily write a function that simply doesn't reference the argument, and it will work fine. Additionally, there are other ways to check whether an argument has been initialized from inside the body of a function. For example, the following function works identically to the function g shown above (which included a default value for y in its definition):

```
> h <- function(x,y) {
+   args <- as.list(match.call())
+   if (is.null(args$y)) {
+     y <- 10
+   }
+   x + y
+ }
```

In practice, you should specify default values in the function signature to make your functions as clear and easy to read as possible.

[†] You might remember from Chapter 7 that "..." is a special type of object in R. The only place you can manipulate this object is inside the body of a function. In this context, it means "all the other arguments for the function."

It is also possible to read the arguments from the variable-length argument list. To do this, you can convert the object **...** to a list within the body of the function. As an example, let's create a function that simply sums all its arguments:

```
> addemup <- function(x,...) {
+     args <- list(...)
+     for (a in args) x <- x + a
+     x
+ }
> addemup(1,1)
[1] 2
> addemup(1,2,3,4,5)
[1] 15
```

You can also directly refer to items within the list **...** through the variables **..1**, **..2**, to **..9**. Use ..1 for the first item, ..2 for the second, and so on. Named arguments are valid symbols within the body of the function. For more information about the scope within which variables are defined, see Chapter 8.

Return Values

In an R function, you may use the **return** function to specify the value returned by the function. For example:

```
> f <- function(x) {return(x^2 + 3)}
> f(3)
[1] 12
```

However, R will simply return the last evaluated expression as the result of a function. So, it is common to omit the **return** statement:

```
> f <- function(x) {x^2 + 3}
> f(3)
[1] 12
```

In some cases, an explicit return value may lead to cleaner code.

Functions As Arguments

Many functions in R can take other functions as arguments. For example, many modeling functions accept an optional argument that specifies how to handle missing values; this argument is usually a function for processing the input data.

As an example of a function that takes another function as an argument, let's look at **sapply**. The **sapply** function iterates through each element in a vector, applying another function to each element in the vector, and returning the results. Here is a simple example:

```
> a <- 1:7
> sapply(a, sqrt)
[1] 1.000000 1.414214 1.732051 2.000000 2.236068 2.449490 2.645751
```

This is a toy example; you could have calculated the same quantity with the expression sqrt(1:7). However, there are many useful functions that don't work properly on a vector with more than one element; sapply provides a simple way to extend such a function to work on a vector. Related functions allow you to summarize every element in a data structure or to perform more complicated calculations. See "Summarizing Functions" on page 194 for information on related functions.

Anonymous Functions

So far, we've mostly seen named functions in R. However, because functions are just objects in R, it is possible to create functions that do not have names. These are called *anonymous functions*. Anonymous functions are usually passed as arguments to other functions. If you're new to functional languages, this concept might seem strange, so let's start with a very simple example.

We will define a function that takes another function as its argument and then applies that function to the number 3. Let's call the function apply.to.three, and we will call the argument f:

```
> apply.to.three <- function(f) {f(3)}
```

Now, let's call apply.to.three with an anonymous function assigned to argument f. As an example, let's create a simple function that takes one argument and multiplies that argument by 7:

```
> apply.to.three(function(x) {x * 7})
[1] 21
```

Here's how this works. When the R interpreter evaluates the expression apply.to.three(function(x) {x * 7}), it assigns the argument f to the anonymous function function(x) {x * 7}. The interpreter then begins evaluating the expression f(3). The interpreter assigns 3 to the argument x for the anonymous function. Finally, the interpreter evaluates the expression 3 * 7 and returns the result.

Anonymous functions are a very powerful concept that is used in many places in R. Above, we used the sapply function to apply a named function to every element in an array. You can also pass an anonymous function as an argument to sapply:

```
> a <- c(1, 2, 3, 4, 5)
> sapply(a, function(x) {x+1})
[1] 2 3 4 5 6
```

This family of functions is a good alternative to control structures.

By the way, it is possible to define an anonymous function and apply it directly to an argument. Here's an example:

```
> (function(x) {x+1})(1)
[1] 2
```

Notice that the function object needs to be enclosed in parentheses. This is because function calls, expressions of the form $f(arguments)$, have very high precedence in R.‡

Properties of Functions

R includes a set of functions for getting more information about function objects. To see the set of arguments accepted by a function, use the `args` function. The `args` function returns a function object with `NULL` as the body. Here are a few examples:

```
> args(sin)
function (x)
NULL
> args(`?`)
function (e1, e2)
NULL
> args(args)
function (name)
NULL
> args(lm)
function (formula, data, subset, weights, na.action, method = "qr",
    model = TRUE, x = FALSE, y = FALSE, qr = TRUE, singular.ok = TRUE,
    contrasts = NULL, offset, ...)
NULL
```

If you would like to manipulate the list of arguments with R code, then you may find the `formals` function more useful. The `formals` function will return a pairlist object, with a pair for every argument. The name of each pair will correspond to each argument name in the function. When a default value is defined, the corresponding value in the pairlist will be set to that value. When no default is defined, the value will be `NULL`. The `formals` function is only available for functions written in R (objects of type `closure`) and not for built-in functions.

Here is a simple example of using `formals` to extract information about the arguments to a function:

```
> f <- function(x,y=1,z=2) {x+y+z}
> f.formals <- formals(f)
> f.formals
$x
```

---

‡ If you omit the parentheses in this example, you will not initially get an error:
```
> function(x) {x+1}(1)
function(x) {x+1}(1)
```

This is because you will have created an object that is a function taking one argument (x) with the body {x+1}(1). There is no error generated because the body is not evaluated. If you were to assign this object to a symbol (so that you can easily apply it to an argument and see what it does), you will find that this function attempts to call a function returned by evaluating the expression {x + 1}. In order not to get an error or an input of class c, you would need to register a generic function that took as input an object of class c (x in this expression) and a numerical value (1 in this expression) and returned a function object. So, omitting the parentheses is not wrong; it is a valid R expression. However, this is almost certainly not what you meant to write.

```
$y
[1] 1

$z
[1] 2

> f.formals$x

> f.formals$y
[1] 1
> f.formals$z
[1] 2
```

You may also use `formals` on the lefthand side of an assignment statement to change the formal argument for a function. For example:

```
> f.formals$y <- 3
> formals(f) <- f.formals
> args(f)
function (x, y = 3, z = 2)
NULL
```

R provides a convenience function called `alist` to construct an argument list. You simply specify the argument list as if you were defining a function. (Note that for an argument with no default, you do not need to include a value but still need to include the equals sign.)

```
> f <- function(x,y=1,z=2) {x + y + z}
> formals(f) <- alist(x=,y=100,z=200)
> f
function (x, y = 100, z = 200)
{
    x + y + z
}
```

R provides a similar function called `body` that can be used to return the body of a function:

```
> body(f)
{
    x + y + z
}
```

Like the `formals` function, the `body` function may be used on the lefthand side of an assignment statement:

```
> f
function (x, y = 3, z = 2)
{
    x + y + z
}
> body(f) <- expression({x * y * z})
> f
function (x, y = 3, z = 2)
{
```

```
        x * y * z
}
```

Note that the body of a function has type `expression`, so when you assign a new value it must have the type `expression`.

Argument Order and Named Arguments

When you specify a function in R, you assign a name to each argument in the function. Inside the body of the function, you can access the arguments by name. For example, consider the following function definition:

```
> addTheLog <- function(first, second) {first + log(second)}
```

This function takes two arguments, called `first` and `second`. Inside the body of the function, you can refer to the arguments by these names.

When you call a function in R, you can specify the arguments in three different ways (in order of priority):

1. Exact names. The arguments will be assigned to *full* names explicitly given in the argument list. Full argument names are matched first:

   ```
   > addTheLog(second=exp(4),first=1)
   [1] 5
   ```

2. Partially matching names. The arguments will be assigned to *partial* names explicitly given in the arguments list:

   ```
   > addTheLog(s=exp(4),f=1)
   [1] 5
   ```

3. Argument order. The arguments will be assigned to names in the order in which they were given:

   ```
   > addTheLog(1,exp(4))
   [1] 5
   ```

When you are using generic functions, you cannot specify the argument name of the object on which the generic function is being called. You can still specify names for other arguments.

When possible, it's a good practice to use exact argument names. Specifying full argument names does require extra typing, but it makes your code easier to read and removes ambiguity.

Partial names are a deprecated feature because they can lead to confusion. As an example, consider the following function:

```
> f <- function(arg1=10,arg2=20) {
+     print(paste("arg1:",arg1))
+     print(paste("arg2:",arg2))
+ }
```

When you call this function with one ambiguous argument, it will cause an error:

```
> f(arg=1)
Error in f(arg = 1) : argument 1 matches multiple formal arguments
```

However, when you specify two arguments, the ambiguous argument could refer to either of the other arguments:

```
> f(arg=1,arg2=2)
[1] "arg1: 1"
[1] "arg2: 2"
> f(arg=1,arg1=2)
[1] "arg1: 2"
[1] "arg2: 1"
```

Side Effects

All functions in R return a value. Some functions also do other things: change variables in the current environment (or in other environments), plot graphics, load or save files, or access the network. These operations are called *side effects*.

Changes to Other Environments

We have already seen some examples of functions with side effects. In Chapter 8, we showed how to directly access symbols and objects in an environment (or in parent environments). We also showed how to access objects on the call stack.

An important function that causes side effects is the <<- operator. This operator takes the following form: *var* <<- *value*. This operator will cause the interpreter to first search through the current environment to find the symbol *var*. If the interpreter does not find the symbol *var* in the current environment, then the interpreter will next search through the parent environment. The interpreter will recursively search through environments until it either finds the symbol *var* or reaches the global environment. If it reaches the global environment before the symbol *var* is found, then R will assign *var* to *value* in the global environment.

Here is an example that compares the behavior of the <- assignment operator and the <<- operator:

```
> x
Error: object "x" not found
> doesnt.assign.x <- function(i) {x <- i}
> doesnt.assign.x(4)
> x
Error: object "x" not found
> assigns.x <- function(i) {x <<- i}
> assigns.x(4)
> x
[1] 4
```

Input/Output

R does a lot of stuff, but it's not completely self-contained. If you're using R, you'll probably want to load data from external files (or from the Internet) and save data to files. These input/output (I/O) actions are side effects, because they do things other than just return an object. We'll talk about these functions extensively in Chapter 12.

Graphics

Graphics functions are another example of side effects in R. Graphics functions may return objects, but they also plot graphics (either on screen or to files). We'll talk about these functions in Chapters 14 and 15.

10

Object-Oriented Programming

The R system includes some support for object-oriented programming (OOP). OOP has become the preferred paradigm for organizing computer software; it's used in almost every modern programming language (Java, C#, Ruby, and Objective C, among others) and in quite a few old ones (Smalltalk, C++). It's easy to understand why: OOP methods lead to code that is faster to write, easier to maintain, and less likely to contain errors. Many R packages are written using OOP mechanisms.

If all you plan to do with R is to load some data, build some statistical models, and plot some charts, you can probably skim this chapter. On the other hand, if you want to write your own code for loading data, building statistical models, and plotting charts, you probably should read this chapter more carefully.

R includes two different mechanisms for object-oriented programming. As you may recall, the R language is derived from the S language. S's object-oriented programming system evolved over time. Around 1990, S version 3 (thus S3) introduced class attributes that allowed single-argument methods. Many R functions (such as the statistical modeling software) were implemented using S3 methods, so S3 methods are still around today. In S version 4 (hence S4), formal classes and methods were introduced that allowed multiple arguments, more abstract types, and more sophisticated inheritance. Many new packages were implemented using S4 methods (and you can find S4 implementations of many key statistical procedures as well). In particular, formal classes are used extensively in Bioconductor.

In this chapter, we'll begin with the newer mechanism, because it is more robust and flexible. I think that it is wise to use S4 classes and methods for new software that needs to represent abstract concepts, and that it is not a good idea to implement new S3 classes. However, you may want to change code that uses S3 classes and methods or use S3 classes and methods in new software. In "Old-School OOP in R: S3" on page 131, we'll talk about how the S3 system works and how to mix S3 and S4 classes.

Overview of Object-Oriented Programming in R

Object-oriented programming is *not* the same thing as programming with objects. R is a very object-centric language; everything in R is an object. However, there is more to OOP than just objects. Here's a short description of what object-oriented programming means.

Key Ideas

As an example of how object-oriented programming is used in R, we'll consider time series.* A time series is a sequence of measurements of a quantity over time. Measurements are taken at equally spaced intervals. Time series have some properties associated with them: a start time, an end time, a number of measurements, a frequency, and so forth.

In OOP, we would create a "time series" class to capture information about time series. A *class* is a formal definition for an object. Each individual time series object is called an *instance* of the class. A function that operates on a specific class of objects is called a *method*.

As a user of time series, you probably don't care too much about how time series are implemented. All you care about is that you know how to create a time series object and manipulate the object through methods. The time series could be stored as a data frame, a vector, or even a long text field. The process of separating the interface from the implementation is called *encapsulation*.

Suppose that we wanted to track the weight history of people over time. For this application, we'd like to keep all the same information as a time series, plus some additional information on individual people. It would be nice to be able to reuse the code for our time series class for objects in the weight history class. In OOP, it is possible to base one class on another and just specify what is different about the new class. This is called *inheritance*. We would say that the "weight history" class *inherits* from the "time series" class. We might also say that the "time series" class is a *superclass* of the "weight history" class and that the "weight history" class is a *subclass* of the "time series" class.

Suppose that you wanted to ask a question like "what is the period of the measurements in the class?" Ideally, it would be nice to have a single function name for finding this information, maybe called "period." In OOP, allowing the same method name to be used for different objects is called *polymorphism*.

Finally, suppose that we implemented the "weight history" class by creating classes for each of its pieces: time series, personal attributes, and so on. The process of creating a new class from a set of other classes is called *composition*. In some

---

* You may have noticed that I picked an example of a class that is already implemented in R. Time series objects are implemented by the `ts` class in the `stats` package. (I introduced `ts` objects in "Time Series" on page 89.) The implementation in the `stats` package is an example of an S3 class. We'll talk more about what that means, and how to use S3 and S4 classes together, next.

languages (like R), a class can inherit methods from more than one other class. This is called *multiple inheritance*.

Implementation Example

If you're familiar with object-oriented programming in other languages (like Java), you'll find that most of the familiar concepts are included in R. However, the syntax and structure in R are different. In particular, you define a class with a call to a function (`setClass`) and define a method with a call to another function (`setMethod`). Before we describe R's implementation of object-oriented programming in depth, let's look at a quick example.

As an example, let's implement a class representing a time series. We'll want to define a new object that contains the following information:

- A set of data values, sampled at periodic intervals over time
- A start time
- An end time
- The period of the time series

Clearly, some of this information is redundant; given many of the attributes of a time series, we can calculate the remaining attributes. Let's start by defining a new class called "TimeSeries." We'll represent a time series by a numeric vector containing the data, a start time, and an end time. We can calculate units, frequency, and period from the start time, end time, and the length of the data vector. As a user of the class, it shouldn't matter how we represent this information, but it does matter to the implementer.

In R, the places where information is stored in an object are called *slots*. We'll name the slots `data`, `start`, and `end`. To create a class, we'll use the `setClass` function:

```
> setClass("TimeSeries",
+   representation(
+     data="numeric",
+     start="POSIXct",
+     end="POSIXct"
+   )
+ )
```

The representation explains the class of the object contained in each slot. To create a new `TimeSeries` object, we will use the `new` function. (The `new` function is a generic *constructor* method for S4 objects.) The first argument specifies the class name; other arguments specify values for slots:

```
> my.TimeSeries <- new("TimeSeries",
+   data=c(1,2,3,4,5,6),
+   start=as.POSIXct("07/01/2009 0:00:00",tz="GMT",
+                    format="%m/%d/%Y %H:%M:%S"),
+   end=as.POSIXct("07/01/2009 0:05:00",tz="GMT",
+                  format="%m/%d/%Y %H:%M:%S")
+ )
```

There is a generic print method for new S4 classes in R that displays the slot names and the contents of each slot:

```
> my.TimeSeries
An object of class "TimeSeries"
Slot "data":
[1] 1 2 3 4 5 6

Slot "start":
[1] "2009-07-01 GMT"

Slot "end":
[1] "2009-07-01 00:05:00 GMT"
```

Not all possible slot values are valid. We want to make sure that end occurs after start and that the lengths of start and end are both exactly 1. We can write a function to check the validity of a TimeSeries object. R allows you to specify a function that will be used to validate a specific class. We can specify this with the setValidity function:

```
> setValidity("TimeSeries",
+     function(object) {
+         object@start <= object@end &&
+         length(object@start) == 1 &&
+         length(object@end) == 1
+     }
+ )
Class "TimeSeries" [in ".GlobalEnv"]

Slots:

Name:      data    start     end
Class: numeric POSIXct POSIXct
```

You can now check that a TimeSeries object is valid with the validObject function:

```
> validObject(my.TimesSeries)
[1] TRUE
```

When we try to create a new TimeSeries object, R will check the validity of the new object and reject bad objects:

```
> good.TimeSeries <- new("TimeSeries",
+     data=c(7,8,9,10,11,12),
+     start=as.POSIXct("07/01/2009 0:06:00",tz="GMT",
+                     format="%m/%d/%Y %H:%M:%S"),
+     end=as.POSIXct("07/01/2009 0:11:00",tz="GMT",
+                     format="%m/%d/%Y %H:%M:%S")
+ )
> bad.TimeSeries <- new("TimeSeries",
+     data=c(7,8,9,10,11,12),
+     start=as.POSIXct("07/01/2009 0:06:00",tz="GMT",
+                     format="%m/%d/%Y %H:%M:%S"),
+     end=as.POSIXct("07/01/1999 0:11:00",tz="GMT",
+                     format="%m/%d/%Y %H:%M:%S")
+ )
Error in validObject(.Object) : invalid class "TimeSeries" object: FALSE
```

(You can also specify the validity method at the time you are creating a class; see the full definition of setClass for more information.)

Now that we have defined the class, let's create some methods that use the class. One property of a time series is its period. We can create a method for extracting the period from the time series. This method will calculate the duration between observations based on the length of the vector in the data slot, the start time, and the end time:

```
> period.TimeSeries <- function(object) {
+    if (length(object@data) > 1) {
+      (object@end - object@start) / (length(object@data) - 1)
+    } else {
+      Inf
+    }
+ }
```

Suppose that you wanted to create a set of functions to derive the data series from other objects (when appropriate), regardless of the type of object (i.e., polymorphism). R provides a mechanism called *generic functions* for doing this.[†] You can define a generic name for a set of functions (like "series"). When you call "series" on an object, R will find the correct method to execute based on the class of the object. Let's create a function for extracting the data series from a generic object:

```
> series <- function(object) {object@data}
> setGeneric("series")
[1] "series"
> series(my.TimesSeries)
[1] 1 2 3 4 5 6
```

The call to setGeneric redefined series as a generic function whose default method is the old body for series:

```
> series
standardGeneric for "series" defined from package ".GlobalEnv"

function (object)
standardGeneric("series")
<environment: 0x19ac4f4>
Methods may be defined for arguments: object
Use  showMethods("series")  for currently available ones.
> showMethods("series")
Function: series (package .GlobalEnv)
object="ANY"
object="TimeSeries"
    (inherited from: object="ANY")
```

As a further example, suppose we wanted to create a new generic function called "period" for extracting a period from an object and wanted to specify that the function period.TimeSeries should be used for TimeSeries objects, but the generic method should be used for other objects. We could do this with the following commands:

† In object-oriented programming terms, this is called *overloading a function*.

```
> period <- function(object) {object@period}
> setGeneric("period")
[1] "period"
> setMethod(period, signature=c("TimeSeries"), definition=period.TimeSeries)
[1] "period"
attr(,"package")
[1] ".GlobalEnv"
> showMethods("period")
Function: period (package .GlobalEnv)
object="ANY"
object="TimeSeries"
```

Now, we can calculate the period of a TimeSeries object by just calling the generic function period:

```
> period(my.TimeSeries)
Time difference of 1 mins
```

It is also possible to define your own methods for existing generic functions, such as summary. Let's define a summary method for our new class:

```
> setMethod("summary",
+    signature="TimeSeries",
+    definition=function(object) {
+       print(paste(object@start,
+                   " to ",
+                   object@end,
+                   sep="",collapse=""))
+       print(paste(object@data,sep="",collapse=","))
+    }
+ )
Creating a new generic function for "summary" in ".GlobalEnv"
[1] "summary"
> summary(my.TimeSeries)
[1] "2009-07-01 to 2009-07-01 00:05:00"
[1] "1,2,3,4,5,6"
```

You can even define a new method for an existing operator:

```
> setMethod("[",
+    signature=c("TimeSeries"),
+    definition=function(x, i, j, ...,drop) {
+       x@data[i]
+    }
+ )
[1] "["
> my.TimesSeries[3]
[1] 3
```

(As a quick side note, this only works for some built-in functions. For example, you can't define a new print method this way. See the help file for S4groupGeneric for a list of generic functions that you can redefine this way, and "Old-School OOP in R: S3" on page 131 for an explanation on why this doesn't always work.)

Now, let's show how to implement a WeightHistory class based on the TimeSeries class. One way to do this is to create a WeightHistory class that inherits from the TimeSeries class but adds extra fields to represent a person's name and height. We can do this with the setClass command by stating that the new class inherits from the TimeSeries class and specifying the extra slots in the WeightHistory class:

```
> setClass(
+   "WeightHistory",
+   representation(
+     height = "numeric",
+     name = "character"
+   ),
+   contains = "TimeSeries"
+ )
```

Now, we can create a WeightHistory object, populating slots named in TimeSeries and the new slots for WeightHistory:

```
> john.doe <- new("WeightHistory",
+   data=c(170, 169, 171, 168, 170, 169),
+   start=as.POSIXct("02/14/2009 0:00:00",tz="GMT",
+     format="%m/%d/%Y %H:%M:%S"),
+   end=as.POSIXct("03/28/2009 0:00:00",tz="GMT",
+     format="%m/%d/%Y %H:%M:%S"),
+   height=72,
+   name="John Doe")
> john.doe
An object of class "WeightHistory"
Slot "height":
[1] 72

Slot "name":
[1] "John Doe"

Slot "data":
numeric(0)

Slot "start":
[1] "2009-02-14 GMT"

Slot "end":
[1] "2009-03-28 GMT"
```

R will validate that the new TimeSeries object contained within WeightHistory is valid. (You can test this yourself.)

Let's consider an alternative way to construct a weight history. Suppose that we had created a Person class containing a person's name and height:

```
> setClass(
+   "Person",
+   representation(
+     height = "numeric",
+     name = "character"
+   )
+ )
```

Now, we can create an alternative weight history that inherits from both a TimeSeries object and a Person object:

```
> setClass(
+   "AltWeightHistory",
+   contains = c("TimeSeries", "Person")
+ )
```

This alternative implementation works identically to the original implementation, but the new implementation is slightly cleaner. This implementation inherits methods from both the TimeSeries and the Person classes.

Suppose that we also had created a class to represent cats:

```
> setClass(
+   "Cat",
+   representation(
+     breed = "character",
+     name = "character"
+   )
+ )
```

Notice that both Person and Cat objects contain a name attribute. Suppose that we wanted to create a method for both classes that checked if the name was "Fluffy." An efficient way to do this in R is to create a virtual class that is a superclass of both the Person and the Cat classes and then write an is.fluffy method for the superclass. (You can write methods for a virtual class but can't create objects from that class because the representation of those objects is ambiguous.)

```
> setClassUnion(
+   "NamedThing",
+   c("Person","Cat")
+ )
```

We could then create an is.fluffy method for the NamedThing class that would apply to both Person and Cat objects. (Note that if we were to define a method of is.fluffy for the Person class, this would override the method from the parent class.) An added benefit is that we could now check to see if an object was a NamedThing:

```
> jane.doe <- new("AltWeightHistory",
+   data=c(130, 129, 131, 128, 130, 129),
+   start=as.POSIXct("02/14/2009 0:00:00",tz="GMT",
+     format="%m/%d/%Y %H:%M:%S"),
+   end=as.POSIXct("03/28/2009 0:00:00",tz="GMT",
+     format="%m/%d/%Y %H:%M:%S"),
+   height=67,
+   name="Jane Doe")
> is(jane.doe,"NamedThing")
[1] TRUE
> is(john.doe,"TimeSeries")
[1] TRUE
```

Object-Oriented Programming in R: S4 Classes

Now that we've seen a quick introduction to object-oriented programming in R, let's talk about the functions for building classes in more depth.

Defining Classes

To create a new class in R, you use the `setClass` function:

```
setClass(Class, representation, prototype, contains=character(),
         validity, access, where, version, sealed, package,
         S3methods = FALSE)
```

Here is a description of the arguments to `setClass`.

Argument	Description	Default
Class	A character value specifying the name for the new class. (Only required argument.)	
representation	A named list of the different slots in the class and the object name associated with each one. (You can specify "ANY" if you want to allow arbitrary objects to be stored in the slot.)	
prototype	An object containing the default object for slots in the class.	
contains	A character vector containing the names of the classes that this class extends (usually called *superclasses*).	character()
validity	A function that checks the validity of an object of this class. (Default is no validity check.) May be changed later with `setValidity`.	
access	Not used; included for compatibility with S-PLUS.	
where	The environment in which to store the object definition.	Default is the environment in which `setClass` was called.
version	Not used; included for compatibility with S-PLUS.	
sealed	A logical value to indicate if this class can be redefined by calling `setClass` again with the same class name.	
package	A character value specifying the package name for this class. (Default is the name of the package in which `setClass` was called.)	
S3methods	A logical value specifying whether S3 methods may be written for this class.	FALSE

To simplify the creation of new classes, the `methods` package includes two functions for creating the representation and prototype arguments, called `representation` and `prototype` (respectively). These functions are very helpful when defining classes that extend other classes as a data part, have multiple superclasses, or combine extending a class and slots.

Some slot names are prohibited in R because they are reserved for attributes. (By the way, objects can have both slots and attributes.) Forbidden names include `"class"`, `"comment"`, `"dim"`, `"dimnames"`, `"names"`, `"row.names"` and `"tsp"`.

If a class extends one of the basic R types (as described in Table 10-1), there will be a slot called .Data containing the data from the basic object type. R code that works on the built-in class will work with objects of the new class; they will just act on the .Data part of the object.

You can explicitly define an inheritance relationship with the setIs function. (This is an alternative to using the contains argument for setClass.)

```
setIs(class1, class2, test=NULL, coerce=NULL, replace=NULL,
      by = character(), where = topenv(parent.frame()), classDef =,
      extensionObject = NULL, doComplete = TRUE)
```

To explicitly set a validation function for a class, you use the setValidity function:

```
setValidity(Class, method, where = topenv(parent.frame()) )
```

If you want to call a function after a new object is created, you may want to define an initialize method for the new object. See "Methods" on page 128 for information on how to add a method for the generic function initialize.

R also allows you to define a virtual class that is a superclass of several other classes. This can be useful if the virtual class does not contain any data by itself, but you want to create a set of methods that can be used by a set of other classes. To do this, you would use the setClassUnion function:

```
setClassUnion(name, members, where)
```

This function takes the following arguments.

Argument	Description
name	A character value specifying the name of the new superclass
members	A character vector specifying the names of the subclasses
where	The environment in which to create the new superclass

New Objects

You can create a new object in R through a call to the class's new method. (In object-oriented programming lingo, this is called a *constructor*.) Calling:

```
new(c,...)
```

returns a new object of class c. It is possible to fill data into the slots of the new object by specifying named arguments in the call to new; each slot will be set to the value specified by the corresponding named argument. If a method named initialize exists for class c, then the function initialize will be called after the new object is created (and the slots are filled in with the optional arguments).

Accessing Slots

You can fetch the value stored in slot *slot_name* of object *object_name* through a call to the function slot(*slot_name, object_name*). R includes a special operator for accessing the objects stored inside another object that is a shorthand for the slot function: the @ operator. This operator takes the form *object_name@slot_name*.

It is also possible to set the object stored in a slot with the familiar assignment operator. For example, to set the "month" slot of a "birthdate" object to the value "June," you would call:

```
> birthdate@month <- "June"
```

or, alternatively:

```
> slot(month, birthdate) <- "June"
```

By default, when changing a value in an object, R will check the validity of the new object. However, it is possible to override this check by using the check=FALSE option when calling slot:

```
> slot(month, birthdate, check=FALSE) <- "June"
```

Doing so is usually unwise and unnecessary.

Working with Objects

To test whether an object *o* is a member of a class *c*, you can use the function is(*o*, *c*). To test whether a class *c1* extends a second class *c2*, you can use the function extends(*c1*, *c2*).

To get a list of the slots associated with an object *o*, you can use the function slot Names(*o*). To get the classes associated with those slots, use getSlots(*o*). To determine the names of the slots in a class *c*, you can use the function slotNames(*c*). Somewhat nonintuitively, getSlots(*c*) returns the set of classes associated with each slot.

Creating Coercion Methods

It is possible to convert an object *o* to class *c* by calling as(*o*, *c*).

To enable coercion for a class that you define, make sure to register coercion methods with the setAs function:

```
setAs(from, to, def, replace, where = topenv(parent.frame()))
```

This function takes the following arguments.

Argument	Description	Default
from	A character value specifying the class name of the input object.	
to	A character value specifying the class name of the output object.	
def	A function that takes an argument of type from and returns a value of type to. In other words, a function that performs the conversion.	
replace	A second function that may be used in a replacement method (that is, the method to use if the as function is used as the destination in an assignment statement). This is a function of two arguments: from and value.	
where	The environment in which to store the definition.	topenv(parent.frame())

Methods

In Chapter 9, we showed how to use functions in R. An important part of a function definition in R is the set of arguments for a function. As you may recall, a function only accepts one set of arguments. When you assign a function directly to a symbol, you can only call that function with a single set of arguments.

Generic functions are a system for allowing the same name to be used for many different functions, with many different sets of arguments, from many different classes.

Suppose that you define a class called meat and a class called dairy and a method called serve. In R, you could assign one function to serve a meat object and another function to serve a dairy object. You could even assign a third function that took both a meat object and a dairy object as arguments and allowed you to serve both of them together. This would not be kosher in some other languages, but it's OK in R.‡

The first step in assigning methods is to create an appropriate generic function (if the function doesn't already exist). To do this, you use the setGeneric function to create a generic method:

```
setGeneric(name, def= , group=list(), valueClass=character(),
          where= , package= , signature= , useAsDefault= ,
          genericFunction= , simpleInheritanceOnly = )
```

This function takes the following arguments.

Argument	Description
name	A character value specifying the name of the generic function.
def	An optional function defining the generic function.
group	An optional character value specifying the group generic to which this function belongs. See the help file for S4groupGeneric for more information.
valueClass	An optional character value specifying the name of the class (or classes) to which objects returned by this function must belong.
where	The environment in which to store the new generic function.
package	A character value specifying the package name with which the generic function is associated.
signature	An optional character vector specifying the names of the formal arguments (as labels) and classes for the arguments to the function (as values). The class name "ANY" can be used to mean that arguments of any type are allowed.
useAsDefault	A logical value or function specifying the function to use as the default method. See the help file for more information.
genericFunction	Not currently used.
simpleInheritanceOnly	A logical value specifying whether to require that methods be inherited through simple inheritance only.

‡ In technical terms, R's implementation is called *parametric polymorphism*.

To associate a method with a class (or, more specifically, a signature with a generic function), you use the setMethod function:

```
setMethod(f, signature=character(), definition,
          where = topenv(parent.frame()),
          valueClass = NULL, sealed = FALSE)
```

Here is a description of the arguments for setMethod.

Argument	Description	Default
f	A generic function *or* the name of a generic function.	
signature	A vector containing the names of the formal arguments (as labels) and classes for the arguments to the function (as values). The class name "ANY" can be used to mean that arguments of any type are allowed.	character()
definition	The function to be called when the method is evaluated.	
where	The environment in which the method was defined.	topenv(parent.frame())
valueClass	Not used; included for backward compatibility.	NULL
sealed	Used to indicate if this class can be redefined by calling setClass again with the same class name.	FALSE

Managing Methods

The methods package includes a number of functions for managing generic methods.

Function	Description
isGeneric	Checks if there is a generic function with the given name.
isGroup	Checks if there is a group generic function with the given name.
removeGeneric	Removes all the methods for a generic function and the generic function itself.
dumpMethod	Dumps the method for this generic function and signature.
findFunction	Returns a list of either the positions on the search list or the current top-level environment on which a function object for a given name exists.
dumpMethods	Dumps all the methods for a generic function.
signature	Returns the names of the generic functions that have methods defined on a specific path.
removeMethods	Removes all the methods for a generic function.
setGeneric	Creates a new generic function of the given name.

The methods package also includes functions for managing methods.

Function	Description
getMethod, selectMethod	Returns the method for a particular function and signature.
existsMethod, hasMethod	Tests if a method (specified by a specific name and signature) exists.
findMethod	Returns the package(s) that contain a method for this function and signature.
showMethods	Shows the set of methods associated with an S4 generic.

For more information on these functions, see the corresponding help files.

Basic Classes

Classes for built-in types are shown in Table 10-1; these are often called *basic classes*. All classes are built on top of these classes. Additionally, it is possible to write new methods for these classes that override the defaults.

Table 10-1. Classes of built-in types

Category	Object Type	Class
Vectors	integer	integer
	double	numeric
	complex	complex
	character	character
	logical	logical
	raw	raw
Compound	list	list
	pairlist	pairlist
	S4	
	environment	environment
Special	any	
	NULL	NULL
	...	
R language	symbol	name
	promise	
	language	call
	expression	expression
	externalptr	externalptr
Functions	closure	function
	special	function
	builtin	function

The vector classes (integer, numeric, complex, character, logical, and raw) all extend the vector class. The vector class is a virtual class.

More Help

Many tools for working with classes are included in the methods package, so you can find additional help on classes with the command library(help="methods").

Old-School OOP in R: S3

This section is about S3 classes and methods. Although S4 classes and methods are much more capable than S3 classes and methods, many important R functions were written before S4 objects were implemented (such as the statistical modeling software). In order to understand, modify, or extend this software, you have to know how S3 classes they are implemented.

S3 Classes

As we saw above, S4 classes implement most features of modern object-oriented programming languages: formal class definitions, simple and multiple inheritance, parameteric polymorphism, and encapsulation. Unfortunately, S3 classes don't implement most of these features.

S3 classes are implemented through an object attribute. An S3 object is simply a primitive R object with additional attributes, including a class name. There is no formal definition for an S3 object; you can manually change the attributes, including the class. Above, we used time series as an example of an S4 class. There is an existing S3 class for representing time series, called "ts" objects. Let's create a sample time series object and look at how it is implemented. Specifically, we'll look at the attributes of the object and then use `typeof` and `unclass` to examine the underlying object:

```
> my.ts <- ts(data=c(1,2,3,4,5),start=c(2009,2),frequency=12)
> my.ts
      Feb Mar Apr May Jun
2009    1   2   3   4   5
> attributes(my.ts)
$tsp
[1] 2009.083 2009.417   12.000

$class
[1] "ts"

> typeof(my.ts)
[1] "double"
> unclass(my.ts)
[1] 1 2 3 4 5
attr(,"tsp")
[1] 2009.083 2009.417   12.000
```

As you can see, a `ts` object is just a numeric vector (of doubles), with two attributes: `class` and `tsp`. The `class` attribute is just the name "ts," and the `tsp` attribute is just a vector with a start time, end time, and frequency. You can't access attributes in an S3 object using the same operator that you use to access slots in an S4 object:

```
> my.ts@tsp
Error: trying to get slot "tsp" from an object (class "ts")
        that is not an S4 object
```

S3 classes lack a lot of the structure of S3 objects. Inheritance is implemented informally, and encapsulation is not enforced by the language.§ S3 classes also don't allow parametric polymorphism. S3 classes do, however, allow simple polymorphism. It is possible to define S3 generic functions and to dispatch by object type.

S3 Methods

S3 generic functions work by naming convention, not by explicitly registering methods for different classes. Here is how to create a generic function using S3 classes:

1. Pick a name for the generic function. We'll call this *gname*.
2. Create a function named *gname*. In the body for *gname*, call UseMethod("*gname*").
3. For each class that you want to use with *gname*, create a function called *gname.classname* whose first argument is an object of class *classname*.

Rather than fabricating an example, let's look at an S3 generic function in R: plot:

```
> plot
function (x, y, ...)
{
    if (is.function(x) && is.null(attr(x, "class"))) {
        if (missing(y))
            y <- NULL
        hasylab <- function(...) !all(is.na(pmatch(names(list(...)),
            "ylab")))
        if (hasylab(...))
            plot.function(x, y, ...)
        else plot.function(x, y, ylab = paste(deparse(substitute(x)),
            "(x)"), ...)
    }
    else UseMethod("plot")
}
<environment: namespace:graphics>
```

Here's how plot works. The function takes a look at the arguments on which it was called. If the first argument is a function, plot does something special. Otherwise, plot calls UseMethod("plot"). UseMethod looks at the class of the object x. It then looks for a function named plot.*class* and calls plot.*class*(x, y, ...).

For example, we defined a new TimeSeries class above. To add a plot method for TimeSeries objects, we simply create a function named plot.TimeSeries:

```
> plot.TimeSeries <- function(object, ...) {
+    plot(object@data, ...)
+ }
```

So, we could now call:

```
plot(my.TimeSeries)
```

§ If the attribute class is a vector with more than one element, then the first element is interpreted as the class of the object, and other elements name classes that the object "inherits" from. That makes inheritance a property of objects, not classes.

and R would, in turn, call `plot.TimeSeries(my.TimeSeries)`.

The function `UseMethod` dispatches to the appropriate method, depending on the class of the first argument's calling function. `UseMethod` iterates through each class in the object's class vector, until it finds a suitable method. If it finds no suitable method, `UseMethod` looks for a function for the class "default." (A closely related function, `NextMethod`. `NextMethod`, is used in a method called by `UseMethod`; it calls the next available method for an object. See the help file for more information.)

Using S3 Classes in S4 Classes

You can't specify an S3 class for a slot in an S4 class. To use an S3 class as a slot in an S4 class, you need to create an S4 class based on the S3 class. A simple way to do this is through the function `setOldClass`:

```
setOldClass(Classes, prototype, where, test = FALSE, S4Class)
```

This function takes the following arguments.

Argument	Description	Default
Classes	A character vector specifying the names of the old-style classes.	
prototype	An object to use as a prototype; this will be used as the default object for the S4 class.	
where	An environment specifying where to store the class definition.	The top-level environment
test	A logical value specifying whether to explicitly test inheritance for the object. Specify test=TRUE if there can be multiple inheritance.	FALSE
S4Class	A class definition for an S4 class or a class name for an S4 class. This will be used to define the new class.	

Finding Hidden S3 Methods

Sometimes, you may encounter cases where individual methods are hidden. The author of a package may choose to hide individual methods in order to encapsulate details of the implementation within the package; hiding methods encourages you to use the generic functions. For example, individual methods for the generic method `histogram` (in the `lattice` package) are hidden:

```
> methods(histogram)
[1] histogram.factor*  histogram.formula* histogram.numeric*

    Nonvisible functions are asterisked > histogram.factor()
Error: could not find function "histogram.factor"
```

Sometimes, you might want to retrieve the hidden methods (for example, to view the R code). To retrieve the hidden method, use the function `getS3method`. For example, to fetch the code for `histogram.formula`, try the following command:

```
> getS3method(f="histogram", class="formula")
```

Alternatively, you can use the function `getAnywhere`:

```
> getAnywhere(histogram.formula)
```

11

High-Performance R

In my experience, R runs very well on modern computers and moderate-size data sets, returning results in seconds or minutes. If you're dealing with small data sets and doing normal calculations, you probably don't have to worry too much about performance. However, if you are dealing with big data sets or doing very complex calculations, then you could run into trouble.

This chapter includes some tips for making R run well when tackling unusually large or complicated problems.

Use Built-in Math Functions

When possible, try to use built-in functions for mathematical computations instead of writing R code to perform those computations. Many common math functions are included as native functions in R. In most cases, these functions are implemented as calls to external math libraries. As an obvious example, if you want to multiply two matrices together, you should probably use the %*% operator and not write your own matrix multiplication code in R.

Often, it is possible to use built-in functions by transforming a problem. As an example, let's consider an example from queueing theory. Queueing theory is the study of systems where "customers" arrive, wait in a "queue" for service, are served, and then leave. As an example, picture a cafeteria with a single cashier. After customers select their food, they proceed to the cashier for payment. If there is no line, they pay the cashier and then leave. If there is a line, they wait in the line until the cashier is free. If we suppose that customers arrive according to a Poisson process and that the time required for the cashier to finish each transaction is given by an exponential distribution, then this is called an M/M/1 queue. (This means "memoryless" arrivals, "memoryless" service time, and one server.)

A very useful formula for queueing theory is Erlang's B formula. Picture a call center with n operators but no queue: if a customer calls the center and there is a free operator, then the operator will answer the customer's call. However, if every operator is busy, the customer will get a busy signal. Further, let's make the same assumptions as above: customers arrive according to a Poisson process, and the time required to service each call is given by an exponential distribution. This is called an M/M/n/n queue. Erlang's B formula tells us the probability that all operators are busy at a given time; it is the probability that a customer who calls into the data center will get a busy signal:

$$p_l = \frac{r^c/c!}{\sum_{i=0}^{c} r^i/i!}$$

Unfortunately, you'll find that it's hard to calculate this value directly for more than a handful of operators because R can't handle numbers as big (or as small) as it needs to handle. One trick to perform this calculation is to transform this formula into formulas that are already implemented as R functions: Poisson distribution calculations:[*]

$$p_l = \frac{r^c/c!}{\sum_{i=0}^{c} r^i/i!} = \frac{e^{-r} r^c/c!}{e^{-r}\sum_{i=0}^{c} r^i/i!} = \frac{e^{-r} r^c/c!}{\sum_{i=0}^{c} e^{-r} r^i/i!}$$

So, to calculate Erlang's B formula in R, you could use an R function like this:

```
erlangb <- function(c, r) {dpois(c,r)/ppois(c,r)}
```

By using the built-in function, we are using compiled code written in a low-level language (usually C or FORTRAN, depending on the function). This code is typically faster than interpreted R code.

Use Environments for Lookup Tables

If you need to store a big lookup table, consider implementing the table using an environment. Environment objects are implemented using hash tables. Vectors and lists are not. This means that looking up a value in a vector with n elements or a list can take $O(n)$ time. Looking up the value in an environment object takes $O(1)$ time (on average).

To make it less awkward, consider defining an S4 class that implements the interface for the lookup table.

Use a Database to Query Large Data Sets

If you need to query large tables of data, you should consider storing the values in a database. You don't need to use an external database; the RSQLite package provides an interface to the SQLite library that allows you to store data in files and query

---

[*] Another alternative is to notice that Erlang's B formula can be rewritten as a recurrence and write a program to iteratively calculate the probability. For more details on this method, see a book like *Fundamentals of Queueing Theory* by Donald Gross et al. (Wiley-Interscience).

the files using SQL. (This is the strategy that is used by Bioconductor to store annotation databases.) See "DBI" on page 173 for more information on how to use this package.

Preallocate Memory

In R, you don't have to explicitly allocate memory before you use it. For example, you could fill an array with numbers using the following code:

```
v <- c()
for (i in 1:100000) {v[i] <- i;}
```

This code works correctly; however, it takes a long time to finish (about 30 seconds on my computer). You can speed up this code substantially by preallocating memory to the vector. You can do this by setting the length, nrow, ncol, or dim attributes for an object. Here is an example:

```
v2 <- c(NA)
length(v2) <- 100000
for (i in 1:100000) {v2[i] <- i;}
```

This code works identically but performs much, much faster.

Monitor How Much Memory You Are Using

The function gc serves two purposes. First, it causes garbage collection to occur immediately, potentially freeing up storage space. Second, it displays statistics on free memory:

```
> gc()
            used  (Mb) gc trigger   (Mb)  max used   (Mb)
Ncells    774900  20.7     919870   24.6   3032449   81.0
Vcells 53549840 408.6  176511395 1346.7 380946917 2906.4
> # remove a big object
> rm(audioscrobbler)
> gc()
            used  (Mb) gc trigger   (Mb)  max used   (Mb)
Ncells    328394   8.8     919870   24.6   3032449   81.0
Vcells 50049839 381.9  141209116 1077.4 380946917 2906.4
```

Monitoring Memory Usage

To check on the (approximate) size of a specific object, use the function object.size:

```
> object.size(1)
32 bytes
> object.size("Hello world!")
72 bytes
> object.size(audioscrobbler)
39374504 bytes
```

The function memory.profile displays information on memory usage by object type:

```
> memory.profile()
       NULL      symbol     pairlist     closure environment
```

```
            1        9479      160358        3360        1342
      promise    language     special     builtin        char
         8162       44776         138        1294       48872
      logical     integer      double     complex   character
         4727        8373        2185           4       29761
          ...         any        list  expression    bytecode
            0           0        3488           2           0
  externalptr     weakref         raw          S4
          993         272         273        1008
```

To monitor how much memory R is using on a Microsoft Windows system, you can use the function memory.size. (On other platforms, this function returns the value Inf with a warning.) On startup, here is how much memory R used:

```
> memory.size()
[1] 10.58104
```

This function reports memory usage in MB. You can check the maximum amount of memory used so far through the memory.size(max=TRUE) option:

```
> memory.size(max=TRUE)
[1] 12.3125
```

Increasing Memory Limits

If you are running out of storage space on a Microsoft Windows platform, you can get or set the memory limit on a system with the function memory.limit:

```
> memory.limit()
[1] 1023.484
> memory.limit(size=1280)
NULL
> memory.limit()
[1] 1280
```

On other platforms, this function will return Inf and print a warning message. On these platforms, you can use the function mem.limits to get or set memory limits:

```
mem.limits(nsize = NA, vsize = NA)
```

The argument nsize specifies the number of cons cells (basic units of storage).

If there are no explicit limits, this function may return NA:

```
> mem.limits()
nsize vsize
   NA    NA
```

Cleaning Up Objects

In R, you usually don't have to manually manage memory; the system automatically allocates and unallocates memory as needed. However, you can get some information on the process (and control it a little) through the function gc, as described earlier.

If you're running out of storage space, you might want to try removing objects from the workspace. You can remove an object (or a set of objects) from an environment

---

with the `rm` function. By default, this function removes objects from the current environment:

```
> # remove a big object
> rm(audioscrobbler)
> gc()
            used  (Mb) gc trigger    (Mb)  max used    (Mb)
Ncells    328394   8.8     919870    24.6   3032449    81.0
Vcells  50049839 381.9  141209116  1077.4 380946917  2906.4
```

Functions for Big Data Sets

If you're working with a very large data set, you may not have enough memory to use the standard regression functions. Luckily, R includes an alternative set of regression functions for working with big data sets. These functions are slower than the standard regression functions, but will work when there is not enough memory to use the standard regression functions:

```
library(biglm)
# substitute for lm, works in dataframes
biglm(formula, data, weights=NULL, sandwich=FALSE)
# substitute for glm, works in data frames
bigglm(formula, data, family=gaussian(),
    weights=NULL, sandwich=FALSE, maxit=8, tolerance=1e-7,
    start=NULL,quiet=FALSE,...)
```

It's even possible to use `bigglm` on data sets inside a database. To do this, you would open a database connection using RODBC or RSQLite and then call `bigglm` with the `data` argument specifying the database connection and `tablename` specifying the table in which to evaluate the formula:

```
bigglm(formula, data, family=gaussian(),
    tablename, ..., chunksize=5000)
```

Parallel Computation with R

One of the best techniques for speeding up large computing problems is to break them into lots of little pieces, solve the pieces separately on different processors and then put the pieces back together. This is called *parallel computing*, because it enables you to solve problems in parallel. For example, suppose that you had a lot of laundry: enough to fill 10 washing machines. Suppose each wash took 45 minutes, and each drying took 45 minutes. If you had only one washing machine and dryer, it would take 495 minutes to finish all the laundry. However, if you had 10 washing machines and ten dryers, you could finish the laundry in 90 minutes.

In Chapters 20 and 21, we will show some cutting-edge techniques for statistical modeling. Many of these problems are very computationally intensive and could take a long time to finish. Luckily, many of them are very parallelizable. For example, we will show several algorithms that build models by fitting a large number of tree models to the underlying data (such as boosting, bagging, and random forests). Each of these algorithms could be run in parallel if more processors were available.

In "Looping Extensions" on page 71, we showed some extensions to R's built-in looping functions. Revolution Computing developed these extensions to help facilitate parallel computation. Revolution Computing has also released a package called doMC that facilitates running R code on multiple cores.

To write code that takes advantage of multiple cores, you need to initialize the doMC package:

```
> library(doMC)
> registerDoMC()
```

This will allow the %dopar% operator (and related functions) in the **foreach** package to run in parallel. There are additional tools available to running R code in parallel on a cluster; see packages doMPI, doSNOW, or doNWS for more information.

Revolution Computing has additional tools available in its enterprise version. See its website for more information.

High-Performance R Binaries

On some platforms (like Mac OS X), R is compiled with high-quality math libraries. However, the default libraries on other platforms (like Windows) can be sluggish. If you're working with large data sets or complicated mathematical operations, you might find it worthwhile to build an optimized version of R with better math libraries.

Revolution R

Revolution Computing is a software company that makes a high-performance version of R. It offers both free and commercial versions, including a 64-bit build of R for Windows. For the latest version, check out its website: *http://www.revolution -computing.com/*.

Revolution R looks a lot like the standard R binaries (although a little outdated; at the time I was writing this book, Revolution was shipping Revolution R 1.3.0 included R 2.7.2, while the current version from CRAN was 2.10.0). The key difference is the addition of improved math libraries. These are multithreaded and can take advantage of multiple cores when available. There are two helper functions included with Revolution R that can help you set and check the number of cores in use. To check the number of cores, use:

```
getMKLthreads()
```

Revolution R guesses the number of threads to use, but you can change the number yourself if it guesses wrong (or if you want to experiment). To set the number of cores explicitly, use:

```
setMKLthreads(n)
```

The help file suggests not setting the number of threads higher than the number of available cores.

Building Your Own

Building your own R can be useful if you want to compile it to run more efficiently. For example, you can compile a 64-bit version of R if you want to work with data sets that require much more than 4 GB of memory. This section explains how to build R yourself.

Building on Microsoft Windows

The easiest way to build your own R binaries on Microsoft Windows is to use the Rtools software. The R compilation process is very sensitive to the tools that you use. So, the Rtools software bundles together a set of tools that are known to work correctly with R. Even if you plan to use your own compiler, math libraries, or other components, you should probably start with the standard toolkit and incrementally modify it. That will help you isolate problems in the build process.

Here is how to successfully build your own R binaries (and installer!) on Microsoft Windows:

1. Download the R source code from *http://cran.r-project.org/src/base/*.

2. Download the "Rtools" software from *http://www.murdoch-sutherland.com/Rtools/*.

3. Run the Rtools installer application. Follow the directions to install Rtools. You can select most default options, but I do not suggest installing all components at this stage. (The "Extras to build R" needs to be installed in the source code directory to be useful. However, we don't install those until steps 4 and 5. Unfortunately, you need other tools from the RTools software in order to execute steps 4 and 5, so we can't change the order of the steps to avoid running the installer twice.) As shown in Figure 11-1, you should select everything except "Extras to build R." We'll install that stuff later, so don't throw out the tools installer yet. Also, if you use Cygwin, be sure to read the notes about conflicts with Cygwin DLLs (dynamic-link libraries). Be sure to select the option allowing Rtools to modify your PATH variable (or make sure to change it yourself).

4. Move the source code file to a build directory, open a command-line window (possibly with cmd), and change to the build directory. (Be sure to open the command shell after installing the Rtools and modifying your PATH. This will guarantee that the commands in the next few steps are available.)

5. Run the following command to unpack the source code into the directory R-2.9.2:

   ```
   tar xvfz R-2.9.2.tar.gz
   ```

   (Note that I used R-2.9.2.tar.gz. Change the command as needed for the R version you are installing.)

6. Rerun the Rtools setup program. This time, select only the "Extras to build R" component, and no other components. Install the components into the source code directory that you just unpacked. (For example, if you have installed R into *C:\stuff\things*, then select *C:\stuff\things\R-2.9.2*.)

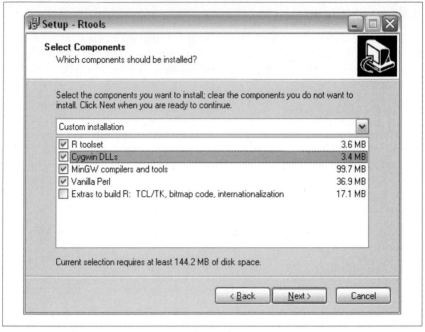

Figure 11-1. Selecting components in Rtools

7. At this point, you may install several additional pieces of software:

   a. (Optional) If you want to build Microsoft HTML help files, then download and install the Microsoft HTML Help Workshop from *http://www.micro soft.com/downloads/details.aspx?FamilyID=00535334-c8a6-452f-9aa0 -d597d16580cc*. Make sure the location where it is installed (for example, *C:\Program Files\HTML Help Workshop*) is included in the PATH.

   b. (Optional) If you want to build your own R installer, then download and install Inno Setup from *http://www.jrsoftware.org/isinfo.php*. After you have done this, edit the file *src\gnuwin32\MkRules in the R-2.9.2* directory. Change ISDIR to the location where Inno Setup was installed. (By default, this location is *C:\Program Files\Inno Setup 5*.)

   c. (Optional) Download and install LaTeX if you want to build PDF versions of the help files. A suitable version is MiKTeX, from *http://www.miktex .org/*.

8. Return to the command window and change directories to the *src\gnuwin32* directory in the R sources (for example, *C:\stuff\things\R-2.9.2\src\gnuwin32*). Run the following command to build R:

   ```
   make all recommended
   ```

9. To check that the build was successful, you can run the command:

   ```
   make check
   ```

   Or, for more comprehensive checks:

   ```
   make check-all
   ```

   I found that the checks failed due to a silly error. (The checks included testing examples in libraries, so the test application tried to open a network connection to *http://foo.bar*, a hostname that could not be resolved.) Use your own discretion about whether the tests were successful or not.

10. If everything worked correctly, you can now try your own build of R. The executables will be located in the *R-2.9.2\bin* directory. The full GUI version is named `Rgui.exe`; the command-line version is `R.exe`.

11. If you would like to build your own installer, then execute the following command in the *src\gnuwin32* directory:

    ```
    make distribution
    ```

    (I got some errors late in the install process. The standard makefiles try to delete content when they're done. If you don't make it past building `rinstaller`, manually run `make cran`.) To check if the process worked, look for the installer in the *gnuwin32\cran* directory.

For more information about how to build R on Microsoft Windows platforms, see the directions in the R Installation and Administration Manual. (You can read the manual online at *http://cran.r-project.org/doc/manuals/R-admin.html*, or you can download a PDF from *http://cran.r-project.org/doc/manuals/R-admin.pdf*.)

Building R on Unix-like systems

Unix-like systems are by far the easiest systems on which to build R. Here is how to do it:

1. Install the standard development tools: gcc, make, perl, binutiles, and LaTeX. (If you don't know if you have all the tools and are using a standard Linux version such as Fedora, you have probably already installed all the components you need. Unfortunately, it's outside the scope of this book to explain how to find and install missing components. Try using the precompiled binaries or find a good book on Unix system administration.)

2. Download the R source code from *http://cran.r-project.org/src/base/*.

3. Run the following command to unpack the source code into the directory R-2.10.0:

   ```
   tar xvfz R-2.10.0.tar.gz
   ```

   (Note that I used R-2.10.0.tar.gz. Change the command as needed for the R version you are installing.)

4. Change to the R-2.10.0 directory. Run the following commands to build R:

```
./configure
make
```

5. To check that the build was successful, you can run the command:

```
make check
```

Or, for more comprehensive checks:

```
make check-all
```

6. Finally, if everything is OK, run the following command to install R:

```
make install
```

These directions will work on Mac OS X if you want to build a command-line version of R or a version of R that works through X Windows. They will not build the full Mac OS X GUI.

Building R on Mac OS X

Building R on Mac OS X is a little trickier than building it on Windows or Linux systems because you have to fetch more individual pieces. For directions on how to compile R on Mac OS X, see *http://cran.r-project.org/doc/manuals/R-admin.html*. You may also want to read the FAQ file at *http://cran.cnr.Berkeley.edu/bin/macosx/RMacOSX-FAQ.html*, which gives some hints on how to build.

Working with Data

This part of the book explains how to accomplish some common tasks with R: loading data, transforming data, and visualizing data. These techniques are useful for any type of data that you want to work with in R.

12

Saving, Loading, and Editing Data

This chapter explains how to load data into R, save data objects from R, and edit data using R.

Entering Data Within R

If you are entering a small number of observations, entering the data directly into R might be a good approach. There are a couple of different ways to enter data into R.

Entering Data Using R Commands

Many of the examples in Parts I and II show how to create new objects directly on the R console. If you are entering a small amount of data, this might be a good approach.

As we have seen before, to create a vector, use the c function:

```
> salary <- c(18700000,14626720,14137500,13980000,12916666)
> position <- c("QB","QB","DE","QB","QB")
> team <- c("Colts","Patriots","Panthers","Bengals","Giants")
> name.last <- c("Manning","Brady","Pepper","Palmer","Manning")
> name.first <- c("Peyton","Tom","Julius","Carson","Eli")
```

It's often convenient to put these vectors together into a data frame. To create a data frame, use the `data.frame` function to combine the vectors:

```
> top.5.salaries <- data.frame(name.last,name.first,team,position,salary)
> top.5.salaries
  name.last name.first    team position    salary
1   Manning     Peyton   Colts       QB  18700000
2     Brady        Tom Patriots       QB  14626720
3    Pepper     Julius Panthers       DE  14137500
4    Palmer     Carson Bengals        QB  13980000
5   Manning        Eli  Giants        QB  12916666
```

Using the Edit GUI

Entering data using individual statements can be awkward for more than a handful of observations. (That's why my example above only included five observations.) Luckily, R provides a nice GUI for editing tabular data: the data editor.

To edit an object with the data editor, use the `edit` function. The `edit` function will open the data editor and return the edited object. For example, to edit the `top.5.salaries` data frame, you would use the following command:

```
> top.5.salaries <- edit(top.5.salaries)
```

Notice that you need to assign the output of the `edit` function to a symbol; otherwise, the edits will be lost. The data editor is designed to edit tabular data objects, specifically data frames and matrices. The `edit` function can be used with other types of objects such as vectors, functions, and lists, but it will open a text editor.

Alternatively, you can use the `fix` function. The `fix` function calls `edit` on its argument and then assigns the result to the same symbol in the calling environment. For the example above, here is how you would use `fix`:

```
> fix(top.5.salaries)
```

On Microsoft Windows, there is a menu item "Data Editor…" under the Edit menu that allows you to enter the name of an object into a dialog box and then calls `fix` on the object.

Windows Data Editor

The data editor on Microsoft Windows is very intuitive. To edit a value, simply click in the cell. To change the name of a column (or to change it from numeric to character), click on the column name and a window will pop up allowing you to make those changes. You may add additional rows and columns simply by entering values into empty cells (see Figure 12-1).

Mac OS X Data Editor

On Mac OS X, the edit window looks (and works) subtly differently. You may use the data editor with data frames or matrices (see Figure 12-2).

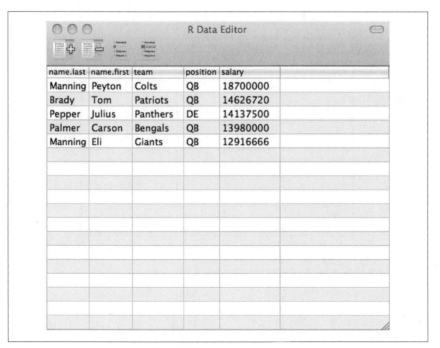

Figure 12-1. Editor window on Windows

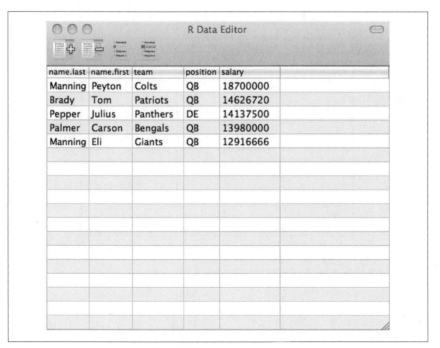

Figure 12-2. Editor window on Mac OS X

You can click on a data cell to edit the value. The buttons on the top have the following effects (from left to right): add a column, delete a column, add a row and delete a row. You can change a column's width by clicking on the lines separating that column from its neighbor and dragging it. You cannot change variable types or names from this editor.

X Windows (Linux) Data Editor

A data editor GUI is also available on X Windows systems. Like the Microsoft Windows version, you can edit the column names. For convenience, this editor includes copy, paste, and quit buttons (see Figure 12-3).

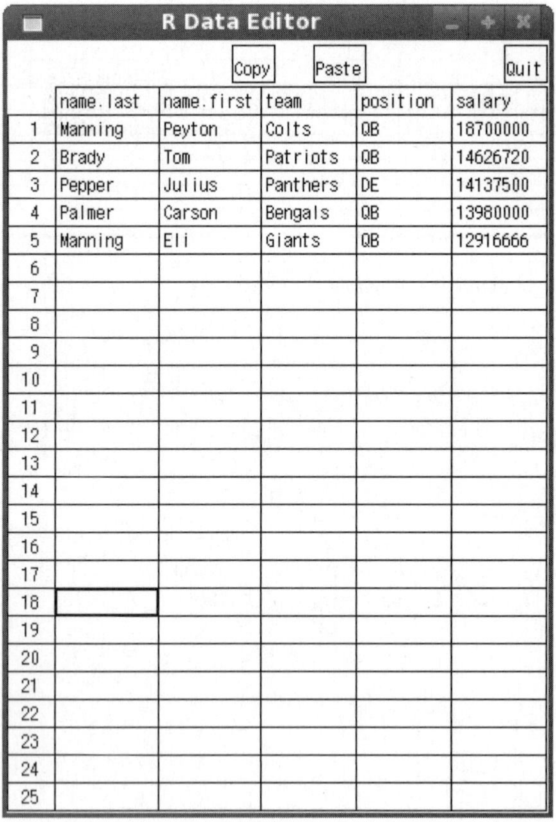

Figure 12-3. Data Editor on X Windows

R Data Editor Versus Spreadsheets

The R data editor can be convenient for inspecting a data frame or a matrix or maybe for editing a couple of values, but I don't recommend using it for doing serious work. If you have a lot of data to enter, I recommend using a real spreadsheet or desktop database program. There are a few reasons for this.

First, the R data editor doesn't provide an Undo or Redo function.

Second, the R data editor doesn't make it very easy to save your work. There is no "save" button. To save, you need to periodically close the editor, save your work, and then reopen the editor. Doing that is awkward and error prone; I would worry about losing my work if I used this editor.

Finally, spreadsheet programs often include data entry forms. (Desktop database programs also often have data entry forms.) If you're entering a complicated set of data, filling out a form for each observation can be much easier than typing the results into a form.

Saving and Loading R Objects

R allows you to save and load R data objects to external files.

Saving Objects with save

The simplest way to save an object is with the **save** function. For example, we could use the following command to save the object `top.5.salaries` to the file *~/top.5.salaries.RData*:

```
> save(top.5.salaries,file="~/top.5.salaries.RData")
```

In R, file paths are *always* specified with forward slashes ("/"), even on Microsoft Windows. So, to save this file to the directory *C:\Documents and Settings\me\My Documents\top.5.salaries.rda*, you would use the following command:

```
> save(top.5.salaries,
+   file="C:/Documents and Settings/me/My Documents/top.5.salaries.rda")
```

Note that the `file` argument must be explicitly named. (Nine out of ten times, I forget to do so.) Now, you can easily load this object back into R with the `load` function:

```
> load("~/top.5.salaries")
```

Incidentally, files saved in R will work across platforms. (For example, the data files for this book were produced on Mac OS X but work on Windows and Linux.) You can save multiple objects to the same file by simply listing them in the same save command. If you want to save every object in the workspace, you can use the **save.image** function. (When you quit R, you will be asked if you want to save your current workspace. If you say yes, the workspace will be saved the same way as this function.)

The **save** function is very flexible and can be used in many different ways. You can save multiple objects, save to files or connections, and save in a variety of formats:

```
save(..., list =, file =, ascii =, version = , envir =,
    compress =, eval.promises = , precheck = )
```

You can omit any argument except the filename. The defaults for **save** are very sensible: objects will be saved in a compressed binary format, and existing files won't be overwritten.

Here is a detailed description of the arguments to **save**.

Argument	Description
...	A set of symbols that name the objects to be saved. (This is a variable-length argument.)
list	Alternatively, you may specify the objects to be saved in a character vector.
file	Specifies where to save the file. Both connections and filenames can be used.
ascii	A logical value that indicates whether to write a human-readable representation of the data (ascii=TRUE) or a binary representation (ascii=FALSE). Default is ascii=FALSE.
version	A numeric value that indicates the file version. For R 0.99.0 through 1.3.1, use version=1. For R 1.4.0 through (at least) 2.8.1, use version=2. Default is version=2.
envir	Specifies the environment in which to find the objects to be saved. Default is the environment in which save was called (to be precise, parent.frame()).
compress	A logical value that indicates whether to compress the file when saving it. (The effect is the same as running gzip on an uncompressed file.) Default is compress=TRUE for binary files (ascii=FALSE) and compress=FALSE for human-readable files (ascii=TRUE).
eval.promises	A logical value that indicates whether promise objects should be forced before saving. Default is eval.promises=TRUE.
precheck	A logical value that indicates whether the save function should check if the object exists before saving (and raise an error if it is). Default is precheck=TRUE.

Importing Data from External Files

One of the nicest things about R is how easy it is to pull in data from other programs. R can import data from text files, other statistics software, and even spreadsheets. You don't even need a local copy of the file: you can specify a file at a URL, and R will fetch the file for you over the Internet.

Text Files

Most text files containing data are formatted similarly: each line of a text file represents an observation (or record). Each line contains a set of different variables associated with that observation. Sometimes, different variables are separated by a special character called the *delimiter*. Other times, variables are differentiated by their location on each line.

Delimited files

R includes a family of functions for importing delimited text files into R, based on the `read.table` function:

```
read.table(file, header, sep = , quote = , dec = , row.names, col.names,
           as.is = , na.strings , colClasses , nrows =, skip = ,
           check.names = , fill = , strip.white = , blank.lines.skip = ,
           comment.char = , allowEscapes = , flush = , stringsAsFactors = ,
           encoding = )
```

The `read.table` function reads a text file into R and returns a `data.frame` object. Each row in the input file is interpreted as an observation. Each column in the input file represents a variable. The `read.table` function expects each field to be separated by a delimiter.

For example, suppose that you had a file called *top.5.salaries.csv* that contained the following text (and only this text):

```
name.last,name.first,team position,salary
"Manning","Peyton","Colts","QB",18700000
"Brady","Tom","Patriots","QB",14626720
"Pepper","Julius","Panthers","DE",14137500
"Palmer","Carson","Bengals","QB",13980000
"Manning","Eli","Giants","QB",12916666
```

This file contains the same data frame that we entered in "Entering Data Using R Commands" on page 147. Notice how this data is encoded:

- The first row contains the column names.
- Each text field is encapsulated in quotes.
- Each field is separated by commas.

To load this file into R, you would specify that the first row contained column names (`header=TRUE`), that the delimiter was a comma (`sep=","`), and that quotes were used to encapsulate text (`quote="\""`). Here is an R statement that loads in this file:

```
> top.5.salaries <- read.table("top.5.salaries.csv",
+                               header=TRUE,
+                               sep=",",
+                               quote="\"")
```

The `read.table` function is very flexible and allows you to load files with many different properties. Here is a brief description of the options for `read.table`.

Argument	Description	Default
file	The name of the file to open or, alternatively, the name of a connection containing the data. You can even use a URL. (This is the one required argument for `read.table`.)	None
header	A logical value indicating whether the first row of the file contains variable names.	FALSE
sep	The character (or characters) separating fields. When "" is specified, any whitespace is used as a separator.	""

Argument	Description	Default
quote	If character values are enclosed in quotes, this argument should specify the type of quotes.	""
dec	The character used for decimal points.	.
row.names	A character vector containing row names for the returned data frame.	None
col.names	A character vector containing column names for the returned data frame.	None
as.is	A logical vector (the same length as the number of columns) that specifies whether or not to convert character values to factors.	!stringsAsFactors
na.strings	A character vector specifying values that should be interpreted as NA.	NA
colClasses	A character vector of class names to be assigned to each column.	NA
nrows	An integer value specifying the number of rows to read. (Invalid values, such as negatives, are ignored.)	-1
skip	An integer value specifying the number of rows in the text file to skip before beginning to read data.	0
check.names	A logical value that specifies whether `read.table` should check if the column names are valid symbol names in R.	TRUE
fill	Sometimes, a file might contain rows of unequal length. This argument is a logical value that specifies whether `read.table` should implicitly add blank fields at the end of rows where some values were missing.	!blank.lines.skip
strip.white	When `sep !=""`, this logical value specifies whether `read.table` should remove extra leading and trailing white space from character fields.	FALSE
blank.lines.skip	A logical value that specifies whether `read.table` should ignore blank lines.	TRUE
comment.char	`read.table` can ignore comment lines in input files if the comment lines begin with a single special character. This argument specifies the character used to delineate these lines.	#
allowEscapes	A logical value that indicates whether escapes (such as "\n" for a new line) should be interpreted or if character strings should be read literally.	FALSE
flush	A logical value that indicates whether `read.table` should skip to the next line when all requested fields have been read in from a line.	FALSE
stringsAsFactors	A logical value indicating whether text fields should be converted to factors.	default.stringsAsFactors()
encoding	The encoding scheme used for the source file.	"unknown"

The most important options are sep and header. You almost always have to know the field separator and know if there is a header field. R includes a set of convenience functions that call read.table with different default options for these values (and a couple of others). Here is a description of these functions.

Function	header	sep	quote	dec	fill	comment.char
read.table	FALSE		\" or \'	.	!blank.lines.skip	#
read.csv	TRUE	,	\"	.	TRUE	
read.csv2	TRUE	;	\"	,	TRUE	
read.delim	TRUE	\t	\"	.	TRUE	
read.delim2	TRUE	\t	\"	,	TRUE	

In most cases, you will find that you can use `read.csv` for comma-separated files or `read.delim` for tab-delimited files without specifying any other options. (Except, I suppose, if you are in Europe, and you use commas to indicate the decimal point in numbers. Then you can use `read.csv2` and `read.delim2`.)

As another example, suppose that you wanted to analyze some historical stock quote data. Yahoo! Finance provides this information in an easily downloadable form on its website; you can fetch a CSV file from a single URL. For example, to fetch the closing price of the S&P 500 index for every month between April 1, 1999, and April 1, 2009, you could use the following URL: *http://ichart.finance.yahoo.com/table.csv ?s=%5EGSPC&a=03&b=1&c=1999&d=03&e=1&f=2009&g=m&ignore=.csv*.

Conveniently, you can use a URL in place of a filename in R. This means that you could load this data into R with the following expression:

```
> sp500 <- read.csv(paste("http://ichart.finance.yahoo.com/table.csv?",
+     "s=%5EGSPC&a=03&b=1&c=1999&d=03&e=1&f=2009&g=m&ignore=.csv"))
> # show the first 5 rows
> sp500[1:5,]
        Date   Open   High    Low  Close      Volume Adj.Close
1 2009-04-01 793.59 813.62 783.32 811.08 12068280000    811.08
2 2009-03-02 729.57 832.98 666.79 797.87  7633306300    797.87
3 2009-02-02 823.09 875.01 734.52 735.09  7022036200    735.09
4 2009-01-02 902.99 943.85 804.30 825.88  5844561500    825.88
5 2008-12-01 888.61 918.85 815.69 903.25  5320791300    903.25
```

We will revisit this example in the next section.

If you're trying to load a really big file, you might find that loading the file takes a long time. It can be very frustrating to wait 15 minutes for a file to load, only to discover that you have specified the wrong separator. A useful technique for testing is to only load a small number of rows into R. For example, to load 20 rows, you would add `nrows=20` as an argument to `read.table`.

Many programs can export data as text files. Here are a few tips for creating text files that you can easily read into R:

- For Microsoft Excel spreadsheets, you can export them as either comma-delimited files (CSV files) or tab-delimited files (TXT files). When possible, you should specify Unix-style line delimiters, not MS-DOS line delimiters. (MS-DOS files end each line with "\n\r," while Unix-style systems end lines with "\n.") There are two things to think about when choosing between CSV and TXT files.

CSV files can be more convenient because (by default) opening these files in Windows Explorer will open these files in Microsoft Excel. However, if you are using CSV files, then you must be careful to enclose text in quotes if the data contains commas (and, additionally, you must escape any quotation marks within text fields). Tab characters occur less often in text, so tab-delimited files are less likely to cause problems.

- If you are exporting data from a database, consider using a GUI tool to query the database and export the results. It is possible to use command-line scripts to export data using tools like sqlplus, pgsql, or mysql, but doing so is often tricky.

  Here are a few options that I have tried. If you are using Microsoft Windows, a good choice is Toad for Data Analysts (available from *http://www.toadsoft .com/tda/tdaindex.html*); this will work with many different databases. If you are exporting from MySQL, MySQL Query Browser is also a good choice; versions are available for Microsoft Windows, Mac OS X, and Linux (you can download it from *http://dev.mysql.com/downloads/gui-tools/5.0.html*). Oracle now produces a free multi-platform query tool called SQL Developer. (You can find it at *http://www.oracle.com/technology/products/database/sql_developer/in dex.html*.)

Fixed-width files

To read a fixed-width format text file into a data frame, you can use the `read.fwf` function:

```
read.fwf(file, widths, header = , sep = ,
         skip = , row.names, col.names, n = ,
         buffersize = , ...)
```

Here is a description of the arguments to `read.fwf`.

Argument	Description	Default
file	The name of the file to open or, alternatively, the name of a connection containing the data. (This is a required argument.)	
widths	An integer vector or a list of integer vectors. If the input file has one record per line, then use an integer vector where each value represents the width of each variable. If each record spans multiple lines, then use a list of integer vectors where each integer vector corresponds to the widths of the variables on that line. (This is a required argument.)	
header	A logical value indicating whether the first line of the file contains variable names. (If it does, the names must be delimited by sep.)	FALSE
sep	The character used to delimit variable names in the header.	\t
skip	An integer specifying the number of lines to skip at the beginning of the file.	A0

Argument	Description	Default
row.names	A character vector used to specify row names in the data frame.	
col.names	A character vector used to specify column names in the data frame.	
n	An integer value specifying the number of rows to records to read into R. (Invalid values, such as negatives, are ignored.)	-1
buffersize	An integer specifying the maximum number of lines to be read at one time. (This value may be tuned to optimize performance.)	2,000

Note that `read.fwf` can also take many arguments used by `read.table`, including `as.is`, `na.strings`, `colClasses`, and `strip.white`.

Saving, Loading, and Editing Data

Using Other Languages to Preprocess Text Files

R is a very good system for numerical calculations and data visualization, but it's not the most efficient choice for processing large text files. For example, the U.S. Centers for Disease Control and Prevention publishes data files containing information on every death in the United States (see *http://www.cdc.gov/nchs/data_ac cess/Vitalstatsonline.htm*). These data files are provided in a fixed-width format. They are very large; the data file for 2006 was 1.1 GB uncompressed. In theory, you could load a subset of data from this file into R using a statement like this:

```
> # data from ftp://ftp.cdc.gov/pub/Health_Statistics/
NCHS/Datasets/DVS/mortality/mort2006us.zip
> mort06 <- read.fwf(file="MORT06.DUSMCPUB",
+    widths= c(19,1,40,2,1,1,2,2,1,4,1,2,2,2,2,1,1,1,16,4,1,1,1,1,
+              34,1,1,4,3,1,3,3,2,283,2,1,1,1,1,33,3,1,1),
+    col.names= c("X0","ResidentStatus","X1","Education1989",
+                 "Education2003","EducationFlag","MonthOfDeath",
+                 "X5","Sex","AgeDetail","AgeSubstitution",
+                 "AgeRecode52","AgeRecode27","AgeRecode12",
+                 "AgeRecodeInfant22","PlaceOfDeath","MaritalStatus",
+                 "DayOfWeekofDeath","X15","CurrentDataYear",
+                 "InjuryAtWork","MannerOfDeath","MethodOfDisposition",
+                 "Autopsy","X20","ActivityCode","PlaceOfInjury",
+                 "ICDCode","CauseRecode358","X24","CauseRecode113",
+                 "CauseRecode130","CauseRecord39","X27","Race",
+                 "BridgeRaceFlag","RaceImputationFlag","RaceRecode3",
+                 "RaceRecord5","X32","HispanicOrigin","X33",
+                 "HispanicOriginRecode","X34")
+ )
```

Unfortunately, this probably won't work very well. First, R processes files less quickly than some other languages. Second, R will try to load the entire table into memory. The file takes up 1.1 GB as a raw text file. Many fields in this file are used to encode categorical values that have a small number of choices (such as race) but show the value as numbers. R will convert these character values from single characters (which take up 1 byte) to integers (which take up 4 bytes). This means that it will take a lot of memory to load this file into your computer.

As an alternative, I'd suggest using a scripting language like Perl, Python, or Ruby to preprocess large, complex text files and turn them into a digestible form. (As a side note, I usually write out lists of field names and lengths in Excel and then use Excel formulas to create the R or Perl code to load them. That's how I generated all of the code shown in this example.) Here's the Perl script that I used to preprocess the raw mortality data file, filtering out fields I didn't need and writing the results to a CSV file:

```perl
#!/usr/bin/perl

# file to preprocess (and filter) mortality data

print "ResidentStatus,Education1989,Education2003,EducationFlag," .
    "MonthOfDeath,Sex,AgeDetail,AgeSubstitution,AgeRecode52," .
    "AgeRecode27,AgeRecode12,AgeRecodeInfant22,PlaceOfDeath," .
    "MaritalStatus,DayOfWeekofDeath,CurrentDataYear,InjuryAtWork," .
    "MannerOfDeath,MethodOfDisposition,Autopsy,ActivityCode," .
    "PlaceOfInjury,ICDCode,CauseRecode358,CauseRecode113," .
    "CauseRecode130,CauseRecord39,Race,BridgeRaceFlag," .
    "RaceImputationFlag,RaceRecode3,RaceRecord5,HispanicOrigin," .
    "HispanicOriginRecode\n";

while(<>) {
    my ($X0,$ResidentStatus,$X1,$Education1989,$Education2003,
    $EducationFlag,$MonthOfDeath,$X5,$Sex,$AgeDetail,
    $AgeSubstitution,$AgeRecode52,$AgeRecode27,$AgeRecode12,
    $AgeRecodeInfant22,$PlaceOfDeath,$MaritalStatus,
    $DayOfWeekofDeath,$X15,$CurrentDataYear,$InjuryAtWork,
    $MannerOfDeath,$MethodOfDisposition,$Autopsy,$X20,$ActivityCode,
    $PlaceOfInjury,$ICDCode,$CauseRecode358,$X24,$CauseRecode113,
    $CauseRecode130,$CauseRecord39,$X27,$Race,$BridgeRaceFlag,
    $RaceImputationFlag,$RaceRecode3,$RaceRecord5,$X32,
    $HispanicOrigin,$X33,$HispanicOriginRecode,$X34)
        = unpack("a19a1a40a2a1a1a2a2a1a4a1a2a2a2a2a1a1a1a16a4a1" .
                "a1a1a1a34a1a1a4a3a1a3a3a2a283a2a1a1a1a1a33a3a1a1",
                $_);
    print "$ResidentStatus,$Education1989,$Education2003,".
        "$EducationFlag,$MonthOfDeath,$Sex,$AgeDetail,".
        "$AgeSubstitution,$AgeRecode52,$AgeRecode27,".
        "$AgeRecode12,$AgeRecodeInfant22,$PlaceOfDeath," .
        "$MaritalStatus,$DayOfWeekofDeath,$CurrentDataYear,".
        "$InjuryAtWork,$MannerOfDeath,$MethodOfDisposition,".
        "$Autopsy,$ActivityCode,$PlaceOfInjury,$ICDCode,".
        "$CauseRecode358,$CauseRecode113,$CauseRecode130,".
        "$CauseRecord39,$Race,$BridgeRaceFlag,$RaceImputationFlag,".
        "$RaceRecode3,$RaceRecord5,$HispanicOrigin," .
        "$HispanicOriginRecode\n";
}
```

I executed this script with the following command in a bash shell:

```
./mortalities.pl < MORT06.DUSMCPUB > MORT06.csv
```

You can now load the data into R with a line like this:

```
> mort06 <- read.csv(file="~/Documents/book/data/MORT06.csv")
```

We'll come back to this data set in the chapters on statistical tests and statistical models.

Other functions to parse data

Most of the time, you should be able to load text files into R with the `read.table` function. Sometimes, however, you might be provided with a file that cannot be read correctly with this function. For example, observations in the file might span multiple lines. To read data into R one line at a time, use the function `readLines`:

```
readLines(con = stdin(), n = -1L, ok = TRUE, warn = TRUE,
          encoding = "unknown")
```

The `readLines` function will return a character vector, with one value corresponding to each row in the file. Here is a description of the arguments to `readLines`.

Argument	Description	Default
con	A character string (specifying a file or URL) or a connection containing the data to read.	stdin()
n	An integer value specifying the number of lines to read. (Negative values mean "read until the end of the file.")	-1L
ok	A logical value specifying whether to trigger an error if the number of lines in the file is less than n.	TRUE
warn	A logical value specifying whether to warn the user if the file does not end with an EOL.	TRUE
encoding	A character value specifying the encoding of the input file.	"unknown"

Note that you can use `readLines` interactively to enter data.

Another useful function for reading more complex file formats is `scan`:

```
scan(file = "", what = double(0), nmax = -1, n = -1, sep = "",
     quote = if(identical(sep, "\n")) "" else "'\"", dec = ".",
     skip = 0, nlines = 0, na.strings = "NA",
     flush = FALSE, fill = FALSE, strip.white = FALSE,
     quiet = FALSE, blank.lines.skip = TRUE, multi.line = TRUE,
     comment.char = "", allowEscapes = FALSE,
     encoding = "unknown")
```

The `scan` function allows you to read the contents of a file into R. Unlike `readLines`, `scan` allows you to read data into a specifically defined data structure using the argument `what`.

Here is a description of the arguments to `scan`.

Argument	Description	Default
file	A character string (specifying a file or URL) or a connection containing the data to read.	""

Argument	Description	Default
what	The type of data to be read. If all fields are the same type, you can specify logical, integer, numeric, complex, character, or raw. Otherwise, specify a list of types to read values into a list. (You can specify the type of each element in the list individually.)	double(0)
nmax	An integer value specifying the number of values to read or the number of records to read (if what is a list). (Negative values mean "read until the end of the file.")	-1
n	An integer value specifying the number of values to read. (Negative values mean "read until the end of the file.")	-1
sep	Character value specifying the separator between values. sep="" means that any whitespace character is interpreted as a separator.	""
quote	Character value used to quote strings.	if(identical(sep, "\n")) "" else "\""
dec	Character value used for decimal place in numbers.	"."
skip	Number of lines to skip at the top of the file.	0
nlines	Number of lines of data to read. Nonpositive values mean that there is no limit.	0
na.strings	Character values specifying how NA values are encoded.	"NA"
flush	A logical value specifying whether to "flush" any remaining text on a line after the last requested item on a line is read into what. (Commonly used to allow comments at the end of lines or to ignore unneeded fields.)	FALSE
fill	Specifies whether to add empty fields to lines with fewer fields than specified by what.	FALSE
strip.white	Specifies whether to strip leading and trailing whitespace from character fields. Only applies when sep is specified.	FALSE
quiet	If quiet=FALSE, scan will print a message showing how many lines were read. If quiet=TRUE, this message is suppressed.	FALSE
blank.lines.skip	Specifies whether to ignore blank lines.	TRUE
multi.line	If what is a list, allows records to span multiple lines.	TRUE
comment.char	Notes a character to be used to specify comment lines.	""
allowEscapes	Specifies whether C-style escapes (such as \t for tab character or \n for newlines) should be interpreted by scan or read verbatim. If allowEscapes=FALSE, they are interpreted as special characters; if allowEscapes=TRUE, they are read literally.	FALSE
encoding	A character value specifying the encoding of the input file.	"unknown"

Like readLines, you can also use scan to enter data directly into R.

Other Software

Although many software packages can export data as text files, you might find it more convenient to read their data files directly. R can read files in many other formats. Table 12-1 shows a list of functions for reading (and writing) files in other formats. You can find more information about these functions in the help files.

Table 12-1. Functions to read and write data

File format	Reading	Writing
ARFF	read.arff	write.arff
DBF	read.dbf	write.dbf
Stata	read.dta	write.dta
Epi Info	read.epiinfo	
Minitab	read.mtp	
Octave	read.octave	
S3 binary files, data.dump files	read.S	
SPSS	read.spss	
SAS Permanent Dataset	read.ssd	
Systat	read.sysstat	
SAS XPORT File	read.xport	

Saving, Loading, and Editing Data

Exporting Data

R can also export R data objects (usually data frames and matrices) as text files. To export data to a text file, use the `write.table` function:

```
write.table(x, file = "", append = FALSE, quote = TRUE, sep = " ",
            eol = "\n", na = "NA", dec = ".", row.names = TRUE,
            col.names = TRUE, qmethod = c("escape", "double"))
```

There are wrapper functions for `write.table` that call `write.table` with different defaults. These are useful if you want to create a file of comma-separated values, for example, to import into Microsoft Excel:

```
write.csv(...)
write.csv2(...)
```

Here is a description of the arguments to `write.table`.

Argument	Description	Default
x	Object to export.	
file	Character value specifying a filename or a connection object to which you would like to write the output.	""
append	A logical value indicating whether to append the output to the end of an existing file (append=TRUE) or replace the file (append=FALSE).	FALSE

Argument	Description	Default
quote	A logical value specifying whether to surround any character or factor values with quotes, or a numeric vector specifying which columns to surround with quotes.	TRUE
sep	A character value specifying the value that separates values within a row.	""
eol	A character value specifying the value to append on the end of each line.	"\n"
na	A character value specifying how to represent NA values.	"NA"
dec	A character value specifying the decimal separator in numeric values.	"."
row.names	A logical value indicating whether to include row names in the output or a numeric vector specifying the rows from which row names should be included.	TRUE
col.names	A logical value specifying whether to include column names or a character vector specifying alternate names to include.	TRUE
qmethod	Specifies how to deal with quotes inside quoted character and factor fields. Specify qmethod="escape" to escape quotes with a backslash (as in C) or qmethod="double" to escape quotes as double quotes (i.e., " is transformed to "").	"escape"

Importing Data from Databases

It is very common for large companies, healthcare providers, and academic institutions to keep data in relational databases. This section explains how to move data from databases into R.

Export Then Import

One of the best approaches for working with data from a database is to export the data to a text file and then import the text file into R. In my experience dealing with very large data sets (1 GB or more), I've found that you can import data into R at a much faster rate from text files than you can from database connections.

For directions on how to import these files into R, see "Text Files" on page 152.

If you plan to extract a large amount of data once and then analyze the data, this is often the best approach. However, if you are using R to produce regular reports or to repeat an analysis many times, then it might be better to import data into R directly through a database connection.

Database Connection Packages

In order to connect directly to a database from R, you will need to install some optional packages. The packages you need depend on the database(s) to which you want to connect, and the connection method that you want to use.

There are two sets of database interfaces available in R:

- **RODBC**. The RODBC package allows R to fetch data from ODBC (Open DataBase Connectivity) connections. ODBC provides a standard interface for different programs to connect to databases.

- **DBI**. The DBI package allows R to connect to databases using native database drivers or JDBC drivers. This package provides a common database abstraction for R software. You must install additional packages to use the native drivers for each database.

Often, you can choose from either option. You might wonder which package is the better choice: RODBC or DBI? Here are a few features to consider.

- **Driver availability**. On Windows and Linux, you can easily find free ODBC drivers for most common databases. On Mac OS X, it can be difficult to find free ODBC drivers for a database. However, JDBC drivers are readily available for each platform.

- **Special features and performance**. A native database interface might take advantage of unique product features and be faster than a generic driver.

- **Package availability**. Not all packages will work on all platforms.

- **Code quality**. The DBI package is written using S4 objects and methods. Using the DBI package can help you write better code.

In this section, I'll show how to configure an ODBC connection to an SQLite database on Microsoft Windows and Mac OS X. SQLite is a tool for storing databases in files. It's completely contained in a C library. This means that you can try the examples in this section without installing a full database system.

For this example, we will use an SQLite database containing the Baseball Databank database. You do not need to install any additional software to use this database. This file is included in the `nutshell` package. To access it within R, use the following expression as a filename: `system.file("data", "bb.db", package = "nutshell")`.

RODBC

The R package for accessing databases through ODBC is the RODBC package. Microsoft and Simba Technologies jointly developed ODBC in the late 1990s based on a design from the SQL Access Group. In ODBC, different data sources are labeled by database source names (DSNs).

Getting RODBC working

Before you can use RODBC, you need to configure the ODBC connection. You only need to do this once; after you have configured R to communicate with your database, you are ready to use RODBC inside R.

1. Install the RODBC package in R.
2. If needed, install the ODBC drivers for your platform.
3. Configure an ODBC connection to your database.

Here are directions for completing each step.

Installing the RODBC package. A quick way to install the RODBC package (if it is not already installed) is with the `install.packages` function:

```
> install.packages("RODBC")
trying URL 'http://cran.cnr.Berkeley.edu/bin/macosx/universal/contrib/2.8/
    RODBC_1.2-5.tgz'
Content type 'application/x-gzip' length 120902 bytes (118 Kb)
opened URL
==================================================
downloaded 118 Kb


The downloaded packages are in
    /var/folders/gj/gj60srEiEVq4hTWB5lvMak+++TM/-Tmp-//Rtmp2UFF7o/
    downloaded_packages
```

You will get slightly different output when you run this command. Don't worry about the output unless you see an error message. If you want to make sure that the package was installed correctly, try loading it in R:

```
> library(RODBC)
```

If there is no error message, then the package is now locally installed (and available). For information about other methods for installing RODBC, see Chapter 4.

Installing ODBC drivers. If you already have the correct ODBC drivers installed (for example, to access a database from Microsoft Excel), then you can skip this step. Table 12-2 shows some sources for ODBC drivers. (I haven't used most of these products, and am not endorsing any of them.)

Table 12-2. Where to find ODBC drivers

Provider	Database	Platforms	Website
MySQL	MySQL	Microsoft Windows, Linux, Mac OS X, Solaris, AIX, FreeBSD, others	*http://dev .mysql.com/ downloads/con nector/odbc/*
Oracle	Oracle	Microsoft Windows, Linux, Solaris	*http://www .oracle.com/ technology/ tech/windows/ odbc/index .html*
PostgreSQL	PostgreSQL	Microsoft Windows, Linux, other Unix-like platforms	*http://www .postgresql .org/ftp/odbc/ versions/*
Microsoft	SQL Server	Microsoft Windows	*http://msdn.mi crosoft.com/en -us/data/ aa937730.aspx*
Data Direct	Oracle, SQL Server, DB2, Sybase, Teradata, MySQL, PostgreSQL, others	Microsoft Windows, Linux, other Unix platforms	*http://www.da tadirect.com/ products/odbc/ index.ssp*

Provider	Database	Platforms	Website
Easysoft	Oracle, SQL Server, others	Microsoft Windows, Linux	*http://www .easysoft.com/ products/data _access/index .html*
Actual Technologies	Oracle, SQL Server, Sybase, MySQL, PostgreSQL, SQLite	Mac OS X	*http://www.ac tualtechnolo gies.com/*
OpenLink Software	Oracle, SQL Server, DB2, Sybase, MySQL, PostgreSQL, others	Microsoft Windows, Mac OS X, Linux, others	*http://uda .openlinksw .com/odbc/*
Christian Werner Software	SQLite	Microsoft Windows, Mac OS X, Linux	*http://www.ch -werner.de/sqli teodbc/*

Follow the directions for the driver that you are using. For the example in this section, I used the SQLite ODBC driver.

Example: SQLite ODBC on Mac OS X. To use this free driver, you'll need to compile and install the driver yourself. Luckily, this process works flawlessly on Mac OS X 10.5.6.[*] Here is how to install the drivers on Mac OS X:

1. Download the latest sources from *http://www.ch-werner.de/sqliteodbc/*. (Do not download the precompiled version.) I used sqliteodbc-0.80.tar.gz. You can do this with this command:

   ```
   % wget http://www.ch-werner.de/sqliteodbc/sqliteodbc-0.80.tar.gz
   ```

2. Unpack and unzip the archive. You can do this with this command:

   ```
   % tar xvfz sqliteodbc-0.80.tar.gz
   ```

3. Change to the directory of sources files:

   ```
   % cd sqliteodbc-0.80
   ```

4. Configure the driver for your platform, compile the driver, and then install it. You can do this with these commands:

   ```
   % ./configure
   % make
   % sudo make install
   ```

Now, you need to configure your Mac to use this driver.

1. Open the ODBC Administrator program (usually in /Applications/Utilities).
2. Select the "Drivers" tab and click "Add...."

---

[*] You may have to install Apple's development tools to build this driver. (It's a good idea to install Apple's developer tools anyway so that you can build R packages from source.) You can download these from *http://developer.apple.com/Tools/*.

3. Enter a name for the driver (like "SQLite ODBC Driver") in the "Description" field. Enter "/usr/local/lib/libsqlite3odbc.dylib" in the "Driver File" and "Setup File" fields, as shown in Figure 12-4. Click the "OK" button.

4. Now, select the "User DSN" tab or "System DSN" tab (if you want this database to be available for all users). Click the "Add..." button to specify the new database.

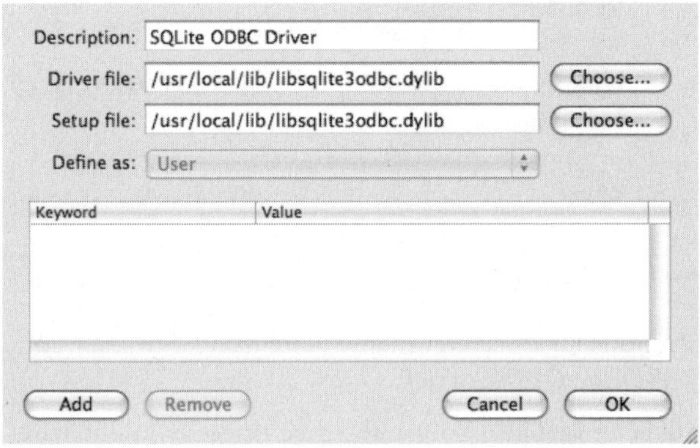

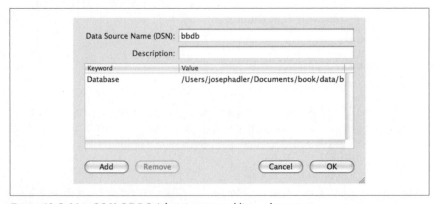

Figure 12-4. Mac OS X ODBC Administrator: driver settings

5. You will be prompted to choose a driver. Choose "SQLite ODBC Driver" (or whatever name you entered above) and click the "OK" button.

6. Enter a name for the data source such as "bbdb." You need to add a keyword that specifies the database location. Click the "Add" button at the bottom of the window. Select the "Keyword" field in the table and enter "Database." Select the "Value" field and enter the path to the database. (I entered "/Library/Frameworks/R.framework/Resources/library/nutshell/bb.db" to use the example in the nutshell package.) Figure 12-5 shows how this looks. Click "OK" when you are done.

Figure 12-5. Mac OS X ODBC Administrator: adding a data source

The ODBC connection is now configured. You can test this with a couple of simple commands in R (we'll explain what these mean below):

```
> bbdb <- odbcConnect("bbdb")
> odbcGetInfo(bbdb)
```

```
                                                          DBMS_Name
                                                            "SQLite"
                                                           DBMS_Ver
                                                             "3.4.0"
                                                    Driver_ODBC_Ver
                                                             "03.00"
                                                   Data_Source_Name
                                                             "bbdb"
                                                        Driver_Name
                                                   "sqlite3odbc.so"
                                                         Driver_Ver
                                                             "0.80"
                                                           ODBC_Ver
                                                       "03.52.0000"
                                                        Server_Name
"/Library/Frameworks/R.framework/Resources/library/nutshell/bb.db"
```

Example: SQLite ODBC on Windows. On Windows, you don't have to build the drivers from source. Here is how to get it working:

1. Download the SQLite ODBC installer package from *http://www.ch-werner.de/sqliteodbc/sqliteodbc.exe*.

2. Run the installer, using the default options in the wizard.

3. Open the ODBC Data Source Administrator application. On Microsoft Windows XP, you can find this in Administrative Tools (under Control Panels). You can click the "Drivers" tab to make sure that the SQLite ODBC drivers are installed, as shown in Figure 12-6.

4. Next, you need to configure a DSN for your database. Go to the "User DSN" tab (or "System DSN" if you want to share the database among multiple users) and click the "Add..." button. Select "SQLite3 ODBC Driver" and click "Finish."

5. You will be prompted for configuration information as shown in Figure 12-7. Enter a data source name of your choice (I used "bbdb"). Enter the path of the database file or use the "Browse..." button to browse for the file. (You can find the path for the file in R using the expression `system.file("data", "bb.db", package="nutshell")`.) Enter 200 ms for the Lock Timeout, select "NORMAL" as the Sync Mode, and click "Don't Create Database." When you are done, click "OK."

You should now be able to access the bbdb file through ODBC. You can check that everything worked correctly by entering a couple of commands in R:

```
> bbdb <- odbcConnect("bbdb")
> odbcGetInfo(bbdb)
```

```
                                                          DBMS_Name
                                                            "SQLite"
                                                           DBMS_Ver
```

```
                                "3.6.10"
                         Driver_ODBC_Ver
                                "03.00"
                        Data_Source_Name
                                 "bbdb"
                             Driver_Name
       "C:\\WINDOWS\\system32\\sqlite3odbc.dll"
                              Driver_Ver
                                 "0.80"
                                ODBC_Ver
                             "03.52.0000"
                             Server_Name
 "C:\\Program Files\\R\\R-2.10.0\\library\\nutshell\\data\\bb.db"
```

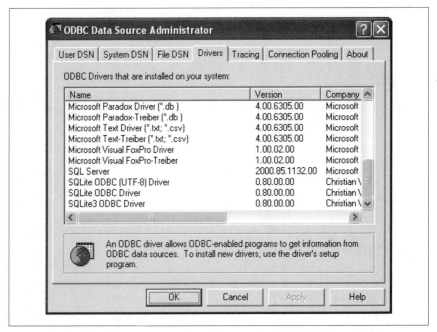

Figure 12-6. ODBC Data Source Administrator: "Drivers" tab

Using RODBC

Connecting to a database in R is like connecting to a file. First, you need to connect to a database. Next, you can execute any database queries. Finally, you should close the connection.

Opening a channel. To establish a connection, use the odbcConnect function:

```
odbcConnect(dsn, uid = "", pwd = "", ...)
```

You need to specify the DSN for the database to which you want to connect. If you did not specify a username and password in the DSN, you may specify a username with the uid argument and a password with the pwd argument. Other arguments are

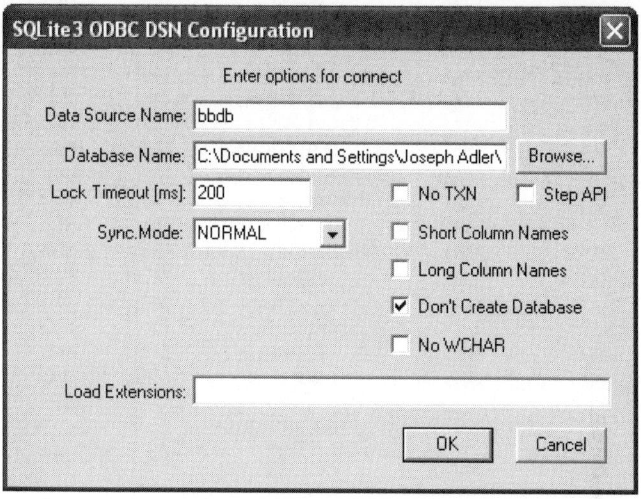

Figure 12-7. SQLite3 ODBC configuration

passed to the underlying `odbcDriverConnect` function. The `odbcConnect` function returns an object of class `RODBC` that identifies the connection. This object is usually called a *channel*.

Here is how you would use this function for the example DSN, "bbdb":

```
> library(RODBC)
> bbdb <- odbcConnect("bbdb")
```

Getting information about the database. You can get information about an ODBC connection using the `odbcGetInfo` function. This function takes a channel (the object returned by `odbcConnect`) as its only argument. It returns a character vector with information about the driver and connection; each value in the vector is named. Example output from this function is shown in "Example: SQLite ODBC on Mac OS X" on page 165 and "Example: SQLite ODBC on Windows" on page 167.

To get a list of the tables in the underlying database that the connected user can read, use the `sqlTables` function. This function returns a data frame with information about the available tables:

```
> sqlTables(bbdb)
   TABLE_CAT TABLE_SCHEM        TABLE_NAME TABLE_TYPE REMARKS
1       <NA>        <NA>           Allstar      TABLE    <NA>
2       <NA>        <NA>       AllstarFull      TABLE    <NA>
3       <NA>        <NA>       Appearances      TABLE    <NA>
4       <NA>        <NA>    AwardsManagers      TABLE    <NA>
5       <NA>        <NA>     AwardsPlayers      TABLE    <NA>
6       <NA>        <NA> AwardsShareManagers   TABLE    <NA>
7       <NA>        <NA>  AwardsSharePlayers    TABLE    <NA>
8       <NA>        <NA>           Batting      TABLE    <NA>
9       <NA>        <NA>       BattingPost      TABLE    <NA>
10      <NA>        <NA>          Fielding      TABLE    <NA>
11      <NA>        <NA>         FieldingOF      TABLE    <NA>
```

```
12      <NA>      <NA>      FieldingPost    TABLE    <NA>
13      <NA>      <NA>             HOFold    TABLE    <NA>
14      <NA>      <NA>         HallOfFame    TABLE    <NA>
15      <NA>      <NA>           Managers    TABLE    <NA>
16      <NA>      <NA>       ManagersHalf    TABLE    <NA>
17      <NA>      <NA>             Master    TABLE    <NA>
18      <NA>      <NA>            Pitching    TABLE    <NA>
19      <NA>      <NA>        PitchingPost    TABLE    <NA>
20      <NA>      <NA>            Salaries    TABLE    <NA>
21      <NA>      <NA>             Schools    TABLE    <NA>
22      <NA>      <NA>      SchoolsPlayers    TABLE    <NA>
23      <NA>      <NA>          SeriesPost    TABLE    <NA>
24      <NA>      <NA>               Teams    TABLE    <NA>
25      <NA>      <NA>     TeamsFranchises    TABLE    <NA>
26      <NA>      <NA>           TeamsHalf    TABLE    <NA>
27      <NA>      <NA>          xref_stats    TABLE    <NA>
```

To get detailed information about the columns in a specific table, use the `sqlColumns` function:

```
> sqlColumns(bbdb,"Allstar")
  TABLE_CAT TABLE_SCHEM TABLE_NAME COLUMN_NAME DATA_TYPE   TYPE_NAME
1                           Allstar    playerID        12  varchar(9)
2                           Allstar      yearID         5 smallint(4)
3                           Allstar        lgID        12     char(2)
  COLUMN_SIZE BUFFER_LENGTH DECIMAL_DIGITS NUM_PREC_RADIX NULLABLE
1           9             9             10              0        0
2           4             4             10              0        0
3           2             2             10              0        0
  REMARKS COLUMN_DEF SQL_DATA_TYPE SQL_DATETIME_SUB CHAR_OCTET_LENGTH
1    <NA>                       12               NA             16384
2    <NA>          0             5               NA             16384
3    <NA>                       12               NA             16384
  ORDINAL_POSITION IS_NULLABLE
1                1          NO
2                2          NO
3                3          NO
```

You can also discover the primary keys for a table using the `sqlPrimaryKeys` function.

Getting data. Finally, we've gotten to the interesting part: executing queries in the database and returning results. RODBC provides some functions that let you query a database even if you don't know SQL.

To fetch a table (or view) from the underlying database, you can use the `sqlFetch` function. This function returns a data frame containing the contents of the table:

```
sqlFetch(channel, sqtable, ..., colnames = , rownames = )
```

You need to specify the ODBC channel with the `channel` argument and the table name with the `sqtable` argument. You can specify whether the column names and row names from the underlying table should be used in the data frame with the `colnames` and `rownames` arguments. The column names from the table will be used in the returned data frame (this is enabled by default). If you choose to use row names, the first column in the returned data is used for column names in the data frame (this is disabled by default). You may pass additional arguments to this function, which are, in turn, passed to `sqlQuery` and `sqlGetResults` (described below).

As an example, let's load the content of the Teams table into a data frame called t:

```
> t <- sqlFetch(bbdb,"Teams")
> names(t)
 [1] "yearID"          "lgID"           "teamID"          "franchID"
 [5] "divID"           "Rank"           "G"               "Ghome"
 [9] "W"               "L"              "DivWin"          "WCWin"
[13] "LgWin"           "WSWin"          "R"               "AB"
[17] "H"               "2B"             "3B"              "HR"
[21] "BB"              "SO"             "SB"              "CS"
[25] "HBP"             "SF"             "RA"              "ER"
[29] "ERA"             "CG"             "SHO"             "SV"
[33] "IPouts"          "HA"             "HRA"             "BBA"
[37] "SOA"             "E"              "DP"              "FP"
[41] "name"            "park"           "attendance"      "BPF"
[45] "PPF"             "teamIDBR"       "teamIDlahman45"  "teamIDretro"
> dim(t)
[1] 2595   48
```

After loading the table into R, you can easily manipulate the data using R commands:

```
> # show wins and losses for American League teams in 2008
> subset(t,
+             subset=(t$yearID==2008 & t$lgID=="AL"),
+             select=c("teamID","W","L"))
     teamID   W   L
2567    LAA 100  62
2568    KCA  75  87
2571    DET  74  88
2573    CLE  81  81
2576    CHA  89  74
2577    BOS  95  67
2578    BAL  68  93
2582    MIN  88  75
2583    NYA  89  73
2585    OAK  75  86
2589    SEA  61 101
2592    TBA  97  65
2593    TEX  79  83
2594    TOR  86  76
```

There are related functions for writing a data frame to a database (`sqlSave`) or for updating a table in a database (`sqlUpdate`); see the help files for these functions for more information.

You can also execute an arbitrary SQL query in the underlying database. SQL is a very powerful language; you can use SQL to fetch data from multiple tables, to fetch a summary of the data in one (or more) tables, or to fetch specific rows or columns from the database. You can do this with the `sqlQuery` function:

```
sqlQuery(channel, query, errors = , max =,  ..., rows_at_time = )
```

This function returns a data frame containing the rows returned by the query. As an example, we could use an SQL query to select only the data shown above (wins and losses by team in the American League in 2008):

```
> sqlQuery(bbdb,
+     "SELECT teamID, W, L FROM Teams where yearID=2008 and lgID='AL'")
   teamID   W   L
1     BAL  68  93
2     BOS  95  67
3     CHA  89  74
4     CLE  81  81
5     DET  74  88
6     KCA  75  87
7     LAA 100  62
8     MIN  88  75
9     NYA  89  73
10    OAK  75  86
11    SEA  61 101
12    TBA  97  65
13    TEX  79  83
14    TOR  86  76
```

If you want to fetch data from a very large table, or from a very complicated query, you might not want to fetch all of the data at one time. The RODBC library provides a mechanism for fetching results piecewise. To do this, you begin by calling sqlQuery (or sqlFetch), but specify a value for max, telling the function the maximum number of rows that you want to retrieve at one time. You can fetch the remaining rows with the sqlGetResults function:

```
sqlGetResults(channel, as.is = ,  errors = , max = , buffsize = ,
              nullstring = , na.strings = , believeNRows = , dec = ,
              stringsAsFactors = )
```

The sqlQuery function actually calls the sqlGetResults function to fetch the results of the query. Here is a list of the arguments for these two functions. (If you are using sqlFetch, the corresponding function to fetch additional rows is sqlFetchMore.)

Argument	Description	Default
channel	Specifies the channel for the underlying database.	
query	A character value specifying the SQL query to execute.	
errors	A logical value specifying what to do when an error is encountered. When errors=TRUE, the function will stop and display the error if an error is encountered. When errors=FALSE, a value of -1 is returned.	TRUE
max	An integer specifying the maximum number of rows to return. Specify 0 for no maximum.	0 (meaning, no maximum)
rows_at_time	An integer specifying the number of rows to fetch from the ODBC connection on each call to the underlying driver; not all drivers allow values greater than 1. (Note that this is a performance optimization; it doesn't mean the same thing as the max argument. For modern drivers, the package documentation suggests a value of 1,024.)	1

Argument	Description	Default
as.is	A logical vector specifying which columns should be converted to factors.	FALSE
buffsize	An integer used to specify the buffer size for the driver. (If you know the approximate number of rows that a query will return, you can specify that value to optimize performance.)	1,000
nullstring	Character values to be used for null values.	NA
na.strings	Character values to be mapped to NA values.	"NA"
believeNRows	A logical value that tells this function whether the row counts returned by the ODBC driver are correct. (This is a performance optimization.)	TRUE
dec	The character used as the decimal point in decimal values.	getOption("dec")
stringsAsFactors	A logical value that specifies whether character value columns not explicitly included in as.is should be converted to factors.	default.stringsAsFactors()

By the way, notice that the `sqlQuery` function can be used to execute any valid query in the underlying database. It is most commonly used to just query results (using SELECT queries), but you can enter any valid data manipulation language query (including SELECT, INSERT, DELETE, and UPDATE queries) and data definition language query (including CREATE, DROP, and ALTER queries).

Underlying Functions

There is a second set of functions in the RODBC package. The functions `odbcQuery`, `odbcTables`, `odbcColumns`, and `odbcPrimaryKeys` are used to execute queries in the database but not to fetch results. A second function, `odbcFetchResults`, is used to get the results. The first four functions return status codes as integers, which is not very R-like. (It's more like C.) The `odbcFetchResults` function returns its results in list form, which can also be somewhat cumbersome. If there is an error, you can retrieve the message by calling `odbcGetErrMsg`.

Sometimes, it might be convenient to use these functions because they give you greater control over how data is fetched from the database. However, the higher-level functions described in this section are usually much more convenient.

Closing a channel.   When you are done using an RODBC channel, you can close it with the `odbcClose` function. This function takes the connection name as its only argument:

```
> odbcClose(bbdb)
```

Conveniently, you can also close all open channels using the `odbcCloseAll` function. It is generally a good practice to close connections when you are done, because this frees resources locally and in the underlying database.

DBI

As described above, there is a second set of packages for accessing databases in R: DBI. DBI is not a single package, but instead is a framework and set of packages for accessing databases. Table 12-3 shows the set of database drivers available through this interface. One important difference between the DBI packages and the RODBC package is in the objects they use: DBI uses S4 objects to represent drivers, connections, and other objects.

Table 12-3. DBI packages

Database	Package
MySQL	RMySQL
SQLite	RSQLite
Oracle	ROracle
PostgreSQL	RPostgreSQL
Any database with a JDBC driver	RJDBC

As an example, let's use the RSQLite package. You can install this package with the following command:

```
> install.packages("RSQLite")
```

When you load this package, it will automatically load the DBI package as well:

```
> library(RSQLite)
Loading required package: DBI
```

If you are familiar with SQL, but new to SQLite, you may want to review what SQL commands are supported by SQLite. You can find this list at *http://www.sqlite.org/lang.html*.

Opening a connection

To open a connection with DBI, use the dbConnect function:

```
dbConnect(drv, ...)
```

The argument drv can be a DBIDriver object or a character value describing the driver to use. You can generate a DBIDriver object with a call to the DBI driver. The dbConnect function can take additional options, depending on the type of database you are using. For SQLite databases, the most important argument is dbname (which specifies the database file). Check the help files for the database you are using for more options. Even arguments for parameters like usernames are *not* the same between databases.

For example, to create a driver for SQLite, you can use a command like this:

```
> drv <- dbDriver("SQLite")
```

To open a connection to the example database, we could use the following command:

```
> con <- dbConnect(drv,
+   dbname=system.file("data", "bb.db", package="nutshell"))
```

Alternatively, we could skip creating the driver object and simply create the connection:

```
> con <- dbConnect("SQLite,
+     dbname=system.file("data", "bb.db", package="nutshell"))
```

There are several reasons why it can be better to explicitly create a driver object. First, you can get information about open connections if you can identify the driver. Additionally, if you are concerned with resource consumption, it may be wise to explicitly create a driver object, because you can free the object later. (See "Cleaning up" on page 178 for more details.)

Getting DB information

There are several ways to get information about an open database connection object. As noted above, DBI objects are S4 objects, so they have meaningful classes:

```
> class(drv)
[1] "SQLiteDriver"
attr(,"package")
[1] "RSQLite"
> class(con)
[1] "SQLiteConnection"
attr(,"package")
[1] "RSQLite"
```

To get the list of connection objects associated with a driver object, use the dbList Connections function:

```
> dbListConnections(drv)
[[1]]
<SQLiteConnection:(4580,0)>
```

You can get some basic information about a connection object, such as the database name and username, through the dbGetInfo function:

```
> dbGetInfo(con)
$host
[1] "localhost"

$user
[1] "NA"

$dbname
[1] "/Library/Frameworks/R.framework/Resources/library/nutshell/data/bb.db"

$conType
[1] "direct"

$serverVersion
[1] "3.6.4"

$threadId
[1] -1

$rsId
integer(0)
```

```
$loadableExtensions
[1] "off"
```

To find the set of tables that you can access from a database connection, use the dbListTables function. This function returns a character vector of table names:

```
> dbListTables(con)
 [1] "Allstar"            "AllstarFull"     "Appearances"
 [4] "AwardsManagers"     "AwardsPlayers"   "AwardsShareManagers"
 [7] "AwardsSharePlayers" "Batting"         "BattingPost"
[10] "Fielding"           "FieldingOF"      "FieldingPost"
[13] "HOFold"             "HallOfFame"      "Managers"
[16] "ManagersHalf"       "Master"          "Pitching"
[19] "PitchingPost"       "Salaries"        "Schools"
[22] "SchoolsPlayers"     "SeriesPost"      "Teams"
[25] "TeamsFranchises"    "TeamsHalf"       "xref_stats"
```

To find the list of columns, use the List dbListFields function. This function takes a connection object and a table name as arguments and returns a character vector of column names:

```
> dbListFields(con,"Allstar")
[1] "playerID" "yearID"   "lgID"
```

Querying the database

To query a database using DBI and return a data frame with the results, use the dbGetQuery function. This function requires a connection object and SQL statement as arguments. Check the help files for your database for additional arguments.

For example, to fetch a list of the wins and losses for teams in the American League in 2008, you could use the following query:

```
> wlrecords.2008 <- dbGetQuery(con,
+     "SELECT teamID, W, L FROM Teams where yearID=2008 and lgID='AL'")
```

To get information on all batters in 2008, you might use a query like this:

```
> batting.2008 <- dbGetQuery(con,
+     paste("SELECT m.nameLast, m.nameFirst, m.weight, m.height, ",
+             "m.bats, m.throws, m.debut, m.birthYear, b.* ",
+             "from Master m inner join Batting b ",
+             "on m.playerID=b.playerID where b.yearID=2008"))
> names(batting.2008)
 [1] "nameLast"  "nameFirst" "weight"    "height"    "bats"
 [6] "throws"    "debut"     "birthYear" "playerID"  "yearID"
[11] "stint"     "teamID"    "lgID"      "G"         "G_batting"
[16] "AB"        "R"         "H"         "2B"        "3B"
[21] "HR"        "RBI"       "SB"        "CS"        "BB"
[26] "SO"        "IBB"       "HBP"       "SH"        "SF"
[31] "GIDP"      "G_old"
> dim(batting.2008)
[1] 1384   31
```

This data set is used in other sections of this book as an example. For convenience, it is included in the nutshell package.

You might find it more convenient to separately submit an SQL query and fetch the results. To do this, you would use the **dbSendQuery** function to send a query and then use **fetch** to get the results. The **dbSendQuery** function returns a **DBIResult** object (actually, it returns an object from a class that inherits from **DBIResult**). You then use the **fetch** function to extract data from the results object.

The **dbSendQuery** function takes the same arguments as **dbGetQuery**. The **fetch** function takes a result object **res** as an argument, an integer value **n** representing the maximum number of rows to return, and additional arguments passed to the methods for a specific database driver. To fetch all records, you can omit n, or use n=-1.

For example, the following R statements are equivalent to the **dbGetQuery** statements shown above:

```
> res <- dbSendQuery(con,
+    "SELECT teamID, W, L FROM Teams where yearID=2008 and lgID='AL'")
> wlrecords.2008 <- fetch(res)
```

You can clear pending results using the **dbClearResult** function:

```
> # query to fetch a lot of results
> res <- dbSendQuery(con,"SELECT * from Master")
> # function to clear the results
> dbClearResult(res)
[1] TRUE
```

If an error occurred, you can get information about the error with the dbGetException function:

```
> # SQL statement that will generate an error.
> # Notice that an error message is printed.
> res <- dbSendQuery(con,"SELECT * from non_existent_table")
Error in sqliteExecStatement(conn, statement, ...) :
  RS-DBI driver: (error in statement: no such table: non_existent_table)
> # now, manually get the error message
> dbGetException(con)
$errorNum
[1] 1

$errorMsg
[1] "error in statement: no such table: non_existent_table"
```

Finally, DBI provides some functions for reading whole tables from a database or writing whole data frames to a database. To read a whole table, use the dbReadTable function:

```
> batters <- dbReadTable(con,"Batting")
> dim(batters)
[1] 91457    24
```

To write a data frame to a table, you can use the **dbWriteTable** function. You can check if a table exists with the **dbExistsTable** function, and you can delete a table with the **dbRemoveTable** function.

Cleaning up

To close a database connection, use the dbDisconnect function:

```
> dbDisconnect(con)
[1] TRUE
```

You can also explicitly unload the database driver, freeing system resources, by using the dbUnloadDriver function. With some databases, you can pass additional arguments to this driver; see the help files for the database you are using for more information.

```
> dbUnloadDriver(drv)
```

TSDBI

There is one last database interface in R that you might find useful: TSDBI. TSDBI is an interface specifically designed for time series data. There are TSDBI packages for many popular databases, as show in Table 12-4.

Table 12-4. TSDBI packages

Database	Package
MySQL	TSMySQL
SQLite	TSSQLite
Fame	TSFame
PostgreSQL	TSPostgreSQL
Any database with an ODBC driver	TSODBC

13

Preparing Data

Back in my freshman year of college, I was planning to be a biochemist. I spent hours and hours of time in the lab: mixing chemicals in test tubes, putting samples in different machines, and analyzing the results. Over time, I grew frustrated because I found myself spending weeks in the lab doing manual work and just a few minutes planning experiments or analyzing results. After a year, I gave up on chemistry and became a computer scientist, thinking that I would spend less time on preparation and testing and more time on analysis.

Unfortunately for me, I chose to do data mining work professionally. Everyone loves building models, drawing charts, and playing with cool algorithms. Unfortunately, most of the time you spend on data analysis projects is spent on preparing data for analysis. I'd estimate that 80% of the effort on a typical project is spent on finding, cleaning, and preparing data for analysis. Less than 5% of the effort is devoted to analysis. (The rest of the time is spent on writing up what you did.)

If you're new to data analysis, you're probably wondering what the big deal is about preparing data. Suppose that you are getting some data off of your company's web servers, or out of a financial database, or from electronic patient records. It all came from computers, so it's perfect, right?

In practice, data is almost never stored in the right form for analysis. Even when data is in the right form, there are often surprises in the data. It takes a lot of work to pull together a usable data set. This chapter explains how to prepare data for analysis with R.

Combining Data Sets

Let's start with one of the most common obstacles to data analysis: working with data that's stored in two different places. For example, suppose that you wanted to look at batting statistics for baseball players by age. In most baseball data sources (like the Baseball Databank data), player information (like ages) is kept in different files from performance data (like batting statistics). So, you would need to combine two files to do this analysis. This section discusses several tools in R used for combining data sets.

Pasting Together Data Structures

R provides several functions that allow you to paste together multiple data structures into a single structure.

Paste

The simplest of these functions is paste. The paste function allows you to concatenate multiple character vectors into a single vector. (If you concatenate a vector of another type, it will be coerced to a character vector first.)

```
> x <- c("a","b","c","d","e")
> y <- c("A","B","C","D","E")
> paste(x,y)
[1] "a A" "b B" "c C" "d D" "e E"
```

By default, values are separated by a space; you can specify another separator (or none at all) with the sep argument:

```
> paste(x,y,sep="-")
[1] "a-A" "b-B" "c-C" "d-D" "e-E"
```

If you would like all of the values in the returned vector to be concatenated with each other (to return just a single value), then specify a value for the collapse argument. The value of collapse will be used as the separator in this value:

```
> paste(x,y,sep="-",collapse="#")
[1] "a-A#b-B#c-C#d-D#e-E"
```

rbind and cbind

Sometimes, you would like to bind together multiple data frames or matrices. You can do this with the rbind and cbind functions. The cbind function will combine objects by adding columns. You can picture this as combining two tables horizontally. As an example, let's start with the data frame for the top five salaries in the NFL in 2008:[*]

```
> top.5.salaries
  name.last name.first    team position   salary
1   Manning     Peyton   Colts       QB 18700000
2     Brady        Tom Patriots       QB 14626720
3    Pepper     Julius Panthers       DE 14137500
4    Palmer     Carson Bengals        QB 13980000
5   Manning        Eli  Giants        QB 12916666
```

Now, let's create a new data frame with two more columns (a year and a rank):

```
> year <- c(2008,2008,2008,2008,2008)
> rank <- c(1,2,3,4,5)
```

---

[*] Salary data is from *http://sportsillustrated.cnn.com/football/nfl/salaries/2008/all.html*. The salary numbers are cap numbers, not cash salaries.

```
> more.cols <- data.frame(year,rank)
> more.cols
  year rank
1 2008    1
2 2008    2
3 2008    3
4 2008    4
5 2008    5
```

Finally, let's put together these two data frames:

```
> cbind(top.5.salaries,more.cols)
  name.last name.first    team position   salary year rank
1   Manning     Peyton   Colts       QB 18700000 2008    1
2     Brady        Tom Patriots       QB 14626720 2008    2
3    Pepper     Julius Panthers       DE 14137500 2008    3
4    Palmer     Carson Bengals       QB 13980000 2008    4
5   Manning        Eli  Giants       QB 12916666 2008    5
```

The `rbind` function will combine objects by adding rows. You can picture this as combining two tables vertically.

As an example, suppose that you had a data frame with the top five salaries (as shown above) and a second data frame with the next three salaries:

```
> top.5.salaries
  name.last name.first    team position   salary
1   Manning     Peyton   Colts       QB 18700000
2     Brady        Tom Patriots       QB 14626720
3    Pepper     Julius Panthers       DE 14137500
4    Palmer     Carson Bengals       QB 13980000
5   Manning        Eli  Giants       QB 12916666
> next.three
  name.last name.first    team position   salary
6     Favre      Brett Packers       QB 12800000
7    Bailey      Champ Broncos       CB 12690050
8  Harrison     Marvin   Colts       WR 12000000
```

You could combine these into a single data frame using the `rbind` function:

```
> rbind(top.5.salaries,next.three)
  name.last name.first    team position   salary
1   Manning     Peyton   Colts       QB 18700000
2     Brady        Tom Patriots       QB 14626720
3    Pepper     Julius Panthers       DE 14137500
4    Palmer     Carson Bengals       QB 13980000
5   Manning        Eli  Giants       QB 12916666
6     Favre      Brett Packers       QB 12800000
7    Bailey      Champ Broncos       CB 12690050
8  Harrison     Marvin   Colts       WR 12000000
```

An extended example

To show how to fetch and combine together data and build a data frame for analysis, we'll use an example from the previous chapter: stock quotes. Yahoo! Finance allows you to download CSV files with stock quotes for a single ticker.

Suppose that you wanted a single data set with stock quotes for multiple securities (say, the 30 stocks in the Dow Jones Industrial Average). You would need a way to bind together the data returned by the query into a single data set. Let's write a function that can return historical stock quotes for multiple securities in a single data frame. First, let's write a function that assembles the URL for the CSV file and then fetches a data frame with the contents.

Here is what this function will do. First, it will define the URL. (I determined the format of the URL by trial and error: I tried fetching CSV files from Yahoo! Finance with different ticker symbols and different date ranges until I knew how to construct the queries.) We will use the `paste` function to put together all of these different character values. Next, we will fetch the URL with the `read.csv` function, assigning the data frame to the symbol `tmp`. The data frame has most of the information we want but doesn't include the ticker symbol. So, we will use the `cbind` function to attach a vector of ticker symbols to the data frame. (By the way, the function uses `Date` objects to represent the date. I also used the current date as the default value for `to`, and the date one year ago as the default value for `from`.)

Here is the function:

```
get.quotes <- function(ticker,
                       from=(Sys.Date()-365),
                       to=(Sys.Date()),
                       interval="d") {

    # define parts of the URL
    base       <- "http://ichart.finance.yahoo.com/table.csv?";
    symbol     <- paste("s=", ticker, sep="");
    # months are numbered from 00 to 11, so format the month correctly
    from.month <- paste("&a=",
        formatC(as.integer(format(from,"%m"))-1,width=2,flag="0"),
        sep="");
    from.day   <- paste("&b=", format(from,"%d"), sep="");
    from.year  <- paste("&c=", format(from,"%Y"), sep="");
    to.month <- paste("&a=",
        formatC(as.integer(format(to,"%m"))-1,width=2,flag="0"),
        sep="");
    to.day     <- paste("&e=", format(to,"%d"), sep="");
    to.year    <- paste("&f=", format(to,"%Y"), sep="");
    inter      <- paste("&g=", interval, sep="");
    last       <- "&ignore=.csv";

    # put together the URL
    url <- paste(base, symbol, from.month, from.day, from.year,
                 to.month, to.day, to.year, inter, last, sep="");

    # get the file
    tmp <- read.csv(url);

    # add a new column with ticker symbol labels
    cbind(symbol=ticker,tmp);
}
```

Now, let's write a function that returns a data frame with quotes from multiple securities. This function will simply call get.quotes once for every ticker in a vector of tickers and bind together the results using rbind:

```
get.multiple.quotes <- function(tkrs,
                                 from=(Sys.Date()-365),
                                 to=(Sys.Date()),
                                 interval="d") {
    tmp <- NULL;
    for (tkr in tkrs) {
        if (is.null(tmp))
                tmp <- get.quotes(tkr,from,to,interval)
        else tmp <- rbind(tmp,get.quotes(tkr,from,to,interval))
        }
    tmp
}
```

Finally, let's define a vector with the set of ticker symbols in the Dow Jones Industrial Average and then build a data frame with data from all 30 tickers:

```
> dow30.tickers <- c("MMM", "AA", "AXP", "T", "BAC", "BA", "CAT", "CVX", "C",
+                     "KO", "DD", "XOM", "GE", "GM", "HPQ", "HD", "INTC", "IBM",
+                     "JNJ", "JPM", "KFT", "MCD", "MRK", "MSFT", "PFE", "PG",
+                     "UTX", "VZ", "WMT", "DIS")
> # date on which I ran this code
> Sys.Date()
[1] "2009-09-22"
> dow30 <- get.multiple.quotes(dow30.tickers)
```

We'll return to this data set below.

Merging Data by Common Fields

As an example, let's return to the Baseball Databank database that we used in "Importing Data from Databases" on page 162. In this database, player information is stored in the Master table. Players are uniquely identified by the column playerID:

```
> dbListFields(con,"Master")
 [1] "lahmanID"      "playerID"      "managerID"     "hofID"
 [5] "birthYear"     "birthMonth"    "birthDay"      "birthCountry"
 [9] "birthState"    "birthCity"     "deathYear"     "deathMonth"
[13] "deathDay"      "deathCountry"  "deathState"    "deathCity"
[17] "nameFirst"     "nameLast"      "nameNote"      "nameGiven"
[21] "nameNick"      "weight"        "height"        "bats"
[25] "throws"        "debut"         "finalGame"     "college"
[29] "lahman40ID"    "lahman45ID"    "retroID"       "holtzID"
[33] "bbrefID"
```

Batting information is stored in the Batting table. Players are uniquely identified by playerID in this table as well:

```
> dbListFields(con,"Batting")
 [1] "playerID" "yearID"    "stint"   "teamID"  "lgID"
 [6] "G"        "G_batting" "AB"      "R"       "H"
[11] "2B"       "3B"        "HR"      "RBI"     "SB"
[16] "CS"       "BB"        "SO"      "IBB"     "HBP"
[21] "SH"       "SF"        "GIDP"    "G_old"
```

Suppose that you wanted to show batting statistics for each player along with his name and age. To do this, you would need to merge data from the two tables. In R, you can do this with the merge function:

```
> batting <- dbGetQuery(con, "SELECT * FROM Batting")
> master <- dbGetQuery(con, "SELECT * FROM Master")
> batting.w.names <- merge(batting,master)
```

In this case, there was only one common variable between the two tables: playerID:

```
> intersect(names(batting),names(master))
[1] "playerID"
```

By default, merge uses common variables between the two data frames as the merge keys. So, in this case, we did not have to specify any more arguments to merge. Let's take a closer look at the arguments to merge (for data frames):

```
merge(x, y, by = , by.x = , by.y = , all = , all.x = , all.y = ,
      sort = , suffixes = , incomparables = , ...)
```

Here is a description of the arguments to merge.

Argument	Description	Default
x	One of the two data frames to combine.	
y	One of the two data frames to combine.	
by	A vector of character values corresponding to column names.	intersect(names(x), names(y))
by.x	A vector of character values corresponding to column names in x. Overrides the list given in by.	by
by.y	A vector of character values corresponding to column names in y. Overrides the list given in by.	by
all	A logical value specifying whether rows from each data frame should be included even if there is no match in the other data frame. This is equivalent to an OUTER JOIN in a database. (Equivalent to all.x=TRUE and all.y=TRUE.)	FALSE
all.x	A logical value specifying whether rows from data frame x should be included even if there is no match in the other data frame. This is equivalent to x LEFT OUTER JOIN y in a database.	all
all.y	A logical value specifying whether rows from data frame x should be included even if there is no match in the other data frame. This is equivalent to x RIGHT OUTER JOIN y in a database.	all
sort	A logical value that specifies whether the results should be sorted by the by columns.	TRUE
suffixes	A character vector with two values. If there are columns in x and y with the same name that are not used in the by list, they will be renamed with the suffixes given by this argument.	suffixes = c(".x", ".y")
incomparables	A list of variables that cannot be matched.	NULL

By default, merge is equivalent to a NATURAL JOIN in SQL. You can specify other columns to make it use merge like an INNER JOIN. You can specify values of ALL to get the same results as OUTER or FULL joins. If there are no matching field names, of if by is of length 0 (or by.x and by.y) are of length 0, then merge will return the full Cartesian product of x and y.

Transformations

Sometimes, there will be some variables in your source data that aren't quite right. This section explains how to change a variable in a data frame.

Reassigning Variables

One of the most convenient ways to redefine a variable in a data frame is to use the assignment operator. For example, suppose that you wanted to change the type of a variable in the dow30 data frame that we created above. When read.csv imported this data, it interpreted the "Date" field as a character string and converted it to a factor:

```
> class(dow30$Date)
[1] "factor"
```

Factors are fine for some things, but we could better represent the date field as a Date object. (That would create a proper ordering on dates and allow us to extract information from them.) Luckily, Yahoo! Finance prints dates in the default date format for R, so we can just transform these values into Date objects using as.Date (see the help file for as.Date for more information). So, let's change this variable within the data frame to use Date objects:

```
> dow30$Date <- as.Date(dow30$Date)
> class(dow30$Date)
[1] "Date"
```

It's also possible to make other changes to data frames. For example, suppose that we wanted to define a new midpoint variable that is the mean of the high and low price. We can add this variable with the same notation:

```
> dow30$mid <- (dow30$High + dow30$Low) / 2
> names(dow30)
[1] "symbol"    "Date"      "Open"      "High"     "Low"
[6] "Close"     "Volume"    "Adj.Close" "mid"
```

The Transform Function

A convenient function for changing variables in a data frame is the transform function. Formally, transform is defined as:

```
transform(`_data`, ...)
```

Notice that there aren't any named arguments for this function. To use transform, you specify a data frame (as the first argument) and a set of expressions that use variables within the data frame. The transform function applies each expression to the data frame and then returns the final data frame.

For example, suppose that we wanted to perform the two transformations listed above: changing the Date column to a Date format, and adding a new midpoint variable. We could do this with transform using the following expression:

```
> dow30.transformed <- transform(dow30, Date=as.Date(Date),
+     mid = (High + Low) / 2)
> names(dow30.transformed)
[1] "symbol"   "Date"      "Open"      "High"      "Low"
[6] "Close"    "Volume"    "Adj.Close" "mid"
> class(dow30.transformed$Date)
[1] "Date"
```

Applying a Function to Each Element of an Object

When transforming data, one common operation is to apply a function to a set of objects (or each part of a composite object) and return a new set of objects (or a new composite object). The base R library includes a set of different functions for doing this.

Applying a function to an array

To apply a function to parts of an array (or matrix), use the apply function:

```
apply(X, MARGIN, FUN, ...)
```

Apply accepts three arguments: X is the array to which a function is applied, FUN is the function, and MARGIN specifies the dimensions to which you would like to apply a function. Optionally, you can specify arguments to FUN as addition arguments to apply arguments to FUN.) To show how this works, here's a simple example. Let's create a matrix with four rows of five elements, corresponding to the numbers between 1 and 20:

```
> x <- 1:20
> dim(x) <- c(5,4)
> x
     [,1] [,2] [,3] [,4]
[1,]   1    6   11   16
[2,]   2    7   12   17
[3,]   3    8   13   18
[4,]   4    9   14   19
[5,]   5   10   15   20
```

Now, let's show how apply works. We'll use the function max because it's easy to look at the matrix above and see where the results came from.

First, let's select the maximum element of each row. (These are the values in the rightmost column: 16, 17, 18, 19, and 20.) To do this, we will specify X=x, MARGIN=1 (rows are the first dimension), and FUN=max:

```
> apply(X=x,MARGIN=1,FUN=max)
[1] 16 17 18 19 20
```

To do the same thing for columns, we simply have to change the value of `MARGIN`:

```
> apply(X=x,MARGIN=2,FUN=max)
[1]  5 10 15 20
```

As a slightly more complex example, we can also use `MARGIN` to apply a function over multiple dimensions. (We'll switch to the function **paste** to show which elements were included.) Consider the following three-dimensional array:

```
> x <- 1:27
> dim(x) <- c(3,3,3)
> x
, , 1

     [,1] [,2] [,3]
[1,]    1    4    7
[2,]    2    5    8
[3,]    3    6    9

, , 2

     [,1] [,2] [,3]
[1,]   10   13   16
[2,]   11   14   17
[3,]   12   15   18

, , 3

     [,1] [,2] [,3]
[1,]   19   22   25
[2,]   20   23   26
[3,]   21   24   27
```

Let's start by looking at which values are grouped for each value of `MARGIN`:

```
> apply(X=x,MARGIN=1,FUN=paste,collapse=",")
[1] "1,4,7,10,13,16,19,22,25" "2,5,8,11,14,17,20,23,26"
[3] "3,6,9,12,15,18,21,24,27"
> apply(X=x,MARGIN=2,FUN=paste,collapse=",")
[1] "1,2,3,10,11,12,19,20,21" "4,5,6,13,14,15,22,23,24"
[3] "7,8,9,16,17,18,25,26,27"
> apply(X=x,MARGIN=3,FUN=paste,collapse=",")
[1] "1,2,3,4,5,6,7,8,9"        "10,11,12,13,14,15,16,17,18"
[3] "19,20,21,22,23,24,25,26,27"
```

Let's do something more complicated. Let's select `MARGIN=c(1, 2)` to see which elements are selected:

```
> apply(X=x,MARGIN=c(1,2),FUN=paste,collapse=",")
       [,1]       [,2]       [,3]
[1,] "1,10,19" "4,13,22" "7,16,25"
[2,] "2,11,20" "5,14,23" "8,17,26"
[3,] "3,12,21" "6,15,24" "9,18,27"
```

This is the equivalent of doing the following: for each value of i between 1 and 3 and each value of j between 1 and 3, calculate `FUN` of `x[i][j][1]`, `x[i][j][2]`, `x[i][j][3]`.

Applying a function to a list or vector

To apply a function to each element in a vector or a list and return a list, you can use the function `lapply`. The function `lapply` requires two arguments: an object X and a function FUNC. (You may specify additional arguments that will be passed to FUNC.) Let's look at a simple example of how to use `lapply`:

```
> x <- as.list(1:5)
> lapply(x,function(x) 2^x)
[[1]]
[1] 2

[[2]]
[1] 4

[[3]]
[1] 8

[[4]]
[1] 16

[[5]]
[1] 32
```

You can apply a function to a data frame, and the function will be applied to each vector in the data frame. For example:

```
> d <- data.frame(x=1:5,y=6:10)
> d
  x  y
1 1  6
2 2  7
3 3  8
4 4  9
5 5 10
> lapply(d,function(x) 2^x)
$x
[1]  2  4  8 16 32

$y
[1]   64  128  256  512 1024
> lapply(d,FUN=max)
$x
[1] 5

$y
[1] 10
```

Sometimes, you might prefer to get a vector, matrix, or array instead of a list. To do this, use the `sapply` function. This function works exactly the same way as `apply`, except that it returns a vector or matrix (when appropriate):

```
> sapply(d,FUN=function(x) 2^x)
     x   y
[1,] 2  64
[2,] 4 128
```

```
[3,]  8  256
[4,] 16  512
[5,] 32 1024
```

Another related function is `mapply`, the "multivariate" version of `sapply`:

```
mapply(FUN, ..., MoreArgs = , SIMPLIFY = , USE.NAMES = )
```

Here is a description of the arguments to `mapply`.

Argument	Description	Default
FUN	The function to apply.	
...	A set of vectors over which FUN should be applied.	
MoreArgs	A list of additional arguments to pass to FUN.	
SIMPLIFY	A logical value indicating whether to simplify the returned array.	TRUE
USE.NAMES	A logical value indicating whether to use names for returned values. Names are taken from the values in the first vector (if it is a character vector) or from the names of elements in that vector.	TRUE

This function will apply `FUN` to the first element of each vector, then to the second, and so on, until it reaches the last element.

Here is a simple example of `mapply`:

```
> mapply(paste,
+        c(1,2,3,4,5),
+        c("a","b","c","d","e"),
+        c("A","B","C","D","E"),
+        MoreArgs=list(sep="-"))
[1] "1-a-A" "2-b-B" "3-c-C" "4-d-D" "5-e-E"
```

Binning Data

Another common data transformation is to group a set of observations into bins based on the value of a specific variable. For example, suppose that you had some time series data where time was measured in days, but you wanted to summarize the data by month. There are several functions available for binning numeric data in R.

Shingles

We briefly mentioned shingles in "Shingles" on page 91. Shingles are a way to represent intervals in R. They can be overlapping, like roof shingles (hence the name). They are used extensively in the `lattice` package, when you want to use a numeric value as a conditioning value.

To create shingles in R, use the `shingle` function:

```
shingle(x, intervals=sort(unique(x)))
```

To specify where to separate the bins, use the `intervals` argument. You can use a numeric vector to indicate the breaks or a two-column matrix, where each row represents a specific interval.

To create shingles where the number of observations is the same in each bin, you can use the `equal.count` function:

```
equal.count(x, ...)
```

Cut

The function `cut` is useful for taking a continuous variable and splitting it into discrete pieces. Here is the default form of `cut` for use with numeric vectors:

```
# numeric form
cut(x, breaks, labels = NULL,
    include.lowest = FALSE, right = TRUE, dig.lab = 3,
    ordered_result = FALSE, ...)
```

There is also a version of cut for manipulating `Date` objects:

```
# Date form
cut(x, breaks, labels = NULL, start.on.monday = TRUE,
    right = FALSE, ...)
```

The `cut` function takes a numeric vector as input and returns a factor. Each level in the factor corresponds to an interval of values in the input vector. Here is a description of the arguments to `cut`.

Argument	Description	Default
x	A numeric vector (to convert to a factor).	
breaks	Either a single integer value specifying the number of break points or a numeric vector specifying the set of break points.	
labels	Labels for the levels in the output factor.	NULL
include.lowest	A logical value indicating if a value equal to the lowest point in the range (if `right=TRUE`) in a range should be included in a given bucket. If `right=FALSE` indicates whether a value equal to the highest point in the range should be included.	FALSE
right	A logical value that specifies whether intervals should be closed on the right and open on the left. (For `right=FALSE`, intervals will be open on the right and closed on the left.)	TRUE
dig.lab	Number of digits used when generating labels (if labels are not explicitly specified).	3
ordered_results	A logical value indicating whether the result should be an ordered factor.	FALSE

For example, suppose that you wanted to count the number of players with batting averages in certain ranges. To do this, you could use the `cut` function and the `table` function:

```
> # load in the example data
> library(nutshell)
> data(batting.2008)
> # first, add batting average to the data frame:
> batting.2008.AB <- transform(batting.2008, AVG = H/AB)
> # now, select a subset of players with over 100 AB (for some
```

```
> # statistical significance):
> batting.2008.over100AB <- subset(batting.2008.AB, subset=(AB > 100))
> # finally, split the results into 10 bins:
> battingavg.2008.bins <- cut(batting.2008.over100AB$AVG,breaks=10)
> table(battingavg.2008.bins)
battingavg.2008.bins
(0.137,0.163] (0.163,0.189] (0.189,0.215]  (0.215,0.24]  (0.24,0.266]
            4             6            24            67           121
(0.266,0.292] (0.292,0.318] (0.318,0.344]  (0.344,0.37]  (0.37,0.396]
          132            70            11             5             2
```

Combining Objects with a Grouping Variable

Sometimes, you would like to combine a set of similar objects (either vectors or data frames) into a single data frame, with a column labeling the source. You can do this with the make.groups function in the lattice package:

```
library(lattice)
make.groups(...)
```

For example, let's combine three different vectors into a data frame:

```
> hat.sizes <- seq(from=6.25, to=7.75, by=.25)
> pants.sizes <- c(30,31,32,33,34,36,38,40)
> shoe.sizes <- seq(from=7, to=12)
> make.groups(hat.sizes, pants.sizes, shoe.sizes)
              data        which
hat.sizes1    6.25    hat.sizes
hat.sizes2    6.50    hat.sizes
hat.sizes3    6.75    hat.sizes
hat.sizes4    7.00    hat.sizes
hat.sizes5    7.25    hat.sizes
hat.sizes6    7.50    hat.sizes
hat.sizes7    7.75    hat.sizes
pants.sizes1 30.00  pants.sizes
pants.sizes2 31.00  pants.sizes
pants.sizes3 32.00  pants.sizes
pants.sizes4 33.00  pants.sizes
pants.sizes5 34.00  pants.sizes
pants.sizes6 36.00  pants.sizes
pants.sizes7 38.00  pants.sizes
pants.sizes8 40.00  pants.sizes
shoe.sizes1   7.00   shoe.sizes
shoe.sizes2   8.00   shoe.sizes
shoe.sizes3   9.00   shoe.sizes
shoe.sizes4  10.00   shoe.sizes
shoe.sizes5  11.00   shoe.sizes
shoe.sizes6  12.00   shoe.sizes
```

Subsets

Often, you'll be provided with too much data. For example, suppose that you were working with patient records at a hospital. You might want to analyze healthcare records for patients between 5 and 13 years of age who were treated for asthma during the past 3 years. To do this, you need to take a subset of the data and not examine the whole database.

Other times, you might have too much relevant data. For example, suppose that you were looking at a logistics operation that fills billions of orders every year. R can only hold a certain number of records in memory and might not be able to hold the entire database. In most cases, you can get statistically significant results with a tiny fraction of the data; even millions of orders might be too many.

Bracket Notation

One way to take a subset of a data set is to use the bracket notation. As you may recall, you can select rows in a data frame by providing a vector of logical values. If you can write a simple expression describing the set of rows to select from a data frame, you can provide this as an index.

For example, suppose that we wanted to select only batting data from 2008. The column batting.w.names$yearID contains the year associated with each row, so we could calculate a vector of logical values describing which rows to keep with the expression batting.w.names$yearID==2008. Now, we just have to index the data frame batting.w.names with this vector to select only rows for the year 2008:

```
> batting.w.names.2008 <- batting.w.names[batting.w.names$yearID==2008,]
> summary(batting.w.names.2008$yearID)
   Min. 1st Qu.  Median    Mean 3rd Qu.    Max.
   2008    2008    2008    2008    2008    2008
```

Similarly, we can use the same notation to select only certain columns. Suppose that we only wanted to keep some variables nameFirst, nameLast, AB, H, BB. We could provide these in the brackets as well:

```
> batting.w.names.2008.short <-
+     batting.w.names[batting.w.names$yearID==2008,
+     c("nameFirst","nameLast","AB","H","BB")]
```

subset Function

As an alternative, you can use the subset function to select a subset of rows and columns from a data frame (or matrix):

```
subset(x, subset, select, drop = FALSE, ...)
```

There isn't anything you can do with subset that you can't do with the bracket notation, but using subset can lead to more readable code. Subset allows you to use variable names from the data frame when selecting subsets, saving some typing. Here is a description of the arguments to subset.

Argument	Description	Default
x	The object from which to calculate a subset.	
subset	A logical expression that describes the set of rows to return.	
select	An expression indicating which columns to return.	
drop	Passed to ` [ `.	FALSE

As an example, let's re-create the same data sets we created above using `subset`:

```
> batting.w.names.2008 <- subset(batting.w.names, yearID==2008)
> batting.w.names.2008.short <- subset(batting.w.names, yearID==2008,
+     c("nameFirst","nameLast","AB","H","BB"))
```

Random Sampling

Often, it is desirable to take a random sample of a data set. Sometimes, you might have too much data (for statistical reasons or performance reasons). Other times, you simply want to split your data into different parts for modeling (usually into training, testing, and validation subsets).

One of the simplest ways to extract a random sample is with the `sample` function. The `sample` function returns a random sample of the elements of a vector:

```
sample(x, size, replace = FALSE, prob = NULL)
```

Argument	Description	Default
x	The object from which the sample is taken	
size	An integer value specifying the sample size	
replace	A logical value indicating whether to sample with, or without, replacement	FALSE
prob	A vector of probabilities for selecting each item	NULL

Somewhat nonintuitively, when applied to a data frame, `sample` will return a random sample of the columns. (Remember that a data frame is implemented as a list of vectors, so `sample` is just taking a random sample of the elements of the list.) So, you need to be a little more clever when you use `sample` with a data frame.

To take a random sample of the observations in a data set, you can use `sample` to create a random sample of row numbers and then select these row numbers using an index operator. For example, let's take a random sample of five elements from the `batting.2008` data set:

```
> batting.2008[sample(1:nrow(batting.2008),5),]
         playerID yearID stint teamID lgID   G G_batting  AB   R  H 2B 3B
90648 izturma01   2008     1    LAA   AL   79             79 290 44 78 14  2
90280 benoijo01   2008     1    TEX   AL   44              3   0  0  0  0  0
90055 percitr01   2008     1    TBA   AL   50              4   0  0  0  0  0
91085  getzch01   2008     1    CHA   AL   10             10   7  2  2  0  0
90503 willijo03   2008     1    FLO   NL  102            102 351 54 89 21  5
      HR RBI SB CS BB SO IBB HBP SH SF GIDP G_old
90648  3  37 11  2 26 27   0   1  2  2    9    79
90280  0   0  0  0  0  0   0   0  0  0    0     3
90055  0   0  0  0  0  0   0   0  0  0    0     4
91085  0   1  1  1  0  1   0   0  0  0    0    10
90503 15  51  3  2 48 82   2  14  1  2    7   102
```

You can also use this technique to select a more complicated random subset. For example, suppose that you wanted to randomly select statistics for three teams. You could do this as follows:

```
> batting.2008$teamID <- as.factor(batting.2008$teamID)
> levels(batting.2008$teamID)
 [1] "ARI" "ATL" "BAL" "BOS" "CHA" "CHN" "CIN" "CLE" "COL" "DET" "FLO"
[12] "HOU" "KCA" "LAA" "LAN" "MIL" "MIN" "NYA" "NYN" "OAK" "PHI" "PIT"
[23] "SDN" "SEA" "SFN" "SLN" "TBA" "TEX" "TOR" "WAS"
> # example of sample
> sample(levels(batting.2008$teamID),3)
[1] "ATL" "TEX" "DET"
> # usage example (note that it's a different random sample of teams)
> batting.2008.3teams <- batting.2008[is.element(batting.2008$teamID,
+       sample(levels(batting.2008$teamID),3)),]
> # check to see that sample only has three teams
> summary(batting.2008.3teams$teamID)
ARI ATL BAL BOS CHA CHN CIN CLE COL DET FLO HOU KCA LAA LAN MIL MIN
  0   0   0   0   0   0  48   0   0   0   0   0   0  41   0  44   0
NYA NYN OAK PHI PIT SDN SEA SFN SLN TBA TEX TOR WAS
  0   0   0   0   0   0   0   0   0   0   0   0   0
```

This function is good for data sources where you simply want to take a random sample of all the observations, but often you might want to do something more complicated like stratified sampling, cluster sampling, maximum entropy sampling, or other more sophisticated methods. You can find many of these methods in the sampling package. For an example using this package to do stratified sampling, see "Machine Learning Algorithms for Classification" on page 445.

Summarizing Functions

Often, you are provided with data that is too fine grained for your analysis. For example, you might be analyzing data about a website. Suppose that you wanted to know the average number of pages delivered to each user. To find the answer, you might need to look at every HTTP transaction (every request for content), grouping together requests into sessions and counting the number of requests. R provides a number of different functions for summarizing data, aggregating records together to build a smaller data set.

tapply, aggregate

The tapply function is a very flexible function for summarizing a vector X. You can specify which subsets of X to summarize as well as the function used for summarization:

```
tapply(X, INDEX, FUN = , ..., simplify = )
```

Here are the arguments to tapply.

Argument	Description	Default
X	The object on which to apply the function (usually a vector).	
INDEX	A list of factors that specify different sets of values of X over which to calculate FUN, each the same length as X.	
FUN	The function applied to elements of X.	NULL

Argument	Description	Default
...	Optional arguments are passed to FUN.	
simplify	If simplify=TRUE, then if FUN returns a scalar, then tapply returns an array with the mode of the scalar. If simplify=FALSE, then tapply returns a list.	TRUE

For example, we can use `tapply` to sum the number of home runs by team:

```
> tapply(X=batting.2008$HR,INDEX=list(batting.2008$teamID),FUN=sum)
ARI ATL BAL BOS CHA CHN CIN CLE COL DET FLO HOU KCA LAA LAN MIL MIN
159 130 172 173 235 184 187 171 160 200 208 167 120 159 137 198 111
NYA NYN OAK PHI PIT SDN SEA SFN SLN TBA TEX TOR WAS
180 172 125 214 153 154 124  94 174 180 194 126 117
```

You can also apply a function that returns multiple items, such as `fivenum` (which returns a vector containing minimum, lower-hinge, median, upper-hinge, maximum) to the data. For example, here is the result of applying `fivenum` to the batting averages of each player, aggregated by league:

```
> tapply(X=(batting.2008$H/batting.2008$AB),
+    INDEX=list(batting.2008$lgID),FUN=fivenum)
$AL
[1] 0.0000000 0.1758242 0.2487923 0.2825485 1.0000000

$NL
[1] 0.0000000 0.0952381 0.2172524 0.2679739 1.0000000
```

You can also use `tapply` to calculate summaries over multiple dimensions. For example, we can calculate the mean number of home runs per player by league and batting hand:

```
> tapply(X=(batting.2008$HR),
+    INDEX=list(batting.w.names.2008$lgID,
+             batting.w.names.2008$bats),
+    FUN=mean)
           B        L        R
AL 3.058824 3.478495 3.910891
NL 3.313433 3.400000 3.344902
```

A function closely related to `tapply` is `by`. The `by` function works the same way as `tapply`, except that it works on data frames. The `INDEX` argument is replaced by an `INDICES` argument. Here is an example:

```
> by(batting.2008[,c("H","2B","3B","HR")],
+    INDICES=list(batting.w.names.2008$lgID,
+       batting.w.names.2008$bats),FUN=mean)
: AL
: B
           H         2B        3B        HR
29.0980392  5.4901961  0.8431373  3.0588235
----------------------------------------------------------
: NL
: B
           H         2B        3B        HR
29.2238806  6.4776119  0.6865672  3.3134328
----------------------------------------------------------
```

```
: AL
: L
           H          2B          3B           HR
32.4301075   6.7258065   0.5967742   3.4784946
----------------------------------------------------
: NL
: L
           H          2B          3B           HR
31.888372   6.283721   0.627907   3.400000
----------------------------------------------------
: AL
: R
           H          2B          3B           HR
34.2549505   7.0495050   0.6460396   3.9108911
----------------------------------------------------
: NL
: R
           H          2B          3B           HR
29.9414317   6.1822126   0.6290672   3.3449024
```

Another option for summarization is the function **aggregate**. Here is the form of **aggregate** when applied to data frames:

```
aggregate(x, by, FUN, ...)
```

Aggregate can also be applied to time series and takes slightly different arguments:

```
aggregate(x, nfrequency = 1, FUN = sum, ndeltat = 1,
          ts.eps = getOption("ts.eps"), ...)
```

Here is a description of the arguments to **aggregate**.

Argument	Description	Default
x	The object to aggregate	
by	A list of grouping elements, each as long as x	
FUN	A scalar function used to compute the summary statistic	no default for data frames; for time series, FUN=SUM
nfrequency	Number of observations per unit of time	1
ndeltat	Fraction of the sampling period between successive observations	1
ts.eps	Tolerance used to decide if nfrequency is a submultiple of the original frequency	getOption("ts.eps")
...	Further arguments passed to FUN	

For example, we can use **aggregate** to summarize batting statistics by team:

```
> aggregate(x=batting.2008[,c("AB","H","BB","2B","3B","HR")],
+     by=list(batting.2008$teamID),FUN=sum)
   Group.1   AB    H   BB   2B  3B   HR
1      ARI 5409 1355  587  318  47  159
2      ATL 5604 1514  618  316  33  130
3      BAL 5559 1486  533  322  30  172
4      BOS 5596 1565  646  353  33  173
5      CHA 5553 1458  540  296  13  235
```

```
6    CHN 5588 1552 636 329 21 184
7    CIN 5465 1351 560 269 24 187
8    CLE 5543 1455 560 339 22 171
9    COL 5557 1462 570 310 28 160
10   DET 5641 1529 572 293 41 200
11   FLO 5499 1397 543 302 28 208
12   HOU 5451 1432 449 284 22 167
13   KCA 5608 1507 392 303 28 120
14   LAA 5540 1486 481 274 25 159
15   LAN 5506 1455 543 271 29 137
16   MIL 5535 1398 550 324 35 198
17   MIN 5641 1572 529 298 49 111
18   NYA 5572 1512 535 289 20 180
19   NYN 5606 1491 619 274 38 172
20   OAK 5451 1318 574 270 23 125
21   PHI 5509 1407 586 291 36 214
22   PIT 5628 1454 474 314 21 153
23   SDN 5568 1390 518 264 27 154
24   SEA 5643 1498 417 285 20 124
25   SFN 5543 1452 452 311 37  94
26   SLN 5636 1585 577 283 26 174
27   TBA 5541 1443 626 284 37 180
28   TEX 5728 1619 595 376 35 194
29   TOR 5503 1453 521 303 32 126
30   WAS 5491 1376 534 269 26 117
```

Aggregating Tables with rowsum

Sometimes, you would simply like to calculate the sum of certain variables in an object, grouped together by a grouping variable. To do this in R, use the rowsum function:

```
rowsum(x, group, reorder = TRUE, ...)
```

For example, we can use rowsum to summarize batting statistics by team:

```
> rowsum(batting.2008[,c("AB","H","BB","2B","3B","HR")],
+     group=batting.2008$teamID)
      AB    H  BB X2B X3B  HR
ARI 5409 1355 587 318  47 159
ATL 5604 1514 618 316  33 130
BAL 5559 1486 533 322  30 172
BOS 5596 1565 646 353  33 173
CHA 5553 1458 540 296  13 235
CHN 5588 1552 636 329  21 184
CIN 5465 1351 560 269  24 187
CLE 5543 1455 560 339  22 171
COL 5557 1462 570 310  28 160
DET 5641 1529 572 293  41 200
FLO 5499 1397 543 302  28 208
HOU 5451 1432 449 284  22 167
KCA 5608 1507 392 303  28 120
LAA 5540 1486 481 274  25 159
LAN 5506 1455 543 271  29 137
MIL 5535 1398 550 324  35 198
MIN 5641 1572 529 298  49 111
```

```
NYA 5572 1512 535 289  20 180
NYN 5606 1491 619 274  38 172
OAK 5451 1318 574 270  23 125
PHI 5509 1407 586 291  36 214
PIT 5628 1454 474 314  21 153
SDN 5568 1390 518 264  27 154
SEA 5643 1498 417 285  20 124
SFN 5543 1452 452 311  37  94
SLN 5636 1585 577 283  26 174
TBA 5541 1443 626 284  37 180
TEX 5728 1619 595 376  35 194
TOR 5503 1453 521 303  32 126
WAS 5491 1376 534 269  26 117
```

Counting Values

Often, it can be useful to count the number of observations that take on each possible value of a variable. R provides several functions for doing this.

The simplest function for counting the number of observations that take on a value is the `tabulate` function. This function counts the number of elements in a vector that take on each integer value and returns a vector with the counts.

As an example, suppose that you wanted to count the number of players who hit 0 HR, 1 HR, 2 HR, 3 HR, and so on. You could do this with the `tabulate` function:

```
> HR.cnts <- tabulate(batting.w.names.2008$HR)
> # tabulate doesn't label results, so let's add names:
> names(HR.cnts) <- 0:(length(HR.cnts)-1)
> HR.cnts
 0  1  2  3  4  5  6  7  8  9 10 11 12 13 14 15 16 17 18 19 20 21 22
92 63 45 20 15 26 23 21 22 15 15 18 12 10 12  4  9  3  3 13  9  7 10
23 24 25 26 27 28 29 30 31 32 33 34 35 36 37 38 39 40 41 42 43 44 45
 4  8  2  5  2  4  0  1  6  6  3  1  2  4  1  0  0  0  0  0  0  0  0
46 47
 0  1
```

A related function (for categorical values) is `table`. Suppose that you are presented with some data that includes a few categorical values (encoded as factors in R) and wanted to count how many observations in the data had each categorical value. To do this, you can use the `table` function:

```
table(..., exclude = if (useNA == "no") c(NA, NaN), useNA = c("no",
    "ifany", "always"), dnn = list.names(...), deparse.level = 1)
```

`Table` returns a table object showing the number of observations that have each possible categorical value.[†] Here are the arguments to `table`.

Argument	Description	Default
...	A set of factors (or objects that can be coerced into factors).	
exclude	Levels to remove from factors.	if (useNA == "no") c(NA, NaN)

[†] If you are familiar with SAS, you can think of `table` as the equivalent to PROC FREQ.

Argument	Description	Default
useNA	Indicates whether to include NA values in the table.	c("no", "ifany", "always")
dnn	Names to be given to dimensions in the result.	list.names(...)
deparse.level	As noted in the help file: "If the argument dnn is not supplied, the internal function list.names is called to compute the 'dimname names'. If the arguments in ... are named, those names are used. For the remaining arguments, deparse.level = 0 gives an empty name, deparse.level = 1 uses the supplied argument if it is a symbol, and deparse.level = 2 will deparse the argument."	1

For example, suppose that we wanted to count the number of left-handed batters, right-handed batters, and switch hitters in 2008. We could use the data frame `batting.w.names.2008` defined above to provide the data and `table` to tabulate the results:

```
> table(batting.w.names.2008$bats)

  B   L   R
118 401 865
```

To make this a little more interesting, we could make this a two-dimensional table showing the number of players who batted and threw with each hand:

```
> table(batting.2008[,c("bats", "throws")])
     throws
bats   L   R
   B  10 108
   L 240 161
   R  25 840
```

We could extend the results to another dimension, adding league ID:

```
, , lgID = AL

     throws
bats   L   R
   B   4  47
   L 109  77
   R  11 393

, , lgID = NL

     throws
bats   L   R
   B   6  61
   L 131  84
   R  14 447
```

Another useful function is `xtabs`, which creates contingency tables from factors using formulas:

```
xtabs(formula = ~., data = parent.frame(), subset, na.action,
      exclude = c(NA, NaN), drop.unused.levels = FALSE)
```

`xtabs` works the same as `table`, but allows you to specify the groupings by specifying a formula and a data frame. In many cases, this can save you some typing. For example, here is how to use `xtabs` to tabulate batting statistics by batting arm and league:

```
> xtabs(~bats+lgID,batting.2008)
     lgID
bats  AL  NL
   B  51  67
   L 186 215
   R 404 461
```

`Table` only works on factors, but sometimes you might like to calculate tables with numeric values as well. For example, suppose that you wanted to count the number of players with batting averages in certain ranges. To do this, you could use the `cut` function and the `table` function:

```
> # first, add batting average to the data frame:
> batting.w.names.2008 <- transform(batting.w.names.2008, AVG = H/AB)
> # now, select a subset of players with over 100 AB (for some
> # statistical significance):
> batting.2008.over100AB <- subset(batting.2008, subset=(AB > 100))
> # finally, split the results into 10 bins:
> battingavg.2008.bins <- cut(batting.2008.over100AB$AVG,breaks=10)
> table(battingavg.2008.bins)
battingavg.2008.bins
(0.137,0.163] (0.163,0.189] (0.189,0.215]  (0.215,0.24]  (0.24,0.266]
            4             6            24            67           121
(0.266,0.292] (0.292,0.318] (0.318,0.344]  (0.344,0.37]  (0.37,0.396]
          132            70            11             5             2
```

Reshaping Data

Very often, you are presented with data that is in the wrong "shape." Sometimes, you might find that a single observation is stored across multiple lines in a data frame. This happens very often in data warehouses. In these systems, a single table might be used to represent many different "facts." Each fact might be associated with a unique identifier, a timestamp, a concept, and an observed value. To build a statistical model or to plot results, you might need to create a version of the data where each line contained a unique identifier, a timestamp, and a column for each concept. So, you might want to transform this "narrow" data set to a "wide" format.

Other times, you might be presented with a sparsely populated data frame that has a large number of columns. Although this format might make analysis straightforward, the data set might also be large and difficult to store. So, you might want to transform this wide data set into a narrow one.

Transposing matrices and data frames

A very useful function is `t`, which transposes objects. The `t` function takes one argument: an object to transpose. The object can be a matrix, vector, or data frame. Here is an example with a matrix:

```
> m <- matrix(1:10, nrow=5)
> m
     [,1] [,2]
[1,]    1    6
[2,]    2    7
[3,]    3    8
[4,]    4    9
[5,]    5   10
> t(m)
     [,1] [,2] [,3] [,4] [,5]
[1,]    1    2    3    4    5
[2,]    6    7    8    9   10
```

When you call t on a vector, the vector is treated as a single row of a matrix. So, the value returned by t will be a matrix with a single column:

```
> v <- 1:10
> v
 [1]  1  2  3  4  5  6  7  8  9 10
> t(v)
     [,1] [,2] [,3] [,4] [,5] [,6] [,7] [,8] [,9] [,10]
[1,]    1    2    3    4    5    6    7    8    9    10
```

Reshaping data frames and matrices

R includes several functions that let you change data between narrow and wide formats. Let's use a small table of stock data to show how these functions work. First, we'll define a small portfolio of stocks. Then we'll get monthly observation for the first three months of 2009:

```
> my.tickers <- c("GE","GOOG","AAPL","AXP","GS")
> my.quotes <- get.multiple.quotes(my.tickers, from=as.Date("2009-01-01"),
+     to=as.Date("2009-03-31"), interval="m")
> my.quotes
   symbol       Date   Open   High    Low  Close     Volume Adj.Close
1      GE 2009-03-02   8.29  11.35   5.87  10.11  277426300     10.11
2      GE 2009-02-02  12.03  12.90   8.40   8.51  194928800      8.51
3      GE 2009-01-02  16.51  17.24  11.87  12.13  117846700     11.78
4    GOOG 2009-03-02 333.33 359.16 289.45 348.06    5346800    348.06
5    GOOG 2009-02-02 334.29 381.00 329.55 337.99    6158100    337.99
6    GOOG 2009-01-02 308.60 352.33 282.75 338.53    5727600    338.53
7    AAPL 2009-03-02  88.12 109.98  82.33 105.12   25963400    105.12
8    AAPL 2009-02-02  89.10 103.00  86.51  89.31   27394900     89.31
9    AAPL 2009-01-02  85.88  97.17  78.20  90.13   33487900     90.13
10    AXP 2009-03-02  11.68  15.24   9.71  13.63   31136400     13.45
11    AXP 2009-02-02  16.35  18.27  11.44  12.06   24297100     11.90
12    AXP 2009-01-02  18.57  21.38  14.72  16.73   19110000     16.51
13     GS 2009-03-02  87.86 115.65  72.78 106.02   30196400    106.02
14     GS 2009-02-02  78.78  98.66  78.57  91.08   28301500     91.08
15     GS 2009-01-02  84.02  92.20  59.13  80.73   22764300     80.29
```

Now, let's keep only the Date, Symbol, and Close columns:

```
> my.quotes.narrow <- my.quotes[,c("symbol","Date","Close")]
> my.quotes.narrow
   symbol       Date Close
1      GE 2009-03-02 10.11
```

```
 2      GE 2009-02-02   8.51
 3      GE 2009-01-02  12.13
 4    GOOG 2009-03-02 348.06
 5    GOOG 2009-02-02 337.99
 6    GOOG 2009-01-02 338.53
 7    AAPL 2009-03-02 105.12
 8    AAPL 2009-02-02  89.31
 9    AAPL 2009-01-02  90.13
10     AXP 2009-03-02  13.63
11     AXP 2009-02-02  12.06
12     AXP 2009-01-02  16.73
13      GS 2009-03-02 106.02
14      GS 2009-02-02  91.08
15      GS 2009-01-02  80.73
```

We can use the unstack function to change the format of this data from a stacked form to an unstacked form:

```
> unstack(my.quotes.narrow,form=Close~symbol)
     GE   GOOG   AAPL   AXP     GS
1 10.11 348.06 105.12 13.63 106.02
2  8.51 337.99  89.31 12.06  91.08
3 12.13 338.53  90.13 16.73  80.73
```

The first argument to unstack specifies the data frame. The second argument, form, uses a formula to specify how to unstack the data frame. The left side of the formula represents the vector to be unstacked (in this case, symbol). The right side indicates the groups to create (in this case Close).

Notice that the unstack operation retains the order of observations, but loses the Date column. (It's probably best to use unstack with data in which there are only two variables that matter.) You can also transform data the other way, stacking observations to create a long list:

```
> unstacked <- unstack(my.quotes.narrow,form=Close~symbol)
> stack(unstacked)
   values  ind
1   10.11   GE
2    8.51   GE
3   12.13   GE
4  348.06 GOOG
5  337.99 GOOG
6  338.53 GOOG
7  105.12 AAPL
8   89.31 AAPL
9   90.13 AAPL
10  13.63  AXP
11  12.06  AXP
12  16.73  AXP
13 106.02   GS
14  91.08   GS
15  80.73   GS
```

R includes a more powerful function for changing the shape of a data frame: the reshape function. Before explaining how to use this function (it's a bit complicated), let's use a couple of examples to show what it does.

First, suppose that we wanted each row to represent a unique date and each column to represent a different stock. We can do this with the **reshape** function:

```
> my.quotes.wide <- reshape(my.quotes.narrow, idvar="Date",
+     timevar="symbol", direction="wide")
> my.quotes.wide
       Date Close.GE Close.GOOG Close.AAPL Close.AXP Close.GS
1 2009-03-02    10.11     348.06     105.12     13.63   106.02
2 2009-02-02     8.51     337.99      89.31     12.06    91.08
3 2009-01-02    12.13     338.53      90.13     16.73    80.73
```

Parameters for **reshape** are stored as attributes of the created data frame:

```
> attributes(my.quotes.wide)
$row.names
[1] 1 2 3

$names
[1] "Date"       "Close.GE"   "Close.GOOG" "Close.AAPL" "Close.AXP"
[6] "Close.GS"

$class
[1] "data.frame"

$reshapeWide
$reshapeWide$v.names
NULL

$reshapeWide$timevar
[1] "symbol"

$reshapeWide$idvar
[1] "Date"

$reshapeWide$times
[1] GE   GOOG AAPL AXP  GS
Levels: GE GOOG AAPL AXP GS

$reshapeWide$varying
     [,1]       [,2]         [,3]         [,4]        [,5]
[1,] "Close.GE" "Close.GOOG" "Close.AAPL" "Close.AXP" "Close.GS"
```

Alternatively, we could have each row represent a stock, and each column represent a different date:

```
> reshape(my.quotes.narrow,idvar="symbol",timevar="Date",direction="wide")
   symbol Close.2009-03-02 Close.2009-02-02 Close.2009-01-02
1      GE            10.11             8.51            12.13
4    GOOG           348.06           337.99           338.53
7    AAPL           105.12            89.31            90.13
10    AXP            13.63            12.06            16.73
13     GS           106.02            91.08            80.73
```

We could even go in the opposite direction:

```
> reshape(my.quotes.wide)
                    Date symbol Close.GE
2009-03-02.GE 2009-03-02     GE    10.11
```

```
2009-02-02.GE    2009-02-02    GE      8.51
2009-01-02.GE    2009-01-02    GE     12.13
2009-03-02.GOOG  2009-03-02  GOOG    348.06
2009-02-02.GOOG  2009-02-02  GOOG    337.99
2009-01-02.GOOG  2009-01-02  GOOG    338.53
2009-03-02.AAPL  2009-03-02  AAPL    105.12
2009-02-02.AAPL  2009-02-02  AAPL     89.31
2009-01-02.AAPL  2009-01-02  AAPL     90.13
2009-03-02.AXP   2009-03-02   AXP     13.63
2009-02-02.AXP   2009-02-02   AXP     12.06
2009-01-02.AXP   2009-01-02   AXP     16.73
2009-03-02.GS    2009-03-02    GS    106.02
2009-02-02.GS    2009-02-02    GS     91.08
2009-01-02.GS    2009-01-02    GS     80.73
```

By the way, you can also use reshape to create columns for multiple data values at once:

```
> my.quotes.oc <- my.quotes[,c("symbol","Date","Close","Open")]
> my.quotes.oc
   symbol       Date  Close   Open
1      GE 2009-03-02  10.11   8.29
2      GE 2009-02-02   8.51  12.03
3      GE 2009-01-02  12.13  16.51
4    GOOG 2009-03-02 348.06 333.33
5    GOOG 2009-02-02 337.99 334.29
6    GOOG 2009-01-02 338.53 308.60
7    AAPL 2009-03-02 105.12  88.12
8    AAPL 2009-02-02  89.31  89.10
9    AAPL 2009-01-02  90.13  85.88
10    AXP 2009-03-02  13.63  11.68
11    AXP 2009-02-02  12.06  16.35
12    AXP 2009-01-02  16.73  18.57
13     GS 2009-03-02 106.02  87.86
14     GS 2009-02-02  91.08  78.78
15     GS 2009-01-02  80.73  84.02
> # now, let's change the shape of this data frame:
> reshape(my.quotes.oc,timevar="Date",idvar="symbol",direction="wide")
   symbol Close.2009-03-02 Open.2009-03-02 Close.2009-02-02
1      GE            10.11            8.29             8.51
4    GOOG           348.06          333.33           337.99
7    AAPL           105.12           88.12            89.31
10    AXP            13.63           11.68            12.06
13     GS           106.02           87.86            91.08
   Open.2009-02-02 Close.2009-01-02 Open.2009-01-02
1            12.03            12.13           16.51
4           334.29           338.53          308.60
7            89.10            90.13           85.88
10           16.35            16.73           18.57
13           78.78            80.73           84.02
```

The tricky thing about reshape is that it is actually two functions in one: a function that transforms long data to wide data and a function that transforms wide data to long data. The direction argument specifies whether you want a data frame that is "long" or "wide."

When transforming to wide data, you need to specify the `idvar` and `timevar` arguments. When transforming to long data, you need to specify the `varying` argument.

By the way, calls to `reshape` are reversible. If you have an object d that was created by a call to `reshape`, you can call `reshape(d)` to get back the original data frame:

```
reshape(data, varying = , v.names = , timevar = , idvar = , ids = , times = ,
        drop = , direction, new.row.names = , sep = , split = )
```

Here are the arguments to `reshape`.

Argument	Description	Default
data	A data frame to reshape.	
varying	A list of variables in the wide format that should be assigned to unique rows in the long format. Usually given as a list of variable names, but can be a matrix of names or a vector of names. (You can also use integers in this argument, which are used to index names (data).)	NULL
v.names	Names of variables in the long format that should be assigned to columns in the wide format.	NULL
timevar	The variable in the long format that identifies unique observations for the same group or individual (when going from the long to the wide format).	"time"
idvar	The variable in the long format that identifies unique groups or individuals (when going from the long to the wide format).	"id"
ids	The values to use for a new `idvar` variable.	1:NROW(data)
times	The values to use for a new `timevar` variable.	seq_along(varying[[1]])
drop	A vector of variable names to exclude from reshaping.	NULL
direction	A character value that specifies the reshaping direction: "wide" reshapes long data to wide data, and "long" reshapes wide data to long data.	
new.row.names	A logical value. When reshaping long data to wide data, specifies whether to create new row names from the values of the id and time variables.	NULL
sep	A character value. The reshape function will attempt to guess values for v.names and v.times when moving from wide to long data. This variable specifies the separator that is used in the variable names.	"."
split	As noted in the description for sep, reshape will attempt to split variable names into v.names and v.times. If the relationship between the variables is more complicated than just concatenation with a single value, reshape can still automatically guess values for v.names and v.times. See the help file for more information.	if (sep=="") { list(regexp= "[A-Za-z][0-9]", include=TRUE) } else { list(regexp=sep, include=FALSE, fixed=TRUE) }

Data Cleaning

Even when data is in the right form, there are often surprises in the data. For example, I used to work with credit data in a financial services company. Valid credit scores (specifically, FICO credit scores) always fall between 340 and 840. However, our data often contained values like 997, 998, and 999. These values did not mean that the customer had really super credit; instead, they had special meanings like "insufficient data."

Or, there might be duplicate records in the data. Again, suppose that you were analyzing data on patients at a hospital. Often, the same doctor might see multiple patients with the same first *and* last names, so multiple patients may be rolled up into a single record incorrectly. However, sometimes the same patient might see multiple doctors, creating multiple records in the database for the same patient.

Data cleaning doesn't mean changing the meaning of data. It means identifying problems caused by data collection, processing, and storage processes and modifying the data so that these problems don't interfere with analysis.

Finding and Removing Duplicates

Data sources often contain duplicate values. Depending on how you plan to use the data, the duplicates might cause problems. It's a good idea to check for duplicates in your data (if they aren't supposed to be there).

R provides some useful functions for detecting duplicate values.

Suppose that you accidentally included one stock ticker twice (say, GE) when you fetched stock quotes:

```
> my.tickers.2 <- c("GE","GOOG","AAPL","AXP","GS","GE")
> my.quotes.2 <- get.multiple.quotes(my.tickers.2, from=as.Date("2009-01-01"),
+ to=as.Date("2009-03-31"), interval="m")
```

R provides some useful functions for detecting duplicate values such as the `duplicated` function. This function returns a logical vector showing which elements are duplicates of values with lower indices. Let's apply `duplicated` to the data frame `my.quotes.2`:

```
> duplicated(my.quotes.2)
 [1] FALSE FALSE FALSE FALSE FALSE FALSE FALSE FALSE FALSE FALSE FALSE
[12] FALSE FALSE FALSE FALSE  TRUE  TRUE  TRUE
```

As expected, `duplicated` shows that the last three rows are duplicates of earlier rows. You can use the resulting vector to remove duplicates:

```
> my.quotes.unique <- my.quotes.2[!duplicated(my.quotes.2),]
```

Alternatively, you could use the `unique` function to remove the duplicate values:

```
my.quotes.unique <- unique(my.quotes.2)
```

Sorting

One last set of operations that you might find useful for analysis are sorting and ranking functions.

To sort the elements of an object, use the **sort** function:

```
> w <- c(5,4,7,2,7,1)
> sort(w)
[1] 1 2 4 5 7 7
```

Add the decreasing=TRUE option to sort in reverse order:

```
> sort(w,decreasing=TRUE)
[1] 7 7 5 4 2 1
```

You can control the treatment of NA values by setting the na.last argument:

```
> length(w)
[1] 6
> length(w) <- 7
> # note that by default, NA.last=NA and NA values are not shown
> sort(w)
[1] 1 2 4 5 7 7
> # set NA.last=TRUE to put NA values last
> sort(w,na.last=TRUE)
[1]  1  2  4  5  7  7  7 NA
> # set NA.last=FALSE to put NA values first
> sort(w,na.last=FALSE)
[1] NA  1  2  4  5  7  7
```

Sorting data frames is somewhat nonintuitive. To sort a data frame, you need to create a permutation of the indices from the data frame and use these to fetch the rows of the data frame in the correct order. You can generate an appropriate permutation of the indices using the **order** function:

```
order(..., na.last = , decreasing = )
```

The **order** function takes a set of vectors as arguments. It sorts recursively by each vector, breaking ties by looking at successive vectors in the argument list. At the end, it returns a permutation of the indices of the vector corresponding to the sorted order. (The arguments na.last and decreasing work the same way as they do for sort.) To see what this means, let's use a simple example. First, we'll define a vector with two elements out of order:

```
> v <- c(11, 12, 13, 15, 14)
```

You can see that the first three elements (11, 12, 13) are in order, and the last two (15, 14) are reversed. Let's call **order** to see what it does:

```
> order(v)
[1] 1 2 3 5 4
```

This means "move row 1 to row 1, move row 2 to row 2, move row 3 to row 3, move row 4 to row 5, move row 5 to row 4." We can return a sorted version of v using an indexing operator:

```
> v[order(v)]
[1] 11 12 13 14 15
```

Suppose that we created the following data frame from the vector v and a second vector u:

```
> u <- c("pig","cow","duck","horse","rat")
> w <- data.frame(v, u)
> w
   v     u
1 11   pig
2 12   cow
3 13  duck
4 15 horse
5 14   rat
```

We could sort the data frame w by v using the following expression:

```
> w[order(w$v),]
   v     u
1 11   pig
2 12   cow
3 13  duck
5 14   rat
4 15 horse
```

As another example, let's sort the my.quotes data frame (that we created earlier) by closing price:

```
> my.quotes[order(my.quotes$Close),]
    symbol       Date   Open    High    Low   Close     Volume Adj.Close
2       GE 2009-02-02  12.03   12.90   8.40    8.51  194928800      8.51
1       GE 2009-03-02   8.29   11.35   5.87   10.11  277426300     10.11
11     AXP 2009-02-02  16.35   18.27  11.44   12.06   24297100     11.90
3       GE 2009-01-02  16.51   17.24  11.87   12.13  117846700     11.78
10     AXP 2009-03-02  11.68   15.24   9.71   13.63   31136400     13.45
12     AXP 2009-01-02  18.57   21.38  14.72   16.73   19110000     16.51
15      GS 2009-01-02  84.02   92.20  59.13   80.73   22764300     80.29
8     AAPL 2009-02-02  89.10  103.00  86.51   89.31   27394900     89.31
9     AAPL 2009-01-02  85.88   97.17  78.20   90.13   33487900     90.13
14      GS 2009-02-02  78.78   98.66  78.57   91.08   28301500     91.08
7     AAPL 2009-03-02  88.12  109.98  82.33  105.12   25963400    105.12
13      GS 2009-03-02  87.86  115.65  72.78  106.02   30196400    106.02
5     GOOG 2009-02-02 334.29  381.00 329.55  337.99    6158100    337.99
6     GOOG 2009-01-02 308.60  352.33 282.75  338.53    5727600    338.53
4     GOOG 2009-03-02 333.33  359.16 289.45  348.06    5346800    348.06
```

You could sort by symbol and then by closing price using the following expression:

```
> my.quotes[order(my.quotes$symbol, my.quotes$Close),]
    symbol       Date   Open    High    Low   Close     Volume Adj.Close
2       GE 2009-02-02  12.03   12.90   8.40    8.51  194928800      8.51
1       GE 2009-03-02   8.29   11.35   5.87   10.11  277426300     10.11
3       GE 2009-01-02  16.51   17.24  11.87   12.13  117846700     11.78
5     GOOG 2009-02-02 334.29  381.00 329.55  337.99    6158100    337.99
6     GOOG 2009-01-02 308.60  352.33 282.75  338.53    5727600    338.53
4     GOOG 2009-03-02 333.33  359.16 289.45  348.06    5346800    348.06
8     AAPL 2009-02-02  89.10  103.00  86.51   89.31   27394900     89.31
9     AAPL 2009-01-02  85.88   97.17  78.20   90.13   33487900     90.13
7     AAPL 2009-03-02  88.12  109.98  82.33  105.12   25963400    105.12
11     AXP 2009-02-02  16.35   18.27  11.44   12.06   24297100     11.90
```

```
10    AXP 2009-03-02   11.68   15.24    9.71  13.63  31136400     13.45
12    AXP 2009-01-02   18.57   21.38   14.72  16.73  19110000     16.51
15     GS 2009-01-02   84.02   92.20   59.13  80.73  22764300     80.29
14     GS 2009-02-02   78.78   98.66   78.57  91.08  28301500     91.08
13     GS 2009-03-02   87.86  115.65   72.78 106.02  30196400    106.02
```

Sorting a whole data frame is a little strange. You can create a suitable permutation using the order function, but you need to call order using do.call for it to work properly. (The reason for this is that order expects a list of vectors and interprets the data frame as a single vector, not as a list of vectors.) Let's try sorting the my.quotes table we just created:

```
> # what happens when you call order on my.quotes directly: the data
> # frame is interpreted as a vector
> order(my.quotes)
  [1]   61  94  96  95  31  62  77 107  70  76 106  46  71  40 108  63
 [17]  116  32  86  78  47 115  85  72  55  41  33 117  87  48  56  42
 [33]  102  57 105 101  97  98 104 103 100  99  75  73  69  74  44 120
 [49]   90  67  45  39  68  43  37  38  83 113  84 114  89 119  60  54
 [65]   59  53  82 112  88 118  52  58  93  92  18  21  24  27  30  17
 [81]   20  23  26  29  16  19  22  25  28  91  66  64  36  65  34  35
 [97]   80 110  81 111  79 109  51  49  50   7   8   9  10  11  12   1
[113]    2   3   4   5   6  13  14  15
> # what you get when you use do.call:
> do.call(order,my.quotes)
 [1]  3  2  1  6  5  4  9  8  7 12 11 10 15 14 13
> # now, return the sorted data frame using the permutation:
> my.quotes[do.call(order, my.quotes),]
     symbol       Date   Open   High    Low  Close     Volume Adj.Close
3        GE 2009-01-02  16.51  17.24  11.87  12.13  117846700     11.78
2        GE 2009-02-02  12.03  12.90   8.40   8.51  194928800      8.51
1        GE 2009-03-02   8.29  11.35   5.87  10.11  277426300     10.11
6      GOOG 2009-01-02 308.60 352.33 282.75 338.53    5727600    338.53
5      GOOG 2009-02-02 334.29 381.00 329.55 337.99    6158100    337.99
4      GOOG 2009-03-02 333.33 359.16 289.45 348.06    5346800    348.06
9      AAPL 2009-01-02  85.88  97.17  78.20  90.13   33487900     90.13
8      AAPL 2009-02-02  89.10 103.00  86.51  89.31   27394900     89.31
7      AAPL 2009-03-02  88.12 109.98  82.33 105.12   25963400    105.12
12      AXP 2009-01-02  18.57  21.38  14.72  16.73   19110000     16.51
11      AXP 2009-02-02  16.35  18.27  11.44  12.06   24297100     11.90
10      AXP 2009-03-02  11.68  15.24   9.71  13.63   31136400     13.45
15       GS 2009-01-02  84.02  92.20  59.13  80.73   22764300     80.29
14       GS 2009-02-02  78.78  98.66  78.57  91.08   28301500     91.08
13       GS 2009-03-02  87.86 115.65  72.78 106.02   30196400    106.02
```

14

Graphics

R includes two different packages for plotting data: `graphics` and `lattice`. The `graphics` package contains a wide variety of functions for plotting data. It is easy to customize or modify charts with the `graphics` package, or to interact with plots on the screen. The `lattice` package contains an alternative set of functions for plotting data. Lattice graphics are well suited for splitting data by a conditioning variable. This chapter gives an overview of the `graphics` package. We'll explain how to use lattice graphics in Chapter 15.

An Overview of R Graphics

R includes tools for drawing most common types of charts, including bar charts, pie charts, line charts, and scatter plots. Additionally, R can also draw some less familiar charts like quantile-quantile (Q-Q) plots, mosaic plots, and contour plots. The following table shows many of the charts included in the `graphics` package.

Graphics package function	Description
barplot	Bar and column charts
dotchart	Cleveland dot plots
hist	Histograms
density	Kernel density plots
stripchart	Strip charts
qqnorm (in stats package)	Quantile-quantile plots
xplot	Scatter plots
smoothScatter	Smooth scatter plots
qqplot (in stats package)	Quantile-quantile plots
pairs	Scatter plot matrices
image	Image plots
contour	Contour plots

Graphics package function	Description
persp	Perspective charts of three-dimensional data
interaction.plot	Summary of the response for two-way combinations of factors
sunflowerplot	Sunflower plots

You can show R graphics on the screen or save them in many different formats. "Graphics Devices" on page 243 explains how to choose output methods. R gives you an enormous amount of control over graphics. You can control almost every aspect of a chart. "Customizing Charts" on page 244 explains how to tweak the output of R to look the way you want. This section shows how to use many common types of R charts.

Scatter Plots

To show how to use scatter plots, we will look at cases of cancer in 2008 and toxic waste releases by state in 2006. Data on new cancer cases (and deaths from cancer) are tabulated by the American Cancer Society; information on toxic chemicals released into the environment is tabulated by the U.S. Environmental Protection Agency (EPA).[*]

The sample data is included in the nutshell package:

```
> library(nutshell)
> data(toxins.and.cancer)
```

To show a scatter plot, use the plot function. Plot is a generic function (you can "plot" many different types of objects); plot can draw many types of objects, including vectors, tables, and time series. For simple scatter plots with two vectors, the function that is called is plot.default:

```
plot(x, y = NULL, type = "p",  xlim = NULL, ylim = NULL,
     log = "", main = NULL, sub = NULL, xlab = NULL, ylab = NULL,
     ann = par("ann"), axes = TRUE, frame.plot = axes,
     panel.first = NULL, panel.last = NULL, asp = NA, ...)
```

Here is a description of the arguments to plot.

Argument	Description	Default
x, y	The data to be plotted. You may specify two separate vectors x and y. Otherwise, you may specify a time series, formula, list, or matrix with two or more columns; see the help file for xy.coords for more details.	
type	A character value that specifies the type of plot: type="p" for points, type="l" for lines, type="o" for overplotted points and lines, type="b" for points joined by lines, type="s" for stair steps, type="h" for	"p"

---

[*] Data from both can be found in the *Statistical Abstract of the United States*, available online at *http://www.census.gov/compendia/statab/*.

Argument	Description	Default
	histogram-style vertical lines, or `type="n"` for no points or lines.	
xlim	A numeric vector with two values specifying the x limits of the plot.	NULL
log	A character value that specifies which axes should be plotted with a logarithmic scale. Use `log=""` for neither, `log="x"` for the x-axis, `log="y"` for the y-axis, and `log="xy"` for both.	""
main	The main title for the plot.	NULL
sub	The subtitle for the plot.	NULL
xlab	The label for the x-axis.	NULL
ylab	The label for the y-axis.	NULL
ann	If `ann=TRUE`, then axis titles and overall titles are included with plots. If `ann=FALSE`, these annotations are not included.	par("ann")
axes	A logical value that specifies whether axes should be drawn.	TRUE
frame.plot	A logical value that specifies whether a box should be drawn around the plot.	axes
panel.first	An expression that is evaluated after the axes are drawn but before points are plotted.	NULL
panel.last	An expression that is evaluated after the points are plotted.	NULL
asp	A numeric value that specifies the aspect ratio of the plot (as y/x).	NA
...	Additional graphical parameters. (See "Graphical Parameters" on page 244 for more information.)	

Now, let's try our first plot. Let's compare the overall cancer rate (number of cancer deaths divided by state population) to the presence of toxins (total toxic chemicals release divided by state area):

```
> # use attach so that we don't have to keep typing the
> # data frame name
> attach(toxins.and.cancer)
> plot(total_toxic_chemicals/Surface_Area,deaths_total/Population)
```

The chart is shown in Figure 14-1. Perhaps there is a stronger correlation between airborne toxins and lung cancer:

```
> plot(air_on_site/Surface_Area,deaths_lung/Population)
```

This chart is shown in Figure 14-2. Suppose that you wanted to know which states were associated with which points. R provides some interactive tools for identifying points on plots. You can use the `locator` function to tell you the coordinates of a specific point (or set of points). To do this, first plot the data. Next, type `locator(1)`. Then click on a point in the open graphics window. As an example, suppose that you plotted the data above, typed `locator(1)`, and then clicked on the point in the upper-right corner. You would see output like this in the R console:

```
> locator(1)
$x
[1] 0.002499427

$y
[1] 0.0008182696
```

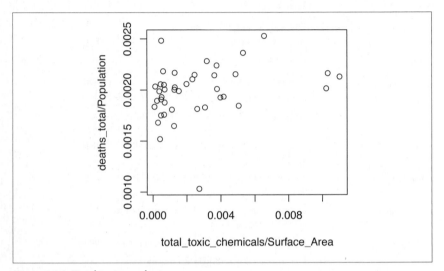

Figure 14-1. Total toxins and new cancer cases

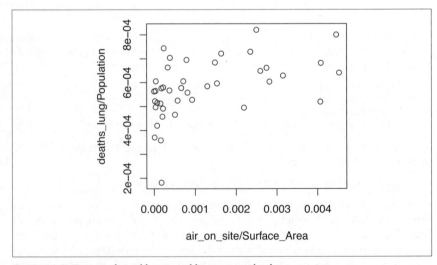

Figure 14-2. Toxins released by air and lung cancer deaths per capita

Another useful function for identifying points is `identify`. This function can be used to interactively label points on a plot. To use `identify` with the data above, you would enter:

```
> identify(air_on_site/Surface_Area, deaths_lung/Population,
+   State_Abbrev)
```

While this command is running, you can click on individual points on the chart, and R will label those points with state names.

If you wanted to label all of the points at once, you could use the `text` function to add labels to the plot. Here is how I drew the plot shown in Figure 14-3:

```
> plot(air_on_site/Surface_Area, deaths_lung/Population,
+   xlab="Air Release Rate of Toxic Chemicals",
+   ylab="Lung Cancer Death Rate")
> text(air_on_site/Surface_Area, deaths_lung/Population,
+   labels=State_Abbrev,
+   cex=0.5,
+   adj=c(0,-1))
```

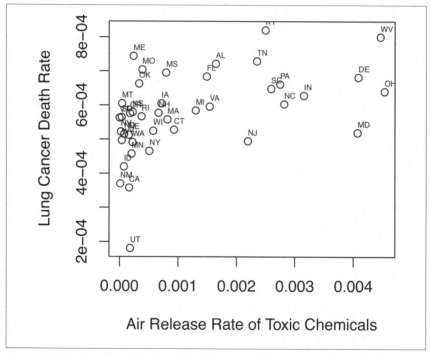

Figure 14-3. Toxins released by air and lung cancer deaths per capita, cleaned up

Notice that I have added some extra arguments to refine the appearance of the plot. The `xlab` and `ylab` arguments are used to add labels to the x- and y-axes, respectively. The `text` function draws a label next to each point. I tweaked the size placement of the labels using the `cex` and `adj` arguments; see "Graphical Parameters" on page 244 for more information.

Is this relationship significant? It is actually statistically significant (see "Correlation tests" on page 356), but we don't have enough information to make a good argument that there is a causal relationship.

The `plot` function is a good choice if you only want to plot two columns of data on one chart. However, suppose that you have more columns of data to plot, perhaps split into different categories. Or, suppose that you want to plot all the columns of one matrix against all the columns of another matrix. To plot multiple sets of columns against each other, you can use the `matplot` function:

```
matplot(x, y, type = "p", lty = 1:5, lwd = 1, pch = NULL,
        col = 1:6, cex = NULL, bg = NA,
        xlab = NULL, ylab = NULL, xlim = NULL, ylim = NULL,
        ..., add = FALSE, verbose = getOption("verbose"))
```

Matplot accepts the following arguments.

Argument	Description	Default
x, y	Vectors or matrices containing the data to be plotted. The number of rows and columns should match.	If x is not specified, then x=1:ncol(y). If y is not specified, then y=x; x=1:ncol(y).
type	A character vector specifying the types of plots to generate. Use type="p" for points, type="l" for lines, type="b" for both, type="c" for the lines part alone of "b", type="o" for both overplotted points and lines, type="h" for histogram-like (or high-density) vertical lines, type="s" for stair steps, type="S" for other steps, or type="n" for no plotting.	"p"
lty	A vector of line types. See "Graphical parameter by name" on page 250 for more details.	1:5
lwd	A vector of line widths. See "Graphical parameter by name" on page 250 for more details.	1
pch	A vector of plotting characters. See "Graphical parameter by name" on page 250 for more details.	NULL
col	A vector of colors. See "Graphical parameter by name" on page 250 for more details.	1:6
cex	A vector of character expansion sizes. See "Graphical parameter by name" on page 250 for more details.	NULL
bg	A vector of background colors for plot symbols. See "Graphical parameter by name" on page 250 for more details.	NA
xlab, ylab	Character values specifying x- and y-axis labels.	NULL
xlim, ylim	Numeric values specifying ranges for the x- and y-axes.	NULL
...	Additional graphical parameters that are passed to par.	NULL
add	A logical value indicating whether to add to the current plot or to generate a new plot.	FALSE
verbose	A logical value indicating whether to write information to the console describing what matplot did.	getOption("verbose")

Many arguments to `matplot` have the same names as standard arguments to `par`. However, because `matplot` generates multiple plots at the same time, these arguments can be specified as vectors of multiple values when called by `matplot`. For more details on standard arguments, see "Graphical Parameters" on page 244.

If you are plotting a very large number of points, then you may prefer the function `smoothScatter`, which plots the density of points by shading different regions of the plot different shades, depending on the density of points in each region:

```
smoothScatter(x, y = NULL, nbin = 128, bandwidth,
              colramp = colorRampPalette(c("white", blues9)),
              nrpoints = 100, pch = ".", cex = 1, col = "black",
              transformation = function(x) x^.25,
              postPlotHook = box,
              xlab = NULL, ylab = NULL, xlim, ylim,
              xaxs = par("xaxs"), yaxs = par("yaxs"), ...)
```

For an example of `smoothScatter`, see "Correlation and Covariance" on page 325.

If you have a data frame with *n* different variables and you would like to generate a scatter plot for each pair of values in the data frame, try the `pairs` function. As an example, let's plot the hits, runs, strikeouts, walks, and home runs for each Major League Baseball (MLB) player who had more than 100 at bats in 2008. To do this, we would use the following R statement:

```
> library(nutshell)
> data(batting.2008)
> pairs(batting.2008[batting.2008$AB>100,c("H","R","SO","BB","HR")])
```

The result is shown in Figure 14-4.

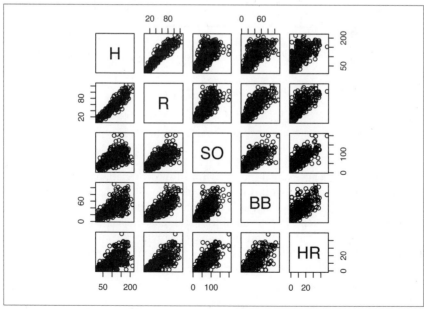

Figure 14-4. Pairs example

Plotting Time Series

R includes tools for plotting time series data. The `plot` function has a method for time series:

```
plot(x, y = NULL, plot.type = c("multiple", "single"),
        xy.labels, xy.lines, panel = lines, nc, yax.flip = FALSE,
        mar.multi = c(0, 5.1, 0, if(yax.flip) 5.1 else 2.1),
        oma.multi = c(6, 0, 5, 0), axes = TRUE, ...)
```

The arguments x and y specify `ts` objects, `panel` specifies how to plot the time series (by default, lines), and other arguments specify how to break time series into different plots (as in lattice). As an example, we'll plot the turkey price data:

```
> library(nutshell)
> data(turkey.price.ts)
> plot(turkey.price.ts)
```

The results are shown in Figure 14-5. As you can see, turkey prices are very seasonal. There are huge sales in November and December (for Thanksgiving and Christmas) and minor sales in spring (probably for Easter).

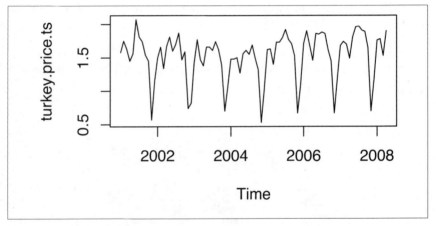

Figure 14-5. Time series plot

Another way to look at seasonal effects is with an autocorrelation plot (which is also called a `correlogram`; see Figure 14-6). This plot shows how correlated points are with each other, by difference in time. You can also plot the autocorrelation function for a time series (which can be helpful for looking at cyclical effects). The plot is generated by default when you call `acf`, which computes the autocorrelation function. (Alternatively, you can generate the autocorrelation function with `acf` and then plot it separately.) Here is how to generate the plot for the turkey price data:

```
> acf(turkey.price.ts)
```

As you can see, points are correlated over 12-month cycles (and inversely correlated over 6-month cycles). Time series analysis is discussed further in Chapter 23.

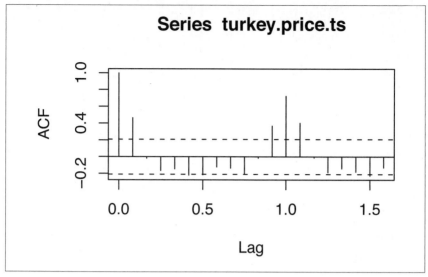

Figure 14-6. Autocorrelation function plot

Bar Charts

To draw bar (or column) charts in R, use the `barplot` function.

As an example, let's look at doctoral degrees awarded in the United States between 2001 and 2006:[†]

```
> doctorates <- data.frame (
+    year=c(2001,2002,2003,2004,2005,2006),
+    engineering=c(5323,5511,5079,5280,5777,6425),
+    science=c(20643,20017,19529,20001,20498,21564),
+    education=c(6436,6349,6503,6643,6635,6226),
+    health=c(1591,1541,1654,1633,1720,1785),
+    humanities=c(5213,5178,5051,5020,5013,4949),
+    other=c(2159,2141,2209,2180,2480,2436)
+ )
```

Or, if you prefer, you can just load the data from the `nutshell` package:

```
> library(nutshell)
> data(doctorates)
```

Now, let's transform this into a matrix for plotting:

```
> # make this into a matrix:
> doctorates.m <- as.matrix(doctorates[2:7])
> rownames(doctorates.m) <- doctorates[,1]
> doctorates.m
          engineering science education health humanities other
```

---

† As with many other examples in this book, this data was taken from the *Statistical Abstract of the United States, 2009*. This data comes from here: *http://www.census.gov/compendia/statab/tables/09s0785.xls.*

2001	5323	20643	6436	1591	5213	2159
2002	5511	20017	6349	1541	5178	2141
2003	5079	19529	6503	1654	5051	2209
2004	5280	20001	6643	1633	5020	2180
2005	5777	20498	6635	1720	5013	2480
2006	6425	21564	6226	1785	4949	2436

The `barplot` function can't work with a data frame, so we've created a matrix object for this problem with the data.

Let's start by just showing a bar plot of doctorates in 2001 by type:

```
> barplot(doctorates.m[1,])
```

As you can see from Figure 14-7, by default R shows the *y*-axis along with the size of each bar, but it does not show the *x*-axis. R also automatically uses column names to name the bars. Suppose that we wanted to show all of the different years as bars stacked next to each other. Suppose that we also wanted the bars plotted horizontally and wanted to show a legend for the different years. To do this, we could use the following expression to generate the chart shown in Figure 14-8:

```
> barplot(doctorates.m,beside=TRUE,horiz=TRUE,legend=TRUE,cex.names=.75)
```

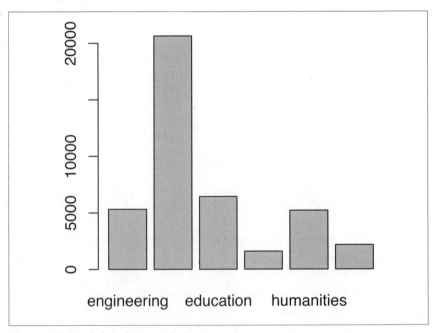

Figure 14-7. Simple bar plot example

Finally, suppose that we wanted to show doctorates by year as stacked bars. To do this, we need to transform the matrix so that each column is a year and each row is a discipline. We also need to make sure that there is enough room to see the legend, so we'll extend the limits on the *y*-axis:

```
> barplot(t(doctorates.m),legend=TRUE,ylim=c(0,66000))
```

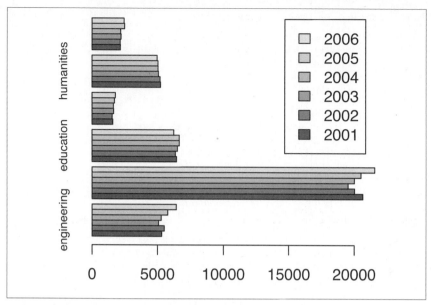

Figure 14-8. Horizontal juxtaposed bar plot example

The chart generated by this expression is shown in Figure 14-9.

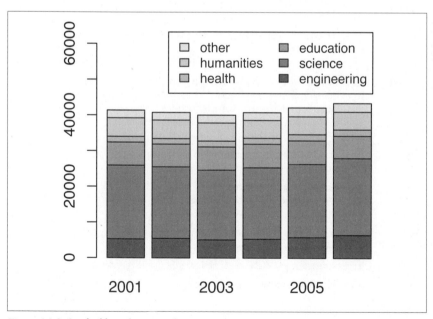

Figure 14-9. Stacked bar plot example

Here is a detailed description of `barplot`:

```
barplot(height, width = 1, space = NULL,
        names.arg = NULL, legend.text = NULL, beside = FALSE,
        horiz = FALSE, density = NULL, angle = 45,
        col = NULL, border = par("fg"),
        main = NULL, sub = NULL, xlab = NULL, ylab = NULL,
        xlim = NULL, ylim = NULL, xpd = TRUE, log = "",
        axes = TRUE, axisnames = TRUE,
        cex.axis = par("cex.axis"), cex.names = par("cex.axis"),
        inside = TRUE, plot = TRUE, axis.lty = 0, offset = 0,
        add = FALSE, args.legend = NULL, ...)
```

The `barplot` function is very flexible; here is a description of the arguments to barplot.

Argument	Description	Default
height	Either a numeric vector or a numeric matrix representing the values to be plotted. If values are given as a matrix and beside=FALSE, then the bars are stacked. If beside=TRUE, then the bars are plotted next to each other.	
width	A numeric vector representing the widths of the bars.	1
space	If beside=FALSE, a numeric value indicating the amount of space between bars. You specify the space as a fraction of the average column width. If beside=TRUE, then you can specify a two-element vector, where the first element specifies the space within a group and the second represents the space between groups.	if (is.matrix(height) & beside=TRUE) c(0, 1) else 0.2
names.arg	A character vector specifying the names to be plotted for each bar (or group of bars).	if (is.matrix(height)) col.names(height) else names(height)
legend.text	A character vector or a logical value. If a logical value is given, then a legend is generated using the row names of height. If a character vector is given, then those character values are used instead. This function is mostly useful when height is a matrix (and there are two different dimensions that need labels).	NULL
beside	A logical value indicating whether columns should be stacked or drawn beside each other. Only meaningful when height is a matrix.	FALSE
horiz	A logical value specifying the direction to draw the bars. If horiz=FALSE, bars are drawn vertically from left to right. If horiz=TRUE, bars are drawn horizontally from bottom to top.	FALSE
density	A numeric value that specifies the density of shading lines in lines per inch. density=NULL means that no lines are drawn.	NULL
angle	A numeric value that specifies the slope of the shading lines (in degrees).	45
col	A vector of colors to use for the bars (or bar components).	Gray is used if height is a vector; a gamma-corrected gray palette if height is a matrix.
border	The color to be used for the border of the bars.	par("fg")
main	A character value to be used as the overall title.	NULL

Argument	Description	Default
sub	A character value to to be used as a subtitle.	NULL
xlab	A character value to use as the label for the *x*-axis.	NULL
ylab	A character value to use as the label for the *y*-axis.	NULL
xlim	Limits for the *x*-axis.	NULL
ylim	Limits for the *y*-axis.	NULL
xpd	A logical value indicating if bars should be allowed to go outside the region.	TRUE
log	A character value specifying whether axis scales should be logarithmic.	""
axes	A logical value indicating whether axes should be drawn.	TRUE
axisnames	If `axisnames=TRUE` and `names.arg` is not null, the second axis is drawn and labeled.	TRUE
cex.axis	A numeric value specifying the size of numeric axis labels relative to other text (`cex.axis=1` means full size).	par("cex.axis")
cex.names	A numeric value specifying the size of axis names relative to other text (`cex.axis=1` means full size).	par("cex.axis")
inside	A logical value that indicates whether to draw lines separating bars when `beside=TRUE`.	TRUE
plot	A logical value indicating whether to plot the chart.	TRUE
axis.lty	The line type for the axis.	0
offset	A numeric vector indicating how much bars should be shifted relative to the *x*-axis.	0
add	A logical value indicating if bars should be added to an existing plot.	FALSE
args.legend	A list of arguments to pass to legend (if `legend.text` is used).	NULL
...	Additional arguments passed to other graphical routines used inside `barplot` (typically, arguments to `par`).	

Pie Charts

One of the most popular ways to plot data is the pie chart. Pie charts can be an effective way to compare different parts of a quantity, though there are lots of good reasons not to use pie charts.[‡] You can draw pie charts in R using the `pie` function:

```
pie(x, labels = names(x), edges = 200, radius = 0.8,
    clockwise = FALSE, init.angle = if(clockwise) 90 else 0,
    density = NULL, angle = 45, col = NULL, border = NULL,
    lty = NULL, main = NULL, ...)
```

[‡] A lot of people dislike pie charts. I think that they are good for saying "look how much bigger this number is than this number," and they are very good at taking up lots of space on a page. Pie charts are not good at showing subtle differences between the size of different slices; search for "why pie charts are bad" on Google, and you'll come up with dozens of sites explaining what's wrong with them. Or, just check the help file for `pie`, which says, "Pie charts are a very bad way of displaying information. The eye is good at judging linear measures and bad at judging relative areas. A bar chart or dot chart is a preferable way of displaying this type of data."

Here is a description of the arguments to `pie`.

Argument	Description	Default
x	A vector of nonnegative numeric values that will be plotted.	
labels	An expression to generate labels, a vector of character strings, or another object that can be coerced to a `graphicsAnnot` object and used as labels.	names(x)
edges	A numeric value indicating the number of segments used to draw the outside of the pie.	200
radius	A numeric value that specifies how big the pie should be. (Parts of the pie are cut off for values over 1.)	0.8
clockwise	A logical value indicating whether slices are drawn clockwise or counterclockwise.	FALSE
init.angle	A numeric value specifying the starting angle for the slices (in degrees).	if (clockwise) 90 else 0
density	A numeric value that specifies the density of shading lines in lines per inch. `density=NULL` means that no lines are drawn.	NULL
angle	A numeric value that specifies the slope of the shading lines (in degrees).	45
col	A numeric vector that specifies the colors to be used for slices. If `col=NULL`, then a set of six pastel colors is used.	NULL
border	Arguments passed to the polygon function to draw each slice.	NULL
lty	The line type used to draw each slice.	NULL
main	A character string that represents the title.	NULL

As a simple example, let's use pie charts to show what happened to fish caught in the United States in 2006:

```
> # 2006 fishery data from
> #   http://www.census.gov/compendia/statab/tables/09s0852.xls
> # units are millions of pounds of live fish
> domestic.catch.2006 <- c(7752,1166,463,108)
> names(domestic.catch.2006) <- c("Fresh and frozen",
+    "Reduced to meal, oil, etc.","Canned","Cured")
> # note: cex.6 setting shrinks text size by 40% so you can see the labels
> pie(acres.harvested, init.angle=100, cex=.6)
```

As shown in Figure 14-10, most of the fish (by weight) was sold fresh or frozen.

Plotting Categorical Data

The `graphics` package includes some very useful, and possibly unfamiliar, tools for looking at categorical data.

Suppose that you want to look at the *conditional density* of a set of categories dependent on a numeric value. You can do this with a conditional density plot, generated by the `cdplot` function:

```
cdplot(x, y,
    plot = TRUE, tol.ylab = 0.05, ylevels = NULL,
    bw = "nrd0", n = 512, from = NULL, to = NULL,
```

```
col = NULL, border = 1, main = "", xlab = NULL, ylab = NULL,
yaxlabels = NULL, xlim = NULL, ylim = c(0, 1), ...)
```

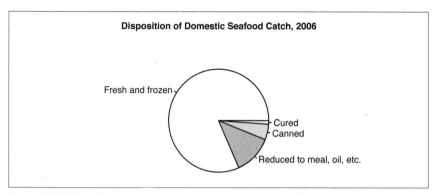

Figure 14-10. Pie chart

Here is the form of `cdplot` when called with a formula:

```
cdplot(formula, data = list(),
    plot = TRUE, tol.ylab = 0.05, ylevels = NULL,
    bw = "nrd0", n = 512, from = NULL, to = NULL,
    col = NULL, border = 1, main = "", xlab = NULL, ylab = NULL,
    yaxlabels = NULL, xlim = NULL, ylim = c(0, 1), ...,
    subset = NULL)
```

The `cdplot` function uses the `density` function to compute kernel density estimates across the range of numeric values and then plots these estimates. Here is the list of arguments to `cdplot`.

Argument	Description	Default
x, y, formula, data	Arguments used to specify the data to plot. You may specify either a numeric vector x containing data to plot and a factor vector y containing grouping information or a `formula` and a data frame (`data`) in which to evaluate the formula.	
subset	A vector specifying the subset of values to be used when plotting. (Only applies when using a formula and a data frame.)	NULL
plot	Logical value specifying whether the conditional densities should be plotted.	TRUE
tol.ylab	A numeric vector that specifies a "tolerance parameter" for y-axis labels. If the difference between two labels is less than this parameter, they are plotted equidistantly.	0.05
ylevels	A character or numeric vector that specifies the order in which levels should be plotted.	NULL
bw	The "smoothing bandwidth" to use when plotting. See the help file for `density` for more details.	"nrd0"
n	A numeric value specifying the number of points at which the density is estimated.	512
from	A numeric value specifying the lowest point at which the density is estimated.	NULL
to	A numeric value specifying the highest point at which the density is estimated.	NULL
col	A vector of fill colors for the different conditional values.	NULL
border	Border color for shaded polygons.	1

Argument	Description	Default
main	Main title.	""
xlab	x-axis label.	NULL
ylab	y-axis label.	NULL
yaxlabels	Character vector for labeling different conditional variables.	NULL
xlim	Range of x variables to plot.	NULL
ylim	Range of y variables to plot.	c(0, 1)
...	Other arguments passed to density.	

As an example, let's look at how the distribution of batting hand varies by batting average among MLB players in 2008:

```
> batting.w.names.2008 <- transform(batting.w.names.2008,
+     bat=as.factor(bats), throws=as.factor(throws))
> cdplot(bats~AVG,data=batting.w.names.2008,
+     subset=(batting.w.names.2008$AB>100))
```

The results are shown in Figure 14-11. As you can see, the proportion of switch hitters (bats=="B") increases with higher batting average.

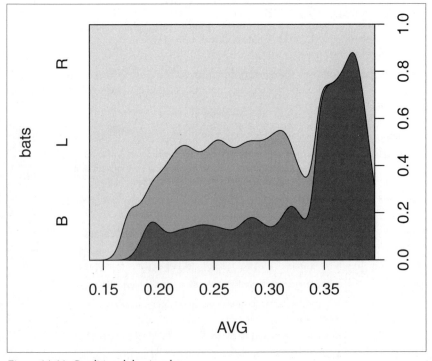

Figure 14-11. Conditional density plot

Suppose, instead, that you simply wanted to plot the proportion of observations for two different categorical variables. R also provides tools for visualizing this type of data. One of the most interesting charts available in R for showing the number of observations with certain properties is the mosaic plot. A mosaic plot shows a set of boxes corresponding to different factor values. The *x*-axis corresponds to one factor, and the *y*-axis to another factor. To create a mosaic plot, use the `mosaicplot` function. Here is the form of the `mosaicplot` function for a contingency table:

```
mosaicplot(x, main = deparse(substitute(x)),
           sub = NULL, xlab = NULL, ylab = NULL,
           sort = NULL, off = NULL, dir = NULL,
           color = NULL, shade = FALSE, margin = NULL,
           cex.axis = 0.66, las = par("las"),
           type = c("pearson", "deviance", "FT"), ...)
```

There is also a method for `mosaicplot` that allows you to specify the data as a formula and data frame:

```
mosaicplot(formula, data = NULL, ...,
           main = deparse(substitute(data)), subset,
           na.action = stats::na.omit)
```

Here is a description of the arguments to `mosaicplot`.

Argument	Description	Default
x, formula, data	Specifies the data to be plotted. You may specify either a contingency table x or a formula and a data frame (data). (If the variables in formula are defined in the current environment, you may omit data.)	
subset	A vector that specifies which values in data to plot.	
main	A character value specifying the main title for the plot.	deparse(substitute(x))
sub	A character value specifying the subtitle for the plot.	NULL
xlab	A character value specifying the label for the *x*-axis.	NULL
ylab	A character value specifying the label for the *y*-axis.	NULL
sort	An integer vector that describes how to sort the variables in x. Specified as a permutation of 1:length(dim(x)).	NULL
off	A numeric vector that specifies the spacing between each level of the mosaic as a percentage.	NULL
dir	A character vector that specifies which direction to plot each vector in x. Use "v" for vertical and "h" for horizontal.	NULL
color	A logical value or character vector specifying colors to use for color shading. You may use color=TRUE for a gamma-corrected color palette, color=NULL for grayscale, or color=FALSE for unfilled boxes.	NULL
shade	A logical value (or numeric vector) specifying whether to produce "extended mosaic plots" to visualize standardized residuals of a log-linear model for the table by color and outline of the mosaic's tiles. You may specify shade=FALSE for standard plots, shade=TRUE for extended plots, or a numeric vector with up to five elements specifying cut points of the residuals.	FALSE
margin	A list of vectors containing marginal totals to fit in a log-linear model. See the help file for loglin for more information.	NULL

Argument	Description	Default
cex.axis	A numeric value specifying the magnification factor to use for axis annotation text.	0.66
las	Specifies the style of the axis labels.	par("las")
type	A character string indicating the type of residuals to plot. Use `type="pearson"` for components of Pearson's chi-squared, `type="deviance"` for components of the likelihood ratio chi-squared, or `type="FT"` for the Freeman-Tukey residuals.	c("pearson", "deviance", "FT")
na.action	A function that specifies what mosaicplot should do if the data contains variables to be cross-tabulated that contain NA values.	A function that omits NA values (specifically, stats::na.omit)
...	Additional graphical parameters passed to other methods.	

As an example, let's create a mosaic plot showing the number of batters in MLB in 2008. On the *x*-axis, we'll show batting hand (left, right, or both), and on the *y*-axis we'll show throwing hand (left or right). This function can accept either a matrix of values or a formula and a data frame. In this example, we'll use a formula and a data frame. The plot is shown in Figure 14-12:

```
> png("~/Documents/book/current/figs/incoming/rian_1413.pdf",
+       width=4.3,height=4.3,units="in",res=72)
> mosaicplot(formula=bats~throws,data=batting.w.names.2008,color=TRUE)
> dev.off()
```

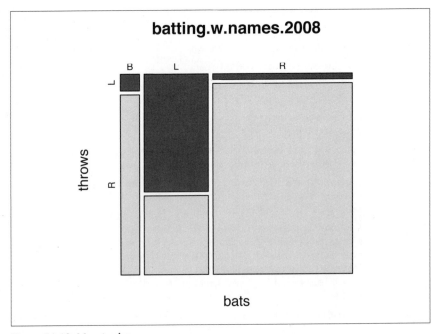

Figure 14-12. Mosaic plot

Another chart that is very similar to a mosaic plot is a spine plot. A spine plot shows different boxes corresponding to the number of observations associated with two factors. Figure 14-13 shows an example of a spine plot using the same batting data we used in the mosaic example:

```
> spineplot(formula=bats~throws,data=batting.w.names.2008)
```

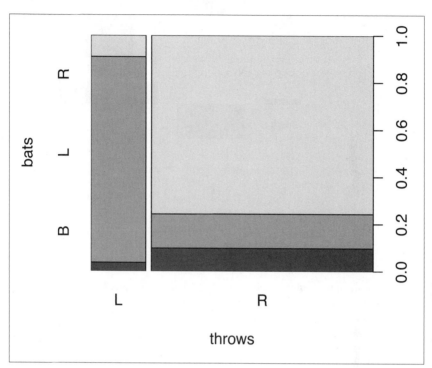

Figure 14-13. Spine plot

Another function for looking at tables of data is `assocplot`. This function plots a set of bar charts, showing the deviation of each combination of factors from independence. (These are also called *Cohen-Friendly association plots*.) As an example, let's look at the same data for batting and throwing hands:

```
> assocplot(table(batting.w.names.2008$bats,batting.w.names.2008$throws),
+     xlab="Throws",ylab="Bats")
```

The resulting plot is shown in Figure 14-14. Other useful plotting functions include `stars` and `fourfoldplot`. See the help files for more information.

Three-Dimensional Data

R includes a few functions for visualizing three-dimensional data. All of these functions can be used to plot a matrix of values. (Row indices correspond to x values, column indices to y values, and values in the matrix to z values.)

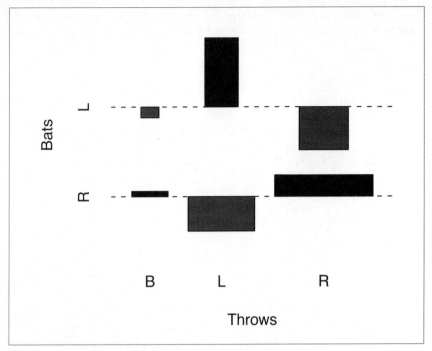

Figure 14-14. Association plot example

As an example of multidimensional data, I used elevation data for Yosemite Valley in Yosemite National Park (you can find a map at *http://www.nps.gov/yose/planyour visit/upload/yosevalley2008.pdf*). The sample data I used for my examples is included in the nutshell library.

Getting Elevation Data

I downloaded the Yosemite data from the U.S. Geological Survey. Specifically, I used the National Map Seamless Server (available at *http://seamless.usgs.gov/web site/seamless/viewer.htm*). This service allows you to search for a specific location and select a region from which to obtain elevation data. After you select the area that you want to export, a window will pop up called the "Request Summary Page." There will be a link on this page to "Modify Data Request." Click this link to modify the defaults, choose to export the data in GridFloat format, save the options, and download the file. The name of the file that I downloaded was *NED_09216343.zip*, though the name of your file will be different.

Unzip the downloaded file. There are many different files inside the archive, including a lot of information about the request. The most important files are the *.hdr* file (which contains information you need to load the data) and the *.flt* file (which contains the data). Here is what was contained in the *ned_09216343.hdr* file that I downloaded:

```
ncols       562
nrows       253
```

```
xllcorner      -119.68111111082
yllcorner      37.699166665986
cellsize       0.00027777777779647
NODATA_value   -9999
byteorder      LSBFIRST
```

The GridFloat format saves the topological data as a stream of 4-byte floating-point values. You can load this into R using the **readBin** function. As noted in this file, there were 562 * 253 four-byte values encoded in little-endian format. So, I loaded the data with the following statement:

```
> yosemite <- readBin(
+   "~/Documents/book/data/NED_09216343/ned_09216343.flt",
+   what="numeric", n=562*253, size=4, endian="little")
```

I then assigned dimensions with this statement:

```
> dim(yosemite) <- c(562,253)
```

Feel free to grab your own data samples for experimentation.

To view a three-dimensional surface, use the **persp** function. This function draws a plot of a three-dimensional surface for a specific perspective. (It does, of course, only draw in two dimensions.) If you want to show your nonstatistician friends that you are doing really cool math stuff with R, this is the function that draws the coolest plots:

```
persp(x = seq(0, 1, length.out = nrow(z)),
      y = seq(0, 1, length.out = ncol(z)),
      z, xlim = range(x), ylim = range(y),
      zlim = range(z, na.rm = TRUE),
      xlab = NULL, ylab = NULL, zlab = NULL,
      main = NULL, sub = NULL,
      theta = 0, phi = 15, r = sqrt(3), d = 1,
      scale = TRUE, expand = 1,
      col = "white", border = NULL, ltheta = -135, lphi = 0,
      shade = NA, box = TRUE, axes = TRUE, nticks = 5,
      ticktype = "simple", ...)
```

Here is a description of the values to **persp**.

Argument	Description	Default
x, y	Numeric vectors that explain what each dimension of z represents. (Specifically, x is a numeric vector representing the x values for each row in z, and y is a numeric vector representing the y values for each column in z.)	x = seq(0, 1, length.out = nrow(z)) sy = seq(0, 1, length.out = ncol(z))
z	A matrix of values to plot.	
xlim, ylim, zlim	Numeric vectors with two values, representing the range of values to plot for x, y, and z, respectively.	xlim = range(x) ylim = range(y), zlim = range(z, na.rm = TRUE)

Argument	Description	Default
xlab, ylab, zlab	Character values specifying titles to plot for the x-, y-, and z-axes.	NULL
main	A character value specifying the main title for the plot.	NULL
sub	A character value specifying the subtitle for the plot.	NULL
theta	A numeric value that specifies the azimuthal direction of the viewing angle.	0
phi	A numeric value that specifies the colatitude of the viewing angle.	15
r	The distance of the viewing point from the center of the plotting box.	sqrt(3)
d	A numeric value that can be used to increase or decrease the perspective effect.	1
scale	A logical value specifying whether to maintain aspect ratios when plotting.	TRUE
expand	A numeric factor used to expand (when $z > 1$) or shrink (when $z < 1$) the z coordinates.	1
col	The color of the surface facets.	"white"
border	The color of the lines drawn around the surface facets.	NULL
ltheta	If specified, the surface is drawn as if illuminated from the direction specified by azimuth `ltheta` and colatitude `lphi`.	-135
lphi	See the explanation for `ltheta`.	0
shade	An exponent used to calculate the shade of the surface facets. See the help file for more information.	NA
box	A logical value indicating whether a bounding box for the surface should be drawn.	TRUE
axes	A logical value indicating whether axes should be drawn.	TRUE
nticks	A numeric value specifying the number of ticks to draw on each axis.	5
ticktype	A character value specifying the types of ticks drawn here. Use `ticktype="simple"` for arrows pointing in the direction of increase, `ticktype="detailed"` to show simple tick marks.	"simple"
...	Additional graphical parameters. See "Graphical Parameters" on page 244.	

As an example of three-dimensional data, let's take a look at Yosemite Valley. Specifically, let's look toward Half Dome. To plot this elevation data, I needed to make two transformations. First, I needed to flip the data horizontally. In the data file, values move east to west (or left to right) as x indices increase and from north to south (or top to bottom) as y indices increase. Unfortunately, persp plots y coordinates slightly differently. Persp plots increasing y coordinates from bottom to top. So, I selected y indices in reverse order. Here is an R expression to do this:

```
> # load the data:
> library(nutshell)
> data(yosemite)
> # check dimensions of data
> dim(yosemite)
[1] 562 253
> # select all 253 rows in reverse order
> yosemite.flipped <- yosemite[,seq(from=253,to=1)]
```

Next, I wanted to select only a square subset of the elevation points. To do this, I selected only the rightmost 253 rows of the yosemite matrix using an expression like this:

```
yosemite.rightmost <- yosemite[nrow(yosemite) - ncol(yosemite) + 1,]
```

Note the "+ 1" in this statement; that's to make sure that we take exactly 253 rows. (This is to avoid a fencepost error.)

To plot the figure, I rotated the image by 225° (through theta=225) and changed the viewing angle to 20° (phi=20). I adjusted the light source to be from a 45° angle (ltheta=45) and set the shading factor to 0.75 (shade=.75) to exaggerate topological features. Putting it all together, here is the code I used to plot Yosemite Valley looking toward Half Dome:

```
> # create halfdome subset in one expression:
> halfdome <- yosemite[(nrow(yosemite) - ncol(yosemite) + 1):562,
+                      seq(from=253,to=1)]
> persp(halfdome,col=grey(.25),border=NA,expand=.15,
+   theta=225, phi=20, ltheta=45,lphi=20,shade=.75)
```

The resulting image is shown in Figure 14-15.

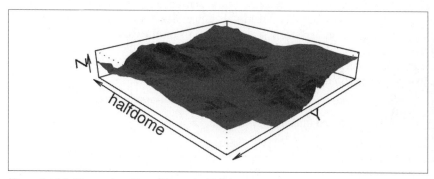

Figure 14-15. Perspective view of Yosemite Valley

Another useful function for plotting three-dimensional data is image. This function plots a matrix of data points as a grid of boxes, color coding the boxes based on the intensity at each location:

```
image(x, y, z, zlim, xlim, ylim, col = heat.colors(12),
      add = FALSE, xaxs = "i", yaxs = "i", xlab, ylab,
      breaks, oldstyle = FALSE, ...)
```

Here is a description of the arguments to image.

Argument	Description	Default
x, y	(Alternatively, you may pass x an argument that is a list containing elements named x, y, and z.)	
z	A matrix of values to plot.	

Argument	Description	Default
xlim, ylim	Two-element numeric vectors that specify the range of values in x and y (respectively) that should be plotted.	
zlim	The range of values for z for which colors should be plotted.	
col	A vector of colors to plot. Typically generated by functions like `rainbow`, `heat.col ors`, `topo.colors`, or `terrain.colors`.	heat.colors(12)
add	A logical value that specifies whether the plot should be added to the existing plot.	FALSE
xaxs, yaxs	Style for the x- and y-axes; see "Graphical parameter by name" on page 250.	xlab="i", ylab="i"
xlab, ylab	Labels for the x and y values.	
breaks	An integer value specifying the number of break points for colors. (There must be at least one more color than break point.)	
oldstyle	If `oldstyle=TRUE`, then the midpoints of the color intervals are equally spaced between the limits. If `oldstyle=FALSE`, then the range is split into color intervals of equal size.	FALSE
...	Additional arguments to par.	

To plot the Yosemite Valley data using `image`, I needed to make several tweaks. First, I needed to specify an aspect ratio that matched the dimensions of the data (`asp=253/562`). Then I specified a range of points on the y dimension to make sure that data was plotted from top to bottom (`y=c(1,0)`). Finally, I specified a set of 32 grayscale colors for this plot (`col=sapply((0:32)/32,gray)`). Here is an expression that generates an image plot from the Yosemite Valley data:

```
> image(yosemite, asp=253/562, ylim=c(1,0), col=sapply((0:32)/32,gray))
```

The results are shown in Figure 14-16.

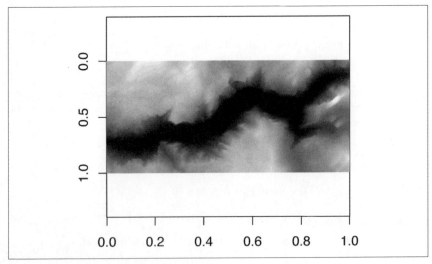

Figure 14-16. Image example: Yosemite Valley

A closely related tool for looking at multidimensional data, particularly in biology, is the heat map. A heat map plots a single variable on two axes, each representing a different factor. The `heatmap` function plots a grid, where each box is encoded with a different color depending on the size of the dependent variable. It may also plot a tree structure (called a *dendrogram*) to the side of each plot showing the hierarchy of values. As you might have guessed, the function for plotting heat maps in R is `heatmap`:

```
heatmap(x, Rowv=NULL, Colv=if(symm)"Rowv" else NULL,
        distfun = dist, hclustfun = hclust,
        reorderfun = function(d,w) reorder(d,w),
        add.expr, symm = FALSE, revC = identical(Colv, "Rowv"),
        scale=c("row", "column", "none"), na.rm = TRUE,
        margins = c(5, 5), ColSideColors, RowSideColors,
        cexRow = 0.2 + 1/log10(nr), cexCol = 0.2 + 1/log10(nc),
        labRow = NULL, labCol = NULL, main = NULL,
        xlab = NULL, ylab = NULL,
        keep.dendro = FALSE, verbose = getOption("verbose"), ...)
```

Another useful function for plotting three-dimensional data is `contour`. The `contour` function plots contour lines, connecting equal values in the data:

```
contour(x = seq(0, 1, length.out = nrow(z)),
        y = seq(0, 1, length.out = ncol(z)),
        z,
        nlevels = 10, levels = pretty(zlim, nlevels),
        labels = NULL,
        xlim = range(x, finite = TRUE),
        ylim = range(y, finite = TRUE),
        zlim = range(z, finite = TRUE),
        labcex = 0.6, drawlabels = TRUE, method = "flattest",
        vfont, axes = TRUE, frame.plot = axes,
        col = par("fg"), lty = par("lty"), lwd = par("lwd"),
        add = FALSE, ...)
```

Here is a table showing the arguments to `contour`.

Argument	Description	Default
x, y	Numeric vectors specifying the location of grid lines at which values in the matrix z are measured. (Alternatively, you may specify a single matrix for x and omit y and z.)	x=seq(0, 1, length.out=nrow(z)) y=seq(0, 1, length.out=ncol(z))
z	A numeric vector containing values to be plotted.	
nlevels	Number of contour levels. (Only used if levels is not specified.)	10
levels	A numeric vector of levels at which to draw lines.	pretty(zlim, nlevels)
labels	A vector of labels for the contour lines.	NULL
xlim, ylim, zlim	Numeric vectors of two elements specifying the range of x, y, and z values (respectively) to include in the plot.	xlim = range(x, finite = TRUE) ylim = range(y, finite = TRUE) zlim = range(z, finite = TRUE)
labcex	Text scaling factor for contour labels.	0.6

Argument	Description	Default
drawlabels	A logical value specifying whether to draw contour labels.	TRUE
method	Character value specifying where to draw contour labels. Options include method="simple", method="edge", and method="flattest".	"flattest"
vfont	A character vector with two elements specifying the font to use for contour labels. vfont[1] specifies a Hershey font family; vfont[2] specifies a type face within the family.	
axes	A logical value indicating whether to print axes.	TRUE
frame.plot	A logical value indicating whether to draw a box around the plot.	axes
col	A color for the contour lines.	par("fg")
lty	A type of lines to draw.	par("lty")
lwd	A width for the lines.	par("lwd")
add	A logical value specifying whether to add the contour lines to an existing plot (add=TRUE) or to create a new plot (add=FALSE).	FALSE
...	Additional arguments passed to plot.window, title, Axis, and box.	

The following expression generates a contour plot using the Yosemite Valley data:

```
> contour(yosemite,asp=253/562,ylim=c(1,0))
```

As with image, we needed to flip the *y*-axis and specify an aspect ratio. The results are shown in Figure 14-17.

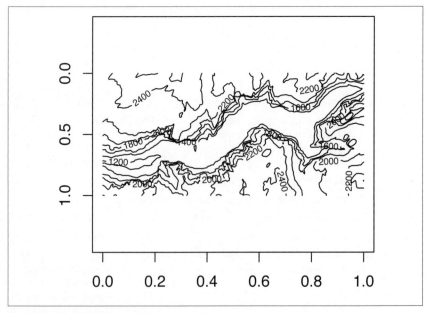

Figure 14-17. Contour example: Yosemite Valley

Contours are commonly added to existing image plots.

Plotting Distributions

When performing data analysis, it's often very important to understand the shape of a data distribution. Looking at a distribution can tell you whether there are outliers in the data, or whether a certain modeling technique will work on your data, or simply how many observations are within a certain range of values.

The best known technique for visualizing a distribution is the histogram. In R, you can plot a histogram with the `hist` function. As an example, let's look at the number of plate appearances (PAs) for batters during the 2008 MLB season. Plate appearances count the number of times that a player had the opportunity to bat; plate appearances include all times a player had a hit, made an out, reached on error, walked, was hit by pitch, hit a sacrifice fly, or hit a sacrifice bunt.

You can load this data set from the `nutshell` package:

```
> library(nutshell)
> data(batting.2008)
```

Let's calculate the plate appearances for each player and then plot a histogram. The resulting histogram is shown in Figure 14-18:

```
> # PA (plate appearances) =
> # AB (at bats) + BB (base on balls) + HBP (hit by pitch) +
> # SF (sacrifice flies) + SH (sacrifice bunts)
> batting.2008 <- transform(batting.2008,
+    PA=AB+BB+HBP+SF+SH)
> hist(batting.2008$PA)
```

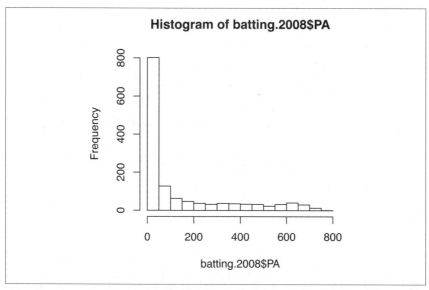

Figure 14-18. Histogram showing the distribution of plate appearances in 2008

The histogram shows that there were a large number of players with fewer than 50 plate appearances. If you were to perform further analysis on this data (for example, looking at the average on-base percentage [OBP]), you might want to exclude these players from your analysis. As we will show in "Proportion Test Design" on page 371, you will need much larger sample sizes than 50 plate appearances to draw conclusions with the data.

Let's try generating a second histogram, this time excluding players with fewer than 25 at bats. We'll also increase the number of bars, using the breaks argument to specify that we want 50 bins:

```
> hist(batting.2008[batting.2008$PA>25,"PA"],breaks=50, cex.main=.8)
```

The second histogram is shown in Figure 14-19.

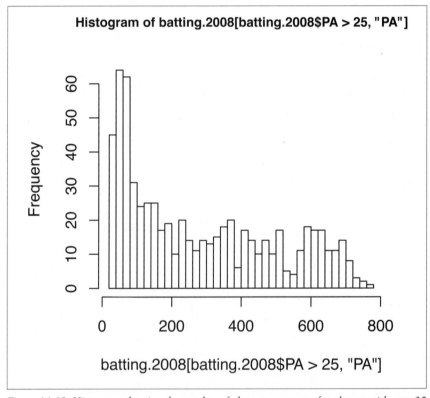

Figure 14-19. Histogram showing the number of plate appearances for players with over 25 plate appearances in 2008

A closely related type of chart is the density plot. Many statisticians recommend using density plots instead of histograms because they are more robust and easier to read. To plot a density plot from the plate appearance data (for batters with more than 25 plate appearances), we use two functions.

First, we use `density` to calculate the kernel density estimates. Next, we use `plot` to plot the estimates. We could plot the diagram with an expression like this:

```
plot(density(batting.2008[batting.2008$PA>25,"PA"]))
```

A common addition to a kernel density plot is a rug. A rug is essentially a strip plot shown along the axis, with each point represented by a short line segment. You can add a rug to the kernel density plot with an expression like:

```
rug(batting.2008[batting.2008$PA>25,"PA"])
```

The final version of the density plot is shown in Figure 14-20.

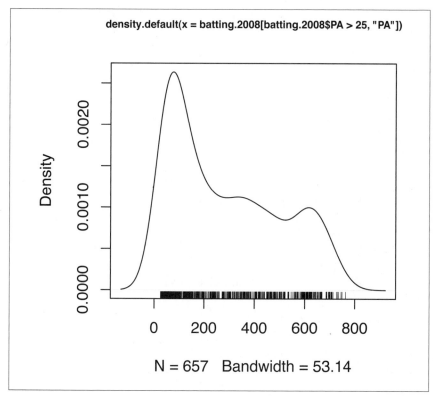

Figure 14-20. Density plot of plate appearances with rug

Another way to view a distribution is the quantile-quantile (Q-Q) plot. Quantile-quantile plots compare the distribution of the sample data to the distribution of a theoretical distribution (often a normal distribution). As the name implies, they plot the quantiles from the sample data set against the quantiles from a theoretical distribution. If the sample data is distributed the same way as the theoretical distribution, all points will be plotted on a 45° line from the lower-left corner to the upper-right corner. Quantile-quantile plots provide a very efficient way to tell how a distribution deviates from an expected distribution.

You can generate these plots in R with the qqnorm function. Without arguments, this function will plot the distribution of points in each quantile, assuming a theoretical normal distribution. The plot is shown in Figure 14-21:

```
> qqnorm(batting.2008$AB)
```

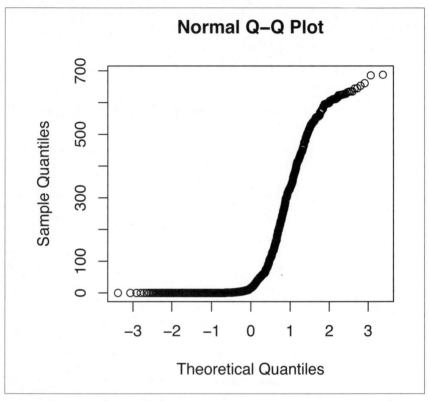

Figure 14-21. Quantile-quantile plot

If you would like to compare two actual distributions, or compare the data distribution to a different theoretical distribution, try the function qqplot.

Box Plots

Another very useful way to visualize a distribution is a box plot. A box plot is a compact way to show the distribution of a variable. The box shows the interquartile range. The *interquartile range* contains values between the 25th and 75th percentile; the line inside the box shows the median. The two "whiskers" on either side of the box show the *adjacent values*. A box plot is shown in Figure 14-22.

The adjacent values are intended to show extreme values, but they don't always extend to the absolute maximum or minimum value. When there are values far outside the range we would expect for normally distributed data, those outlying values are plotted separately. Specifically, here is how the adjacent values are

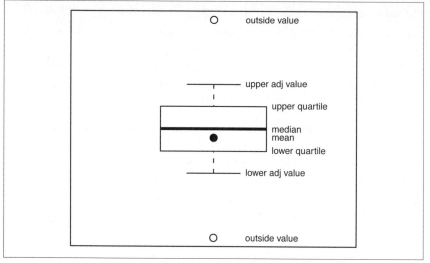

Figure 14-22. Box plot components

calculated: the upper adjacent value is the value of the largest observation that is less than or equal to the upper quartile plus 1.5 times the length of the interquartile range; the lower adjacent value is the value of the smallest observation that is greater than or equal to the lower quartile less 1.5 times the length of the interquartile range. Values outside the range of the whiskers are called *outside values* and are plotted individually.

To plot a box plot, use the `boxplot` function. Here is the default method of `boxplot` for vectors:

```
boxplot(x, ..., range = 1.5, width = NULL, varwidth = FALSE,
        notch = FALSE, outline = TRUE, names, plot = TRUE,
        border = par("fg"), col = NULL, log = "",
        pars = list(boxwex = 0.8, staplewex = 0.5, outwex = 0.5),
        horizontal = FALSE, add = FALSE, at = NULL)
```

And here is the form of `boxplot` when a formula is specified:

```
boxplot(formula, data = NULL, ..., subset, na.action = NULL)
```

Here is a description of the arguments to `boxplot`.

Argument	Description	Default
formula	A formula of the form y ~ grp, where y is a variable to be plotted and grp is a variable describing a set of different plotting groups.	
data	A data frame (or list) in which the variables used in formula are defined.	
subset	A vector specifying a subset of observations to use in plotting.	
x	A vector specifying values to plot.	
...	Additional vectors to plot (or graphical parameters to pass to bxp). Each additional vector is plotted as an additional box.	

Graphics

Argument	Description	Default
range	A numeric value that determines the maximum amount that the whiskers extend from the boxes.	1.5
width	A numeric vector specifying the widths of the boxes being plotted.	NULL
varwidth	If varwidth=TRUE, each box is drawn with a width proportional to the square root of the number of observations represented by the box. If varwidth=FALSE, boxes are plotted with the same width.	FALSE
notch	If notch=TRUE, a "notch" is drawn in the boxes. Notches are drawn at +/-1.58 IQR/ sqrt(n); see the help file for boxplot.stats for an explanation of what they mean.	FALSE
outline	A logical value specifying whether outliers should be drawn.	TRUE
names	A character vector specifying the group names used to label each box plot.	
plot	If plot=TRUE, the box plots are plotted. If plot=FALSE, boxplot returns a list of statistics that could be used to draw a box plot; see the help file for boxplot for more details.	TRUE
border	A character vector specifying the color to use for the outline of each box plot.	par("fg")
col	A character vector specifying the color to use for the background of each box plot.	NULL
log	A character value indicating whether the x-axis (log="x"), y-axis (log="y"), both axes (log="xy"), or neither axis (log="") should be plotted with a logarithmic scale.	""
pars	A list of additional graphical parameters passed to bxp.	list(boxweb = 0.8, stapleweb = 0.5, outwex = 0.5)
horizontal	A logical value indicating whther the boxes should be drawn horizontally (horizontal=TRUE) or vertically (horizontal=FALSE).	FALSE
add	A logical value specifying whether the box plot should be added to an existing chart (add=TRUE) or if a new chart should be drawn (add=FALSE).	FALSE
at	A numeric vector specifying the locations at which each box plot should be drawn.	1:n, where n is the number of boxes

As an example, let's look at the team batting data from 2008. We'll restrict the data to include only American League teams (it's too hard to read a plot with 30 boxes, so this cuts it to 16) and include only players with over 100 plate appearances (to cut out marginal players with a small number of plate appearances). Finally, let's adjust the text size on the axis so that all the labels fit. Here is the expression:

```
> batting.2008 <- transform(batting.2008,
+   OBP=(H+BB+HBP)/(AB+BB+HBP+SF))
> boxplot(OBP~teamID,
+   data=batting.2008[batting.2008$PA>100 & batting.2008$lgID=="AL",],
+   cex.axis=.7)
```

The results are shown in Figure 14-23.

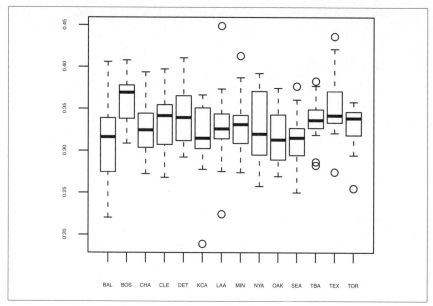

Figure 14-23. Box plot showing on-base percentage for players in the AL in 2008

Graphics Devices

Graphics in R are plotted on a *graphics device*. You can manually specify a graphics device or let R use the default device. In an interactive R environment, the default is to use the device that plots graphics on the screen. On Microsoft Windows, the `windows` device is used. On most Unix systems, the `X11` device is used. On Mac OS X, the `quartz` device is used. You can generate graphics in common formats using the `bmp`, `jpeg`, `png`, and `tiff` devices. Other devices include `postscript`, `pdf`, `pictex` (to generate LaTeX/PicTeX), `xfig`, and `bitmap`.

Most devices allow you to specify the width, height, and point size of the output (with the `width`, `height`, and `pointsize` arguments, of course). For devices that generate files, you can usually use the argument name `file`.[§] When you are done writing a graphic to a file, call the `dev.off` function to close and save the file.

In writing this book, I used the `png` function to generate the graphics printed in this book. For example, I used the following code to produce the first plot in "Scatter Plots" on page 212:

```
> png("scatter.1.pdf", width=4.3, height=4.3, units="in", res=72)
> attach(toxins.and.cancer)
```

[§] For `postscript`, `pdf`, `pictex`, `xfig`, and `bitmap`, the name of the argument is `file`. For `bmp`, `jpeg`, `png`, and `tiff`, the name of the argument is `filename`. However, you can safely use the argument name `file` because of the way R's argument matching rules work. In general, this isn't a good practice, but it's easier than trying to remember the difference between the different devices.

```
> plot(total_toxic_chemicals/Surface_Area,deaths_total/Population)
> dev.off()
```

Customizing Charts

There are many ways to change how R plots charts. The most intuitive is through arguments to a charting function. Another way to customize charts is by setting session parameters. An additional way to change a chart is through a function that modifies a chart (for example, adding titles, trend lines, or more points). Finally, it is possible to write your own charting functions from scratch.

This section describes common arguments and parameters for controlling how charts are plotted.

Common Arguments to Chart Functions

Conveniently, most charting functions in R share some arguments. Here is a table of common arguments for charting functions.

Argument	Description
add	Should this plot be added to the existing plots on the device, or should the device be cleaned first?
axes	Controls whether axes will be plotted on the chart.
log	Controls whether points are plotted on a logarithmic scale.
type	Controls the type of graph being plotted.
xlab, ylab	Labels for x- and y-axes, respectively.
main	Main title for the plot.
sub	Subtitle for the plot.

Graphical Parameters

This section describes the graphical parameters available in the `graphics` package. In most cases, you can specify these parameters as arguments to graphics functions. However, you can also use the `par` function to set graphics parameters. The `par` function sets the graphics functions for a specific graphics device. These new settings will be the defaults for any new plot until you close the device.

The `par` function can be useful if you want to set parameters once and then plot multiple charts. It can also be useful if you want to use the same set of parameters many times. You could write a function to set the right parameters and then call it each time you want to plot some charts:

```
> my_graphics_params <- function () {
    par(some graphics parameters)
}
```

You can check or set the values of these parameters for the active device through the `par` function. If there is no active device, `par` will open the default device.

To check the value of a parameter with `par`, use a character string to specify the value name. To set a parameter's value, use the parameter name as an argument name. To get a vector showing all graphical parameters, simply call `par` with no arguments. Almost all parameters can be read or written. (The only exceptions are `cin`, `cra`, `csi`, `cxy`, and `din`, which can only be read.)

For example, the parameter `bg` specifies the background color for plots. By default, this parameter is set to "transparent":

```
> par("bg")
[1] "transparent"
```

You could use the `par` function to change the `bg` parameter to "white":

```
> par(bg="white")
> par("bg")
[1] "white"
```

"Graphical parameter by name" on page 250 gives details about each graphical parameter by name. However, check the help file for each function to make sure that the parameter means what you think it means. Sometimes, plotting functions have arguments with the same name as graphics parameters to `par` that do different things. For example, the function `points` has an argument named `bg` that means "the background color used in points drawn with this function."

Annotation

Titles and axis labels are called *chart annotation*. You can control chart annotation with the `ann` parameter. (If you set `ann=FALSE`, then titles and axis labels are not printed.)

Margins

R allows you to control the size of the margin around a plot. Figure 14-24 shows how this works. The whole graphics device is called the *device region*. The area where data is plotted is called the *plot region*.

Use the `mai` argument to specify the margin size in inches and use `mar` to specify the margin in lines of text. If you are using `mar`, you can use `mex` to control how big a line of text is in the margin (compared to the rest of the plot). To control the margins around titles and labels, use the `mgp` parameter. To check the overall dimensions of a device (in inches), you can use the read-only parameter `din`.

By default, R maximizes the use of available space out to the margins (`pty="m"`), but you can easily ask R to use a square region by setting `pty="s"`.

Multiple plots

In R, you can plot multiple charts within the same chart area. You can do this with the standard graphics functions by setting the `mfcol` parameter for a device. For example, to plot six figures within the plot area in three rows of two columns, you would set `mfcol` as follows:

```
> par(mfcol=c(3,2))
```

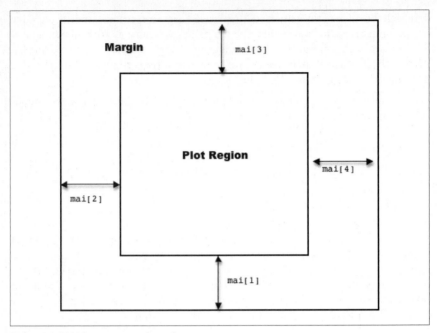

Figure 14-24. Margins around graphics area

Each time a new figure is plotted, it will be plotted in a different row or column within the device, starting with the top-left corner. Plots are then added one at a time, first filling each column from top to bottom, and moving to the next column to the right when each column is filled. For example, let's plot six different figures:

```
> png("~/Documents/book/current/figs/multiplefigs.1.pdf",
+     width=4.3,height=6.5,units="in",res=72)
> par(mfcol=c(3,2))
> pie(c(5,4,3))
> plot(x=c(1,2,3,4,5),y=c(1.1,1.9,3,3.9,6))
> barplot(c(1,2,3,4,5))
> barplot(c(1,2,3,4,5),horiz=TRUE)
> pie(c(5,4,3,2,1))
> plot(c(1,2,3,4,5,6),c(4,3,6,2,1,1))
> dev.off()
```

The result of these commands is shown in Figure 14-25.

If a matrix of subplots is being drawn on a graphics device, you can specify the next plot location using the argument mfg=c(row, column, nrows, ncolumns).

Figure 14-26 shows an example of how margins and plotting areas are defined when using multiple figures. Within the device region are a set of *figure regions* corresponding to each individual figure. Within each figure region, there is a plot region. There is an outer margin that surrounds all of the figure area; you may control these with the parameters omi, oma, and omd. Within each figure, as with all plots, there is a second margin area, controlled by mai, mar, and mex. (If you are writing your own

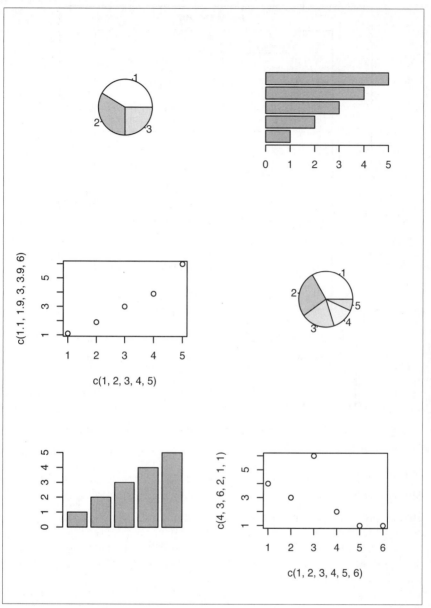

Figure 14-25. Multiple-figure example

graphics functions, you may find it useful to use the xpd parameter to control where graphics are clipped.)

To find the size of the current plot area (within the grid), check the parameter pin. To get the coordinates of the plot region, check the parameter plt. To find the

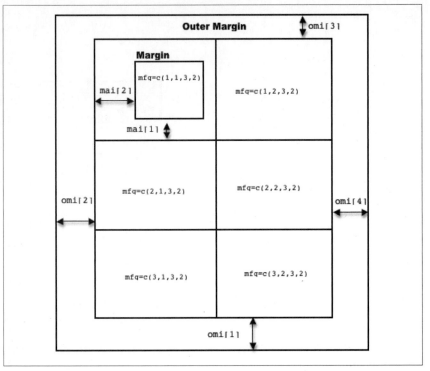

Figure 14-26. Multiple figure layout

dimensions of the current plot area using normalized device coordinates, use the parameter `fig`.||

You may find it easier to use the functions `layout` or `split.screen`. Better still, use the packages `grid` or `lattice`.

Text properties

Many parameters control the way text is shown within a plot.

Text size. The parameter `ps` specifies the default point size of text. A second parameter, `cex`, specifies a default scaling factor for text. You may specify additional scaling factors for different types of text: `cex.axis` for axis annotation, `cex.lab` for *x* and *y* labels, `cex.main` for main titles, and `cex.sub` for subtitles. In many cases, all three parameters are used to find the size of a line of text. Here is an example of how this works. To determine the point size for a chart title, multiply `ps * cex * cex.main`.

You may use the read-only parameters `cin`, `cra`, `csi`, and `cxy` to check the size of characters.

|| Normalized device coordinates map the overall chart space onto a 1 × 1 area. (So, *x* coordinates vary between 0 and 1, and *y* coordinates between 0 and 1.)

Typeface. The font is specified through the `family` argument. Somewhat confusingly, the text style is specified through the `font` argument. You can specify the style for the axis with `font.axis`, for labels with `font.lab`, for main titles with `font.main`, and for subtitles with `font.sub`.

Alignment and spacing. To control how text is aligned, use the `adj` parameter. To change the spacing between lines of text, use the `lheight` parameter.

Rotation. To rotate each character, use the `crt` parameter. To rotate whole strings, use the `srt` parameter.

Line properties

You can also change the way that lines are drawn. To change the line end style, use `lend`. To change the line join style, use `ljoin` and `lmiter`. Line type is specified by `lty` and line width by `lwd`. To change the way boxes are drawn around plots, use the `bty` parameter.

Colors

You can change the default background color with `bg` and the default foreground color with `fg`. The default plotting color is specified by `col`. Use `col.axis` to change the color of axes, `col.lab` to change the color of labels, `col.main` to change the color of the main title, and `col.sub` to change the color of the subtitle.

You can specify colors in many different ways: as a string, using RGB (red/green/blue) components, or referencing a palette by integer index. To get a list of valid color names, use the `colors` function. To specify a color using RGB components, use a string of the form `"#RRGGBB"`, where *RR*, *GG*, and *BB* are hexadecimal values specifying the amount of red, green, and blue, respectively. To view or change a color palette, use the `palette` function. Other functions are available for specifying colors, including `rgb`, `hsv`, `hcl`, `gray`, and `rainbow`.

Axes

The argument `lab` controls how axes are annotated. To change the style of axis labels, use `las`. To change the margin for the axis title, labels, and lines, use `mgp`.

You can specify the size of tick marks in lines of text with `tcl`, or as a fraction of the plot area with `tck`. To change the minimum and maximum tick mark locations, use `xaxp` and `yaxp`. To change the way intervals are calculated, use `xaxs` and `yaxs`. To remove the *x*-axis or *y*-axis, use `xaxt="n"` or `yaxt="n"`.

You can also change the orientation of axis labels with the `las` parameter.

Points

You can change the symbol used for points with the `pch` argument. To get a list of point types, use the `points` function.

Graphical parameter by name

Here is a table showing all the graphical parameters available in R that can be set with par.#

Parameter	Description	Mnemonic	Default
adj	Controls how text is justified in text, mtext, and title strings. Set adj=0 for left-justified text, adj=1 for right-justified text, and adj=0.5 for centered text.		0.5
ann	If ann=TRUE, then axis titles and overall titles are included with plots. If ann=FALSE, these annotations are not included. Used by high-level functions that call plot.default.	ANNotation	TRUE
ask	Within an interactive session, if ask=TRUE, then the user is asked for input before a new chart is drawn.		FALSE
bg	The background color for the device region.		transparent
bty	The type of box to draw around a plot. Use bty="o" for a box on all sides, bty="l" for the left and bottom only, bty="7" for top and right only, bty="]" for right side, bottom, and top only, bty="c" for left side, bottom, and top only, bty="u" for left, right, and bottom only, and bty="n" for no box. To draw the box, use the box function.	Box TYpe (values correspond to the shape of the letter)	o
cex	This parameter controls the size of text and plotted points. cex=1 means "normal size," cex=0.75 means "shrink the text and points to 75% of normal size." This parameter is reset when the device size changes.	Character EXpansion	1
cex.axis	Text magnification for axis notations, relative to cex.		1
cex.lab	Text magnification for x and y labels, relative to cex.		1
cex.main	Text magnification for main titles, relative to cex.		1.2
cex.sub	Text magnification for subtitles, relative to cex.		1
cin	Character size in inches. (Equivalent to cra, just with different units.)	Character size in INches	c(0.15, 0.2)
col	Default plotting color.		black
col.axis	Color for axis annotation.		black
col.lab	Color for axis labels.		black
col.main	Color for main titles.		black

#Incidentally, I generated this table with R code like this:

```
print_pars <- function() {
    for (n in names(par())) {
        p <- par(n);
        if (length(p) == 1) {
            print(paste(n,p,sep="="));
        } else {
            print(paste(n,"=c(",paste(p,collapse=","),sep="",")"));
        }
    }
}
```

Parameter	Description	Mnemonic	Default
col.sub	Color for subtitles.		black
cra	Character size in pixels.	Character size in RAsters	c(10.8, 14.4)
crt	A numeric value that specifies (in degrees) how individual characters should be rotated. Use srt for whole strings.	Character RoTation	0
csi	Height of default size characters in inches. Same as par("cin")[2].		0.2
cxy	Default character size in user coordinate units.		c(0.02604167, 0.03875969)
din	Device dimensions in inches.	Dimensions in INches	c(7, 7)
err	Degree of error reporting. Currently does nothing.		0
family	Name of font family used to draw text. Common values include "serif", "sans", "mono", and Hershey fonts. (See help file for Hershey for more information.)		
fg	Color for foreground of plots.		black
fig	A numeric vector that specifies the coordinates of the figure area.		c(0, 1, 0, 1)
fin	The figure area dimensions in inches.		c(7, 7)
font	An integer that specifies what "font" to use in text (though it sounds like this really means "style"). Use font=1 for normal, font=2 for bold, font=3 for italic, font=4 for bold and italic, and font=5 to substitute the Adobe Symbol font.		1
font.axis	The "font" to be used for axis annotation.		1
font.lab	The "font" to be used for x and y labels.		1
font.main	The "font" to be used for the main plot titles.		2
font.sub	The "font" to be used for plot subtitles.		1
lab	A numeric vector with three elements (x, y, len) that specifies the way that axes are annotated. x, y specify the approximate number of tick marks on the x- and y-axes, and len specifies the label length.	Controls whether to plot a black labrador retriever in the middle of the plot instead of your data	c(5, 5, 7)
las	Specifies the style of axis labels. Use las=0 for parallel to the axis, las=1 for horizontal, las=2 for perpendicular, and las=3 for vertical. (Also used in mtext.)		0
lend	Specifies the style of line ends. Use lend=0 or lend="round" for rounded ends, len=1 or lend="butt" for butt line caps and lend=2 or lend="square" for square caps.	Line END	round
lheight	Specifies the height of a line when spacing multiple lines of text.	Line HEIGHT multiplier	1

Parameter	Description	Mnemonic	Default
ljoin	Specifies the style for joining lines. Use ljoin=0 or ljoin="round" for round line joins, ljoin=1 or ljoin="mitre" for mitred line joins and, ljoin=2 or ljoin="bevel" for beveled line joins.	Line JOIN style	round
lmitre	Controls when mitred line joins are converted to beveled joins.	Line MITRE limit	10
lty	Line type. You can specify a numeric value (0=blank, 1=solid, 2=dashed, 3=dotted, 4=dotdash, 5=longdash, 6=twodash) or a character value ("blank", "solid", "dashed", "dotted", "dotdash", "longdash", or "twodash").	Line TYpe	solid
lwd	A positive number specifying line width.	Line WiDth	1
mai	A numeric vector c(bottom, left, top, right) that specifies margin size in inches.	MArgin size in Inches	c(1.02, 0.82, 0.82, 0.42)
mar	A numeric vector c(bottom, left, top, right) that specifies margin size, in number of lines.	MARgin size in lines	c(5.1, 4.1, 4.1, 2.1)
mex	Specifies the expansion factor used for line size in mar. (The exact relationship is mai=mar*mex*csi.)	Margin EXpansion factor	1
mfcol	This parameter allows you to split a graphics device into a "matrix" of subplots. This parameter is a numeric vector with two values: c(nrows, ncols). nrows represents a number of rows, and ncols represents a number of columns. When more than one row and column are specified, R will split the device area into the specified number of rows and columns.		c(1, 1)
mfg	If a matrix of subplots is being drawn on a graphics device, you can use mfg to specify the next plot to be drawn. Specified as c(row, column, nrows, ncolumns). (When queried, this returns the location of the last figure plotted.)		c(1, 1, 1, 1)
mfrow	Identical to mfcol.		c(1, 1)
mgp	A numeric vector with three values that controls the margin line for an axis title. mgp[1] is used for the title; mgp[2:3] is used for the axis.		c(3, 1, 0)
mkh	The height of symbols when pch is an integer. (As of R 2.9.0, this parameter has no effect.)		0.001
new	A logical value that indicates whether the plotting routine should pretend that the graphics device has been freshly initialized (and is thus empty). Used to plot a figure on top of another one.		>FALSE
oma	A numeric vector c(bottom, left, top, right) that specifies the outer margin in lines.	Outer MArgin in lines	c(0, 0, 0, 0)
omd	A numeric vector c(bottom, left, top, right) that specifies the outer margin as a fraction of the size of the whole device. (For example, a value of 0 means the leftmost or top value, and a value of 0.5 means dead center.)	Outer Margin in Device coordinates	c(0, 1, 0, 1)
omi	A numeric vector c(bottom, left, top, right) that specifies the outer margin in inches.	Outer Margin in Inches	c(0, 0, 0, 0)

Parameter	Description	Mnemonic	Default
pch	Specifies the default point type. Symbols can be specified as numbers (see the help file for the points function to get a complete list of values) or as a character. Some common values are pch=19: solid circle, pch=20: bullet (smaller circle), pch=21: filled circle, pch=22: filled square, pch=23: filled diamond, pch=24: filled triangle point up, pch=25: filled triangle point down.	Point CHaracter	1
pin	The dimensions of the current plot in inches.	Plot in INches	c(5.76, 5.16)
plt	A vector that specifies the coordinates of the plot region as fractions of the figure region.		c(0.1171429, 0.9400000, 0.1457143, 0.8828571)
ps	An integer value specifying the point size of text (not symbols).	Point Size	12
pty	Specifies the type of plotting region. Use pty="s" for square, pty="m" to maximize the use of space.		m
smo	Specifies how smooth circles and arcs should be. (Currently ignored.)		1
srt	Specifies the rotation of strings in degrees. (Only used by text.)		0
tck	The length of tick marks. Specified as a fraction of the width or height of the plotting region (whichever is smallest).		NA
tcl	The length of tick marks as a fraction of the height of a line of text.		-0.5
usr	A vector c(x1, x2, y1, y2) that specifies the extreme values of user coordinates in the plotting region. (Note that these values are scaled exponentially when a logarithmic scale is used.)		c(0, 1, 0, 1)
xaxp	Controls how tick marks are shown in the x-axis. Specified as a vector c(x1, x2, n). When a linear scale is being used, specifies the minimum (x1) and maximum (x2) tick mark locations and the number of tick marks. When a logarithmic scale is being used, these parameters mean something else: x1 is the lowest power of 10, x2 is the highest power of 10, and n specifies the number of tick marks plotted for each power of 10.		c(0, 1, 5)
xaxs	Controls the calculation method used to find axis intervals on the x-axis. The regular method, xaxs="r", extends the data range by 4% on each side and then tries to find pretty labels. The internal method, xaxs="i", tries to find labels within the data range. (There are other valid values, but they aren't currently implemented.)		r
xaxt	Specifies the x-axis type. Use xaxt="n" for no axis; any other value to plot an axis.		s
xlog	A logical value that specifies whether the scale of the x-axis is logarithmic.		FALSE

Graphics

Parameter	Description	Mnemonic	Default
xpd	Controls clipping. Use xpd=FALSE to clip to the plot region, xpd=TRUE to clip to the figure region, and xpd=NA to clip to the device region.		FALSE
yaxp	Controls how tick marks are shown in the y-axis. See xaxp for a full explanation.		c(0, 1, 5)
yaxs	Controls the calculation method used to find axis intervals on the y-axis. See xaxs for a full explanation.		r
yaxt	Specifies the y-axis type. Use xaxt="n" for no axis; any other value to plot an axis.		s
ylog	A logical value that specifies whether the scale of the y-axis is logarithmic.		FALSE

Basic Graphics Functions

It is possible to use these functions to either modify an existing chart or draw a chart yourself from scratch. Many of these functions are called from higher-level graphics functions. These higher-level functions pass extra arguments to these lower-level functions. So, even if you do not plan to use these functions directly, you may find it useful to pass arguments to them to customize charts.

Here is a table of low-level graphics functions called by the higher-level graphics functions listed above. (You can often look at arguments for the low-level graphics functions to determine how to customize the look of plots generated with the high-level functions.)

High-level function	Low-level functions
plot	title, plot.new, plot.xy, plot.window, points, lines, axis, box, xy.coords
matplot	plot
pairs	plot, points
barplot	title, plot.window, title, axis
pie	plot.window, polygon, lines, text, title
dotchart	plot.window, mtext, abline, points, axis, box, title
coplot	axis, plot.new, plot.window, points, grid
cdplot	plot, axis, box
mosaicplot	polygon, text, segments, title
spineplot	axis, plot, rect, axis
persp	title, persp (internal)
image	plot, image (internal)
contour	plot.window, title, Axis, box, contour (internal)
heatmap	image, axis, plot, title

High-level function	Low-level functions
hist	plot
qqnorm	plot
qqplot	plot
boxplot	bxp
bxp	points, polygon, segments, axis, Axis, title, box, plot.new, plot.window
points	plot.xy
lines	plot.xy

points

You can plot points on a chart using the `points` function:

```
points(x, y = NULL, type = "p", ...)
```

This can be very useful for adding an additional set of points to an existing plot (typically a scatter plot), usually with a different color or plot symbol. Most of the same arguments for the `plot` function apply to `points`. The most useful arguments are `col` (to specify the foreground color for plotted points), `bg` (to specify the background color of plotted points), `pch` (to specify the plotting character), `cex` (to specify the size of plotted points), and `lwd` (to specify the line width for plotted symbols).

You can also add points to an existing matrix plot with `matpoints`:

```
matpoints(x, y, type = "p", lty = 1:5, lwd = 1, pch = NULL,
          col = 1:6, ...)
```

lines

A similarly useful function is `lines`:

```
lines(x, y = NULL, type = "l", ...)
```

Like `points`, this is often used to add to an existing plot. The `lines` function plots a set of line segments on an existing plot. (The values in x and y specify the intersections between the line segments.) As with `points`, many arguments for `plot` also apply to `lines`. Some especially useful arguments are `lty` (line type), `lwd` (line width), `col` (line color), `lend` (line end style), `ljoin` (line join style), and `lmitre` (line mitre style).

You can also add lines to an existing plot with `matlines`:

```
matlines (x, y, type = "l", lty = 1:5, lwd = 1, pch = NULL,
          col = 1:6, ..
```

curve

To plot a curve on the current graphical device, you can use the `curve` function:

```
curve(expr, from = NULL, to = NULL, n = 101, add = FALSE,
      type = "l", ylab = NULL, log = NULL, xlim = NULL, ...)
```

Graphics

Here is a description of the arguments to the **curve** function.

Argument	Description	Default
expr	The expression to plot (written as a function of x) or the name of a function to plot.	
from	The lowest x value at which expr is evaluated.	NULL
to	The highest x value at which expr is evaluated.	NULL
n	A positive integer value specifying the number of values at which to evaluate expr between the x limits (specified by xlim).	101
add	A logical value indicating whether to add the curve to the current plot.	FALSE
type	Specifies the plot type. Use type="p" for points, type="l" for lines, type="o" for overplotted points and lines, type="b" for points joined by lines, type="c" for empty points joined by lines, stype="s" or type="S" for stair steps, type="h" for histogram-like vertical lines, or type="n" to plot nothing.	"l"
ylab	A character value specifying the label for the y-axis.	ylab
log	A logical value specifying whether to plot on a logarithmic scale.	log
xlim	A numeric vector with two values specifying the lowest and highest x values to plot.	NULL
...	Additional arguments passed to plot.	

text

You can use the **text** function to add text to an existing plot. (We used the **text** function to label points on a scatter plot in "Scatter Plots" on page 212.)

```
text (x, y = NULL, labels = seq_along(x), adj = NULL,
    pos = NULL, offset = 0.5, vfont = NULL,
    cex = 1, col = NULL, font = NULL, ...)
```

Here are the arguments to **text**.

Argument	Description	Default
x, y	These arguments specify the coordinates at which the text labels will be drawn.	y=NULL
labels	A vector of character values specifying the text values that should be drawn on the chart.	seq_along(x)
adj	A numeric vector with one or two values (each between 0 and 1). If one value is used, it represents the horizontal adjustment. If two values are used, the first represents the horizontal adjustment, and the second represents the vertical adjustment.	NULL
pos	A numeric value that specifies where the text should be positioned. Use pos=1 for below, pos=2 for the left, pos=3 for above, and pos=4 for right. Overrides values specified in adj.	NULL
offset	A numeric value that specifies the offset of the labels in terms of character widths. (Only valid when pos is specified.)	0.5

Argument	Description	Default
vfont	A character vector with two elements specifying the font to use for labels. vfont[1] specifies a Hershey font family; vfont[2] specifies a typeface within the family.	NULL
cex	Numeric value specifying the character expansion factor.	1
col	Specifies the color of plotted text.	NULL
font	Specifies the font to be used for the plotted text.	NULL
...	Additional graphical parameters.	

For an example of how to use the text function, see "Scatter Plots" on page 212.

abline

To plot a single line across the plot area, you can use the abline function:

```
abline(a = NULL, b = NULL, h = NULL, v = NULL, reg = NULL,
       coef = NULL, untf = FALSE, ...)
```

Here is a description of the arguments to abline.

Argument	Description	Default
a	The intercept for the line.	NULL
b	The slope for the line.	NULL
h	A numeric vector of y values for horizontal lines.	NULL
v	A numeric vector of x values for vertical lines.	NULL
reg	Specifies an object with a coef method.	NULL
coef	A numeric vector with two elements specifying the intercept and slope.	NULL
untf	A logical value specifying whether to "untransform" the line; if one or both axes are in logarithmic coordinates and untf=true, then the line is shown in original coordinates. Otherwise, the line is plotted in transformed coordinates.	NULL
...	Additional graphical parameters. See "Graphical Parameters" on page 244 for more details.	

Typically, you would use one call to abline to draw a single line. For example:

```
> # draw a simple plot as a background
> plot(x=c(0,10),y=c(0,10))
> # plot a horizontal line at y=4
> abline(h=4)
> # plot a vertical line at x=3
> abline(v=3)
> # plot a line with a y-inctercept of 1 and slope of 1
> abline(a=1,b=1)
> # plot a line with a y-intercept of 10 and slope of -1,
> # but this time, use the coef argument:
> abline(coef=c(10,-1))
```

However, you can also specify multiple arguments, and `abline` will plot all of the specified lines. For example:

```
> # plot a grid of lines between 1 and 10:
> abline(h=1:10,v=1:10)
```

If you just want to plot a grid on a plot, you might want to use the **grid** function instead:

```
grid(nx = NULL, ny = nx, col = "lightgray", lty = "dotted",
     lwd = par("lwd"), equilogs = TRUE)
```

polygon

To draw a polygon, you can use the `polygon` function:

```
polygon(x, y = NULL, density = NULL, angle = 45,
        border = NULL, col = NA, lty = par("lty"), ..
```

The x and y arguments specify the vertices of the polygon. For example, the following expression draws a 2 × 2 square on a graph centered at (3, 3):

```
> polygon(x=c(2,2,4,4),y=c(2,4,4,2))
```

For the special case where you just need to draw a rectangle, you can use the **rect** function:

```
rect(xleft, ybottom, xright, ytop, density = NULL, angle = 45,
     col = NA, border = NULL, lty = par("lty"), lwd = par("lwd"),
     ...)
```

segments

To draw a set of line segments connecting pairs of points, you can use the segments function:

```
segments(x0, y0, x1, y1,
         col = par("fg"), lty = par("lty"), lwd = par("lwd"),
         ...)
```

This function draws a set of line segments from each pair of vertices specified by (x0[i],y0[i]) to (x1[i], y1[i]).

legend

The `legend` function adds a legend to a chart:

```
legend(x, y = NULL, legend, fill = NULL, col = par("col"),
       lty, lwd, pch,
       angle = 45, density = NULL, bty = "o", bg = par("bg"),
       box.lwd = par("lwd"), box.lty = par("lty"), box.col = par("fg"),
       pt.bg = NA, cex = 1, pt.cex = cex, pt.lwd = lwd,
       xjust = 0, yjust = 1, x.intersp = 1, y.intersp = 1,
       adj = c(0, 0.5), text.width = NULL, text.col = par("col"),
       merge = do.lines && has.pch, trace = FALSE,
       plot = TRUE, ncol = 1, horiz = FALSE, title = NULL,
       inset = 0, xpd, title.col = text.col)
```

Here is a list of arguments to `legend`. (Many of these can also be passed along as arguments to functions that draw legends.)

Argument	Description	Default
x, y	The coordinates at which the legend will be positioned.	y=NULL
legend	A character vector to appear in the legend.	
fill	A character vector specifying a color associated with each legend label. If specified, boxes filled with these colors are shown next to the labels.	NULL
col	The color of lines appearing in the legend.	par("col")
lty	The line type for lines appearing in the legend.	
lwd	The line width for lines appearing in the legend.	
pch	A vector of values specifying point characters appearing in the legend.	
angle	Angle of shading lines.	45
density	Density of shading lines.	NULL
bty	Box type for box drawn around the legend.	"o"
bg	Background color for the legend box.	par("bg")
box.lwd	Line width for the legend box.	par("lwd")
box.lty	Line type for the legend box.	par("lty")
box.col	Line color for the legend box.	par("fg")
pt.bg	Background color for points shown in the legend box (if pch is specified).	NA
cex	Character expansion value for legend relative to par("cex").	1
pt.cex	Expansion factor for points in the legend.	cex
pt.lwd	Line width for points in the legend.	lwd
xjust	Specifies how the legend should be justified relative to the x location. Use xjust=0 for left justification, xjust=0.5 to center, and xjust=1 for right justification.	0
yjust	Specifies how the legend should be justified relative to the y location.	1
x.intersp	Character "interspacing factor" for horizontal spacing.	1
y.intersp	Character "interspacing factor" for vertical spacing.	1
adj	String adjustment for legend text.	c(0, 0.5)
text.width	Width of legend text in user coordinates.	NULL
text.col	Color used for legend text.	par("col")
merge	If merge=TRUE, merge points and lines but not filled boxes.	do.lines && has.pch
trace	Logical value. If trace=TRUE, shows how legend calculates stuff.	FALSE
plot	Logical value. If plot=FALSE, calculations are returned but no legend is drawn.	TRUE
ncol	Specifies the number of columns to draw in the legend.	1

Argument	Description	Default
horiz	Specifies whether the legend should be laid out vertically (horiz=FALSE) or horizontally (horiz=TRUE).	FALSE
title	A character value to be placed at the top of the legend box.	NULL
inset	Inset distance from the margins. Specified as a fraction of the plot region.	0
xpd	Controls clipping while the legend is being drawn. See "Graphical parameter by name" on page 250 for more details.	
title.col	Color for title.	text.col

title

To annotate a plot, use the title function:

```
title(main = NULL, sub = NULL, xlab = NULL, ylab = NULL,
      line = NA, outer = FALSE, ...)
```

This function adds a main title (main), a subtitle (sub), an *x*-axis label (xlab), and a *y*-axis label (ylab). Specify a value of line to move the labels outward from the edge of the plot. Specify outer=TRUE if you would like to place labels in the outer margin.

axis

To add axes to a plot, use the axis function:

```
axis(side, at = NULL, labels = TRUE, tick = TRUE, line = NA,
     pos = NA, outer = FALSE, font = NA, lty = "solid",
     lwd = 1, lwd.ticks = lwd, col = NULL, col.ticks = NULL,
     hadj = NA, padj = NA, ...)
```

Here is a table of arguments to axis. (Many of these arguments can be passed to functions that draw axes.)

Argument	Description	Default
side	An integer value specifying where to draw the axis. Use side=1 for below, side=2 for left, side=3 for above, and side=4 for right.	
at	A numeric vector specifying points at which tick marks are drawn. (If not specified, uses the same method as axTicks to compute "pretty" tick mark locations.)	NULL
labels	Either a logical value or a vector. If logical, specifies whether numeric annotations are added at tick marks. If a vector is specified, each value specifies the label to place at each tick mark.	TRUE
tick	A logical value specifying if tick values and an axis will be drawn.	TRUE
line	The number of lines into the margin at which the axis will be drawn. (Can be used to add space between plotted values and the axis.) Use line=NA for no space.	NA

Argument	Description	Default
pos	The coordinate at which the axis will be drawn. (If not NA, overrides line.)	NA
outer	A logical value specifying whether the axis should be drawn in the outer margin. Use outer=FALSE to draw the axis in the standard margin.	FALSE
font	Font for axis text.	NA
lty	Line type for axis line and tick marks.	"solid"
lwd	Line width for axis line.	1
twd.ticks	Line width for tick marks.	lwd.ticks
col	Color for axis line.	col
col.ticks	Color for tick marks.	col.ticks
hadj	Adjustment for all labels parallel to the reading direction. See "Graphical Parameters" on page 244 for more information on the parameter adj.	NA
padj	Adjustment for all labels perpendicular to the reading direction. See "Graphical Parameters" on page 244 for more information on the parameter adj.	NA
...	Other graphical parameters. See "Graphical Parameters" on page 244 for more information.	

box

The box function can be used to draw a box around the current figure region. This can be useful when plotting multiple figures within a graphics device:

```
box(which = "plot", lty = "solid", ...)
```

The which argument specifies where to draw the box. Values for which include "plot," "figure," "inner," and "outer"). You might find the box argument useful for showing these different regions.

mtext

The mtext function can be used to add text to a margin of a plot:

```
mtext(text, side = 3, line = 0, outer = FALSE, at = NA,
    adj = NA, padj = NA, cex = NA, col = NA, font = NA, ...)
```

Use the side parameter to specify where to plot the text (side = 1 for bottom, side = 2 for left, side = 3 for top, and side = 4 for right). The line argument specifies where to write the text, in terms of "margin lines" (starting at 0 for closest to the plot area).

trans3d

To add lines or points to a perspective plot (from persp), you might find the function trans3d convenient:

```
trans3d(x,y,z, pmat)
```

This function takes vectors of points x, y, and z and translates them into the correct screen position. The argument pmat is a *perspective matrix* that is used for translation. The persp function will return an appropriate perspective matrix object for use by trans3d.

15

Lattice Graphics

The `lattice` package provides a different way to plot graphics in R. Lattice graphics look different from standard R graphics, are created with different functions, and have different options. Lattice functions make it easy to do some things that are hard to do with standard graphics, such as plotting multiple plots on the same page or superimposing plots. Additionally, most lattice functions produce clean, readable output by default. This chapter shows what lattice graphics can do and explains how to use them.

The real strength of the `lattice` package is in splitting a chart into different panels (shown in a grid), or groups (shown with different colors or symbols) using a conditioning or grouping variable. This chapter includes many examples that start with a simple chart and then split it into multiple pieces to answer a question raised by the original plot.

History

In the early 1990s, Richard Becker and William Cleveland (two researchers at Bell Labs) built a revolutionary new system for displaying data called *Trellis graphics*. (You can find more information about the Trellis software at *http://cm.bell-labs.com/cm/ms/departments/sia/project/trellis/*.) Cleveland devised a number of novel plots for visualizing data based on research into how users visualize information.*

The `lattice` package is an implementation of Trellis graphics in R.† You may notice that some functions still contain the Trellis name. The `lattice` package includes many types of charts that will be familiar to most readers such as scatter plots, bar charts, and histograms. But it also includes some plots that you may not have seen before such as dot plots, strip plots, and quantile-quantile plots. This chapter will

---

* See [Cleveland1993] for more information.

† It's not exactly the same as the S version, but unless you want to use old S/S+ code, the differences will probably not matter to you.

show you how to use different types of charts, familiar and unfamiliar, in the lattice package.

An Overview of the Lattice Package

Lattice graphics consist of one or more rectangular drawing areas called *panels*. The data assigned to each panel is referred to as a *packet*. Lattice functions work by calling one or more *panel functions*, which actually plot the packets within panels. To change the appearance of a plot, you can specify arguments to the plotting function or change the panel function.

How Lattice Works

Here is what typically happens in a lattice session:

1. The end user calls a high-level lattice plotting function.
2. The lattice function examines the calling arguments and default parameters, assembles a lattice object, and returns the object. (Note that the class of the object is actually "trellis." This means that many of the methods that act on an object, like print or plot, are named plot.trellis or print.trellis.)
3. The user calls print.lattice or plot.lattice with the lattice object as an argument. (This typically happens automatically on the R console.)
4. The function plot.lattice sets up the matrix of panels, assigns packets to different panels (specified by the argument packet.panel) and then calls the panel function specified in the lattice object to draw the individual panels.

Lattice graphics are extremely modular; they share many high-level functions (like plot.lattice) and low-level functions (like panel.axis, which draws axes). This means that they share many common arguments. It also means that you can customize the appearance of lattice graphics by creating substitute components.

A Simple Example

There are many arguments to lattice functions, but in this section we'll focus on a handful of key arguments for specifying what data to plot.

As you may have noticed, functions in the graphics package don't have completely consistent arguments. Many of them share some common parameters (see "Customizing Charts" on page 244), but many of them have different names for arguments with the same purpose. (For example, data for barplot is specified with the height argument, while data for plot is specified with x and y.) Arguments within the lattice package are much more consistent.

You can always specify the data to plot using a formula and a data frame. Let's create a simple data set and plot a scatter plot with xyplot:

```
> d <- data.frame(x=c(0:9),y=c(1:10),z=c(rep(c("a","b"),times=5)))
> d
   x  y z
1  0  1 a
```

```
 2   1   2 b
 3   2   3 a
 4   3   4 b
 5   4   5 a
 6   5   6 b
 7   6   7 a
 8   7   8 b
 9   8   9 a
10   9  10 b
```

To plot this data frame, we'll use the formula y~x and specify the data frame d. The first argument given is the formula. (The argument used to be called "formula" and is currently named x. The help files for lattice warn not to pass this as a named argument, possibly because the name may change again.) To specify the data frame containing the plotting data, we use the argument data:

```
> xyplot(y~x,data=d)
```

The resulting plot is shown in Figure 15-1. Formulas in the lattice package can also specify a conditioning variable. The conditioning variable is used to assign data points to different panels. For example, we can plot the same data shown above in two panels, split by the conditioning variable z. To do this, we will change the formula to y~x|z:

```
> package(lattice)
> xyplot(y~x|z,data=d)
```

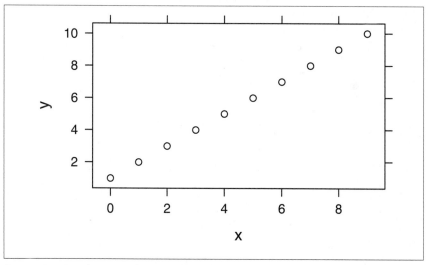

Figure 15-1. Simple scatter plot example

The scatter plot with the conditioning variable is shown in Figure 15-2. As you can see, the data is now split into two panels. If you would prefer to see the two data series superimposed on the same plot, you can specify a grouping variable. To do this, use the argument groups to specify the grouping variable(s):

```
> xyplot(y~x,groups=z,data=d)
```

Lattice Graphics

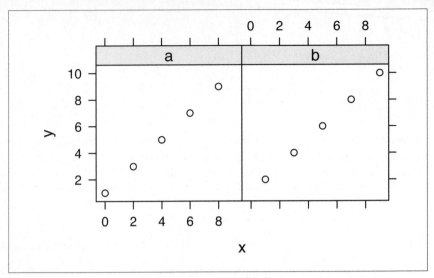

Figure 15-2. Simple scatter plot with conditioning variable

As shown in Figure 15-3, the two data series are represented by different symbols. (If you try this example yourself using the R console, the different groups will be plotted in different colors. To make the charts readable in black and white, I generated the charts using special settings.)

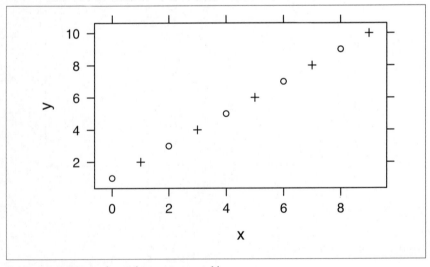

Figure 15-3. Scatter plot with grouping variable

Using Lattice Functions

The easiest way to use lattice graphics is by calling a high-level plotting function. Most of these functions are the equivalent of a similar function in the graphics package. Here's a table showing how standard graphics functions map to lattice functions.

Graphics package function	Trellis package function	Description
barplot	barchart	Bar and column charts
dotchart	dotplot	Cleveland dot plots
hist	histogram	Histograms
Two functions: density and plot.density	densityplot	Kernel density plots
stripchart	stripplot	Strip charts
No function in graphics package; qqnorm in stats package	qqmath	Quantile-quantile plots
xplot	xyplot	Scatter plots
No function in graphics package; qqplot in stats package	qq	Quantile-quantile plots
pairs	splom	Scatter plot matrices
image	levelplot	Image plots
contour	contourplot	Contour plots
persp	cloud, wireframe	Perspective charts of three-dimensional data

When you call a high-level lattice function, it does not actually plot the data. Instead, each of these functions returns a lattice object. To actually show the graphic, you need to use a `print` or `plot` command. If you simply execute a lattice function on the R command line, R runs `print` automatically, so the graphic is shown. However, if you call a lattice function inside another function or inside a script and you want to show the results, make sure that you actually call `print`.

For some (but not all) lattice functions, it is possible to specify the source data in multiple forms. For example, the function `histogram` can also accept data arguments as factors or numeric vectors. These methods are provided for convenience where appropriate. For example, I frequently plot contingency tables as bar charts, so I often use the `table` method of `barchart`. Here is a table of data types accepted by different lattice functions.

Trellis function	Data types
barchart	Array, formula, matrix, numeric vector, table
dotplot	Array, formula, matrix, numeric vector, table
histogram	Factor, formula, numeric vector
densityplot	Formula, numeric vector
stripplot	Formula, numeric vector
qqmath	Formula, numeric vector
xyplot	Formula
qq	Formula

Trellis function	Data types
splom	Data frame formula, matrix
levelplot	Array, formula, matrix, table
contourplot	Array, formula, matrix, table
cloud	Formula, matrix, table
wireframe	Formula, matrix

For more details on arguments to lattice functions, see "Customizing Lattice Graphics" on page 308.

Custom Panel Functions

With standard graphics, you could easily superimpose points, lines, text, and other objects on existing charts. It's possible to do the same thing with lattice graphics, but it's a little trickier.

In order to add extra graphical elements to a lattice plot, you need to use a custom panel function. As we described above, low-level panel functions actually plot graphics. The high-level functions simply specify how data is divided between panels, and how different elements (legends, strips, axes, etc.) need to be added. To add extra elements to a lattice chart, you need to change the panel function.

As a simple example, let's add a diagonal line to Figure 15-2. To do this, we'll create a new custom panel function that calls both `panel.xyplot` and `panel.abline`. The new panel function will pass along its arguments to `panel.xyplot`. We'll specify a line that crosses the y-axis at 1 (through the `a=1` argument to `panel.abline`) and has slope 1 (through the `b=1` argument to `panel.abline`). Here's the code to generate this chart:

```
xyplot(y~x|z, data=d,
    panel=function(...){
        panel.abline(a=1,b=1)
        panel.xyplot(...)
    }
)
```

As you can see, the chart with the custom panel function (Figure 15-4) is identical to the chart we showed above for multiple panels (Figure 15-2, shown previously), except with the addition of the diagonal lines.

High-Level Lattice Plotting Functions

This section describes high-level lattice functions. (We'll cover panel functions in the next section.) We'll start with functions for plotting a single vector of values, then functions for plotting two variables, then functions for plotting three variables, and some other functions that build on these functions.

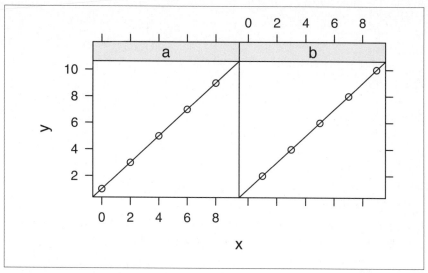

Figure 15-4. Scatter plot showing custom panel function

Univariate Trellis Plots

In this section, I'm going to use the same data set for most of the examples: births in the United States during 2006.[‡] The original data file contains a record for every birth in the United States during 2006, but the version that is included in the nutshell package only contains a 10% sample. Each record includes the following variables:

DOB_MM
   Month of birth

DOB_WK
   Day of week of birth

MAGER
   Mother's age

TBO_REC
   Total birth order

WTGAIN
   Weight gain (by mother)

SEX
   Sex of the child (M or F)

APGAR5
   Apgar score

[‡] This data set is available from *http://www.cdc.gov/nchs/data_access/Vitalstatsonline.htm*. I used the 2006 Birth Data File in this book. The data file is 3.1 GB uncompressed, which is way too big to load easily into R on a machine with only 4 GB. I used a Perl script to parse this file and return a limited number of records in CSV format.

DMEDUC
>   Mother's education

UPREVIS
>   Number of prenatal visits

ESTGEST
>   Estimated weeks of gestation

DMETH_REC
>   Delivery method

DPLURAL
>   "Plural" births (i.e., single, twins, triplets, etc.)

DBWT
>   Birth weight (in grams)

It takes a little while to process the raw data, so I've included a 10% sample of this data set within the nutshell package as births2006.smpl.

Processing the Birth Data

The natality files are gigantic; they're approximately 3.1 GB uncompressed. That's a little larger than R can easily process, so I used Perl to translate these files to a form easily readable by R. (It's possible to read and parse individual lines in R using the function scan, but I found that a little bit cumbersome. Perl is a lot cleaner and easier.) First, I used the following Perl script to process the raw file:

```perl
#! /usr/bin/perl
print "DOB_MM,DOB_WK,MAGER,TBO_REC,WTGAIN,SEX,APGAR5," .
        "DMEDUC,UPREVIS,ESTGEST,DMETH_REC,DPLURAL,DBWT\n";

while(<>) {
    my ($trash1,$DOB_MM,$trash2,$DOB_WK,$trash3,$MAGER,$trash4,
        $DMEDUC,$trash5,$TBO_REC,$trash6,$UPREVIS,$trash7,
        $WTGAIN,$trash8,$DMETH_REC,$trash9,$APGAR5,$trash10,
        $DPLURAL,$trash11,$SEX,$trash12,$ESTGEST,$trash13,$DBWT)
        = unpack("a18a2a8a1a59a2a65a2a59a1a52a2a4a2a125" .
                    "a1a11a2a6a1a12a1a9a2a15a4", $_);
    print "$DOB_MM,$DOB_WK,$MAGER,$TBO_REC,$WTGAIN,$SEX,$APGAR5," .
        "$DMEDUC,$UPREVIS,$ESTGEST,$DMETH_REC,$DPLURAL,$DBWT\n";
}
```

Next, I used the following R code to construct the data set:

```r
births2006.raw <- read.csv("~/Documents/book/data/births2006.csv")

dmeth_rec <- function(X) {
    f <- function(tst) {
        switch(tst,'Vaginal', 'C-section', '', '', '',
            '', '', '', 'Unknown');
    }
    as.factor(as.character(sapply(X,f)));
}
```

```
udmeth_rec <- function(X) {
    f <- function(tst) {
        switch(tst,
            'Vaginal (not VBAC)',  # 1
            'VBAC',                # 2
            'Primary C-section',   # 3
            'Repeat C-section',    # 4
            '',  # 5
            '',  # 6
            '',  # 7
            '',  # 8
            'Unstated'  # 9
            );
    }
    as.factor(as.character(sapply(X,f)));
}

dmeduc <- function(X) {
    f <- function(tst) {
        switch(tst,
            'Not on certificate',
          '0'='No formal education',
          '1'='1 Years of elementary school',
          '2'='2 Years of elementary school',
          '3'='3 Years of elementary school',
          '4'='4 Years of elementary school',
          '5'='5 Years of elementary school',
          '6'='6 Years of elementary school',
          '7'='7 Years of elementary school',
          '8'='8 Years of elementary school',
          '9'='1 year of high school',
          '10'='2 years of high school',
          '11'='3 years of high school',
          '12'='4 years of high school',
          '13'='1 year of college',
          '14'='2 years of college',
          '15'='3 years of college',
          '16'='4 years of college',
          '17'='5 or more years of college',
          '99'='Not stated'
            );
    }
    as.factor(as.character(sapply(X,f)));
}

tbo_rec <- function(x) { ifelse(x==9,NA,x) }

wtgain <- function(x) { ifelse(x==99,NA,x) }

apgar5 <- function(x) { ifelse(x==99,NA,x) }

estgest <- function(x) { ifelse(x==99,NA,x) }

dbwt <- function(x) { ifelse(x==9999,NA,x) }

dplural <- function(X) {
```

```
         f <- function(tst) {
                 switch(tst,'1 Single','2 Twin','3 Triplet',
                   '4 Quadruplet', '5 Quintuplet or higher')
                 }
         as.factor(as.character(sapply(X,f)));
         }

     births2006 <- transform(births2006.raw,
         TBO_REC=tbo_rec(TBO_REC),
         WTGAIN=wtgain(WTGAIN),
         APGAR5=apgar5(APGAR5),
         DMETH_REC=dmeth_rec(DMETH_REC),
         DMEDUC=dmeduc(DMEDUC),
         DPLURAL=dplural(DPLURAL),
         DBWT=dbwt(DBWT)
         );
```

Finally, I took a 10% sample of the original data set so that it would fit in the
nutshell package:

```
> births2006.idx <- sample(1:nrow(births2006),427323)
> births2006.smpl <- births2006[births2006.idx,]
> dim(births2006.smpl)
[1] 427323    13
```

Bar charts

To draw bar charts with Trellis graphics, use the function barchart. The default
method for barchart accepts a formula and a data frame as arguments:

```
barchart(x,
         data,
         panel = lattice.getOption("panel.barchart"),
         box.ratio = 2,
         ...)
```

You specify the formula with the argument x and the data frame with the argument
data. (I'll explain the rest of the arguments below.) However, you can also call
barchart on an object of class table:

```
barchart(x, data, groups = TRUE,
         origin = 0, stack = TRUE, ..., horizontal = TRUE)
```

To call barchart with an object of class table, simply call barchart with the argument
x set to a table. (You shouldn't specify an argument for data; if you do, barchart will
print a warning and ignore the argument.)

By default, the charts are actually drawn by the panel function panel.barchart:

```
panel.barchart(x, y, box.ratio = 1, box.width,
               horizontal = TRUE,
               origin = NULL, reference = TRUE,
               stack = FALSE,
               groups = NULL,
               col = if (is.null(groups)) plot.polygon$col
                     else superpose.polygon$col,
```

```
border = if (is.null(groups)) plot.polygon$border
        else superpose.polygon$border,
lty = if (is.null(groups)) plot.polygon$lty
        else superpose.polygon$lty,
lwd = if (is.null(groups)) plot.polygon$lwd
        else superpose.polygon$lwd,
...)
```

Let's start by calculating a table of the number of births by day of week and then printing a bar chart to show the number of births by day of week. It's the first time that we're using lattice graphics, so let's start by loading the lattice package:

```
> library(lattice)
> births.dow <- table(births2006.smpl$DOB_WK)
> barchart(births.dow)
```

The results are shown in Figure 15-5. This is the default format for the barchart function: horizontal bars, a frame along the outside, tick marks, and turquoise-colored bars (on screen).

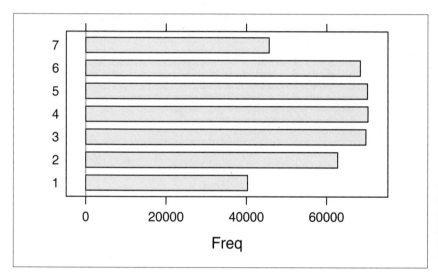

Figure 15-5. Births by day of week

Notice that many more babies are born on weekdays than on weekends. That's a little surprising: you might think that the number of births would be nearly the same, regardless of the day of the week. We'll use lattice graphics to explore this data set further, to see if we can better understand this phenomenon.

You might wonder if there is a difference in the number of births because of the delivery method; maybe doctors just schedule a lot of cesarean sections on weekdays, and natural births occur all the time. This is the type of question that the lattice package is great for answering. Let's start by eliminating records where the delivery method was unknown and then tabulate the number of births by day of week and method:

```
> births2006.dm <- transform(
+    births2006.smpl[births2006.smpl$DMETH_REC != "Unknown",],
+     DMETH_REC=as.factor(as.character(DMETH_REC)))
> dob.dm.tbl <- table(WK=births2006.dm$DOB_WK, MM=births2006.dm$DMETH_REC)
```

Now, let's plot the results:

```
> barchart(dob.dm.tbl)
```

The chart is shown in Figure 15-6. By default, barchart prints stacked bars with no legend. In Trellis terminology, the different colors show different *groups*. It does look like both types of births are less common on weekends, but it's tough to compare the number of each type of birth in this chart. Also, notice that the different shades aren't labeled, so it's not immediately obvious what each shade represents. Let's try to change the way the chart is displayed.

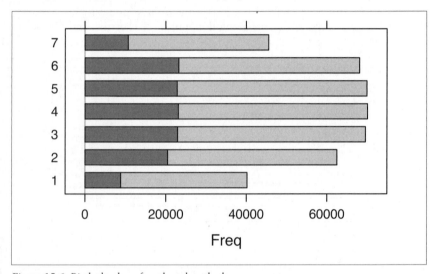

Figure 15-6. Births by day of week and method

As an alternative, let's try unstacking the bars (by specifying stack=FALSE) and adding a legend (by specifying auto.key=TRUE):

```
> # code to create file: remove from final version of book
> trellis.device(device.pdf, color=FALSE,
+    filename="~/Documents/book/current/figs/incoming/rian_1507.pdf",
+    width=4.3,height=4.3,units="in",res=72)
> barchart(dob.dm.tbl,stack=FALSE,auto.key=TRUE)
```

The results are shown in Figure 15-7. It's a little easier to see that both types of births decrease on weekends, but it's still a little difficult to compare values within each group. (When I try to focus on each group, I get distracted by the other group.) Different colored groups aren't the best choice for this data, so let's try a different approach.

First, let's try changing this chart in two ways. We'll split it into two different panels by telling barchart not to group by color, using the groups=FALSE argument. Second,

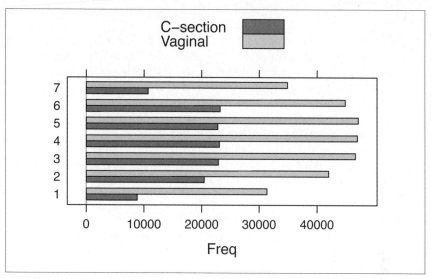

Figure 15-7. Births by day of week and method: unstacked bars

we'll change to columns (using the `horizontal=FALSE` argument), so we can easily compare the different values:

```
> barchart(dob.dm.tbl,horizontal=FALSE,groups=FALSE)
```

The new chart is shown in Figure 15-8. The two different charts are in different *panels*. Now, we can more clearly see what's going on. The number of vaginal births decreases on weekends, by maybe 25 to 30%. However, C-sections drop by 50 to 60%. As you can see, lattice graphics let you quickly try different ways to present information, helping you zero in on the method that best illustrates what is happening in the data.

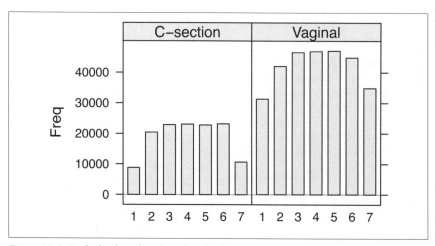

Figure 15-8. Births by day of week and method: two panels of columns

Dot plots

A good alternative to bar charts are Cleveland dot plots. Like bar charts, dot plots are useful for showing data where there is a single point for each category. Visually, they seem a lot less "busy" to me than bar charts, so I like using them to summarize larger data tables. To show dot plots in R, use the function `dotplot`:

```
dotplot(x,
        data,
        panel = lattice.getOption("panel.dotplot"),
        ...)
```

Much like `barchart`, the default method expects you to specify the data in a formula and a data frame, but there is a method for plotting tables as well:

```
## S3 method for class 'table':
dotplot(x, data, groups = TRUE, ..., horizontal = TRUE)
```

As an example of `dotplot`, let's look at a chart of data on births by day of week. Is the pattern we saw above a seasonal pattern? First, we'll create a new table counting births by month, week, and delivery method:

```
> dob.dm.tbl.alt <- table(WEEK=births2006.dm$DOB_WK,
+     MONTH=births2006.dm$DOB_MM,
+     METHOD=births2006.dm$DMETH_REC)
```

Next, we'll plot the results using a dot plot. In this plot, we'll keep on grouping, so that different delivery methods are shown in different colors (`groups=TRUE`). To help highlight differences, we'll disable stacking values (`stack=FALSE`). Finally, we'll print a key so that it's obvious what each symbol represents (`auto.key=TRUE`):

```
> dotplot(dob.dm.tbl.alt,stack=FALSE,auto.key=TRUE,groups=TRUE)
```

The results are shown in Figure 15-9. (To make the results print nicely, I generated these charts with the default black-and-white color scheme. If you try this yourself, the table may look slightly different. Depending on your platform, you'll probably see hollow blue circles for C-section births and hollow purple sections for vaginal births.) As you can see, there are slight seasonal differences, but the overall pattern remains the same.

As another example of dot plots, let's look at the tire failure data. In 2003, the National Highway Traffic Safety Administration (NHTSA) began a study into the durability of radial tires on light trucks. (This was three years after the Firestone recall of tires for Ford Explorers.) The NHTSA performed the tests in Phoenix, because it felt that the hot and dry conditions would be unusually stressful for tires (and because it had noted that many tire failures occur in the American Southwest). Over the next few years, it conducted hundreds of different tests on tires and released the data to the public. (See *http://www.nhtsa.gov/portal/site/nhtsa/menuitem.8027fe7cfb6e727568d07a30343c44cc/* for links to this study.)

Tests were carried out on six different types of tires. Here is a table of the characteristics of the tires.

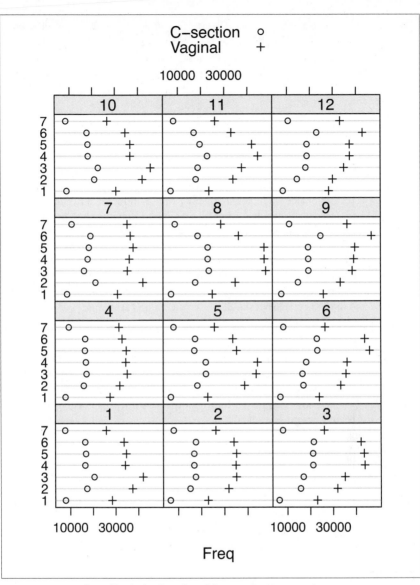

Figure 15-9. Number of births by day of week by month

Tire	Size	Load Index	Speed Rating	Brand	Model	OE Vehicle	OE Model
B	P195/65R15	89	S	BF Goodrich	Touring T/A	Chevy	Cavalier
C	P205/65R15	92	V	Goodyear	Eagle GA	Lexus	ES300
D	P235/75R15	108	S	Michelin	LTX M/S	Ford, Dodge	E 150 Van, Ram Van 1500

Tire	Size	Load Index	Speed Rating	Brand	Model	OE Vehicle	OE Model
E	P265/75R16	114	S	Firestone	Wilderness AT	Chevy/ GMC	Silverado, Tahoe, Yukon
H	LT245/75R16/E	120/116	Q	Pathfinder	ATR A/S OWL	NA	NA
L	255/65R16	109	H	General	Grabber ST A/S	Mercedes	ML320

As an example, we're going to look at one particular batch of tests from this study. The test was called a "Stepped-Up Speed to failure test." In this test, tires were mounted on testing devices. The testing facility then conducted a number of basic tests on the tires to check that they were intact. The test facility then proceeded to test the tires at increasing speeds until the tires failed. Specifically, the testing facility tested each tire at a specific speed for 1 hour, and then it proceeded to increase the speed in 10-km/h increments until either (a) the tire failed or (b) a prescribed limit was reached for each tire. (The limit was dependent on the speed rating for the tire.) After the limit was reached, the test was run continuously until the tire failed. The test data set is in the package nutshell, under the name tires.sus.

The data set contains a lot of information, but we're going to focus on only three variables. Time_To_Failure is the time before each tire failed (in hours), Speed_At_Failure_km_h is the testing speed at which the tire failed, and Tire_Type is the type of tire tested. We know that tests were only run at certain stepped speeds; despite the fact that speed is a numeric variable, we can treat it as a factor. So, we can use dot plots to show the one continuous variable (time to failure) by the speed at failure for each different type of tire:

```
> library(nutshell)
> data(tires.sus)
> dotplot(as.factor(Speed_At_Failure_km_h)~Time_To_Failure|Tire_Type,
+    data=tires.sus)
```

The result is shown in Figure 15-10. This diagram let's us clearly see how quickly tires failed in each of the tests. For example, all type D tires failed quickly at the testing speed of 180 km/h, but some type H tires lasted a long time before failure. We'll revisit this example in "Comparing means" on page 344.

Histograms

A very popular chart for showing the distribution of a variable is the histogram. You can plot histograms in the trellis package with the function histogram:

```
histogram(x,
          data,
          allow.multiple, outer = TRUE,
          auto.key = FALSE,
          aspect = "fill",
          panel = lattice.getOption("panel.histogram"),
          prepanel, scales, strip, groups,
          xlab, xlim, ylab, ylim,
```

```
type = c("percent", "count", "density"),
nint = if (is.factor(x)) nlevels(x)
else round(log2(length(x)) + 1),
endpoints = extend.limits(range(as.numeric(x), finite = TRUE),
                          prop = 0.04),
breaks,
equal.widths = TRUE,
drop.unused.levels = lattice.getOption("drop.unused.levels"),
...,
lattice.options = NULL,
default.scales = list(),
subscripts,
subset)
```

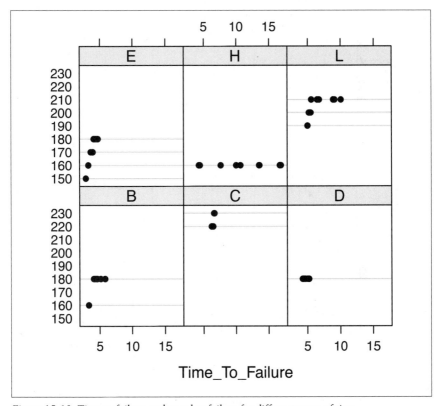

Figure 15-10. Time to failure and speed at failure for different types of tires

By default, histograms are drawn by `panel.histogram`:

```
panel.histogram(x,
                breaks,
                equal.widths = TRUE,
                type = "density",
                nint = round(log2(length(x)) + 1),
                alpha, col, border, lty, lwd,
                ...)
```

As an example of histograms, let's look at average birth weights, grouped by number of births:

```
> histogram(~DBWT|DPLURAL, data=births2006.smpl)
```

The results are shown in Figure 15-11. Notice that the panels are ordered alphabetically by the conditioning variable. (That's why the group names have the numbers at the front.) Also notice that the `histogram` function tries to fill in all the available space with squarish panels. This helps make each chart readable by itself, but makes it difficult to compare the different groups.

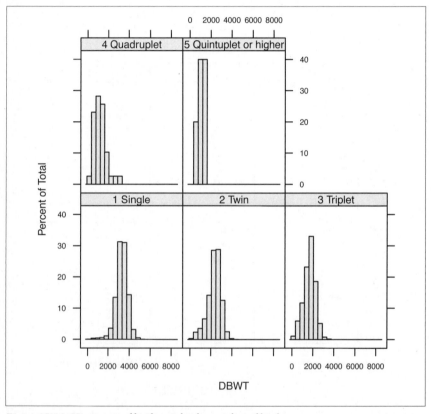

Figure 15-11. Histogram of birth weights by number of births

To make it easier to compare groups, we can explicitly stack the charts on top of each other using the `layout` variable:

```
> histogram(~DBWT|DPLURAL, data=births2006.smpl, layout=c(1, 5))
```

The resulting chart is shown in Figure 15-12. As you can see, birth weights are roughly normally distributed within each group, but the mean weight drops as the number of births increases.

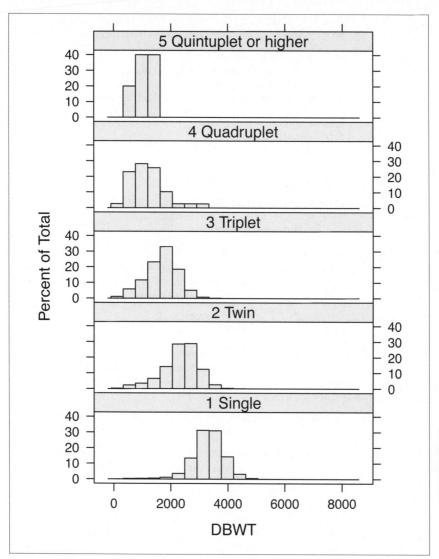

Figure 15-12. Histogram of birth weights by number of births: vertically stacked

Density plots

If you'd like to see a single line showing the distribution, instead of a set of columns representing bins, you can use kernel density plots. To draw them in R, use the function densityplot:

```
densityplot(x,
            data,
            allow.multiple = is.null(groups) || outer,
            outer = !is.null(groups),
            auto.key = FALSE,
```

```
                  aspect = "fill",
                  panel = lattice.getOption("panel.densityplot"),
                  prepanel, scales, strip, groups, weights,
                  xlab, xlim, ylab, ylim,
                  bw, adjust, kernel, window, width, give.Rkern,
                  n = 50, from, to, cut, na.rm,
                  drop.unused.levels = lattice.getOption("drop.unused.levels"),
                  ...,
                  lattice.options = NULL,
                  default.scales = list(),
                  subscripts,
                  subset)
```

By default, panels are drawn by `panel.densityplot`:

```
panel.densityplot(x, darg, plot.points = "jitter", ref = FALSE,
                  groups = NULL, weights = NULL,
                  jitter.amount, type, ...)
```

Let's redraw the example above, replacing the histogram with a density plot. By default, `densityplot` will draw a strip chart under each chart, showing every data point. However, because the data set is so big (there are 427,432 observations), we'll tell `densityplot` not to do this by specifying `plot.points=FALSE`:

```
> densityplot(~DBWT|DPLURAL,data=births2006.smpl,
+    layout=c(1,5),plot.points=FALSE)
```

The results are shown in Figure 15-13. One advantage of density plots over histograms is that you can stack them on top of each other and still read the results. By changing the conditioning variable (`DPLURAL`) to a grouping variable, we can stack these charts on top of each other:

```
> densityplot(~DBWT,groups=DPLURAL,data=births2006.smpl,
+    plot.points=FALSE,auto.key=TRUE)
```

The superimposed density plots are shown in Figure 15-14. As you can see, it's easier to compare distribution shapes (and centers) by superimposing the charts.

Strip plots

A good alternative to histograms are strip plots, especially when there isn't much data to plot. Strip plots look similar to dot plots, but they show different information. Dot plots are designed to show one value per category (often a mean or a sum), while strip plots show many values. You can think of strip plots as one-dimensional scatter plots. To draw strip plots in R, use the `stripplot` function:

```
stripplot(x,
          data,
          panel = lattice.getOption("panel.stripplot"),
          ...)
```

By default, panels are drawn by `panel.stripplot`:

```
panel.stripplot(x, y, jitter.data = FALSE,
                factor = 0.5, amount = NULL,
                horizontal = TRUE, groups = NULL,
                ...)
```

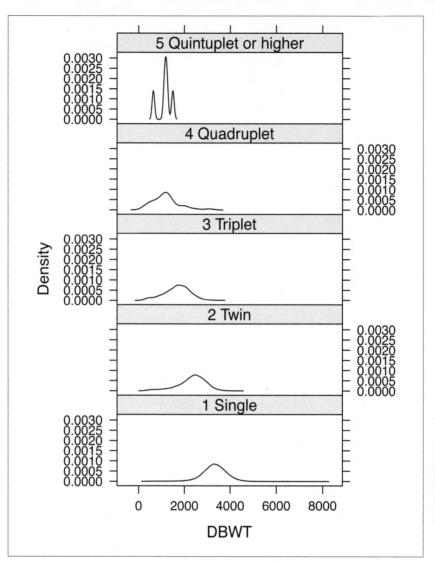

Figure 15-13. Density plots showing birth weight by number of babies

As an example of a strip plot, let's look at the weights of babies born in sets of 4 or more. There were only 44 observations in our data set that match this description, so a strip plot is a reasonable way to show density. In this case, we'll use the `subset` argument to specify the set of observations we want to plot, and add some random vertical noise to make the points easier to read by specifying `jitter.data=TRUE`:

```
> stripplot(~DBWT, data=births2006.smpl,
+   subset=(DPLURAL=="5 Quintuplet or higher" |
+         DPLURAL=="4 Quadruplet"),
+   jitter.data=TRUE)
```

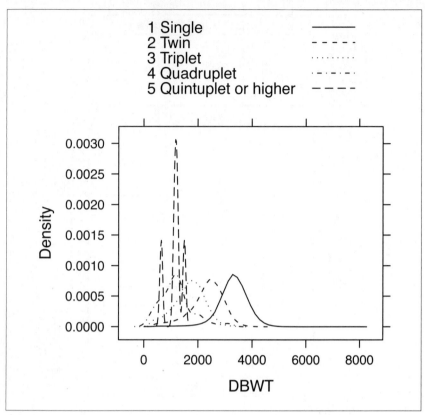

Figure 15-14. Superimposed density plots showing birth weight by number of babies

The resulting chart is shown in Figure 15-15.

Univariate quantile-quantile plots

Another useful plot that you can generate within the lattice package is the quantile-quantile plot. A quantile-quantile plot compares the distribution of actual data values to a theoretical distribution. Specifically, it plots quantiles of the observed data against quantiles of a theoretical distribution. If the plotted points form a straight diagonal line (from top right to bottom left), then it is likely that the observed data comes from the theoretical distribution. Quantile-quantile plots are a very powerful technique for seeing how closely a data set matches a theoretical distribution (or how much it deviates from it).

To plot quantile-quantile plots using lattice graphics, use the function qqmath:

```
qqmath(x,
        data,
        allow.multiple = is.null(groups) || outer,
        outer = !is.null(groups),
        distribution = qnorm,
        f.value = NULL,
```

```
auto.key = FALSE,
aspect = "fill",
panel = lattice.getOption("panel.qqmath"),
prepanel = NULL,
scales, strip, groups,
xlab, xlim, ylab, ylim,
drop.unused.levels = lattice.getOption("drop.unused.levels"),
...,
lattice.options = NULL,
default.scales = list(),
subscripts,
subset)
```

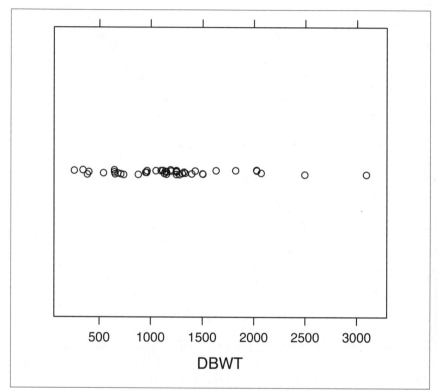

Figure 15-15. Weight of babies born in sets of four or more

By default, panels are drawn by `panel.qqmath`:

```
panel.qqmath(x, f.value = NULL,
            distribution = qnorm,
            qtype = 7,
            groups = NULL, ...
```

By default, the function qqmath compares the sample data to a normal distribution. If the sample data is really normally distributed, you'll see a vertical line. As an example, let's plot 100,000 random values from a normal distribution to show what qqmath does:

```
> qqmath(rnorm(100000))
```

The results are shown in Figure 15-16.

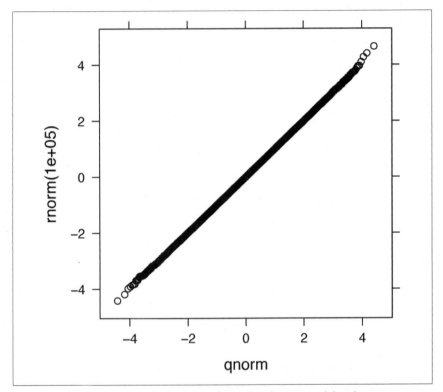

Figure 15-16. Quantile-quantile plot for random values from normal distribution

Let's plot a set of quantile-quantile plots for the birth weight data. Because the data set is rather large, we'll only plot a random sample of 50,000 points:

```
qqmath(~DBWT|DPLURAL,
        data=births2006.smpl[sample(1:nrow(births2006.smpl), 50000), ],
        pch=19,
        cex=0.25,
        subset=(DPLURAL != "5 Quintuplet or higher"))
```

As you can see from Figure 15-17, the distribution of birth weights is not quite normal.

As another example, let's look at real estate prices in San Francisco in 2008 and 2009. This data set is included in the nutshell package as sanfrancisco.home.sales. (See "More About the San Francisco Real Estate Prices

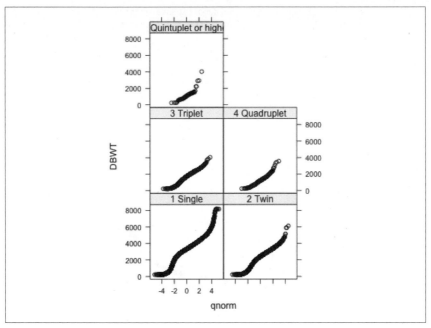

Figure 15-17. Quantile-quantile plots for birth weights

Data Set" on page 290 for more information on this data set.) Here is how to load the data:

```
> library(nutshell)
> data(sanfrancisco.home.sales)
```

Intuitively, it doesn't make sense for real estate prices to be normally distributed. There are far more people with below-average incomes than above-average incomes. The lowest recorded price in the data set is $100,000; the highest is $9,500,000. Let's take a look at this distribution with qqmath:

```
> qqmath(~price,data=sanfrancisco.home.sales)
```

The distribution is shown in Figure 15-18. As expected, the distribution is not normal. It looks exponential, so let's try a log transform:

```
> qqmath(~log(price),data=sanfrancisco.home.sales)
```

A log transform yields a distribution that looks pretty close to normally distributed (see Figure 15-19). Let's take a look at how the distribution changes based on the number of bedrooms. To do this, we'll split the distribution into groups and change the way the points are plotted. Specifically, we'll plot smooth lines instead of individual points. (Point type is actually an argument for panel.xyplot, which is used to draw the chart.) We'll add a key to the plot (using auto.key=TRUE). We'll pass an explicit subset as an argument to the function instead of using the subset argument. (This helps clean up the key, which would show unused factor levels otherwise.)

```
> qqmath(~log(price),groups=bedrooms,
+    data=subset(sanfrancisco.home.sales,
```

```
+                    !is.na(bedrooms)&bedrooms>0&bedrooms<7),
+    auto.key=TRUE,drop.unused.levels=TRUE,type="smooth")
> dev.off()
```

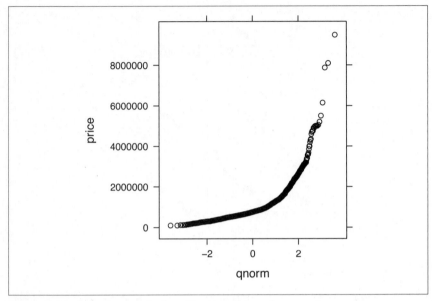

Figure 15-18. Quantile-quantile plot of San Francisco real estate prices

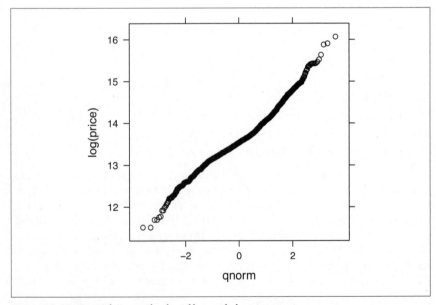

Figure 15-19. Quantile-quantile plot of log-scaled property prices

Notice that the lines are separate, with higher values for higher numbers of bedrooms (see Figure 15-20). We can do the same thing for square footage (see Figure 15-21). (I used the function `cut2` from the package `HMisc` to divide square footages into six even quantiles.)

```
> library(Hmisc)
> qqmath(~log(price),groups=cut2(squarefeet,g=6),
+    data=subset(sanfrancisco.home.sales,!is.na(squarefeet)),
+    auto.key=TRUE, drop.unused.levels=TRUE, type="smooth")
```

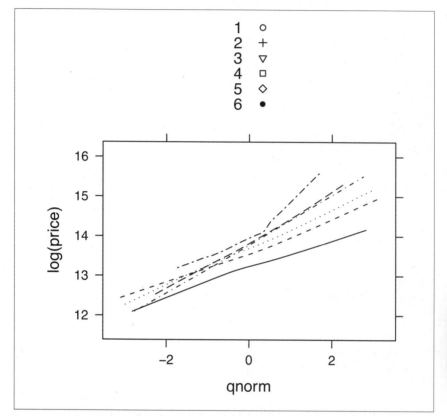

Figure 15-20. Quantile-quantile plots of logs of property prices for different numbers of bedrooms

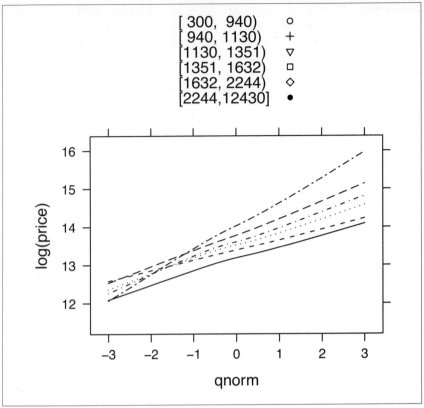

Figure 15-21. Quantile-quantile plots of logs of property prices for different numbers of square feet

Here, the separation is even more clear. We can see the same separation by neighborhood. (We'll come back to this analysis, in Chapter 20.)

More About the San Francisco Real Estate Prices Data Set

In a few places in this chapter (and again in Chapters 20 and 21), we'll use a data set consisting of real estate sale prices in San Francisco between February 13, 2008, and July 14, 2009:

```
> names(sanfrancisco.home.sales)
 [1] "street"    "city"       "zip"        "saledate"   "price"
 [6] "bedrooms"  "squarefeet" "lotsize"    "yearbuilt"  "condolike"
```

In the San Francisco Bay area, real estate sales are published in the newspapers once a week. I put together this data set by compiling information from multiple papers. The reason I'm including 17 months of data is because that was what was available when I wrote this chapter. The data set contains 3,281 observations and 10 variables:

---

street
> Street address for the property.

city
> City in which the property was located. (In this data set, it's `'San Fran cisco'` for every observation.)

zip
> Zip code for the property.

saledate
> Approximate date on which the sale was recorded. (Different papers sometimes disagree by a day or two.)

price
> Sales price for the property.

bedrooms
> A count of the number of bedrooms.

squarefeet
> Interior space in square feet.

lotsize
> Lot size in square feet.

yearbuilt
> Year in which the property was built.

condolike
> Variable derived from `street`, to indicate if the address was qualified by a unit number. (Indicates the presence of a `'#'` in the variable `street`.)

latitude, longitude
> Geographic coordinates for the property.

This is a real data set, so it's not completely clean.[§] It contains data compiled from many sources: real estate listings, self-reported data, government records. So, there may be errors and inconsistencies in the data. Moreover, there are some missing values.

I picked this data set as an example because I had some questions about the way that real estate data is reported in the media. Writers often talk about the number of sales, or the median price, or the price per square foot. I wanted to know a little more about real estate prices. Is there a premium for bedrooms (above square footage)? When the market slowed down in the housing bust, did it slow down across all price points? How sensitive are median prices to one-time events (like a large new condominium building)?

In case you're wondering where this data came from, here's a detailed explanation:

- First, I downloaded real estate sales listings from San Francisco Bay area newspaper web sites. (I wrote a spider to grab and parse the data.) This is how I got sale dates, street addresses, sales prices, bedroom counts, home sizes, lot sizes, and years built.

[§] It's not completely dirty, either. I spent some time cleaning up and correcting the data: removing blatant duplicates, adding years to some recent condo listings, and a few other fixes.

- Next, I got latitude and longitude information for each address from different web services. I merged these files together outside R (using SQLite).

- I downloaded neighborhood information from Zillow.com. You can download neighborhood data from *http://www.zillow.com/howto/api/neighbor hood-boundaries.htm*. (By the way, it's hard to do what Zillow.com does. See "Machine Learning Algorithms for Regression" on page 405 for some examples of price prediction.)

- Finally, I loaded the data into R, merged in neighborhood information, and finished creating the data set.

Here is the code that I used to put the data set together. I used some special packages for reading spacial data in order to load the neighborhood data and determine in which neighborhood each home was located:

```
# load in the shapefile
library(sp)
library(maptools)
# ca.neighborhood.shapes <- read.shape("ZillowNeighborhoods-CA.shp")
ca.neighborhood.shapes <- readShapePoly("ZillowNeighborhoods-CA.shp")
# extract san francisco coordinates
sf.neighborhood.shapes <-
  ca.neighborhood.shapes[ca.neighborhood.shapes$CITY=="San Francisco",]
# function to look up shapes
neighborhood <- function(s, lon, lat) {
    names <- s$NAME;
    for (name in names) {
        lons <- s[s$NAME==name,]@polygons[[1]]@Polygons[[1]]@coords[,1];
        lats <- s[s$NAME==name,]@polygons[[1]]@Polygons[[1]]@coords[,2];
        res <- point.in.polygon(lon,lat,lons,lats);
        if (res==1) {
            return(name);
            }
        }
    return(NA);
}
map_neighborhoods <- function(s, lons, lats) {
    neighborhoods <- rep(NA,length(lons));
    for (i in 1:length(lons)) {
        neighborhoods[i] <- neighborhood(s, lons[i], lats[i]);
        }
    return(neighborhoods);
}
# loading sf data with coordinates
sanfrancisco.home.sales.raw <- read.csv("san_fran_re_sales_wcoors.csv")
# exclude bad coordinates (outside SF)
 sanfrancisco.home.sales.clean <- transform(sanfrancisco.home.sales.raw,
    latitude=ifelse(latitude>37.7&latitude<37.85,latitude,NA),
    longitude=ifelse(latitude>37.7&latitude<37.85,longitude,NA),
    date=as.Date(date,format="%m/%d/%Y"),
    lotsize=ifelse(lotsize<10000,lotsize,NA),
    month=cut(as.Date(date,format="%m/%d/%Y"),"month"),
    lotsize=ifelse(lotsize<15000,lotsize,NA)
 )
# transform date fields
```

```
# finally, build the data set with properly named neighborhoods
sanfrancisco.home.sales <- transform(sanfrancisco.home.sales.clean,
  neighborhood=map_neighborhoods(
    sf.neighborhood.shapes, longitude, latitude))
save(sanfrancisco.home.sales,file="sanfrancisco.home.sales.RData")
```

Bivariate Trellis Plots

This section describes Trellis plots for plotting two variables. Many real data sets (for example, financial data) record relationships between multiple numeric variables. The tools in this section can help you examine those relationships.

Scatter plots

To generate scatter plots with the `trellis` package, use the function `xyplot`:

```
xyplot(x,
       data,
       allow.multiple = is.null(groups) || outer,
       outer = !is.null(groups),
       auto.key = FALSE,
       aspect = "fill",
       panel = lattice.getOption("panel.xyplot"),
       prepanel = NULL,
       scales = list(),
       strip = TRUE,
       groups = NULL,
       xlab,
       xlim,
       ylab,
       ylim,
       drop.unused.levels = lattice.getOption("drop.unused.levels"),
       ...,
       lattice.options = NULL,
       default.scales,
       subscripts = !is.null(groups),
       subset = TRUE)
```

Most of the work is done by the panel function `panel.xyplot`:

```
panel.xyplot(x, y, type = "p",
             groups = NULL,
             pch, col, col.line, col.symbol,
             font, fontfamily, fontface,
             lty, cex, fill, lwd,
             horizontal = FALSE, ...,
             jitter.x = FALSE, jitter.y = FALSE,
             factor = 0.5, amount = NULL)
```

As an example of a scatter plot, let's take a look at the relationship between house size and price. Let's start with a simple scatter plot, showing size and price:

```
> xyplot(price~squarefeet,data=sanfrancisco.home.sales)
```

The results of this command are shown in Figure 15-22. It looks like there is a rough correspondence between size and price (the plot looks vaguely cone shaped). This chart is hard to read, so let's try modifying it. Let's trim outliers (sales prices over 4,000,000 and properties over 6,000 square feet) using the subset argument. Additionally, let's take a look at how this relationship varies by zip code. San Francisco is a pretty big place, and not all neighborhoods are equally in demand. (You probably know the cliché about the first three rules of real estate: location, location, location.) So, additionally, let's try splitting this relationship by zip code.

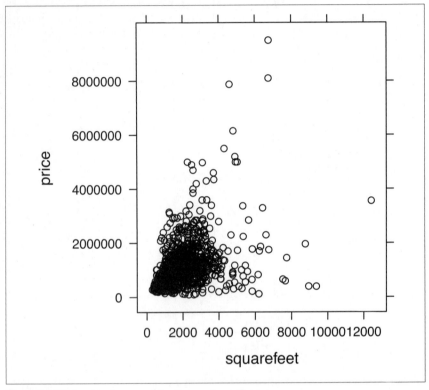

Figure 15-22. Scatter plot comparing house size and price

Before plotting the price data, let's pick a subset of zip codes to plot. A few parts of the city are sparsely populated (like the financial district, 94104) and don't have enough data to make plotting interesting. Also, let's exclude zip codes where square footage isn't available:

```
> table(subset(sanfrancisco.home.sales,!is.na(squarefeet),select=zip))
```

94100	94102	94103	94104	94105	94107	94108	94109	94110	94111	94112
2	52	62	4	44	147	21	115	161	12	192
94114	94115	94116	94117	94118	94121	94122	94123	94124	94127	94131
143	101	124	114	92	92	131	71	85	108	136
94132	94133	94134	94158							
82	47	105	13							

So, we'll exclude 94100, 94104, 94108, 94111, 94133, and 94158 because there are too few sales to be interesting. (Note the **strip** argument. This simply prints the zip codes with the plots.)

```
> trellis.par.set(fontsize=list(text=7))
> xyplot(price~squarefeet|zip, data=sanfrancisco.home.sales,
+    subset=(zip!=94100 & zip!=94104 & zip!=94108 &
+            zip!=94111 & zip!=94133 & zip!=94158 &
+            price<4000000 &
+            ifelse(is.na(squarefeet),FALSE,squarefeet<6000)),
+    strip=strip.custom(strip.levels=TRUE))
```

The resulting plot is shown in Figure 15-23. Now, the linear relationship is much more pronounced. Note the different slopes in different neighborhoods. As you might expect, some up-and-coming neighborhoods (like zip code 94110, which includes the Mission and Bernal Heights) are more shallowly sloped, while ritzy neighborhoods (like zip code 94123, which includes the Marina and Cow Hollow) are more steeply sloped.

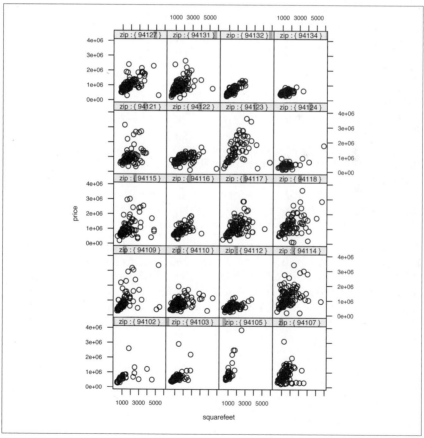

Figure 15-23. Scatter plot comparing house size and price by zip code

We can make this slightly more readable by using neighborhood names. Let's rerun the code, conditioning by neighborhood. We'll also add a diagonal line to each plot (through a custom panel function) to make the charts even easier to read. We'll also change the default points plotted to be solid (through the `pch=19` argument) and shrink them to a smaller size (through the `cex=.2` argument):

```
> trellis.par.set(fontsize=list(text=7))
> dollars.per.squarefoot <- mean(
+    sanfrancisco.home.sales$price / sanfrancisco.home.sales$squarefeet,
+    na.rm=TRUE);
> xyplot(price~squarefeet|neighborhood,
+    data=sanfrancisco.home.sales,
+    pch=19,
+    cex=.2,
+    subset=(zip!=94100 & zip!=94104 & zip!=94108 &
             zip!=94111 & zip!=94133 & zip!=94158 &
             price<4000000 &
             ifelse(is.na(squarefeet),FALSE,squarefeet<6000)),
+    strip=strip.custom(strip.levels=TRUE,
+    horizontal=TRUE,
+    par.strip.text=list(cex=.8)),
+    panel=function(...) {
+    panel.abline(a=0,b=dollars.per.squarefoot);
+    panel.xyplot(...);
+    }
+ )
```

This plot is shown in Figure 15-24.

Box plots in lattice

The San Francisco home sales data set was taken from a particularly interesting time: the housing market crash. (The market fell a little late in San Francisco compared to other cities.) Let's take a look at how prices changed over time during this period. We could plot just the median price or mean price, or the number of sales. However, the lattice package gives us tools that will let us watch how the whole distribution changed over time. Specifically, we can use box plots.

Box plots in the lattice package are just like box plots drawn with the graphics package, as described in "Box Plots" on page 240. The boxes represent prices from the 25th through the 75th percentiles (the interquartile range), the dots represent median prices, and the whiskers represent the minimum or maximum values. (When there are values that stretch beyond 1.5 times the length of the interquartile range, the whiskers are truncated at those extremes.)

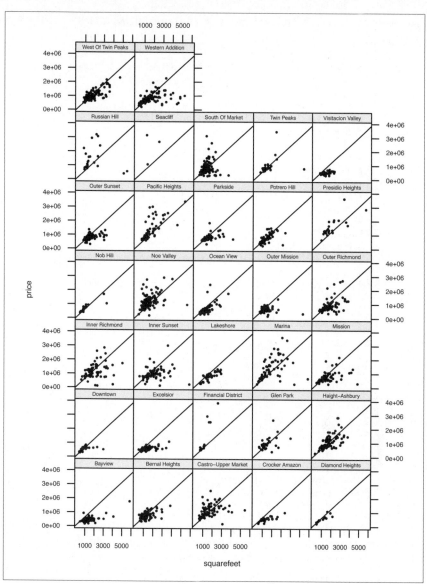

Figure 15-24. Scatter plot comparing house size and price by neighborhood

To show box plots with Trellis graphics, use the function `bwplot`:

```
bwplot(x,
       data,
       allow.multiple = is.null(groups) || outer,
       outer = FALSE,
       auto.key = FALSE,
       aspect = "fill",
       panel = lattice.getOption("panel.bwplot"),
       prepanel = NULL,
       scales = list(),
       strip = TRUE,
       groups = NULL,
       xlab,
       xlim,
       ylab,
       ylim,
       box.ratio = 1,
       horizontal = NULL,
       drop.unused.levels = lattice.getOption("drop.unused.levels"),
       ...,
       lattice.options = NULL,
       default.scales,
       subscripts = !is.null(groups),
       subset = TRUE)
```

This function will, in turn, call `panel.bwplot`:

```
panel.bwplot(x, y, box.ratio = 1,
             box.width = box.ratio / (1 + box.ratio),
             horizontal = TRUE,
             pch, col, alpha, cex,
             font, fontfamily, fontface,
             fill, varwidth = FALSE,
             notch = FALSE, notch.frac = 0.5,
             ...,
             levels.fos,
             stats = boxplot.stats,
             coef = 1.5,
             do.out = TRUE)
```

Let's show a set of box plots, with one plot per month. We'll need to round the date ranges to the nearest month. A convenient way to do this in R is with the cut function. Here's the number of sales by month in this data set:

```
> table(cut(sanfrancisco.home.sales$saledate,"month"))
```

```
2008-02-01 2008-03-01 2008-04-01 2008-05-01 2008-06-01 2008-07-01
       139        230        267        253        237        198
2008-08-01 2008-09-01 2008-10-01 2008-11-01 2008-12-01 2009-01-01
       253        223        272        118        181        114
2009-02-01 2009-03-01 2009-04-01 2009-05-01 2009-06-01 2009-07-01
       123        142        116        180        150         85
```

As you may remember from above, the cutoff dates don't fall neatly on the beginning and ending of each month:

```
> min(sanfrancisco.home.sales$saledate)
[1] "2008-02-13"
> max(sanfrancisco.home.sales$saledate)
[1] "2009-07-14"
```

So don't focus too much on the volumes in February 2008 or July 2009. (Volume was much lower in the spring.) Let's take a look at the distribution of sales prices by month. Here's the code to present this data using the default representation:

```
> bwplot(price~cut(saledate,"month"),data=sanfrancisco.home.sales)
```

Unfortunately, this doesn't produce an easily readable plot, as you can see in Figure 15-25. It's clear that there are a large number of outliers that are making the plot hard to see. Box plots assume a normal distribution, but this doesn't make intuitive sense for real estate prices (as we saw in "Univariate quantile-quantile plots" on page 284). Let's try plotting the box plots again, this time with the log-transformed values. To make it more readable, we'll change to vertical box plots and rotate the text at the bottom:

```
> bwplot(log(price)~cut(saledate,"month"),
+     data=sanfrancisco.home.sales,
+     scales=list(x=list(rot=90)))
```

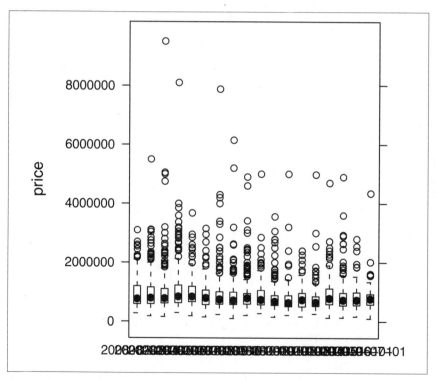

Figure 15-25. Box plot of real esate prices by month: first attempt

Taking a look at the plot (shown in Figure 15-26), we can more clearly see some trends. Median prices moved around a lot during this period, though the interquartile range moved less. Moreover, it looks like sales at the high end of the market slowed down quite a bit (looking at the outliers on the top and the top whiskers). But, interestingly, the basic distribution appears pretty stable from month to month.

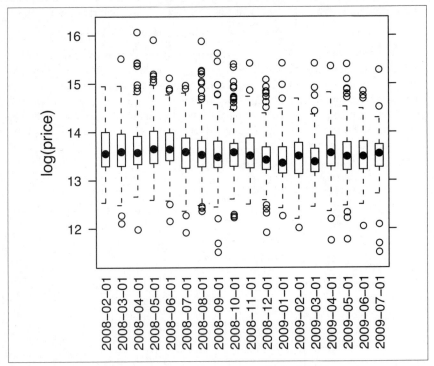

Figure 15-26. Box plot of real estate prices by month (log transformed)

Scatter plots matrices

If you would like to generate a matrix of scatter plots for many different pairs of variables, use the `splom` function:

```
splom(x,
      data,
      auto.key = FALSE,
      aspect = 1,
      between = list(x = 0.5, y = 0.5),
      panel = lattice.getOption("panel.splom"),
      prepanel,
      scales,
      strip,
      groups,
      xlab,
      xlim,
      ylab = NULL,
```

```
       ylim,
       superpanel = lattice.getOption("panel.pairs"),
       pscales = 5,
       varnames,
       drop.unused.levels,
       ...,
       lattice.options = NULL,
       default.scales,
       subset = TRUE)
```

Most of the work is done by `panel.splom`:

```
panel.splom(...)
```

Bivariate quantile-quantile plots

If you would like to generate quantile-quantile plots for comparing two distributions, use the function qq:

```
qq(x, data, aspect = "fill",
    panel = lattice.getOption("panel.qq"),
    prepanel, scales, strip,
    groups, xlab, xlim, ylab, ylim, f.value = NULL,
    drop.unused.levels = lattice.getOption("drop.unused.levels"),
    ...,
    lattice.options = NULL,
    qtype = 7,
    default.scales = list(),
    subscripts,
    subset)
```

Trivariate Plots

If you would like to plot three-dimensional data with Trellis graphics, there are several functions available.

Level plots

To plot three-dimensional data in flat grids, with colors showing different values for the third dimension, use the `levelplot` function:

```
levelplot(x,
          data,
          allow.multiple = is.null(groups) || outer,
          outer = TRUE,
          aspect = "fill",
          panel = lattice.getOption("panel.levelplot"),
          prepanel = NULL,
          scales = list(),
          strip = TRUE,
          groups = NULL,
          xlab,
          xlim,
          ylab,
          ylim,
          at,
          cuts = 15,
```

```
        pretty = FALSE,
        region = TRUE,
        drop.unused.levels = lattice.getOption("drop.unused.levels"),
        ...,
        lattice.options = NULL,
        default.scales = list(),
        colorkey = region,
        col.regions,
        alpha.regions,
        subset = TRUE)
```

Most of the work is done by `panel.levelplot`:

```
panel.levelplot(x, y, z,
                subscripts,
                at = pretty(z),
                shrink,
                labels,
                label.style = c("mixed", "flat", "align"),
                contour = FALSE,
                region = TRUE,
                col = add.line$col,
                lty = add.line$lty,
                lwd = add.line$lwd,
                ...,
                col.regions = regions$col,
                alpha.regions = regions$alpha)
```

As an example of level plots, we will look at the San Francisco home sales data set. Let's start by looking at the number of home sales in different parts of the city. To do this, we'll need to use that coordinate data in the San Francisco home sales data set. Unfortunately, we can't use the coordinates directly; the coordinates are too precise, so the `levelplot` function simply plots a large number of points. (Try executing `levelplot(price~latitude+longitude)` to see what I mean.)

We'll need to break the data into bins and count the number of homes within each bin. To do this, we'll use the `table` and `cut` functions:

```
attach(sanfrancisco.home.sales)
levelplot(table(cut(longitude,breaks=40),
                cut(latitude,breaks=40)),
          scales=list(y=list(cex=.5),
                      x=list(rot=90,cex=.5)))
```

This plot is shown in Figure 15-27. If we were interested in looking at the average sales price by area, we could use a similar strategy. Instead of `table`, we'll use the `tapply` function to aggregate observations. And while we're at it, we'll cut out the axis labels:

```
> levelplot(tapply(price,
+                   INDEX=list(cut(longitude,breaks=40),
+                              cut(latitude, breaks=40)),
+                   FUN=mean),
+           scales=list(draw=FALSE))
```

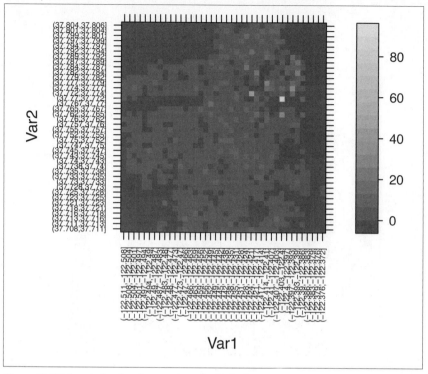

Figure 15-27. Level plot showing number of sales by location

This figure is shown in Figure 15-28. And, of course, you can use conditioning values with level plots. Let's look at the number of home sales, by numbers of bedrooms. We'll simplify the data slightly by looking at houses with zero to four bedrooms and then houses with five bedrooms or more. We'll also cut the number of breaks to keep the charts legible:

```
bedrooms.capped <- ifelse(bedrooms<5,bedrooms,5);
levelplot(table(cut(longitude,breaks=25),
                cut(latitude,breaks=25),
                bedrooms.capped),
          scales=list(draw=FALSE))
```

This figure is shown in Figure 15-29.

Contour plots

If you would like to show contour plots with lattice (which resemble topographic maps), use the `contourplot` function:

```
contourplot(x,
            data,
            panel = lattice.getOption("panel.contourplot"),
            cuts = 7,
            labels = TRUE,
            contour = TRUE,
```

```
            pretty = TRUE,
            region = FALSE,
            ...)
```

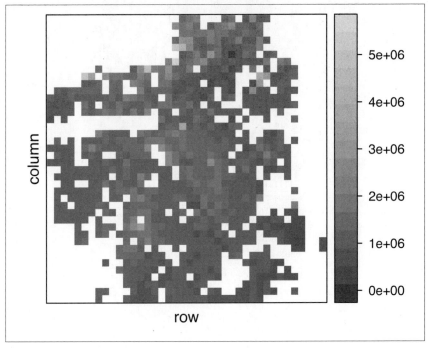

Figure 15-28. Level plot showing mean price by location

Cloud plots

To plot points in three dimensions (technically, projections into two dimensions of the points in three dimensions), use the function `cloud`:

```
cloud(x,
      data,
      allow.multiple = is.null(groups) || outer,
      outer = FALSE,
      auto.key = FALSE,
      aspect = c(1,1),
      panel.aspect = 1,
      panel = lattice.getOption("panel.cloud"),
      prepanel = NULL,
      scales = list(),
      strip = TRUE,
      groups = NULL,
      xlab,
      ylab,
      zlab,
      xlim = if (is.factor(x)) levels(x) else range(x, finite = TRUE),
      ylim = if (is.factor(y)) levels(y) else range(y, finite = TRUE),
      zlim = if (is.factor(z)) levels(z) else range(z, finite = TRUE),
```

```
at,
drape = FALSE,
pretty = FALSE,
drop.unused.levels,
...,
lattice.options = NULL,
default.scales =
list(distance = c(1, 1, 1),
     arrows = TRUE,
     axs = axs.default),
colorkey,
col.regions,
alpha.regions,
cuts = 70,
subset = TRUE,
axs.default = "r")
```

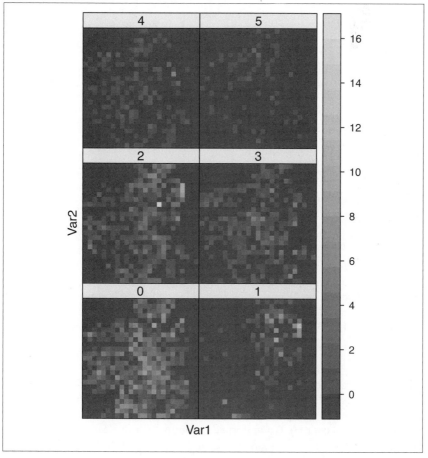

Figure 15-29. Level plot showing number of sales by location for different numbers of bedrooms

By default, plots are drawn with `panel.cloud`:

```
panel.cloud(x, y, subscripts, z,
            groups = NULL,
            perspective = TRUE,
            distance = if (perspective) 0.2 else 0,
            xlim, ylim, zlim,
            panel.3d.cloud = "panel.3dscatter",
            panel.3d.wireframe = "panel.3dwire",
            screen = list(z = 40, x = -60),
            R.mat = diag(4), aspect = c(1, 1),
            par.box = NULL,
            xlab, ylab, zlab,
            xlab.default, ylab.default, zlab.default,
            scales.3d,
            proportion = 0.6,
            wireframe = FALSE,
            scpos,
            ...,
            at)
```

Wire-frame plots

Finally, if you would like to show a three-dimensional surface, use the function `wireframe`:

```
wireframe(x,
          data,
          panel = lattice.getOption("panel.wireframe"),
          ...)
```

Other Plots

If you have fitted a model to a data set, the `rfs` function can help you visualize how well the model fits the data:

```
rfs(model, layout=c(2, 1), xlab="f-value", ylab=NULL,
    distribution = qunif,
    panel, prepanel, strip, ...)
```

The `rfs` function plots residual and fit-spread (RFS) plots. As an example, we'll use the model described in "Example: A Simple Linear Model" on page 373. The example is a linear model for runs scored in baseball games as a function of team offensive statistics. For a full explanation, see Chapter 20; here, we just want to show what charts are plotted for linear models with the `rfs` function:

```
> rfs(runs.mdl)
```

The plot generated by this command is shown in Figure 15-30. Notice that the two curves are S shaped. The residual plot is a quantile-quantile plot of the residuals; we'd expect the plot to be linear if the data fit the assumed distribution. The default distribution choice for `rfs` is a uniform distribution, which clearly isn't right. Let's try generating a second set of plots, assuming a normal distribution for the residuals:

```
> rfs(runs.mdl,distribution=qnorm)
```

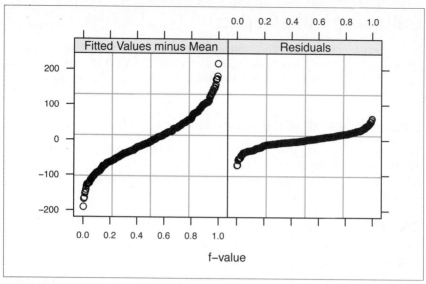

Figure 15-30. RFS plot for runs model (uniformly distributed residuals)

The results are shown in Figure 15-31. Notice that the plots are roughly linear. We expect a normally distributed error function for a linear regression model, so this is a good thing.

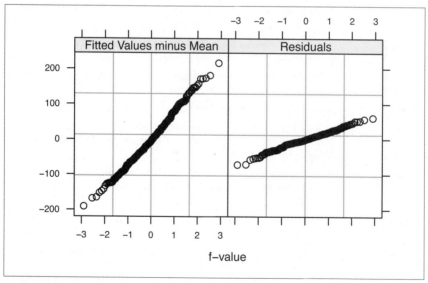

Figure 15-31. RFS plot for runs model (normally distributed residuals)

Customizing Lattice Graphics

Most lattice functions share common arguments; the same argument has a similar effect in multiple functions. This section describes what each of those arguments does. Additionally, this section will explain how to fine-tune the output of lattice functions.

Common Arguments to Lattice Functions

Lattice functions share many common arguments. Instead of explaining what each function does separately I'll explain them in a single table. (Note that the default values for many of these arguments, in particular the panel functions, aren't the same between functions.)

Argument	Description
x	The object to plot. May be a formula, array, numeric vector, or table.
data	When x is a formula, data is a data frame in which the function is evaluated.
allow.multiple	Specifies how to interpret formulas of the form $y_1 + y_2 \sim X \mid Z$ (where X is a function of multiple variables and Z may also be a function of multiple variables). By default, if allow.multiple=TRUE, the lattice function will plot both $y_1 \sim X \mid Z$ and $y_2 \sim X \mid Z$ superimposed on the same panel. However, if you set allow.multiple=FALSE, then the lattice function will plot $I(y_1 + y_2) \sim X \mid Z$ (summing $y_1 + y_2$).
outer	Specifies whether to superimpose plots or not when allow.multiple=TRUE and multiple dependent variables are specified. When outer=FALSE, the plots are superimposed; when outer=TRUE, plots are shown in different panels.
box.ratio	For plots that show data in rectangles (bwplot, barchart, and stripplot), a numeric value that specifies the ratio of the width of the rectangles to the inner rectangle space.
horizontal	For plots that can be laid out vertically or horizontally (bwplot, dotplot, barchart and stripplot), a logical value that specifies the direction to plot.
panel	The panel function used to actually draw the plots.
aspect	Specifies the aspect ratio to use for different panels. Allowable values are aspect="fill" to fill the available space (the default), aspect="xy" to compute aspect ratios based on Cleveland's 45° banking rule, and aspect="iso" for isometric scales.
groups	Specifies a variable (or expression of variables) describing groups of data to pass to the panel function. In most cases, groups specifies the sets of values to show in different colors or with different symbols.
auto.key	A logical value specifying whether to automatically draw a key showing the names of groups corresponding to different colors or symbols. (The variables key and legend override auto.key.)
prepanel	A function that takes the same arguments as panel and returns a list containing values xlim, ylim, dx, and dy (and, less frequently, xat and yat). The prepanel function is used to determine how much space is required to plot a panel. See the help files or [Sarkar2008] for more information.
strip	A logical value specifying whether strips (that label panels) should be drawn.
xlab	A character value specifying the label for the x-axis.
ylab	A character value specifying the label for the y-axis.
scales	A list that specifies how the x- and y-axes should be drawn.

Argument	Description
subscripts	A logical value specifying whether a vector named subscripts should be passed to the panel function. See the help files or [Sarkar2008] for more information.
subset	Specifies the subset of values from `data` to plot. (By default, includes all values.) You can specify a logical vector or an expression that can be evaluated within `data`. (Note: be careful of NA values in subset vectors. Additionally, note that subset does not remove unused levels from plotted factors, so keys may contain these values.)
xlim	Specifies the minimum and maximum values for the *x*-axis.
ylim	Specifies the minimum and maximum values for the *y*-axis.
drop.unused.levels	A logical value (or a list outlining what to do for different components of x) specifying whether to drop unused levels of factors.
default.scales	A list giving the default value of scales. See the help files or [Sarkar2008] for more information.
lattice.options	A list of plotting parameters, similar to `par` values for standard R graphics. See the help file for `lattice.options` for more information.
...	Arguments passed to the internal function `trellis.skeleton`.

trellis.skeleton

Arguments to `trellis.skeleton`, which are effectively arguments to all high-level Trellis functions even when not listed.

Argument	Description
as.table	Specifies the order in which panels are drawn. Use `as.table=FALSE` to draw from left to right, bottom to top or `as.table=TRUE` to draw from left to right, top to bottom.
between	A list with components x and y specifying the space between panels.
key	A list of arguments that define a legend of the components in the plot.
legend	A list specifying a set of grid objects to be used as legends. See the help file for `xyplot` or [Sarkar2008] for more details.
page	A single-argument function to be called after drawing each page. (The argument is the page number.)
main	A character value or expression specifying the main title for the plot.
sub	A character value or expression specifying the subtitle for the plot.
par.strip.text	A list of parameters that control the strip text. (Includes `col`, `cex`, `lines`, `abbreviate`, `minlength`, `dot`.)
layout	A numeric vector specifying the number of rows, columns, and pages. You may specify a value of 0 for a dimension to mean "fit in as many as needed for this dimension to meet my request for the other dimensions." For example, `c(1, 5)` means "one column, five rows," while `c(0, 5)` means "as many columns as are needed with exactly five rows."
skip	A logical vector specifying which panels to skip printing.
strip.left	A function to draw strips on the left side of each panel.
xlab.default	Default label for *x*-axis when `xlab` is not specified.
ylab.default	Default label for *y*-axis when `ylab` is not specified.

Argument	Description
xscale.components	A function to determine axis notation for the *x*-axis. See the help file for `xscale.compo nents.default` for more information.
yscale.components	A function to determine axis notation for the *y*-axis. See the help file for `xscale.compo nents.default` for more information.
axis	A function that draws axis notation. See the help file for `axis.default` for more information.
perm.cond	A numeric vector specifying a permutation of the conditioning variables. By default, the lattice functions draw panels in the order in which the conditioning variables are specified; this variable allows you to change that behavior. See the help file for more information.
index.cond	A list of a function that can be used to subset or reorder the array of conditioning variables. See the help file for `xyplot` for more information.
par.settings	A list of parameters, such as those set with `trellis.par.set`. See below for a list of available parameters.
plot.args	A list of arguments to `plot.trellis`. (See below for a table of arguments.)

Controlling How Axes Are Drawn

You can control how axes are drawn in the `lattice` package by named values in the argument `scales`. You may specify a single list for *x*- and *y*-axes or specify a list of lists with separate *x*- and *y*-axes. (For example, to shrink all text by 50% and just plot the *x*-axis as a base 2 logarithm, use the argument `scales=list(cex=.5, x = list(log = 2))`.) Here is a table of the available arguments.

Argument	Description
relation	Determines how limits are calculated for each panel. Specify `relation="same"` to use the same scale, `relation="free"` to determine different limits in each panel, `relation="sliced"` to keep the length the same in each panel but use different limits.
tick.number	Suggested number of tick marks. Ignored for character values, factors, and shingles.
draw	A logical value specifying whether to draw the axis.
alternating	Specifies whether to alternate axis locations between panels. Specify `alternating=TRUE` to alternate, `alternating=FALSE` not to alternate. Alternatively, you can specify a numeric vector that describes what to do with each panel: 0 not to draw axes, 1 to draw bottom/left, 2 to draw top/right, 3 to draw on both sides.
limits	Limits for each axis; equivalent to `xlim` and `ylim`.
at	A numeric vector describing where to plot tick marks (in native coordinates) or a list describing where to plot tick marks for each panel.
labels	Labels to accompany `at`, specified as a vector (or list of vectors).
cex	A numeric value that controls the size of axis labels ("character expansion" factor). Can specify a vector of length 2 to separately control left/bottom and right/top.
font, fontface, fontfamily	Specify typeface for axis labels.
tck	A numeric value specifying the length of the tick marks.
col	Color of tick marks and labels.

Argument	Description
rot	Angle to rotate axis labels. Can specify a vector of length 2 to separately control left/bottom and right/top.
abbreviate	A logical value specifying whether to abbreviate labels using the function `abbreviate`.
minlength	An argument passed to function `abbreviate` if argument `abbreviate=TRUE`.
log	Specifies whether to transform the data values to log scale prior to drawing and label the axis in log scale. Specify `log=FALSE` not to transform the values, `log="e"` to transform using a natural logarithm, or set log to another numeric value to use that base logarithm.
format	The format to use for data/time variables; see the help file for `strptime` for more information.
axs	Use `axs="r"` to pad date values on each side, `axs="i"` to use exact values.

Parameters

In "Graphical Parameters" on page 244, we talked about the set of graphical parameters available with conventional graphics in R. As you may recall, you could use the function **par** to get or set default parameters. For example, to check the value of the parameter cex:

```
> par("cex")
[1] 1
```

The namespace is not hierarchical; every parameter has a single name. Currently, there are 70 different parameters available in the standard **graphics** package:

```
> length(par())
[1] 70
```

There is a similar mechanism for lattice graphics. It's a little more complicated, but it's also a lot easier to understand than a single list of named items, and it's a lot more flexible.

To check the value of a setting, use the function **trellis.par.get**. As an example, let's check the values of the **"axis.text"** parameter, which controls the look of text printed on axes:

```
> trellis.par.get("axis.text")
$alpha
[1] 1

$cex
[1] 0.8

$col
[1] "#000000"

$font
[1] 1
```

To change a setting, use **trellis.par.set**. To make the text even smaller, we could change the parameter axis.text$cex to 0.5 with the following command:

```
> trellis.par.set(list(axis.text = list(cex = 0.5)))
```

If you'd like a list of all settings, simply call `trellis.par.get` with no arguments. Or, even better, try the function `show.settings`, which shows all the settings graphically:

```
> show.settings()
```

An example of the output of `show.settings` is shown in Figure 15-32. Lattice graphics parameters are hierarchical; you can think of them as a list of lists. There are 34 high-level groups of parameters describing how different components are drawn:

```
> names(trellis.par.get())
 [1] "grid.pars"         "fontsize"          "background"
 [4] "clip"              "add.line"          "add.text"
 [7] "plot.polygon"      "box.dot"           "box.rectangle"
[10] "box.umbrella"      "dot.line"          "dot.symbol"
[13] "plot.line"         "plot.symbol"       "reference.line"
[16] "strip.background"  "strip.shingle"     "strip.border"
[19] "superpose.line"    "superpose.symbol"  "superpose.polygon"
[22] "regions"           "shade.colors"      "axis.line"
[25] "axis.text"         "axis.components"   "layout.heights"
[28] "layout.widths"     "box.3d"            "par.xlab.text"
[31] "par.ylab.text"     "par.zlab.text"     "par.main.text"
[34] "par.sub.text"
```

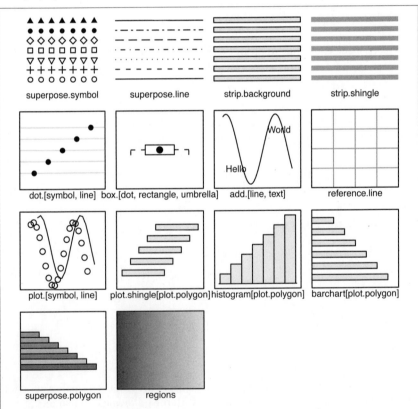

Figure 15-32. Example of show.settings

Here's an explanation of what each of these groups of parameters control:

grid.pars
> A list of global parameters that can't be set elsewhere, such as `lex` and `lineend`.

fontsize
> Base font size for all text on the Trellis device.

background
> Color of plot background.

clip
> Controls clipping for panels and strips.

add.line, add.text
> Specifies the appearance of lines or text plotted by helper functions like `panel.grid` and `panel.text`.

plot.polygon
> Specifies the appearance of bars in panels generated by `panel.barchart` and `panel.histogram`.

box.dot, box.rectangle, box.umbrella
> Specifies the appearance of points, rectangles, and umbrellas in panels plotted by `panel.bwplot`.

dot.line
> Specifies the appearance of lines in panels plotted by `panel.dotplot`.

dot.symbol
> Specifies the appearance of lines in symbols plotted by `panel.dotplot`.

plot.line
> Specifies the appearance of lines plotted by `panel.xyplot`, `panel.densityplot`, and `panel.cloud`.

plot.symbol
> Specifies the appearance of points plotted by `panel.xyplot`, `panel.density plot`, and `panel.cloud`.

reference.line
> Specifies the appearance of reference lines plotted by `panel.grid` and `panel.text`.

strip.background, strip.shingle, strip.border
> Specifies the default appearance of strips.

superpose.line, superpose.symbol, superpose.polygon
> Specifies the appearance of lines, symbols, and polygons on superimposed plots.

regions
> Specifies how regions are plotted by `panel.levelplot` and `panel.wireframe`.

shade.colors
> Specifies colors for plots by `panel.levelplot` and `panel.wireframe`.

axis.line, axis.text
> Specifies how lines and text are plotted in axes.

axis.components
> Controls the appearance of axes.

layout.heights, layout.widths
> Controls the height and width of panels in a lattice.

box.3d
> Specifies the way boxes are drawn by `panel.cloud` and `panel.wireframe`.

par.xlab.text, par.ylab.text, par.zlab.text
> Controls how text labels are plotted.

par.main.text, par.sub.text
> Specifies defaults for main and subtitles.

Within these groups, there are more parameters. There are a total of 378 parameters. However, there are only 46 unique parameters within these groups.‖ Here is an explanation of the most common subparameters (many of which are similar to standard graphical parameters):

alpha
> Controls transparency.

border
> Border color.

cex
> Character expansion factor; size of this type relative to `fontsize`.

col
> Color for lines and points.

fill
> Color for fills.

font
> Font face.

lineheight
> Height of a line, as a multiple of text size.

lty
> Line type.

lwd
> Line width.

pch
> Plotting character.

‖ In case you're curious, here's the code I used to count them:
```
> # count the total number of parameters
> length(names(unlist(trellis.par.get())))
[1] 378
> # count the number of unique parameters
> n <- names(trellis.par.get())
> p <- NA
> for (i in 1:34) {p <- c(p,names(trellis.par.get(n[i])));}
> length(table(p))
```

Here is an explanation of some of the nonstandard subparameters:

palette
Function generating color palette through parameter `shade.colors`.

text, points
Specifies text format through parameter `fontsize`.

panel, strip
Controls clipping for panels and strips in parameter clip.

top.padding, main, main.key.padding, key.top, key.axis.padding, axis.top, strip, panel, axis.panel, between, axis.bottom, axis.xlab.padding, xlab, xlab.key.padding, key.bottom, key.sub.padding, sub, bottom.padding
Parameters for `layout.heights`.

left.padding, key.left, key.ylab.padding, ylab, ylab.axis.padding, axis.left, axis.panel, strip.left, panel, between, axis.right, axis.key.padding, key.right, right.padding
Parameters for `layout.widths`.

For more information, see the help files for `par` (in the `graphics` package), `gpar` (in the `grid` package), or the help files for different panel functions.

plot.trellis

As we noted above, lattice functions do not plot results; they return lattice objects. To plot a lattice object, you need to call `print` or `plot` on the lattice object.

The function that actually does the work is the `plot.trellis` function (which the help file claims is an alias for the `print.trellis` function). It's possible to control how lattice objects are printed through arguments to `plot.trellis`. As shown above, you can also pass these arguments to lattice functions through the `plot.args` argument. Here's a list of arguments for `plot.trellis`.

Argument	Description	Default
x	The Trellis object to plot.	
position	A vector of four numbers, `c(xmin, ymin, xmax, ymax)`, specifying where to plot the object. Coordinates are between 0 and 1 for both dimensions.	
split	A vector of four integers, `c(x, y, nx, ny)`, that says to position the current plot at the x, y position in a regular array of nx by ny plots.	
more	A logical value specifying whether more plots will follow on the current page.	FALSE
newpage	A logical value specifying whether the plot should be on a new page.	FALSE
packet.panel	A function that determines which packet is plotted in which panel.	packet.panel.default
draw.in	A grid viewport in which to draw the plot.	NULL

Argument	Description	Default
panel.height	A list of two components (x and units) specifying the height of each panel in the lattice plot.	lattice.getOption("layout.heights")$panel
panel.width	A list of two components (x and units) specifying the width of each panel in the lattice plot.	lattice.getOption("layout.widths")$panel
save.object	A logical value indicating whether to "save" the last object printed. See the help file for more information.	lattice.getOption("save.object")
panel.error	A function that is executed if an error occurs while plotting the panel.	lattice.getOption("panel.error")
prefix	A character string to use as a prefix in viewport names, to distinguish similar plots. See the help file for more information.	
...	Extra arguments: these are ignored.	

strip.default

To change the way strips are drawn, you can specify your own strip function as an argument to a lattice function. Strip functions are a little complicated to write from scratch, so it is usually best to modify the strips by writing new function that creates a wrapper around the function strip.default:

```
strip.default(which.given,
              which.panel,
              var.name,
              factor.levels,
              shingle.intervals,
              strip.names = c(FALSE, TRUE),
              strip.levels = c(TRUE, FALSE),
              sep = " : ",
              style = 1,
              horizontal = TRUE,
              bg = trellis.par.get("strip.background")$col[which.given],
              fg = trellis.par.get("strip.shingle")$col[which.given],
              par.strip.text = trellis.par.get("add.text"))
```

The simplest way to modify the appearance of the strips is by using the function strip.custom. This function accepts the same arguments as strip.default and returns a new function that can be specified as an argument to a lattice function.

Here's a description of the arguments to strip.default (and, in turn, to strip.custom).

Argument	Description	Default
which.given, which.panel, var.name, factor.levels, shingle.intervals	These arguments contain the data for actually drawing the strip. (Probably not needed for strip.custom.)	

Argument	Description	Default
strip.names	A logical vector with two elements that specifies whether to draw variable names in strips. strip.names[0] is used for factors, and strip.names[1] for shingles.	c(FALSE, TRUE)
strip.levels	A logical vector with two elements that specifies whether to draw variable values in strips. strip.names[0] is used for factors, and strip.names[1] for shingles.	c(TRUE, FALSE)
sep	A character value specifying the separator if both name and level are shown.	
style	An integer value specifying how the current level of a factor is encoded. See the help file for more information.	
horizontal	A logical value specifying whether the labels should be horizontal.	
bg	Specifies the background color.	trellis.par.get("strip.background") $col[which.given]
fg	Specifies the foreground color.	trellis.par.get("strip.shingle") $col[which.given]
par.strip.text	A list of parameters controlling the way text is drawn in the script (such as col, cex, font).	trellis.par.get("add.text")

simpleKey

To customize the way that keys (or legends) are drawn for plots with multiple groups of variables, you may specify a custom function to the key argument, or you may use the auto.key argument to automatically draw a key using the simpleKey function. If you specify autoKey=TRUE, then simpleKey is called with the default arguments to generate the key. Alternatively, you can specify a list of arguments that are, in turn, passed as arguments to simpleKey to draw the legend:

```
simpleKey(text, points = TRUE,
          rectangles = FALSE,
          lines = FALSE,
          col, cex, alpha, font,
          fontface, fontfamily,
          lineheight, ...)
draw.key(key, draw=FALSE, vp=NULL, ...)
```

Here is a description of the arguments to simpleKey.

Argument(s)	Description	Default
text	A character or expression vector specifying the text to be used to describe groups	
points	A logical value specifying whether a key should be provided for points	TRUE
rectangles	A logical value specifying whether a key should be provided as filled rectangles	FALSE
lines	A logical value specifying whether a key should be provided for lines	FALSE

Argument(s)	Description	Default
col, cex, alpha, font, fontface, fontfamily, lineheight	Graphical parameters that control different aspects of the key	

Low-Level Functions

In "Custom Panel Functions" on page 268, we showed how to modify the appearance of a chart through custom panel functions. The `lattice` package includes a variety of different panel functions that you can use to customize your charts. You can start with one of the included panel functions, use another panel function, or even write your own.

Low-Level Graphics Functions

Here is a list of some primitive panel plotting functions available within the `lattice` package. These are functions that are useful for writing your own panel functions from scratch, though they can also be used in conjunction with higher-level functions. (For example, you can use `panel.text` along with `panel.barchart` to plot a bar chart with added text.)

Function(s)	Description
llines, panel.line	Plots lines
lpoints, panel.points	Plots points
ltext, panel.text	Plots text
lsegments, panel.segments	Plots line segments
lpolygon, panel.polygons	Plots polygons
larrows, panel.arrows	Plots arrows
lrect, panel.rect	Plots rectangles
panel.axis	Plots axes
panel.superpose	Superimposes panel functions on top of the same plot (by grouping value)

For more information on how to use these functions, see the help file for any of these functions (such as `llines`).

Panel Functions

Here is a list of some functions for adding to, or customizing the appearance of, other panels. You can use these functions to add lines, text, and other graphical elements to lattice graphics. For an example of using panel functions to modify the appearance of a plot, see "Custom Panel Functions" on page 268.

Function	Description
panel.abline	Adds a line to the chart area of a panel.
panel.curve	Adds a curve (defined by a mathematical expression) to the chart area of a panel.
panel.rug	Adds a "rug" to a panel. (Rugs look a lot like strip plots; you can superimpose a rug to show both exact points and groups in charts like density plots.)
panel.mathdensity	Plots a probability distribution given by a distribution function.
panel.average	Plots average values (grouped by a factor).
panel.fill	Fills the panel with a specified color.
panel.grid	Plots a reference grid.
panel.loess	Adds a smooth curve (fitted by `loess`).
panel.lmline	Plots a line fitted to the underlying data by a linear regression.
panel.refline	Adds a line to the chart area of a panel; just like `panel.abline`, except with different default settings (appropriate for, as you probably guessed, reference lines).
panel.qqmathline	Adds a line through the points at the 25th and 75th percentile points of the sample and theoretical distribution. (Mostly useful for Q-Q plots.)
panel.violin	Draws violin plots. Usually used in place of box-and-whisker plots in box plots.

For more details on these functions, see the corresponding help files.

Lattice Graphics

IV

Statistics with R

This part of the book contains information about statistics in R: statistical tests, statistical modeling, and other analysis tools.

16

Analyzing Data

This chapter describes a number of techniques for analyzing data in R. Many of the functions described in this chapter are useful for preparing data for other analysis, or are the building blocks for other analyses.

Summary Statistics

R includes a variety of functions for calculating summary statistics.

To calculate the mean of a vector, use the `mean` function. You can calculate minima with the `min` function, or maxima with the `max` function. As an example, let's use the `dow30` data set that we created in "An extended example" on page 181. This data set is also available in the `nutshell` package:

```
> library(nutshell)
> data(dow30)
> mean(dow30$Open)
[1] 36.24574
> min(dow30$Open)
[1] 0.99
> max(dow30$Open)
[1] 122.45
```

For each of these functions, the argument `na.rm` specifies how `NA` values are treated. By default, if any value in the vector is `NA`, then the value `NA` is returned. Specify `na.rm=TRUE` to ignore missing values:

```
> mean(c(1,2,3,4,5,NA))
[1] NA
> mean(c(1,2,3,4,5,NA),na.rm=TRUE)
[1] 3
```

Optionally, you can also remove outliers when using the `mean` function. To do this, use the `trim` argument to specify the fraction of observations to filter:

```
> mean(c(-1,0:100,2000))
[1] 68.4369
```

```
> mean(c(-1,0:100,2000),trim=0.1)
[1] 50
```

To calculate the minimum and maximum at the same time, use the range function. This returns a vector with the minimum and maximum value:

```
> range(dow30$Open)
[1]   0.99 122.45
```

Another useful function is quantile. This function can be used to return the values at different percentiles (specified by the probs argument):

```
> quantile(dow30$Open, probs=c(0,0.25,0.5,0.75,1.0))
     0%     25%     50%     75%    100%
  0.990  19.655  30.155  51.680 122.450
```

You can return this specific set of values (minimum, 25th percentile, median, 75th percentile, and maximum) with the fivenum function:

```
> fivenum(dow30$Open)
[1]   0.990  19.650  30.155  51.680 122.450
```

To return the interquartile range (the difference between the 25th and 75th percentile values), use the function IQR:

```
> IQR(dow30$Open)
[1] 32.025
```

Each of the functions above can be useful on its own, but can also be used with apply, tapply, or another aggregation function to calculate statistics for a data frame or subsets of a data frame.

The most convenient function for looking at summary information is summary. It is a generic function that works on data frames, matrices, tables, factors, and other objects. As an example, let's take a look at the output of summary for the dow30 data set that we used above:

```
> summary(dow30)
     symbol              Date              Open              High
 MMM    : 252   2008-09-22:  30   Min.   :  0.99   Min.   :  1.01
 AA     : 252   2008-09-23:  30   1st Qu.: 19.66   1st Qu.: 20.19
 AXP    : 252   2008-09-24:  30   Median : 30.16   Median : 30.75
 T      : 252   2008-09-25:  30   Mean   : 36.25   Mean   : 36.93
 BAC    : 252   2008-09-26:  30   3rd Qu.: 51.68   3rd Qu.: 52.45
 BA     : 252   2008-09-29:  30   Max.   :122.45   Max.   :122.88
 (Other):5970   (Other)   :7302
      Low              Close             Volume             Adj.Close
 Min.   :  0.27   Min.   :  0.75   Min.   :1.336e+06   Min.   :  0.75
 1st Qu.: 19.15   1st Qu.: 19.65   1st Qu.:1.111e+07   1st Qu.: 19.38
 Median : 29.55   Median : 30.10   Median :1.822e+07   Median : 29.41
 Mean   : 35.53   Mean   : 36.24   Mean   :5.226e+07   Mean   : 35.64
 3rd Qu.: 50.84   3rd Qu.: 51.58   3rd Qu.:4.255e+07   3rd Qu.: 50.97
 Max.   :121.62   Max.   :122.11   Max.   :2.672e+09   Max.   :122.11
```

As you can see, summary presents information about each variable in the data frame. For numeric values, it shows the minimum, 1st quartile, median, mean, 3rd quartile, and maximum values. For factors, summary shows the count of the most frequent values. (Less frequent values are grouped into an "Other" category.) Summary doesn't show meaningful information for character values.

A useful (text based) tool for looking at the distribution of a numeric vector is the stem function:

```
stem(x, scale = 1, width = 80, atom = 1e-08)
```

The argument x is a numeric vector, scale controls the length of the plot, width controls the width, and atom is a tolerance factor.

As an example of a stem plot, we'll look at field goal attempts in the NFL during 2005. Specifically, we'll look at the attempted distances for missed field goals. To do this, we'll use the subset function to select only missed field goals and then plot the yards for each attempt:

```
> stem(subset(field.goals,PLTYPE=="FG no")$YARDS)

  The decimal point is at the |

  20 | 0
  22 |
  24 |
  26 | 00
  28 | 0000000
  30 | 0000000
  32 | 00000000
  34 | 000
  36 | 0000
  38 | 00000000000000
  40 | 0000000000
  42 | 0000000000000000
  44 | 00000000000
  46 | 000000000000000000
  48 | 000000000000000000
  50 | 000000000000
  52 | 00000000000000000000
  54 | 0000
  56 | 000
  58 | 00
  60 | 00
  62 | 0
```

Correlation and Covariance

Very often, when analyzing data, you want to know if two variables are *correlated*. Informally, correlation answers the question "when we increase (or decrease) x, does y increase (or decrease), and by how much?" Formally, correlation measures the linear dependence between two random variables. Correlation measures range between –1 and 1; 1 means that one variable is a (positive) linear function of the other, 0 means the two variables aren't correlated at all, and –1 means that one variable is a negative linear function of the other (the two move in completely opposite directions; see Figure 16-1).

Figure 16-1. Correlation

The most commonly used correlation measurement is the Pearson correlation statistic (it's the formula behind the CORREL function in Excel):

$$r = \frac{\sum_{i=1}^{n}(x_i - \bar{x})(y_i - \bar{y})}{\sqrt{\sum_{i=1}^{n}(x_i - \bar{x})^2}\sqrt{\sum_{i=1}^{n}(y_i - \bar{y})^2}}$$

where $\bar{x}$ is the mean of variable x, and $\bar{y}$ is the mean of variable y. The Pearson correlation statistic is rooted in properties of the normal distribution and works best with normally distributed data. An alternative correlation function is the Spearman correlation statistic. Spearman correlation is a nonparametric statistic and doesn't make any assumptions about the underlying distribution:

$$\rho = \frac{n(\sum x_i y_i) - (\sum x_i)(\sum y_i)}{\sqrt{n(\sum x_i^2) - (\sum x_i)^2}\sqrt{n(\sum y_i^2) - (\sum y_i)^2}}$$

Another measurement of how well two random variables are related is Kendall's tau. Kendall's tau formula works by comparing rankings of values in the two random variables, not by comparing the values themselves:

$$\tau = \frac{n_c - n_d}{1/2\,n(n-1)}$$

In this formula, n is the length of the two random variables, n_c counts the number of concordant pairs, and n_d counts the number of discordant pairs.

To compute correlations in R, you can use the function cor. This function can be used to compute each of the correlation measures shown above:

```
cor(x, y = NULL, use = "everything",
    method = c("pearson", "kendall", "spearman"))
```

You can compute correlations on two vectors (assigned to arguments x and y), a data frame (assigned to x with y=NULL), or a matrix (assigned to x with y=NULL). If you specify a matrix or data frame, then cor will compute the correlation between each pair of variables and return a matrix of results.

The method argument specifies the correlation calculation. The use argument specifies how the function should treat NA values. If you want an error raised when values are NA, choose use="all.obs". If you would like the result to be NA when an element is NA, choose use="everything". To omit cases where values are NA, choose use="complete.obs". To omit cases where values are NA, but return NA if all values are NA, specify use="na.or.complete". Finally, to omit pairs where at least one value is NA, choose use="pairwise.complete.obs".

As an example, let's look at the 2006 birth data that we used above. Specifically, let's ask whether the mother's weight gain correlates with the baby's weight. Let's start by selecting only valid birth weights and weight gain values. We'll also only exclude premature births. (I picked gestation age > 35 weeks, though this might not technically be premature.) Finally, we'll only include single births:

```
> births2006.cln <- births2006[
+     !is.na(births2006$WTGAIN) &
+     !is.na(births2006$DBWT) &
+     births2006$DPLURAL=="1 Single" &
+     births2006$ESTGEST>35,]
```

First, we'll take a look at how these two variables are related. Because there are 3,232,884 observations, a normal scatter plot would be hard to read, so we'll use smoothScatter instead:

```
> smoothScatter(births2006.cln$WTGAIN,births2006.cln$DBWT)
```

The plot is shown in Figure 16-2. From this diagram, we'd expect to see a slight correlation. (We wouldn't expect a very strong correlation because of the big blob, but the blob is angled a little bit.) Let's compute the Pearson correlation:

```
> cor(births2006.cln$WTGAIN,births2006.cln$DBWT)
[1] 0.1750655
```

Let's also calculate the Spearman correlation:

```
> cor(births2006.cln$WTGAIN,births2006.cln$DBWT,method="spearman")
[1] 0.1783328
```

As you can see, both measures indicate that there is a modest correlation between the two variables.

A closely related idea is covariance. Covariance is defined as:

$$cov(x, y) = E\big[(x - E[x])(y - E[y])\big]$$

which is the numerator of the Pearson correlation formula. You can compute covariance in R using the cov function, which accepts the same arguments as cor:

```
cov(x, y = NULL, use = "everything",
    method = c("pearson", "kendall", "spearman"))
```

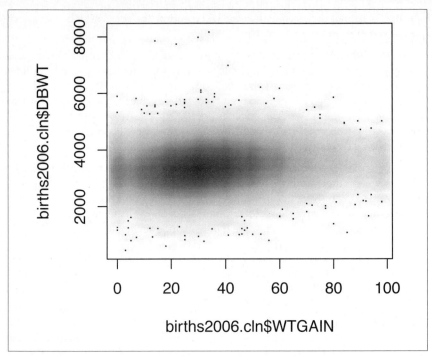

Figure 16-2. Plot of birth weight as a function of mother's weight gain

If you have computed a covariance matrix, you can use the R function `cov2cor` to compute the correlation matrix.

You can also compute weighted covariance measurements using the `cov.wt` formula:

```
cov.wt(x, wt = rep(1/nrow(x), nrow(x)), cor = FALSE, center = TRUE,
       method = c("unbiased", "ML"))
```

Principal Components Analysis

Another technique for analyzing data is principal components analysis. Principal components analysis breaks a set of (possibly correlated) variables into a set of uncorrelated variables.

In R, principal components analysis is available through the function `prcomp` in the `stats` package:

```
## S3 method for class 'formula':
prcomp(formula, data = NULL, subset, na.action, ...)

## Default S3 method:
prcomp(x, retx = TRUE, center = TRUE, scale. = FALSE,
       tol = NULL, ...)
```

Here is a description of the arguments to `prcomp`.

Argument	Description	Default
formula	In the formula method, specifies formula with no response variable, indicating columns of a data frame to use in the analysis.	
data	An optional data frame containing the data specified in formula.	
subset	An (optional) vector specifying observations to include in the analysis.	
na.action	A function specifying how to deal with NA values.	
x	In the default method, specifies a numeric or complex matrix of data for the analysis.	
retx	A logical value specifying whether rotated variables should be returned.	TRUE
center	A logical value specifying whether values should be zero centered.	TRUE
scale.	A logical value specifying whether values should be scaled to have unit variance.	TRUE
tol	A numeric value specifying a tolerance value below which components should be omitted.	NULL
...	Additional arguments passed to other methods.	

As an example, let's try principal components analysis on a matrix of team batting statistics. Let's start by loading the data for every team between 2000 and 2008. We'll use the SQLite database that we used in Chapter 14, and extract the fields we want using an SQL query. (Because this is a book on R and not a book on baseball, I renamed the common abbreviations to more intuitive names for plays.)

```
> library(RSQLite)
> drv <- dbDriver("SQLite")
> con <- dbConnect(drv,
+    dbname=paste(.Library, "/nutshell/data/bb.db", sep=""))
> team.batting.00to08 <- dbGetQuery(con,
+    paste(
+      'SELECT teamID, yearID, R as runs, ',
+      '  H-"2B"-"3B"-HR as singles, ',
+      '  "2B" as doubles, "3B" as triples, HR as homeruns, ',
+      '  BB as walks, SB as stolenbases, CS as caughtstealing, ',
+      '  HBP as hitbypitch, SF as sacrificeflies, ',
+      '  AB as atbats ',
+      '  FROM Teams ',
+      '  WHERE yearID between 2000 and 2008'
+    )
+  )
```

You can also find this data already loaded in the `team.batting.00to08` data set in the nutshell package. Eventually, we'll do some analysis on runs scored. For now, we'll use principal components analysis on the remaining variables in the matrix:

```
> batting.pca <- princomp(~singles+doubles+triples+homeruns+
+        walks+hitbypitch+sacrificeflies+
+        stolenbases+caughtstealing,
+        data=team.batting.00to08)
> batting.pca
Call:
princomp(formula = ~singles + doubles + triples + homeruns +
    walks + hitbypitch + sacrificeflies + stolenbases + caughtstealing,
    data = team.batting.00to08)
```

```
Standard deviations:
   Comp.1    Comp.2    Comp.3    Comp.4    Comp.5    Comp.6    Comp.7
74.900981 61.871086 31.811398 27.988190 23.788859 12.884291  9.150840
   Comp.8    Comp.9
 8.283972  7.060503

 9  variables  and  270 observations.
```

The princomp function returns a princomp object. You can get information on the importance of each component with the summary function:

```
> summary(batting.pca)
Importance of components:
                           Comp.1      Comp.2     Comp.3      Comp.4
Standard deviation     74.9009809  61.8710858 31.8113983 27.98819003
Proportion of Variance  0.4610727   0.3146081  0.0831687  0.06437897
Cumulative Proportion   0.4610727   0.7756807  0.8588494  0.92322841
                           Comp.5      Comp.6     Comp.7
Standard deviation     23.78885885 12.88429066 9.150840397
Proportion of Variance  0.04650949  0.01364317 0.006882026
Cumulative Proportion   0.96973790  0.98338107 0.990263099
                           Comp.8      Comp.9
Standard deviation      8.283972499 7.060503344
Proportion of Variance  0.005639904 0.004096998
Cumulative Proportion   0.995903002 1.000000000
```

To show the contribution of each variable to the components, you can use the loadings method:

```
> loadings(batting.pca)

Loadings:
               Comp.1 Comp.2 Comp.3 Comp.4 Comp.5 Comp.6 Comp.7
singles         0.313  0.929 -0.136         0.136
doubles                      -0.437  0.121 -0.877
triples                                                   0.424
homeruns       -0.235        -0.383  0.825  0.324
walks          -0.914  0.328  0.150 -0.182
hitbypitch                                        -0.989
sacrificeflies                                            0.321
stolenbases            0.131  0.758  0.502 -0.307        -0.232
caughtstealing               0.208  0.104                 0.813
               Comp.8 Comp.9
singles
doubles        -0.100
triples         0.775  0.449
homeruns
walks
hitbypitch
sacrificeflies  0.330 -0.882
stolenbases
caughtstealing -0.521  0.105


               Comp.1 Comp.2 Comp.3 Comp.4 Comp.5 Comp.6 Comp.7
SS loadings     1.000  1.000  1.000  1.000  1.000  1.000  1.000
Proportion Var  0.111  0.111  0.111  0.111  0.111  0.111  0.111
Cumulative Var  0.111  0.222  0.333  0.444  0.556  0.667  0.778
```

```
            Comp.8 Comp.9
SS loadings     1.000  1.000
Proportion Var  0.111  0.111
Cumulative Var  0.889  1.000
```

There is a `plot` method for `princomp` objects that displays a "scree" plot of the variance against each principal component:

```
plot(batting.pca)
```

The results are shown in Figure 16-3. A second useful method for visualizing principal components is the biplot (see Figure 16-4):

```
> biplot(batting.pca,cex=0.5,col=c("gray50","black"))
```

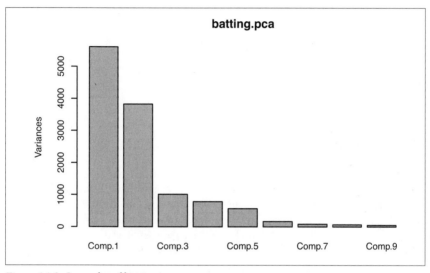

Figure 16-3. Scree plot of batting.pca

A biplot graphically displays the contributions of each of the variables to a pair of principal components and also shows individual observations on the same scale. This example shows the contribution of each variable to components 1 and 2. Individual observations are also plotted on the chart. (I showed these in gray so that you could more clearly see the plot of the projections.) As you can see, `singles` and `walks` are the primary contributors to the first two components.

We'll revisit this data example in more depth in "Example: A Simple Linear Model" on page 373.

Note that there is a `princomp` function that does the same thing, but works differently. It calculates the principal components by using the `eigen` function on the correlation or covariance matrix generated by the `cor` function. This function is included for compatibility with S-PLUS (it produces the same results as the equivalent method in S-PLUS). For more information on `princomp`, see the help file.

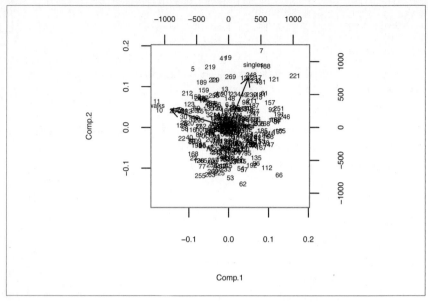

Figure 16-4. Biplot of batting.pca

Factor Analysis

In most data analysis problems, there are some quantities that we can observe, and some that we cannot. The classic examples come from the social sciences. Suppose that you wanted to measure intelligence. It's not possible to directly measure an abstract concept like intelligence, but it is possible to measure performance on different tests. You could use factor analysis to analyze a set of test scores (the observed values) to try to determine intelligence (the hidden value).

Factor analysis is available in R through the function `factanal` in the `stats` package:

```
factanal(x, factors, data = NULL, covmat = NULL, n.obs = NA,
        subset, na.action, start = NULL,
        scores = c("none", "regression", "Bartlett"),
        rotation = "varimax", control = NULL, ...)
```

Here is a description of the arguments to `factanal`.

Argument	Description	Default
x	A formula or a numeric matrix to be used for analysis.	
factors	A numeric value indicating the number of factors to be fitted.	
data	A data frame in which to evaluate x (if x is a formula).	NULL
covmat	A covariance matrix (or a list returned by cov.wt).	NULL
n.obs	The number of observations (if covmat is specified).	NA
subset	Specifies which observations to include in the analysis.	
na.action	A function that specifies how to handle missing observations (if x is a formula).	

Argument	Description	Default
start	A matrix of starting values for the algorithm.	NULL
scores	A character value specifying the type of scores to produce. Use `scores="none"` for no scores, `scores="regression"` for Thompson's scores, or `scores="Bartlett"` for Bartlett's weighted least squares scores.	"none"
rotation	A character value naming the function for rotating the factors.	"varimax"
control	A list of control values for the fit.	

Bootstrap Resampling

When analyzing statistics, analysts often wonder if the statistics are sensitive to a few outlying values. Would we get a similar result if we were to omit a few points? What are the range of values for the statistic? It is possible to answer this question for an arbitrary statistic using a technique called *bootstrapping*.

Formally, bootstrap resampling is a technique for estimating the bias of an estimator. An *estimator* is a statistic calculated from a data sample that provides an estimate of a true underlying value, often a mean, standard deviation, or a hidden parameter.

Bootstrapping works by repeatedly selecting random observations from a data sample (with replacement) and recalculating the statistic. In R, you can use bootstrap resampling through the boot function in the boot package:

```
library(boot)
boot(data, statistic, R, sim="ordinary", stype="i",
     strata=rep(1,n), L=NULL, m=0, weights=NULL,
     ran.gen=function(d, p) d, mle=NULL, simple=FALSE, ...)
```

Arguments to boot include the following.

Argument	Description	Default
data	A vector, matrix, or data frame containing the input data.	
statistic	A function that, when applied to the data, returns a vector containing the statistic of interest. The function takes two arguments: the source data and a vector that specifies which values to select for each bootstrap replicate. The meaning of the second argument is defined by stype.	
R	A numeric value specifying the number of bootstrap replicates.	
sim	A character value specifying the type of simulation. Possible values include `"ordinary"`, `"parametric"`, `"balanced"`, `"permutation"`, and `"antithetic"`.	"ordinary"
stype	A character value that specifies what the second argument to the statistic function represents. Possible values of stype are `"i"` (indices), `"f"` (frequencies), and `"w"` (weights).	"i"
strata	An integer vector or factor specifying the strata for multisample problems.	rep(1, n)
l	A vector of influence values evaluated at the observations (when `sim="antithetic"`).	NULL
m	Specifies the number of predictions at each bootstrap replicate.	0
weights	A numeric vector of weights for data.	NULL

Argument	Description	Default
ran.gen	A function that describes how random values are generated (when `sim="parametric"`).	function(d, p) d
mle	The second argument passed to `ran.gen`; typically, a maximum likelihood estimate (hence the name).	NULL
simple	Specifies the method for generating random values. Specifying `simple=TRUE` causes values to be selected on each iteration, saving storage space but costing time.	FALSE
...	Additional arguments passed to `statistic`.	

As an example of **boot**, let's look at real estate sale prices. Usually, the media reports median sale prices in a region. We can use the bootstrap to look at how biased median is as an estimator:

```
> b <- boot(data=home.sale.prices.june2008,
+           statistic = function(d,i) {median(d[i])},
+           R=1000)
> b

ORDINARY NONPARAMETRIC BOOTSTRAP
Call:
boot(data = home.sale.prices.june2008, statistic = function(d,
    i) {
    median(d[i])
}, R = 1000)
Bootstrap Statistics :
    original   bias    std. error
t1*   845000   -3334    23287.27
```

The **boot** function tells us that the median is a very slightly biased estimator.

17

Probability Distributions

Many statistical tests work by calculating a test statistic and then comparing the test statistic to a value from a theoretical distribution. R provides a set of functions to calculate densities, distributions, and quantiles for common statistical distributions. You can also generate random values from these distributions. This section describes how to use these functions (using the normal distribution as an example) and then lists most functions included with the R stats library.

Normal Distribution

As an example, we'll start with the normal distribution. As you may remember from statistics classes, the probability density function for the normal distribution is:

$$f(x) = \frac{e^{-(x-\mu)^2/(2\sigma^2)}}{\sqrt{2\pi\sigma^2}}$$

To find the probability density at a given value, use the dnorm function:

```
dnorm(x, mean = 0, sd = 1, log = FALSE)
```

The arguments to this function are fairly intuitive: x specifies the value at which to evaluate the density, mean specifies the mean of the distribution, sd specifies the standard deviation, and log specifies whether to return the raw density (log=FALSE) or the logarithm of the density (log=TRUE). As an example, you can plot the normal distribution with the following command:

```
> plot(dnorm, -3, 3, main = "Normal Distribution")
```

The plot is shown in Figure 17-1.

The distribution function for the normal distribution is pnorm:

```
pnorm(q, mean = 0, sd = 1, lower.tail = TRUE, log.p = FALSE)
```

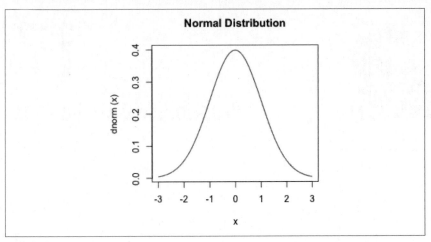

Normal Distribution

Figure 17-1. Normal distribution

You can use the distribution function to tell you the probability that a randomly selected value from the distribution is less than or equal to q. Specifically, it returns $p = \Pr(x \le q)$. The value q is specified by the argument q, the mean by mean, and the standard deviation by sd. If you would like the raw value p, then specify log.p=FALSE; if you would like $\log(p)$, then specify log.p=TRUE. By default, lower.tail=TRUE, so this function returns $\Pr(x \le q)$; if you would prefer $\Pr(x > q)$, then specify lower.tail=FALSE. Here are a few examples of pnorm:

```
> # mean is zero, normal distribution is symmetrical, so
> # probability(q <= 0) is .5
> pnorm(0)
[1] 0.5
> # what is the probability that the value is less than
> # 1 standard deviation below the mean?
> pnorm(-1)
[1] 0.1586553
> # what is the probability that the value is within
> # 1.96 standard deviations of the mean?
> pnorm(-1.96,lower.tail=TRUE) + pnorm(1.96,lower.tail=FALSE)
[1] 0.04999579
```

You can plot the cumulative normal distribution with a command like this:

```
> plot(pnorm, -3, 3, main = "Cumulative Normal Distribution")
```

The plot is shown in Figure 17-2.

The quantile function is the reverse of the distribution function. Specifically, this function returns q where $p = \Pr(x \le q)$. In R, you can calculate the quantile function for the normal distribution with the function qnorm:

```
qnorm(p, mean = 0, sd = 1, lower.tail = TRUE, log.p = FALSE)
```

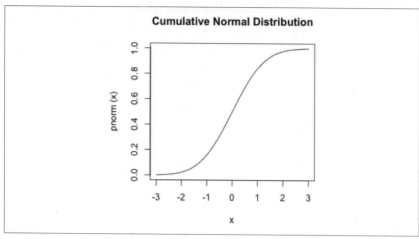

Cumulative Normal Distribution

Figure 17-2. Cumulative normal distribution

As above, p specifies *p* where *p* = Pr(*x* ≤ *q*), mean specifies the mean of the distribution, sd specifies the standard deviation, and lower.tail specifies whether *p* = Pr(*x* ≤ *q*) (lower.tail=TRUE) or *p* = Pr(*x* > *q*) (lower.tail=FALSE). The argument log.p specifies whether the input value is the logarithm of *p* (log.p=TRUE) or just *p* (log.p=FALSE). Here are a few examples:

```
> # find the median of the normal distribution
> qnorm(0.5)
[1] 0
> qnorm(log(0.5),log.p=TRUE)
[1] 0
> # qnorm is the inverse of pnorm
> qnorm(pnorm(-1))
[1] -1
> # finding the left and right sides of a 95% confidence interval
> c(qnorm(.025), qnorm(.975))
[1] -1.959964  1.959964
```

Finally, it is possible to generate random numbers taken from the normal distribution. Selecting random numbers from a specific distribution can be useful in testing statistical functions, running simulations, in sampling methods, and in many other contexts. To do this in R, use the function rnorm:

```
rnorm(n, mean = 0, sd = 1)
```

For example, you could generate 10,000 randomly selected values from a normal distribution with a command like rnorm(10000). You could plot these with an expression like this:

```
> hist(rnorm(10000),breaks=50)
```

The plot is shown in Figure 17-3.

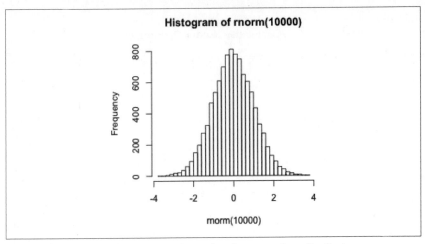

Figure 17-3. Histogram of 10,000 random values from a uniform distribution

Common Distribution-Type Arguments

Almost all the R functions that generate values of probability distributions work the same way. They follow a similar naming convention:

- Probability density functions (PDFs) begin with d.[*]
- Distribution functions begin with p.
- Quantile functions begin with q.
- Random number generators begin with r.

Similarly, most types of functions share certain common arguments:

- For density functions: x, log
- For distribution functions: q, lower.tail, log.p
- For quantile functions: p, lower.tail, log.p
- For random numbers: n (except for hypergeometric distributions, where n is renamed to nn)

This might make it easier to remember which function to use for which application, and which arguments you need to specify. Of course, you can always just look up the right function to use in R's help system. Or in this book.

Distribution Function Families

Here is a table showing the probability distribution functions available in R. In addition to the arguments listed above that are common to each type of function, there are also some arguments that are common to each family.

---

[*] For discrete distributions, these are technically probability mass functions (PMFs), though the function names still begin with "d."

Family	PDF or PMF	R functions	Family arguments
Beta	$f(x) = \dfrac{\Gamma(a+b)}{\Gamma(a)\Gamma(b)} x^{a-1}(1-x)^{b-1}$	dbeta pbeta qbeta rbeta	shape1, shape2, ncp = 0
Binomial	$p(x) = \dbinom{n}{x} p^x (1-p)^{n-x}$	dbinom pbinom qbinom rbinom	size, prob
Birthday		pbirthday qbirthday	classes, coincident
Cauchy	$f(x) = \dfrac{1}{\pi s \left(1 + ((x-l)/s)^2\right)}$	dcauchy pcauchy qcauchy rcauchy	location, scale
Chi-squared	$f(x, df) = \dfrac{x^{\frac{df}{2}-1}\, e^{-\frac{x}{2}}}{\Gamma(1/2\,df)\, 2^{df/2}}$	dchisq pchisq qchisq rchisq	df, ncp=0
Exponential	$f(x) = \lambda e^{-\lambda x}$	dexp pexp qexp rexp	rate
F-distribution	$\sqrt{\dfrac{(df_1 x)^{df_1} df_2^{df_2}}{(df_1 x + df_2)^{df_1 + df_2}}} \Big/ x\, B\!\left(\dfrac{df_1}{2}, \dfrac{df_2}{2}\right)$	df pf qf rf	df1, df2, ncp
Gamma	$(x) = \dfrac{1}{s^a \Gamma(a)} x^{a-1} e^{-x/s}$	dgamma pgamma qgamma rgamma	shape, rate=1, scale=1/rate
Geometric	$p(x) = p(1-p)^x$	dgeom pgeom qgeom rgeom	prob

Family	PDF or PMF	R functions	Family arguments
Hypergeometric	$$p(x)=\dfrac{\dbinom{m}{x}\dbinom{n}{k-x}}{\dbinom{m+n}{k}}$$	dhyper phyper qhyper rhyper	m, n, k (and note that the common variable "n" has been renamed "nn")
Log-normal	$$f(x)=\dfrac{1}{\sqrt{2\pi}\,\sigma x}e^{-((\log x-\mu)^2/(2\sigma^2))}$$	dlnorm plnorm qlnorm rlnorm	meanlog, sdlog
Logistic	$$F(x)=\dfrac{1}{1+e^{-(x-m)/s}}$$	dlogis plogis qlogis rlogis	location, scale
Multinomial distribution	$$p(X)=\dfrac{N!}{x[1]!\cdots x[\text{length}(x)]!}\prod_{j=1,\dots,\text{length}(x)}\text{prob}[j]^{x[j]}$$	dmultinom rmultinom	
Negative binomial	$$p(x)=\dfrac{\Gamma(x+n)}{\Gamma(n)x!}p^n(1-p)^x$$	dnbinom pnbinom qnbinom rnbinom	size, prob, mu
Normal	$$f(x)=\dfrac{1}{\sqrt{2\pi}\,\sigma}e^{-((x-\mu)^2/(2\sigma^2))}$$	dnorm pnorm qnorm rnorm	mean, sd
Poisson	$$p(x)=\dfrac{\lambda^x e^{-\lambda}}{x!}$$	dpois ppois qpois rpois	lambda
Student's T-distribution	$$f(x)=\dfrac{\Gamma((df+1)/2)}{\sqrt{df\,\pi}\,\Gamma(df/2)}\left(1+\dfrac{x^2}{df}\right)^{-(df+1)/2}$$	dt pt qt rt	df, ncp

Family	PDF or PMF	R functions	Family arguments
Studentized range distribution		ptukey qtukey	nmeans, df, nranges
Uniform distribution		dunif punif qunif runif	min, max
Weibull	$f(x) = \left(\dfrac{a}{b}\right)\left(\dfrac{x}{b}\right)^{a-1} e^{-(x/b)^a}$	dweibull pweibull qweibull rweibull	shape, scale
Wilcoxon rank sum		dwilcox pwilcox qwilcox rwilcox	m, n
Wilcoxon signed rank		dsignrank psignrank qsignrank rsignrank	n

18

Statistical Tests

Many data problems boil down to statistical tests. For example, you might want to answer a question like:

- Does this new drug work better than a placebo?
- Does the new web site design lead to significantly more sales than the old design?
- Can this new investment strategy yield higher returns than an index fund?

To answer questions like these, you would formulate a hypothesis, design an experiment, collect data, and use a tool like R to analyze the data. This chapter focuses on the tools available in R for answering these questions.

> To be helpful, I've tried to include enough description of different statistical methods to help remind you when to use each method (in addition to how to find them in R). However, because this is a Nutshell book, I can't describe where these formulas come from, or when they're safe to use. R is a good substitute for expensive, proprietary statistics software packages. However, *R in a Nutshell* isn't a good substitute for a good statistics course or a good statistics book.

I've broken this chapter into two sections: tools for continuous random variables and tools for categorical random variables (or counts).

Continuous Data

This section describes tests that apply to continuous random variables. Many important measurements fall into this category, such as times, dollar amounts, and chemical concentrations.

Normal Distribution-Based Tests

We'll start off by showing how to use some common statistical tests that assume the underlying data is normally distributed. Normal distributions occur frequently in nature, so this is often a good assumption.[*]

Comparing means

Suppose that you designed an experiment to show that some effect is true. You have collected some data and now want to know if the data proves your hypothesis. One common question is to ask if the mean of the experimental data is close to what the experimenter expected; this is called the null hypothesis. Alternately, the experimenter may calculate the probability that an alternative hypothesis was true. Specifically, suppose that you have a set of observations $x_1, x_2, ..., x_n$ with experimental mean μ and want to know if the experimental mean is different from the null hypothesis mean μ_0. Furthermore, assume that the observations are normally distributed. To test the validity of the hypothesis, you can use a t-test. In R, you would use the function t.test:

```
## Default S3 method:
t.test(x, y = NULL,
       alternative = c("two.sided", "less", "greater"),
       mu = 0, paired = FALSE, var.equal = FALSE,
       conf.level = 0.95, ...)
```

Here is a description of the arguments to the t.test function.

Argument	Description	Default
x	A numeric vector of data values.	
y	A numeric vector of data values (use y=NULL for comparing a single vector of values to a null hypothesis mean, mu, or a vector to compare vector x to vector y).	NULL
alternative	A character value specifying the alternative hypothesis. Use alternative="two.sided" for a two-sided distribution, alternative="less" for lower, and alternative="greater" for higher.	c("two.si-ded", "less", "greater")
mu	A numeric value specifying the value of the mean under the null hypothesis (if testing a single vector of values) or the difference between the means (if comparing the means of two vectors).	0
paired	A logical value indicating if the vectors are paired. See the next section for a description of how to use paired.	FALSE
var.equal	A logical value indicating whether the variance of the two vectors is assumed to be the same. If var.equal=TRUE, then the pooled variance is used. If var.equal=FALSE, then the Welch method is used.	FALSE
conf.level	The confidence interval.	0.95
...	Optional values passed to other methods.	

---

[*] One of the most famous results in probability theory is something called the *central limit theorem*. The central limit theorem states, in a nutshell, that if x is the sum of a set of random variables $x_1, x_2, ..., x_n$, then the distribution of x approaches the normal distribution as $n \to \infty$.

Let's take a look at an example of how you would use the `t.test` function. We'll use the same example data that we used in "Dot plots" on page 276. Suppose that we thought, a priori, that tires of type H should last for approximately 8 hours until failure.[†] We'd like to compare the true mean of this data to the hypothetical mean and determine if the difference was statistically significant using a *t*-test.

To load the sample data, use the following command:

```
> library(nutshell)
> data(tires.sus)
```

To begin, let's extract a vector with the set of values in which we are interested and calculate the true mean:

```
> times.to.failure.h <- subset(tires.sus,
+   Tire_Type=="H" &
+   Speed_At_Failure_km_h==160
+ )$Time_To_Failure
> times.to.failure.h
 [1] 10.00 16.67 13.58 13.53 16.83  7.62  4.25 10.67  4.42  4.25
> mean(times.to.failure.h)
[1] 10.182
```

As you can see, the true mean for these 10 tests was slightly longer than expected (10.182). We can use the function `t.test` to check if this difference is statistically significant:

```
> t.test(times.to.failure.h,mu=9)

        One Sample t-test

data:  times.to.failure.h
t = 0.7569, df = 9, p-value = 0.4684
alternative hypothesis: true mean is not equal to 9
95 percent confidence interval:
  6.649536 13.714464
sample estimates:
mean of x
   10.182
```

Here's an explanation of the output from the `t.test` function. First, the function shows us the test statistic ($t = 0.7569$), the degrees of freedom ($df = 9$), and the calculated *p*-value for the test (p-value = 0.4684). The *p*-value means that the probability that the mean value from an actual sample was higher than 10.812 (or lower than 7.288) was 0.4684. This means that it is very likely that the true mean time to failure was 10.

The next line states the alternative hypothesis: the true mean is not equal to 9, which we would reject based on the result of this test. Next, the `t.test` function shows the 95% confidence intervals for this test, and, finally, it gives the actual mean.

[†] This is a slightly contrived example, because I just made up the hypothetical mean value. In reality, the mean value might come from another experiment (perhaps a published experiment, where the raw data was not available). Or, the mean value might have been derived from theory.

Another common situation is when you have two groups of observations, and you want to know if there is a significant difference between the means of the two groups of observations. You can also use a *t*-test to compare the means of the two groups.

Let's pick another example from the tire data. Looking at the characteristics of the different tires that were tested, notice that three of the six tires had the same speed rating: S. Based on this speed rating, we would expect all three tires to last the same amount of time in the test:

```
> times.to.failure.e <- subset(tires.sus,
+   Tire_Type=="E" & Speed_At_Failure_km_h==180)$Time_To_Failure
> times.to.failure.d <- subset(tires.sus,
+   Tire_Type=="D" & Speed_At_Failure_km_h==180)$Time_To_Failure
> times.to.failure.b <- subset(tires.sus,
+   Tire_Type=="B" & Speed_At_Failure_km_h==180)$Time_To_Failure
```

Let's start by comparing the mean times until failure for tires of types D and E:

```
> t.test(times.to.failure.e, times.to.failure.d)

    Welch Two Sample t-test

data:  times.to.failure.e and times.to.failure.d
t = -2.5042, df = 8.961, p-value = 0.03373
alternative hypothesis: true difference in means is not equal to 0
95 percent confidence interval:
 -0.82222528 -0.04148901
sample estimates:
mean of x mean of y
 4.321000  4.752857
```

The results here are similar to the results from the single-sample *t*-test. In this case, notice that the results were statistically significant at the 95% confidence interval; tires of type E lasted longer than tires of type D.

As an another example, let's compare tires of types E and B:

```
> t.test(times.to.failure.e, times.to.failure.b)

    Welch Two Sample t-test

data:  times.to.failure.e and times.to.failure.b
t = -1.4549, df = 16.956, p-value = 0.1640
alternative hypothesis: true difference in means is not equal to 0
95 percent confidence interval:
 -0.5591177  0.1027844
sample estimates:
mean of x mean of y
 4.321000  4.549167
```

In this case, the difference in means was not significant between the two groups (because the calculated *p*-value was 0.1640). Notice that the output in R is otherwise identical; the t.test function doesn't explicitly say if the results were significant or not.

For two-sample *t*-tests, you can also use a formula to specify a *t*-test if the data is included in a data frame, and the two groups of observations are differentiated by a factor:

```
## S3 method for class 'formula':
t.test(formula, data, subset, na.action, ...)
```

The formula specifies the variables to use in the test.

As an example, let's look at data on field goals kicked in the NFL during 2005. Specifically, let's look at the distance of successful field goals kicked in indoor and outdoor stadiums. Many TV commentators talk about the difficulty of kicking field goals outdoors, due to snow, wind, and so on. But does it make a significant difference in the distance of successful field goals? (Or, for that matter, in bad field goals?) We can use a *t*-test to find out.

First, let's put together the data set:

```
> library(nutshell)
> data(field.goals)
> good <- transform(
+    field.goals[field.goals$play.type=="FG good",
+                c("yards","stadium.type")],
+    outside=(stadium.type=="Out"))
> bad  <- transform(
+    field.goals[field.goals$play.type=="FG no",
+                c("yards","stadium.type")],
+      outside=(stadium.type=="Out"))
```

Now, let's use the `t.test` function to compare the distance of field goals in indoor and outdoor stadiums:

```
> t.test(yards~outside,data=good)

        Welch Two Sample t-test

data:  yards by outside
t = 1.1259, df = 319.428, p-value = 0.2610
alternative hypothesis: true difference in means is not equal to 0
95 percent confidence interval:
 -0.685112  2.518571
sample estimates:
mean in group FALSE  mean in group TRUE
           35.31707            34.40034
```

Although the average successful field goal length was about a yard longer, the difference is not significant at a 95% confidence level. The same is true for field goals that missed:

```
> t.test(yards~outside,data=bad)

        Welch Two Sample t-test

data:  yards by outside
t = 1.2016, df = 70.726, p-value = 0.2335
alternative hypothesis: true difference in means is not equal to 0
95 percent confidence interval:
```

```
    -1.097564  4.425985
sample estimates:
mean in group FALSE  mean in group TRUE
           45.18421             43.52000
```

Was there a statistically significant difference in the distances that coaches attempted to kick field goals? Let's take a look:

```
> field.goals.inout <-
+   transform(field.goals,
+               outside=(stadium.type=="Out")
+             )
> t.test(yards~outside, data=field.goals.inout)

        Welch Two Sample t-test

data:  yards by outside
t = 1.5473, df = 401.509, p-value = 0.1226
alternative hypothesis: true difference in means is not equal to 0
95 percent confidence interval:
 -0.3152552  2.6461541
sample estimates:
mean in group FALSE  mean in group TRUE
           37.14625             35.98080
```

Again, the difference does not appear to be statistically significant at a 95% level.

Comparing paired data

Sometimes, you are provided with paired data. For example, you might have two observations per subject: one before an experiment and one after the experiment. In this case, you would use a paired *t*-test. You can use the t.test function, specifying paired=TRUE, to perform this test.

As an example of paired data, we can look at the SPECint2006 results. SPEC is an organization that provides computer performance data using standardized tests. The organization defines a number of different tests for different applications: database performance, web server performance, graphics performance, and so on. For our example, we'll use a simple metric: the integer performance of different computers on typical desktop computing tasks.

SPEC provides two different types of tests: tests with standard settings and tests that are optimized for specific computers. As an example of paired data, we will compare the unoptimized results (called "baseline") to the optimized results, to see if there is a statistically significant difference between the results. This data set is a good example of paired data: we have two different test results for each computer system. As an example, we will look only at single-chip, dual-core systems:

```
> library(nutshell)
> data(SPECint2006)
> t.test(subset(SPECint2006,Num.Chips==1&Num.Cores==2)$Baseline,
+         subset(SPECint2006,Num.Chips==1&Num.Cores==2)$Result,
+         paired=TRUE)

    Paired t-test
```

```
data:  subset(SPECint2006, Num.Chips == 1 & Num.Cores == 2)$Baseline
  and subset(SPECint2006, Num.Chips == 1 & Num.Cores == 2)$Result
t = -21.8043, df = 111, p-value < 2.2e-16
alternative hypothesis: true difference in means is not equal to 0
95 percent confidence interval:
 -1.957837 -1.631627
sample estimates:
mean of the differences
              -1.794732
```

In this case, we can clearly see that the results were significant at the 95% confidence interval. (This isn't a very big surprise. It's well known that optimizing compiler settings and system parameters can make a big difference on system performance. Additionally, submitting optimized results is optional: organizations that could not tune their systems very well probably would not voluntarily share that fact.)

Comparing variances of two populations

To compare the variances of two samples from normal populations, R includes the var.test function which performs an *F*-test:

```
## Default S3 method:
var.test(x, y, ratio = 1,
         alternative = c("two.sided", "less", "greater"),
         conf.level = 0.95, ...)

## S3 method for class 'formula':
var.test(formula, data, subset, na.action, ...)
```

Let's continue with the example from above. Is there a difference in the variance of field goal lengths between indoor and outdoor stadiums? Let's take a look:

```
> var.test(yards~outside, data=field.goals.inout)

        F test to compare two variances

data:  yards by outside
F = 1.2432, num df = 252, denom df = 728, p-value = 0.03098
alternative hypothesis: true ratio of variances is not equal to 1
95 percent confidence interval:
 1.019968 1.530612
sample estimates:
ratio of variances
          1.243157
```

As you can see from the output above, the *p*-value is less than .05, indicating that the difference in variance between the two populations is statistically significant. To test that the variances in each of the groups (samples) are the same, you can use Bartlett's test. In R, this is available through the bartlett.test function:

```
bartlett.test(x, ...)

## Default S3 method:
bartlett.test(x, g, ...)
```

```
## S3 method for class 'formula':
bartlett.test(formula, data, subset, na.action, ...)
```

Using the same example as above, let's compare variances of the two groups using the Bartlett test:

```
> bartlett.test(yards~outside, data=field.goals.inout)

        Bartlett test of homogeneity of variances

data:  yards by outside
Bartlett's K-squared = 4.5808, df = 1, p-value = 0.03233
```

Comparing means across more than two groups

To compare the means across more than two groups, you can use a method called *analysis of variance* (ANOVA).[‡] ANOVA methods are very important for statistics. A full explanation of ANOVA requires an explanation of statistical models in R, which are covered in Chapter 20.

A simple way to perform these tests is through aov:

```
aov(formula, data = NULL, projections = FALSE, qr = TRUE,
    contrasts = NULL, ...)
```

As an example, let's consider the 2006 U.S. Mortality data set. (I showed how to load this data set in "Using Other Languages to Preprocess Text Files" on page 157.) Specifically, we'll look at differences in age at death by cause of death. This is a pretty silly example; clearly, the average age at which people die of natural causes is going to be higher than the age at which they die for other reasons. However, this should help illustrate how the statistic works.

I mapped the disease codes in the original file into readable values and then summarized causes into a small number of reasons. To do this, I created a function to translate the numeric codes into character values. (I grouped together some common causes of death.) The mort06.smpl data set is included in the nutshell package.

Let's take a look at the summary statistics for age by cause:

```
> library(nutshell)
> data(mort06.smpl)
> tapply(mort06.smpl$age,INDEX=list(mort06.smpl$Cause),FUN=summary)
$Accidents
   Min. 1st Qu.  Median    Mean 3rd Qu.    Max.    NA's
   0.00   31.00   48.00   50.88   73.00  108.00    8.00

$`Alzheimer's Disease`
   Min. 1st Qu.  Median    Mean 3rd Qu.    Max.
  40.00   82.00   87.00   86.07   91.00  109.00

$Cancer
   Min. 1st Qu.  Median    Mean 3rd Qu.    Max.
   0.00   61.00   72.00   70.24   81.00  107.00
```

[‡] This process is named for the analysis process, not for the results. It doesn't compare variances; it compares means by analyzing variances.

```
$`Chronic respiratory diseases`
   Min. 1st Qu.  Median   Mean 3rd Qu.    Max.   NA's
   0.00   70.00   78.00  76.37   84.00  106.00   1.00

$Diabetes
   Min. 1st Qu.  Median   Mean 3rd Qu.    Max.   NA's
   0.00   63.00   75.00  72.43   83.00  104.00   1.00

$`Heart Disease`
   Min. 1st Qu.  Median   Mean 3rd Qu.    Max.   NA's
   0.00   70.00   81.00  77.66   88.00  112.00   4.00

$Homicide
   Min. 1st Qu.  Median   Mean 3rd Qu.    Max.   NA's
   0.00   22.00   28.00  32.19   42.00   92.00   2.00

$`Influenza and pneumonia`
   Min. 1st Qu.  Median   Mean 3rd Qu.    Max.   NA's
   0.00   76.00   84.00  80.16   90.00  108.00   1.00

$Other
   Min. 1st Qu.  Median   Mean 3rd Qu.    Max.   NA's
   0.00   60.00   78.00  70.44   87.00  110.00  10.00

$Suicide
   Min. 1st Qu.  Median   Mean 3rd Qu.    Max.   NA's
   8.00   32.00   45.00  46.14   57.00   97.00   2.00
```

Now, let's fit an ANOVA model to the data and show a summary of the model. To do this in R, we simply need to use the aov function:

```
> aov(age~Cause,data=mort06.smpl)
Call:
   aov(formula = age ~ Cause, data = mort06.smpl)

Terms:
                    Cause Residuals
Sum of Squares   15727886  72067515
Deg. of Freedom         9    243034

Residual standard error: 17.22012
Estimated effects may be unbalanced
29 observations deleted due to missingness
```

To get more information on ANOVA results, you can use the model.tables to print information on aov objects:

```
## S3 method for class 'aov':
model.tables(x, type = "effects", se = FALSE, cterms, ...)
```

The argument x specifies the model object, type specifies the type of results to print, se specifies whether to compute standard errors, and cterms specifies which tables should be compared. As an example, here is the output of model.tables for the cause of death example above:

```
> model.tables(aov(age~Cause, data=mort06.smpl))
Tables of effects

 Cause
    Accidents Alzheimer's Disease      Cancer
       -21.41               13.77      -2.056
 rep  12363.00            7336.00   57162.000
    Chronic respiratory diseases  Diabetes Heart Disease Homicide
                           4.077    0.1343        5.371    -40.1
 rep                   12386.000  7271.0000    82593.000   1917.0
    Influenza and pheumonia      Other Suicide
                         7.863    -1.855  -26.15
 rep                  5826.000 52956.000 3234.00
```

As another example of aov, let's consider weight gain by women during pregnancy:

```
> library(nutshell)
> data(births2006.smpl)
> births2006.cln <- births2006.smpl[births2006.smpl$WTGAIN<99 &
                                   !is.na(births2006.smpl$WTGAIN),]
> tapply(X=births2006.cln$WTGAIN,
+        INDEX=births2006.cln$DOB_MM,
+        FUN=mean)
        1        2        3        4        5        6
 30.94405 31.08356 31.29317 31.33610 31.07242 30.92589
        7        8        9       10       11       12
 30.57734 30.54855 30.25546 30.43985 30.79077 30.85564
```

It appears that weight gain increases slightly during winter months, but is this difference statistically significant? Let's take a look:

```
> aov(WTGAIN~DOB_MM, births2006.cln)
Call:
   aov(formula = WTGAIN ~ DOB_MM, data = births2006.cln)

Terms:
                  DOB_MM Residuals
Sum of Squares     14777  73385301
Deg. of Freedom        1    351465

Residual standard error: 14.44986
Estimated effects may be unbalanced
```

Often, it's better to use lm to fit a linear model and then use the anova function to extract information about analysis of variance. For large models, it is often more efficient to use the update function to change an existing model than to create a new model from scratch. See "Example: A Simple Linear Model" on page 373 for more information on the lm function, model objects, and the update function. The anova function presents results slightly differently than the aov function, as you can see in this example:

```
> mort06.smpl.lm <- lm(age~Cause,data=mort06.smpl)
> anova(mort06.smpl.lm)
Analysis of Variance Table

Response: age
```

```
               Df   Sum Sq Mean Sq F value    Pr(>F)
Cause           9 15727886 1747543  5893.3 < 2.2e-16 ***
Residuals 243034 72067515     297
---
Signif. codes:  0 '***' 0.001 '**' 0.01 '*' 0.05 '.' 0.1 ' ' 1
```

ANOVA calculations assume that the variance is equal across groups. When you know this is not true (or suspect this is not true), you can use the oneway.test function to calculate whether two or more samples have the same mean:

```
oneway.test(formula, data, subset, na.action, var.equal = FALSE)
```

This is similar to calling t.test with var.equal=FALSE (and the calculation method was also developed by Welch).

There are other functions for printing information about aov objects: proj returns a list of matrices giving the projections of data onto the linear model, TukeyHSD returns confidence intervals on the differences between the means of the levels of a factor with the specified family-wise probability of coverage, and se.contrast returns the standard errors for one or more contrasts in an aov object.

Pairwise *t*-tests between multiple groups

Sometimes, you're not interested in just whether there is a difference across groups, but would like to know more details about the differences. One way to do this is by performing a *t*-test between every pair of groups. To do this in R, you can use the pairwise.t.test function:

```
pairwise.t.test(x, g, p.adjust.method = p.adjust.methods,
                pool.sd = !paired, paired = FALSE,
                alternative = c("two.sided", "less", "greater"), ...)
```

This function calculates pairwise comparisons between group levels with corrections for multiple testing. The argument x specifies a vector of numeric values, and g specifies a factor that is used to group values. The argument pool.sd specifies whether to calculate a single standard deviation value across all groups and use this for the test.

As an example, let's return to the tire data that we used in the example above. When we looked at the t.test function, we created three different vectors for the different types of tires. Here, we'll just use the pairwise *t*-test to compare all the tires by type:

```
> pairwise.t.test(tires.sus$Time_To_Failure,tires.sus$Tire_Type)

	Pairwise comparisons using t tests with pooled SD

data:  tires.sus$Time_To_Failure and tires.sus$Tire_Type

  B       C       D       E       H
C 0.2219  -       -       -       -
D 1.0000  0.5650  -       -       -
E 1.0000  0.0769  1.0000  -       -
H 2.4e-07 0.0029  2.6e-05 1.9e-08 -
L 0.1147  1.0000  0.4408  0.0291  0.0019
```

```
P value adjustment method: holm
```

As you can see, there is no statistically significant difference between the means of a few pairs of groups (such as C and L, D and L, or D and E), but there is a significant difference between some others (such as B and H, C and H, or E and L).

Testing for normality

To test if a distribution is normally distributed in R, you can use the Shapiro-Wilk test for normality through the `shapiro.test` function:

```
shapiro.test(x)
```

Using the example above, let's look at field goal lengths in the NFL in 2005. Was the distribution of field goal lengths normally distributed? My first instinct is to take a look at the distribution using a histogram or a quantile-quantile plot. Here is some R code to plot both, side by side:

```
par(mfcol=c(1,2),ps=6.5)
hist(fg_attempts$YARDS,breaks=25)
qqnorm(fg_attempts$YARDS,pch=".")
```

The plot is shown in Figure 18-1. It seems plausible that the distribution was normal: the distribution is roughly bell curve shaped, and the quantile-quantile plot is roughly linear. To get a more rigorous answer, we can use the Shapiro-Wilk test. Here's what the Shapiro-Wilk test tells us:

```
> shapiro.test(field.goals$YARDS)

        Shapiro-Wilk normality test

data:  field.goals$YARDS
W = 0.9728, p-value = 1.307e-12
```

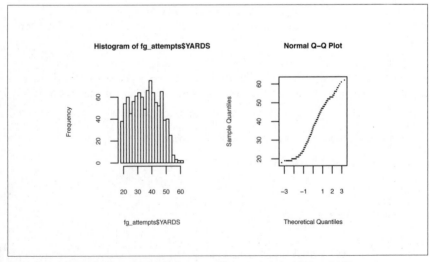

Figure 18-1. Distribution of field goal attempt distances in the NFL in 2005

As you can tell from the *p*-value, it is quite likely that this data came from a normal distribution.

Testing if a data vector came from an arbitrary distribution

You can use the Kolmogorov-Smirnov test to see if a vector came from an arbitrary probability distribution (not just a normal distribution):

```
ks.test(x, y, ...,
        alternative = c("two.sided", "less", "greater"),
        exact = NULL)
```

The argument x specifies the test data. The argument y specifies the arbitrary distribution; it can be a vector of data values, a character name for a probability distribution function, or a distribution function. (You can pass along additional arguments to the distribution function.) The `alternative` argument allows you to specify the alternative hypothesis, and the `exact` value specifies whether to calculate exact values (for large *x* and *y*) or approximations.

Using the example above, we can use the `ks.test` function. We'll specify the normal distribution (using the `pnorm` function):

```
> ks.test(field.goals$YARDS,pnorm)

    One-sample Kolmogorov-Smirnov test

data:  field.goals$YARDS
D = 1, p-value < 2.2e-16
alternative hypothesis: two-sided

Warning message:
In ks.test(field.goals$YARDS, pnorm) :
  cannot compute correct p-values with ties
```

Notice the warning message; ties are extremely unlikely for values from a true normal distribution. If there are ties in the data, that is a good sign that the test data is not actually normally distributed, so the function prints a warning.

Testing if two data vectors came from the same distribution

The Kolmogorov-Smirnov test can also be used to test the probability that two data vectors came from the same distribution. As an example, let's look at the SPECint2006 data that we saw in "Comparing paired data" on page 348. What is the probability that the benchmark data and the optimized data come from the same distribution? We'll compare the benchmark and optimized data using the `ks.test` function, adding some jitter to the values to suppress the warning about ties:

```
> ks.test(jitter(subset(SPECint2006,Num.Chips==1&Num.Cores==2)$Baseline),
+         jitter(subset(SPECint2006,Num.Chips==1&Num.Cores==2)$Result))

    Two-sample Kolmogorov-Smirnov test

data:  jitter(subset(SPECint2006, Num.Chips == 1 & Num.Cores == 2)$Baseline)
  and jitter(subset(SPECint2006, Num.Chips == 1 & Num.Cores == 2)$Result)
```

```
D = 0.2143, p-value = 0.01168
alternative hypothesis: two-sided
```

According to the *p*-value, the two vectors likely came from the same distribution (at a 95% confidence level).

Correlation tests

The functions in "Correlation and Covariance" on page 325 simply compute the degree of correlation between pairs of vectors, but they don't tell you if the correlation is significant. If you'd like to check whether there is a statistically significant correlation between two vectors, you can use the `cor.test` function:

```
## Default S3 method:
cor.test(x, y,
         alternative = c("two.sided", "less", "greater"),
         method = c("pearson", "kendall", "spearman"),
         exact = NULL, conf.level = 0.95, ...)

## S3 method for class 'formula':
cor.test(formula, data, subset, na.action, ...)
```

For example, let's look at how this function works on two obviously correlated vectors:

```
> cor.test(c(1, 2, 3, 4, 5,  6,  7,  8),
+          c(0, 2, 4, 6, 8, 10, 11, 14))

        Pearson's product-moment correlation

data:  c(1, 2, 3, 4, 5, 6, 7, 8) and c(0, 2, 4, 6, 8, 10, 11, 14)
t = 36.1479, df = 6, p-value = 2.989e-08
alternative hypothesis: true correlation is not equal to 0
95 percent confidence interval:
 0.9868648 0.9996032
sample estimates:
      cor
0.997712
```

And two less correlated vectors:

```
> cor.test(c(1, 2, 3, 4, 5, 6, 7, 8),
+          c(5, 3, 8, 1, 7, 0, 0, 3))

        Pearson's product-moment correlation

data:  c(1, 2, 3, 4, 5, 6, 7, 8) and c(5, 3, 8, 1, 7, 0, 0, 3)
t = -1.2232, df = 6, p-value = 0.2671
alternative hypothesis: true correlation is not equal to 0
95 percent confidence interval:
 -0.8757371  0.3764066
sample estimates:
       cor
-0.4467689
```

Let's revisit the data on environmental toxins and lung cancer that we examined in "Scatter Plots" on page 212. This data compared the amount of airborne toxins released in each state to the deaths by lung cancer in each state:

```
> cor.test(air_on_site/Surface_Area, deaths_lung/Population)

        Pearson's product-moment correlation

data:  air_on_site/Surface_Area and deaths_lung/Population
t = 3.4108, df = 39, p-value = 0.001520
alternative hypothesis: true correlation is not equal to 0
95 percent confidence interval:
 0.2013723 0.6858402
sample estimates:
      cor
0.4793273
```

The test shows that there appears to be a positive correlation between these two quantities that is statistically significant. However, don't infer that there is a causal relationship between the rates of toxins released and the rates of lung cancer deaths. There are many alternate explanations for this phenomenon. For example, states with lots of dirty industrial activity may also be states with lower levels of income, which, in turn, correlates with lower-quality medical care. Or, perhaps, states with lots of industrial activity may be states with higher rates of smoking. Or, maybe states with lower levels of industrial activity are less likely to identify cancer as a cause of death. Whatever the explanation, I thought this was a neat result.

Non-Parametric Tests

Although many real data sources can be approximated well by a normal distribution, there are many cases where you know that the data is not normally distributed, or do not know the shape of the distribution. A good alternative to the tests described in "Normal Distribution-Based Tests" on page 344 are non-parametric tests. These tests can be more computationally intensive than tests based on a normal distribution, but they may help you make better choices when the distribution is not normally distributed.

Comparing two means

The Wilcoxon test is the non-parametric equivalent to the *t*-test:

```
## Default S3 method:
wilcox.test(x, y = NULL,
            alternative = c("two.sided", "less", "greater"),
            mu = 0, paired = FALSE, exact = NULL, correct = TRUE,
            conf.int = FALSE, conf.level = 0.95, ...)

## S3 method for class 'formula':
wilcox.test(formula, data, subset, na.action, ...)
```

The Wilcoxon test works by looking at the ranks of different elements in x and y; the exact values don't matter. To get the test statistic for x and y, you can calculate:

$$w = \begin{cases} 1 & \text{if } y[j] < x[i] \\ 0 & \text{otherwise} \end{cases}$$

$$W = \sum_{\forall i,j} w$$

That is, look at all pairs of values $(x[i], y[j])$, counting the number of cases where $y[j] < x[i]$. If the two vectors were both from the same distribution, you'd expect this to be true for roughly half the pairs. The Wilcoxon distribution can be used to estimate the probability of different values for W; the p-value for the Wilcoxon test comes from this distribution. Notice that there is no version of the Wilcoxon test that compares a data sample to a hypothesized mean.

As an example, let's take a look at the same examples we used for t-tests. Let's start by looking at the times to failure for tires. As above, let's start by comparing tires of type E to tires of type D:

```
> wilcox.test(times.to.failure.e, times.to.failure.d)

	Wilcoxon rank sum test with continuity correction

data:  times.to.failure.e and times.to.failure.d
W = 14.5, p-value = 0.05054
alternative hypothesis: true location shift is not equal to 0

Warning message:
In wilcox.test.default(times.to.failure.e, times.to.failure.d) :
  cannot compute exact p-value with ties
```

Here's an explanation of the output. The test function first shows the test statistic ($W = 14.5$) and the p-value for the statistic (0.05054). Notice that this is different from the result for the t-test. With the t-test, there was a significant difference between the means of the two groups, but with the Wilcoxon rank sum test, the difference between the two groups is not significant at a 95% confidence level (though it barely misses).

Also note the warning. The Wilcoxon test statistic is based on the rank order of the observations, not their specific values. In our test data, there are a few ties:

```
> times.to.failure.d
[1] 5.22 4.47 4.25 5.22 4.68 5.05 4.3
> times.to.failure.e
 [1] 4.48 4.70 4.52 4.63 4.07 4.28 4.25 4.10 4.08 4.10
```

Because there was a tie, the function above actually used a normal approximation; see the help file for more information.

As with the standard *t*-test function, there is also a formula method for `wilcox.test`. As above, let's compare the distance of field goals made in indoor stadiums versus outdoor stadiums:

```
> wilcox.test(yards~outside, data=good)

        Wilcoxon rank sum test with continuity correction

data:  YARDS by outside
W = 62045, p-value = 0.3930
alternative hypothesis: true location shift is not equal to 0
```

Comparing more than two means

The Kruskal-Wallis rank sum test is a non-parametric equivalent to ANOVA analysis:

```
kruskal.test(x, ...)

## Default S3 method:
kruskal.test(x, g, ...)

## S3 method for class 'formula':
kruskal.test(formula, data, subset, na.action, ...
```

As an example, here is the output for the mortality data that we used as an example for ANOVA statistics:

```
> kruskal.test(age~Cause,data=mort06.smpl)

    Kruskal-Wallis rank sum test

data:  age by Cause
Kruskal-Wallis chi-squared = 34868.1, df = 9, p-value
< 2.2e-16
```

Comparing variances

To compare the variance between different groups using a nonparametric test, R includes an implementation of the Fligner-Killeen (median) test through the `fligner.test` function:

```
## Default S3 method:
fligner.test(x, g, ...)
## S3 method for class 'formula':
fligner.test(formula, data, subset, na.action, ...)
```

Here is the output of `fligner.test` for the mortality data above:

```
> fligner.test(age~Cause,data=mort06.smpl)

    Fligner-Killeen test of homogeneity of variances

data:  age by Cause
Fligner-Killeen:med chi-squared = 15788, df = 9,
p-value < 2.2e-16
```

Statistical Tests

Difference in scale parameters

There are some tests in R for testing for differences in scale parameters. To use the Ansari-Bradley two-sample test for a difference in scale parameters, use the function ansari.test:

```
## Default S3 method:
ansari.test(x, y,
            alternative = c("two.sided", "less", "greater"),
            exact = NULL, conf.int = FALSE, conf.level = 0.95,
            ...)

## S3 method for class 'formula':
ansari.test(formula, data, subset, na.action, ...)
```

To use Mood's two-sample test for a difference in scale parameters in R, try the function mood.test:

```
## Default S3 method:
mood.test(x, y,
          alternative = c("two.sided", "less", "greater"), ...)

## S3 method for class 'formula':
mood.test(formula, data, subset, na.action, ...)
```

Discrete Data

There is a different set of tests for looking at the statistical significance of discrete random variables (like counts of proportions), and so there is a different set of functions in R for performing those tests.

Proportion Tests

If you have a data set with several different groups of observations and are measuring the probability of success in each group (or the fraction of some other characteristic), you can use the function prop.test to measure whether the difference between groups is statistically significant. Specifically, prop.test can be used for testing the null hypothesis that the proportions (probabilities of success) in several groups are the same or that they equal certain given values:

```
prop.test(x, n, p = NULL,
          alternative = c("two.sided", "less", "greater"),
          conf.level = 0.95, correct = TRUE)
```

As an example, let's revisit the field goal data. Above, we considered the question "is there a difference in the length of attempts indoors and outdoors?" Now, we'll ask the question "is the probability of success the same indoors as it is outdoors?"

First, let's create a new data set containing only good and bad field goals. (We'll eliminate blocked and aborted attempts; there were only 8 aborted attempts and 24 blocked attempts in 2005, but 787 good attempts and 163 bad (no good) attempts.)

```
> field.goals.goodbad <- field.goals[field.goals$play.type=="FG good" |
                                      field.goals$play.type=="FG no", ]
```

Now, let's create a table of successes and failures by stadium type:

```
> field.goals.table <- table(field.goals.goodbad$play.type,
+                             field.goals.goodbad$stadium.type)
> field.goals.table

              Both  In Out
  FG aborted    0   0   0
  FG blocked    0   0   0
  FG good      53 152 582
  FG no        14  24 125
```

The table isn't quite right for `prop.test`; we need a table with two columns (one with a count of successes and one with a count of failures), and we don't want to show empty factor levels. Let's remove the two rows we don't need and transpose the table:

```
> field.goals.table.t <- t(field.goals.table[3:4,])
> field.goals.table.t

      FG good FG no
  Both     53    14
  In      152    24
  Out     582   125
```

Now, we're ready to see if there is a statistically significant difference in success between the three groups. We can simply call `prop.test` on the `field.goals.table.t` object to check:

```
> prop.test(field.goals.table.t)

        3-sample test for equality of proportions without
        continuity correction

data:  field.goals.table
X-squared = 2.3298, df = 2, p-value = 0.3120
alternative hypothesis: two.sided
sample estimates:
   prop 1    prop 2    prop 3
0.7910448 0.8636364 0.8231966
```

As you can see, the results are not significant.

Binomial Tests

Often, an experiment consists of a series of identical trials, each of which has only two outcomes. For example, suppose that you wanted to test the hypothesis that the probability that a coin would land on heads was .5. You might design an experiment where you flipped the coin 50 times and counted the number of heads. Each coin flip is an example of a Bernoulli trial. The distribution of the number of heads is given by the binomial distribution.

R includes a function for evaluating such a trial to determine whether to accept or reject the hypothesis:

```
binom.test(x, n, p = 0.5,
           alternative = c("two.sided", "less", "greater"),
           conf.level = 0.95)
```

The argument x gives the number of successes, n gives the total number of trials, p gives the probability of each success, `alternative` gives the alternative hypothesis, and `conf.level` gives the returned confidence level.

As an example, let's look at David Ortiz's performance during the 2008 season. In 2008, he had a batting average of .264 (110 hits in 416 at bats). Suppose that he was actually a .300 hitter—that the actual probability that he would get a hit in a given at bat was .3. What were the odds that he hit .264 or less in this number of at bats? We can use the function `binom.test` to estimate this probability:

```
> binom.test(x=110,n=416,p=0.3,alternative="less")

        Exact binomial test

data:  110 and 416
number of successes = 110, number of trials = 416, p-value =
0.06174
alternative hypothesis: true probability of success is less than 0.3
95 percent confidence interval:
 0.0000000 0.3023771
sample estimates:
probability of success
              0.2644231
```

Unlike some other test functions, the *p*-value represents the probability that the fraction of successes (.26443431) was at least as far from the hypothesized value (.300) after the experiment. We specified that the alternative hypothesis was "less," meaning that the *p*-value represents the probability that the fraction of successes was less than .26443431, which in this case was .06174.

In plain English, this means that if David Ortiz was a "true" .300 hitter, the probability that he actually hit .264 or worse in a season was .06174.

Tabular Data Tests

A common problem is to look at a table of data and determine if there is a relationship between two categorical variables. If there were no relationship, the two variables would be statistically independent. In these tests, the hypothesis is that the two variables are independent. The alternative hypothesis is that the two variables are not independent.

Tables of data often come up in experimental contexts: there is one column of data from a test population and one from a control population. In this context, the analyst often wants to calculate the probability that the two sets of data could have come from the same population (which would imply the same proportions in each). This is an equivalent problem, so the same test functions can be used.

For small contingency tables (and small values), you can obtain the best results using Fisher's exact test. Fisher's exact test calculates the probability that the deviation from the independence was greater than or equal to the sample quantities. So, a high *p*-value means that the sample data implies that the two variables are likely to be independent. A low *p*-value means that the sample data implies that the two variables are not independent.

In R, you can use the function `fisher.test` to perform Fisher's exact test:

```
fisher.test(x, y = NULL, workspace = 200000, hybrid = FALSE,
            control = list(), or = 1, alternative = "two.sided",
            conf.int = TRUE, conf.level = 0.95,
            simulate.p.value = FALSE, B = 2000)
```

Here is a description of the arguments to `fisher.test`.

Argument	Description	Default
x	Specifies the sample data to use for the test. Either a matrix (representing a two-dimensional contingency table) or a factor.	
y	Specifies the sample data to use for the test. If x is a factor, then y should be a factor. If x is a matrix, then y is ignored.	NULL
workspace	An integer value specifying the size of the workspace to use in the network algorithm (in units of 4 bytes).	200000
hybrid	For tables larger than 2×2, specifies whether exact probabilities should be calculated (hybrid=FALSE) or an approximation should be used (hybrid=TRUE).	FALSE
control	A list of named components for low-level control of `fisher.test`; see the help file for more information.	list()
or	The hypothesized odds ratio for the 2 × 2 case.	1
alternative	The alternative hypothesis. Must be one of "two.sided", "greater", or "less".	"two.sided"
conf.int	A logical value specifying whether to compute and return confidence intervals in the results.	TRUE
conf.level	Specifies the confidence level to use in computing the confidence interval.	0.95
simulate.p.value	A logical value indicating whether to use Monte Carlo simulation to compute *p*-values in tables larger than 2 × 2.	FALSE
B	An integer indicating the number of replicates to use in Monte Carlo simulations.	2000

If you specify x and y as factors, then R will compute a contingency table from these factors. Alternatively, you can specify a matrix for x containing the contingency table.

Fisher's exact test can be very computationally intensive for large tables, so statisticians usually use an alternative test: chi-squared tests. Chi-squared tests are not exactly the same as Fisher's tests. With a chi-squared test, you explicitly state a hypothesis about the probability of each event and then compare the sample distribution to the hypothesis. The *p*-value is the probability that a distribution at least as different from the hypothesized distribution arose by chance.

In R, you can use the function `chisq.test` to calculate a chi-squared contingency table and goodness-of-fit tests:

```
chisq.test(x, y = NULL, correct = TRUE,
           p = rep(1/length(x), length(x)), rescale.p = FALSE,
           simulate.p.value = FALSE, B = 200
```

Here is a description of the arguments to chisq.test.

Argument	Description	Default
x	Specifies the sample data to use for the test. Either a matrix or a vector.	
y	Specifies the sample data to use for the test. If x is a factor, then y should be a vector. If x is a matrix, then y is ignored.	NULL
correct	A logical value specifying whether to apply continuity correction when computing the test statistic for 2×2 tables.	TRUE
p	A vector of probabilities that represent the hypothesis to test. (Note that the default is to assume equal probability for each item.)	rep(1/length(x), length(x))
rescale.p	A logical value indicating whether p needs to be rescaled to sum to 1.	FALSE
simulate.p.value	A logical value indicating whether to compute p-values using Monte Carlo simulation.	FALSE
B	An integer indicating the number of replicates to use in Monte Carlo simulations.	200

If you specify x and y as vectors, then R will compute a contingency table from these vectors (after coercing them to factors). Alternatively, you can specify a matrix for x containing the contingency table.

As an example, let's use the 2006 births data set. (For a detailed description of this data set, see "Univariate Trellis Plots" on page 269.) We will take a look at the number of male and female babies delivered during July 2006, by delivery method. We'll take a subset of births during July where the delivery method was known and then tabulate the results:

```
> births.july.2006 <- births2006.smpl[births2006.smpl$DMETH_REC!="Unknown" &
                            births2006.smpl$DOB_MM==7, ]
> nrow(births2006.smpl)
[1] 427323
> nrow(births.july.2006)
[1] 37060
> method.and.sex <- table(
+     births.july.2006$SEX,
+     as.factor(as.character(births.july.2006$DMETH_REC)))
> method.and.sex

    C-section Vaginal
  F      5326   12622
  M      6067   13045
```

Note that the delivery methods were actually slightly unbalanced by gender during July 2006:

```
> 5325 / (5326 + 6067)
[1] 0.4673923
```

```
> 12622 / (12622 + 13045)
[1] 0.4917598
```

However, there isn't an intuitive reason why this should be true. So, let's check whether this difference is statistically significant: is the difference due to chance or is it likely that these two variables (delivery method and sex) are independent? We can use Fisher's exact test to answer this question:

```
> fisher.test(method.and.sex)

        Fisher's Exact Test for Count Data

data:  method.and.sex
p-value = 1.604e-05
alternative hypothesis: true odds ratio is not equal to 1
95 percent confidence interval:
 0.8678345 0.9485129
sample estimates:
odds ratio
 0.9072866
```

The *p*-value is the probability of obtaining results that were at least as far removed from independence as these results. In this case, the *p*-value is very low, indicating that the results were very far from what we would expect if the variables were truly independent. This implies that we should reject the hypothesis that the two variables are independent.

As a second example, let's look only at twin births. (Note that each record represents a single birth, not a single pregnancy.)

```
> twins.2006 <- births2006.smpl[births2006.smpl$DPLURAL=="2 Twin" &
+                               births2006.smpl$DMETH_REC != "Unknown",]
> method.and.sex.twins <-
+    table(twins.2006$SEX,
+          as.factor(as.character(twins.2006$DMETH_REC)))
> method.and.sex.twins

    C-section Vaginal
  F      4924    1774
  M      5076    1860
```

Now, let's see if there is a statistically significant difference in delivery methods between the two sexes:

```
> fisher.test(method.and.sex.twins)

        Fisher's Exact Test for Count Data

data:  method.and.sex.twins
p-value = 0.67
alternative hypothesis: true odds ratio is not equal to 1
95 percent confidence interval:
 0.9420023 1.0981529
sample estimates:
odds ratio
 1.017083
```

In this case, the *p*-value (0.67) is very high, so it is very likely that the two variables are independent.

We can look at the same table using a chi-squared test:

```
> chisq.test(method.and.sex.twins)

        Pearson's Chi-squared test with Yates' continuity
        correction

data:  method.and.sex.twins
X-squared = 0.1745, df = 1, p-value = 0.6761
```

By the way, we could also have just passed the two factors to `chisq.test`, and `chisq.test` would have calculated the contingency table for us:

```
> chisq.test(twins.2006$DMETH_REC,twins.2006$SEX)

        Pearson's Chi-squared test with Yates' continuity
        correction

data:  twins.2006$DMETH_REC and twins.2006$SEX
X-squared = 0.1745, df = 1, p-value = 0.6761
```

As above, the *p*-value is very high, so it is likely that the two variables are independent for twin births.

Let's ask another interesting question: how many babies are born on weekdays versus weekends? Let's start by tabulating the number of births, by day of week, during 2006:

```
> births2006.byday <- table(births2006.smpl$DOB_WK)
> births2006.byday

    1     2     3     4     5     6     7
40274 62757 69775 70290 70164 68380 45683
```

Curiously, the number of births on days 1 and 7 (Sunday and Saturday, respectively) are sharply lower than the number of births on other days. We can use a chi-squared test to determine what the probability is that this distribution arose by chance. As noted above, by default, we perform a chi-squared test under the assumption that the actual probability of a baby being born on each day is given by the vector `p=rep(1/ length(x), length(x))`, which in this case is 1/7 for every day. So, we're asking what the probability is that a distribution at least as unbalanced as the one above arose by chance:

```
> chisq.test(births2006.byday)

        Chi-squared test for given probabilities

data:  births2006.byday
X-squared = 15873.20, df = 6, p-value < 2.2e-16
```

As you might have guessed, this effect was statistically significant. The *p*-value is very, very small, indicating that it is very unlikely that this effect arose due to chance. (Of course, with a sample this big, it's not hard to find significant effects.)

The `chisq.test` function can also perform tests on multidimensional tables. As an example, let's build a table showing the number of births by day and month:

```
> births2006.bydayandmonth <- table(births2006.smpl$DOB_WK,
+                                     births2006.smpl$DOB_MM)
> births2006.bydayandmonth

    1    2    3    4    5    6    7
1 3645 2930 2965 3616 2969 3036 3976
2 5649 4737 4779 4853 5712 5033 6263
3 6265 5293 5251 5297 6472 5178 5149
4 5131 5280 6486 5173 6496 5540 5499
5 5127 5271 6574 5162 5347 6863 5780
6 4830 5305 6330 5042 4975 6622 5760
7 3295 3392 3408 4185 3364 3464 4751

    8    9   10   11   12
1 3160 3270 3964 2999 3744
2 5127 4850 6167 5043 4544
3 7225 5805 6887 5619 5334
4 7011 5725 5445 6838 5666
5 6945 5822 5538 6165 5570
6 5530 7027 5256 5079 6624
7 3686 4669 3564 3509 4396
```

As above, let's check the probability that this distribution arose by chance under the assumption that the probability of each combination was equal:

```
> chisq.test(births2006.bydayandmonth)

        Pearson's Chi-squared test

data:  births2006.bydayandmonth
X-squared = 4729.620, df = 66,
p-value < 2.2e-16
```

Much like the one-dimensional table, we see that the effects are statistically significant; it is very unlikely that this unbalanced distribution arose due to chance.

For three-way interactions, you can try a Cochran-Mantel-Haenszel test. This is implemented in R through the `mantelhaen.test` function:

```
mantelhaen.test(x, y = NULL, z = NULL,
                alternative = c("two.sided", "less", "greater"),
                correct = TRUE, exact = FALSE, conf.level = 0.95)
```

To test for symmetry in a two-dimensional contingency table, you can use McNemar's chi-squared test. This is implemented in R as `mcnemar.test`:

```
mcnemar.test(x, y = NULL, correct = TRUE)
```

Non-Parametric Tabular Data Tests

The Friedman rank sum test is a non-parametric relative of the chi-squared test. In R, this is implemented through the `friedman.test` function:

```
friedman.test(y, ...)

## Default S3 method:
friedman.test(y, groups, blocks, ...)

## S3 method for class 'formula':
friedman.test(formula, data, subset, na.action, ...)
```

As examples, let's look at some of the same tables we looked at above:

```
> friedman.test(method.and.sex.twins)

        Friedman rank sum test

data:  method.and.sex.twins
Friedman chi-squared = 2, df = 1,
p-value = 0.1573
```

Just like the chi-squared test, the Friedman rank sum test shows that it is very likely that the two distributions are independent.

19

Power Tests

When designing an experiment, it's often helpful to know how much data you need to collect to get a statistically significant sample (or, alternatively, the maximum significance of results that can be calculated from a given amount of data). R provides a set of functions to help you calculate these amounts.

Experimental Design Example

Suppose that you want to test the efficacy of a new drug for treating depression. A common score used to measure depression is the Hamilton Rating Scale for Depression (HAMD). This measure varies from 0 to 48, where higher values indicate increased depression. Let's consider two different experimental design questions. First, suppose that you had collected 50 subjects for the study and split them into two groups of 25 people each. What difference in HAMD scores would you need to observe in order for the results to be considered statistically significant?

We assume a standard deviation of 8.9 for this experiment.[*] We'll also assume that we want a power of .95 for the experiment (meaning that the probability of a Type II error is less than .05). To calculate the minimum statistically significant difference in R, we could use the following expression:

---

[*] Number from *http://www.fda.gov/OHRMS/DOCKETS/ac/07/slides/2007-4273s1_05.pdf*.

```
> power.t.test(power=.95, sig.level=.05, sd=8.9, n=25)

        Two-sample t test power calculation

              n = 25
          delta = 9.26214
             sd = 8.9
      sig.level = 0.05
          power = 0.95
    alternative = two.sided

    NOTE: n is number in *each* group
```

According to the output, the difference in means between the two groups would need to be at least 9.26214 to be significant at this level. Suppose that we doubled the number of subjects. What difference would be considered significant?

```
> power.t.test(power=.95, sig.level=.05, sd=8.9, n=50)

        Two-sample t test power calculation

              n = 50
          delta = 6.480487
             sd = 8.9
      sig.level = 0.05
          power = 0.95
    alternative = two.sided

    NOTE: n is number in *each* group
```

As you can see, the power functions can be very useful for designing an experiment. They can help you to estimate, in advance, how large a difference you need to see between groups to get statistically significant results.

t-Test Design

If you are designing an experiment where you will use a *t*-test to check the significance of the results (typically, an experiment where you calculate the mean value of a random variable for a "test" population and a "control" population), then you can use the power.t.test function to help design the experiment:

```
power.t.test(n = NULL, delta = NULL, sd = 1, sig.level = 0.05,
             power = NULL,
             type = c("two.sample", "one.sample", "paired"),
             alternative = c("two.sided", "one.sided"),
             strict = FALSE)
```

For this function, n specifies the number of observations (per group); delta is the true difference in means between the groups; sd is the true standard deviation of the underlying distribution; sig.level is the significance level (Type I error probability); power is the power of the test (1 – Type II error probability); type specifies whether the test is one sample, two sample, or paired; alternative specifies whether the test is one or two sided; and strict specifies whether to use a strict interpretation in the two-sided case. This function will calculate either n, delta, sig.level, sd, or power, depending on the input. You must specify at least four of these parameters: n, delta, sd, sig.level, power. The remaining argument must be null; this is the value that the function calculates.

Proportion Test Design

If you are designing an experiment where you will be measuring a proportion (using prop.test), you can use the power.prop.test function:

```
power.prop.test(n = NULL, p1 = NULL, p2 = NULL, sig.level = 0.05,
                power = NULL,
                alternative = c("two.sided", "one.sided"),
                strict = FALSE)
```

For this function, n specifies the number of observations (per group), p1 is the probability of success in one group, p2 is the probability of success in the other group, sig.level is the significance level (Type I error probability), power is the power of the test (1 – Type II error probability), alternative specifies whether the test is one or two-sided, and strict specifies whether to use a strict interpretation in the two-sided case. This function will calculate either n, p1, p2, sig.level, or power, depending on the input. You must specify at least four of these parameters: n, p1, p2, sig.level, power. The remaining argument must be null; this is the value that the function calculates.

As an example of power.prop.test, let's consider situational statistics in baseball. Starting in the 2009 season, when ESPN broadcast baseball games, they displayed statistics showing how the batter performed in similar situations. More often than not, the statistics were derived from a very small number of situations. For example, ESPN might show that the hitter had three hits in ten tries when hitting with two men on base and two outs. These statistics sound really interesting, but do they have any meaning? We can use prop.test to help find out.

Suppose that a hitter is batting with two men on base and two outs. The TV broadcaster tells us that the batter's average is .300 in these situations, but only .260 in other situations. Furthermore, let's assume that the true probability that he gets a hit in an at bat in other situations is .260. How many at bats would he need to have in situations with two men on base and two outs in order for the .300 estimate to be statistically significant at a 95% confidence level, with a power of .95?

```
> power.prop.test(p1=.260,p2=.300,sig.level=0.05,
+     power=.95,alternative="one.sided")

        Two-sample comparison of proportions power calculation

              n = 2724.482
             p1 = 0.26
             p2 = 0.3
      sig.level = 0.05
          power = 0.95
    alternative = one.sided

NOTE: n is number in *each* group
```

That's right, the estimate is over 2,724 at bats. So, let's ask the opposite question: what is the confidence we can have in the results? Let's fix **sig.level=0.05** and **power=0.95**:

```
> power.prop.test(n=10, p1=.260, p2=.300, power=.95,
+    sig.level=NULL, alternative="one.sided")

     Two-sample comparison of proportions power calculation

              n = 10
             p1 = 0.26
             p2 = 0.3
      sig.level = 0.9256439
          power = 0.95
    alternative = one.sided

 NOTE: n is number in *each* group

> power.prop.test(n=10,p1=.260,p2=.300,power=NULL,
+    sig.level=.05,alternative="one.sided")

     Two-sample comparison of proportions power calculation

              n = 10
             p1 = 0.26
             p2 = 0.3
      sig.level = 0.05
          power = 0.07393654
    alternative = one.sided

 NOTE: n is number in *each* group
```

With significance levels that low, I think it's safe to say that most of these situational statistics are nonsense.

ANOVA Test Design

If you are designing an experiment where you will be using ANOVA, you can use the **power.anova.test** function:

```
power.anova.test(groups = NULL, n = NULL,
                 between.var = NULL, within.var = NULL,
                 sig.level = 0.05, power = NULL
```

For this function, **groups** specifies the number of groups, **n** specifies the number of observations (per group), **between.var** is the variance between groups, **within.var** is the variance within groups, **sig.level** is the significance level (Type I error probability), and **power** is the power of the test (1 – Type II error probability). This function will calculate either **groups**, **n**, **sig.level**, **between.var**, **power**, **within.var**, or **sig.level**, depending on the input. You must specify exactly six of these parameters, and the remaining argument must be null; this is the value that the function calculates.

20

Regression Models

A *regression model* shows how a continuous value (called the *response variable*, or the *dependent variable*) is related to a set of other values (called the *predictors, stimulus variables*, or *independent variables*). Often, a regression model is used to predict values where they are unknown. For example, warfarin is a drug commonly used as a blood thinner or anticoagulant. A doctor might use a regression model to predict the correct dose of warfarin to give a patient based on several known variables about the patient (such as the patient's weight). Another example of a regression model might be for marketing financial products. An analyst might estimate the average balance of a credit card customer (which, in turn, affects the expected revenue from that customer).

Sometimes, a regression model is simply used to explain a phenomenon, but not to actually predict values. For example, a scientist might suspect that weight is correlated to consumption of certain types of foods, but wants to adjust for a variety of factors, including age, exercise, genetics (and, hopefully, other factors). The scientist could use a regression model to help show the relationship between weight and food consumed by including other variables in the regression. Models can be used for many other purposes, including visualizing trends, analysis of variance tests, and testing variable significance.

This chapter looks at regression models in R; classification models are covered in Chapter 21. To show how to use statistical models in R, I will start with the simplest type of model: linear regression models. (Specifically, I'll use the least squares method to estimate coefficients.) I'll show how to build, evaluate, and refine a model in R. Then I'll describe functions in R for building more sophisticated types of models.

Example: A Simple Linear Model

A linear regression assumes that there is a linear relationship between the response variable and the predictors. Specifically, a linear regression assumes that a response variable y is a linear function of a set of predictor variables $x_1, x_2, ..., x_n$.

As an example, we're going to look at how different metrics predict the runs scored by a baseball team.* Let's start by loading the data for every team between 2000 and 2008. We'll use the SQLite database that we used in Chapter 14 and extract the fields we want using an SQL query:

```
> library(RSQLite)
> drv <- dbDriver("SQLite")
> con <- dbConnect(drv,
+     dbname=paste(.Library, "/nutshell/data/bb.db", sep=""))
> team.batting.00to08 <- dbGetQuery(con,
+     paste(
+         'SELECT teamID, yearID, R as runs, ',
+         '  H-"2B"-"3B"-HR as singles, ',
+         '  "2B" as doubles, "3B" as triples, HR as homeruns, ',
+         '  BB as walks, SB as stolenbases, CS as caughtstealing, ',
+         '  HBP as hitbypitch, SF as sacrificeflies, ',
+         '  AB as atbats ',
+         '  FROM Teams ',
+         '  WHERE yearID between 2000 and 2008'
+     )
+ )
```

Or, if you'd like, you can just load the file from the nutshell package:

```
> library(nutshell)
> data(team.batting.00to08)
```

Because this is a book about R and not a book about baseball, I renamed the common abbreviations to more intuitive names for plays. Let's look at scatter plots of runs versus each other variable, so that we can see which variables are likely to be most important.

We'll create a data frame for plotting, using the make.groups function:

```
> attach(team.batting.00to08);
> forplot <- make.groups(
+     singles        = data.frame(value=singles,        teamID,yearID,runs),
+     doubles        = data.frame(value=doubles,        teamID,yearID,runs),
+     triples        = data.frame(value=triples,        teamID,yearID,runs),
+     homeruns       = data.frame(value=homeruns,       teamID,yearID,runs),
+     walks          = data.frame(value=walks,          teamID,yearID,runs),
+     stolenbases    = data.frame(value=stolenbases,    teamID,yearID,runs),
+     caughtstealing = data.frame(value=caughtstealing, teamID,yearID,runs),
+     hitbypitch     = data.frame(value=hitbypitch,     teamID,yearID,runs),
+     sacrificeflies = data.frame(value=sacrificeflies, teamID,yearID,runs)
+     );
> detach(team.batting.00to08);
```

Now, we'll generate the scatter plots using the xyplot function:

```
> xyplot(runs~value|which, data=forplot,
+     scales=list(relation="free"),
```

---

* This example is closely related to the batter runs formula, which was popularized by Pete Palmer and Jim Thorne in the 1984 book *The Hidden Game of Baseball*. The original batter runs formula worked slightly differently: it predicted the number of runs above or below the mean, and it had no intercept. For more about this problem, see [Adler2006].

```
+    pch=19, cex=.2,
+    strip=strip.custom(strip.levels=TRUE,
+    horizontal=TRUE,
+    par.strip.text=list(cex=.8))
+ )
```

The results are shown in Figure 20-1. Intuitively, teams that hit a lot of home runs score a lot of runs. Interestingly, teams that walk a lot score a lot of runs as well (maybe even more than teams that score a lot of singles).

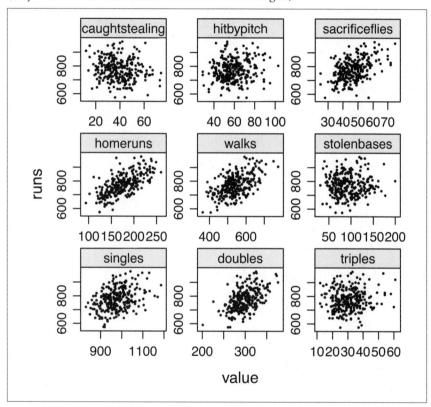

Figure 20-1. Scatter plots: runs as a function of different batter statistics

Fitting a Model

Let's fit a linear model to the data and assign it to the variable `runs.mdl`. We'll use the `lm` function, which fits a linear model using ordinary least squares:

```
> runs.mdl <- lm(
+    formula=runs~singles+doubles+triples+homeruns+
+    walks+hitbypitch+sacrificeflies+
+    stolenbases+caughtstealing,
+    data=team.batting.00to08)
```

R doesn't show much information when you fit a model. (If you don't print the returned object, most modeling functions will not show *any* information, unless there is an error.) To get information about a model, you have to use helper functions.

Helper Functions for Specifying the Model

In a formula object, some symbols have special interpretations. Specifically, "+", "*", "-", and "^" are interpreted specially by R. This means that you need to use some helper functions to represent simple addition, multiplication, subtraction, and exponentiation in a model formula. To interpret an expression literally, and not as formula, use the identity function I(). For example, suppose that you want to include only the product of variables a and b in a formula specification, but not just a or b. If you specify a*b, this is interpreted as a, b, or a*b. To include only a*b, use the identity function I() to protect the expression a*b:

```
lm(y~I(a*b))
```

Sometimes, you would like to fit a polynomial function. Writing out all the terms individually can be tedious, but R provides a short way to specify all the terms at once. To do this, you use the poly function to add all terms up to a specified degree:

```
poly(x, ..., degree = 1, coefs = NULL, raw = FALSE
```

As arguments, the poly function takes a vector x (or a set of vectors), degree to specify a maximum degree to generate, coefs to specify coefficients from a previous fit (when using poly to generate predicted values), and raw to specify whether to use raw and not orthogonal polynomials. For more information on how to specify formulas, see "Formulas" on page 88.

Getting Information About a Model

In R, statistical models are represented by objects; statistical modeling functions return statistical model objects. When you fit a statistical model with most statistical software packages (such as SAS or SPSS) they print a lot of diagnostic information. In R, most statistical modeling function do not print any information.

If you simply call a model function in R, but don't assign the model to a variable, the R console will print the object. (Specifically, it will call the generic method print with the object generated by the modeling function.) R doesn't clutter your screen with lots of information you might not want. Instead, R includes a large set of functions for printing information about model objects. This section describes the functions for getting information about lm objects. Many of these functions may also be used with other types of models; see the help files for more information.

Viewing the model

For most model functions (including lm), the best place to start is with the print method. If you are using the R console, you can simply enter the name of the returned object on the console to see the results of print:

```
> runs.mdl

Call:
```

```
lm(formula = runs ~ singles + doubles + triples + homeruns +
    walks + hitbypitch + sacrificeflies + stolenbases + caughtstealing,
    data = team.batting.00to08)

Coefficients:
  (Intercept)         singles         doubles          triples
    -507.16020         0.56705         0.69110          1.15836
     homeruns           walks      hitbypitch   sacrificeflies
      1.47439         0.30118         0.37750          0.87218
   stolenbases  caughtstealing
      0.04369        -0.01533
```

To show the formula used to fit the model, use the `formula` function:

```
formula(x, ...)
```

Here is the formula on which the model function was called:

```
> formula(runs.mdl)
runs ~ singles + doubles + triples + homeruns + walks + hitbypitch +
    sacrificeflies + stolenbases + caughtstealing
```

To get the list of coefficients for a model object, use the `coef` function:

```
coef(object, ...)
```

Here are the coefficients for the model fitted above:

```
> coef(runs.mdl)
  (Intercept)         singles         doubles          triples
  -507.16019759      0.56704867      0.69110420      1.15836091
     homeruns           walks      hitbypitch   sacrificeflies
   1.47438916      0.30117665      0.37749717      0.87218094
   stolenbases  caughtstealing
   0.04369407     -0.01533245
```

Alternatively, you can use the alias `coefficients` to access the `coef` function.

To get a summary of a linear model object, you can use the `summary` function. The method used for linear model objects is:

```
summary(object, correlation = FALSE, symbolic.cor = FALSE, ...)
```

For the example above, here is the output of the `summary` function:

```
> summary(runs.mdl)

Call:
lm(formula = runs ~ singles + doubles + triples + homeruns +
    walks + hitbypitch + sacrificeflies + stolenbases + caughtstealing,
    data = team.batting.00to08)

Residuals:
    Min      1Q  Median      3Q     Max
-71.9019 -11.8282  -0.4193 14.6576  61.8743

Coefficients:
              Estimate Std. Error t value Pr(>|t|)
(Intercept)  -507.16020   32.34834 -15.678  < 2e-16 ***
singles         0.56705    0.02601  21.801  < 2e-16 ***
```

```
doubles           0.69110    0.05922  11.670  < 2e-16 ***
triples           1.15836    0.17309   6.692 1.34e-10 ***
homeruns          1.47439    0.05081  29.015  < 2e-16 ***
walks             0.30118    0.02309  13.041  < 2e-16 ***
hitbypitch        0.37750    0.11006   3.430 0.000702 ***
sacrificeflies    0.87218    0.19179   4.548 8.33e-06 ***
stolenbases       0.04369    0.05951   0.734 0.463487
caughtstealing   -0.01533    0.15550  -0.099 0.921530
---
Signif. codes:  0 '***' 0.001 '**' 0.01 '*' 0.05 '.' 0.1 ' ' 1

Residual standard error: 23.21 on 260 degrees of freedom
Multiple R-squared: 0.9144,    Adjusted R-squared: 0.9114
F-statistic: 308.6 on 9 and 260 DF,  p-value: < 2.2e-16
```

When you print a summary object, the following method is used:

```
print(x, digits = max(3, getOption("digits") - 3),
      symbolic.cor = x$symbolic.cor,
      signif.stars = getOption("show.signif.stars"), ...)
```

Predicting values using a model

To get the vector of residuals from a linear model fit, use the `residuals` function:

```
residuals(object, ...)
```

To get a vector of fitted values, use the `fitted` function:

```
fitted(object, ...)
```

Suppose that you wanted to use the model object to predict values in another data set. You can use the `predict` function to calculate predicted values using the model object and another data frame:

```
predict(object, newdata, se.fit = FALSE, scale = NULL, df = Inf,
        interval = c("none", "confidence", "prediction"),
        level = 0.95, type = c("response", "terms"),
        terms = NULL, na.action = na.pass,
        pred.var = res.var/weights, weights = 1, ...)
```

The argument `object` specifies the model returned by the fitting function, `newdata` specifies a new data source for predictions, and `na.action` specifies how to deal with missing values in `newdata`. (By default, `predict` ignores missing values. You can choose `na.omit` to simply return `NA` for observations in `newdata` with missing values.) The `predict` function can also return confidence intervals for predictions, in addition to exact values; see the help file for more information.

Analyzing the fit

To get the list of coefficients for a model object, use the `coef` function:

```
coef(object, ...)
```

Here are the coefficients for the model fitted above:

```
> coef(runs.mdl)
    (Intercept)          singles         doubles          triples
   -507.16019759       0.56704867      0.69110420       1.15836091
       homeruns            walks       hitbypitch sacrificeflies
     1.47438916       0.30117665      0.37749717       0.87218094
    stolenbases caughtstealing
     0.04369407      -0.01533245
```

Alternatively, you can use the alias coefficients to access the coef function.

To compute confidence intervals for the coefficients in the fitted model, use the confint function:

```
confint(object, parm, level = 0.95, ...)
```

The argument object specifies the model returned by the fitting function, parm specifies the variables for which to show confidence levels, and level specifies the confidence level. Here are the confidence intervals for the coefficients of the model fitted above:

```
> confint(runs.mdl)
                       2.5 %          97.5 %
(Intercept)    -570.85828008 -443.4621151
singles           0.51583022    0.6182671
doubles           0.57449582    0.8077126
triples           0.81752968    1.4991921
homeruns          1.37432941    1.5744489
walks             0.25570041    0.3466529
hitbypitch        0.16077399    0.5942203
sacrificeflies    0.49451857    1.2498433
stolenbases      -0.07349342    0.1608816
caughtstealing   -0.32152716    0.2908623
```

To compute the influence of different parameters, you can use the influence function:

```
influence(model, do.coef = TRUE, ...)
```

For more friendly output, try influence.measures:

```
influence.measures(model)
```

To get analysis of variance statistics, use the anova function. For linear models, the method used is anova.lmlist, which has the following form:

```
anova.lmlist(object, ..., scale = 0, test = "F")
```

By default, F-test statistics are included in the results table. You can specify test="F" for F-test statistics, test="Chisq" for chi-squared test statistics, test="Cp" for Mallows' C_p statistic, or test=NULL for no test statistics. You can also specify an estimate of the noise variance σ^2 through the scale argument. If you set scale=0 (the default), the anova function will calculate an estimate from the test data. The test statistic and p-values compare the mean square for each row to the residual mean square.

Here are the ANOVA statistics for the model fitted above:

```
> anova(runs.mdl)
Analysis of Variance Table

Response: runs
                Df Sum Sq Mean Sq  F value     Pr(>F)
singles          1 215755  215755  400.4655 < 2.2e-16 ***
doubles          1 356588  356588  661.8680 < 2.2e-16 ***
triples          1    237     237    0.4403 0.5075647
homeruns         1 790051  790051 1466.4256 < 2.2e-16 ***
walks            1 114377  114377  212.2971 < 2.2e-16 ***
hitbypitch       1   7396    7396   13.7286 0.0002580 ***
sacrificeflies   1  11726   11726   21.7643 4.938e-06 ***
stolenbases      1    357     357    0.6632 0.4161654
caughtstealing   1      5       5    0.0097 0.9215298
Residuals      260 140078     539
---
Signif. codes:  0 '***' 0.001 '**' 0.01 '*' 0.05 '.' 0.1 ' ' 1
```

Interestingly, it appears that triples, stolen bases, and times caught stealing are not statistically significant.

You can also view the effects from a fitted model. The effects are the uncorrelated single degree of freedom values obtained by projecting the data onto the successive orthogonal subspaces generated by the QR-decomposition during the fitting process. To obtain a vector of orthogonal effects from the model, use the effects function:

```
effects(object, set.sign = FALSE, ...)
```

To calculate the variance-covariance matrix from the linear model object, use the vcov function:

```
vcov(object, ...)
```

Here is the variance-covariance matrix for the model fitted above:

```
> vcov(runs.mdl)
                 (Intercept)        singles        doubles        triples
(Intercept)     1046.4149572 -6.275356e-01 -6.908905e-01 -0.8115627984
singles           -0.6275356  6.765565e-04 -1.475026e-04  0.0001538296
doubles           -0.6908905 -1.475026e-04  3.506798e-03 -0.0013459187
triples           -0.8115628  1.538296e-04 -1.345919e-03  0.0299591843
homeruns          -0.3190194  2.314669e-04 -3.940172e-04  0.0011510663
walks             -0.2515630  7.950878e-05 -9.902388e-05  0.0004174548
hitbypitch        -0.9002974  3.385518e-04 -4.090707e-04  0.0018360831
sacrificeflies     1.6870020 -1.723732e-03 -2.253712e-03 -0.0051709718
stolenbases        0.2153275 -3.041450e-04  2.871078e-04 -0.0009794480
caughtstealing    -1.4370890  3.126387e-04  1.466032e-03 -0.0016038175
                      homeruns         walks    hitbypitch sacrificeflies
(Intercept)     -3.190194e-01 -2.515630e-01 -0.9002974059    1.6870019518
singles          2.314669e-04  7.950878e-05  0.0003385518   -0.0017237324
doubles         -3.940172e-04 -9.902388e-05 -0.0004090707   -0.0022537124
triples          1.151066e-03  4.174548e-04  0.0018360831   -0.0051709718
homeruns         2.582082e-03 -4.007590e-04 -0.0008183475   -0.0005078943
walks           -4.007590e-04  5.333599e-04  0.0002219440   -0.0010962381
```

```
hitbypitch      -8.183475e-04   2.219440e-04   0.0121132852   -0.0011315622
sacrificeflies  -5.078943e-04  -1.096238e-03  -0.0011315622    0.0367839752
stolenbases     -2.041656e-06  -1.400052e-04  -0.0001197102   -0.0004636454
caughtstealing   3.469784e-04   6.008766e-04   0.0001742039   -0.0024880710
                stolenbases   caughtstealing
(Intercept)      2.153275e-01  -1.4370889812
singles         -3.041450e-04   0.0003126387
doubles          2.871078e-04   0.0014660316
triples         -9.794480e-04  -0.0016038175
homeruns        -2.041656e-06   0.0003469784
walks           -1.400052e-04   0.0006008766
hitbypitch      -1.197102e-04   0.0001742039
sacrificeflies  -4.636454e-04  -0.0024880710
stolenbases      3.541716e-03  -0.0050935339
caughtstealing  -5.093534e-03   0.0241794596
```

To return the deviance of the fitted model, use the `deviance` function:

```
deviance(object, ...)
```

Here is the deviance for the model fitted above (though this value is just the residual sum of squares in this case because `runs.mdl` is a linear model):

```
> deviance(runs.mdl)
[1] 140077.6
```

Finally, to plot a set of useful diagnostic diagrams, use the `plot` function:

```
plot(x, which = c(1:3,5),
     caption = list("Residuals vs Fitted", "Normal Q-Q",
         "Scale-Location", "Cook's distance",
         "Residuals vs Leverage",
         expression("Cook's dist vs Leverage  " * h[ii] / (1 - h[ii]))),
     panel = if(add.smooth) panel.smooth else points,
     sub.caption = NULL, main = "",
     ask = prod(par("mfcol")) < length(which) && dev.interactive(),
     ...,
     id.n = 3, labels.id = names(residuals(x)), cex.id = 0.75,
     qqline = TRUE, cook.levels = c(0.5, 1.0),
     add.smooth = getOption("add.smooth"), label.pos = c(4,2),
     cex.caption = 1)
```

This function shows the following plots:

- Residuals against fitted values
- A scale-location plot of sqrt{| residuals |} against fitted values
- A normal Q-Q plot
- (Not plotted by default) A plot of Cook's distances versus row labels
- A plot of residuals against leverages
- (Not plotted by default) A plot of Cook's distances against leverage/(1 – leverage)

There are many more functions available in R for regression diagnostics; see the help file for `influence.measures` for more information on many of these.

Refining the Model

Often, it is better to use the `update` function to refit a model. This can save you some typing if you are using R interactively. Additionally, this can save on computation time (for large data sets). You can run `update` after changing the formula (perhaps adding or subtracting a term) or even after changing the data frame.

For example, let's fit a slightly different model to the data above. We'll omit the variable `sacrificeflies` and add `0` as a variable (which means to fit the model with no intercept):

```
> runs.mdl2 <- update(runs.mdl,formula=runs ~ singles + doubles +
+    triples + homeruns + walks + hitbypitch +
+    stolenbases + caughtstealing + 0)
> runs.mdl2

Call:
lm(formula = runs ~ singles + doubles + triples + homeruns +
      walks + hitbypitch + stolenbases + caughtstealing - 1,
      data = team.batting.00to08)

Coefficients:
      singles          doubles          triples          homeruns
      0.29823          0.41280          0.95664           1.31945
        walks       hitbypitch      stolenbases    caughtstealing
      0.21352         -0.07471          0.18828          -0.70334
```

Details About the lm Function

Now that we've seen a simple example of how models work in R, let's describe in detail what `lm` does and how you can control it. A linear regression model is appropriate when the response variable (the thing that you want to predict) can be estimated from a linear function of the predictor variables (the information that you know). Technically, we assume that:

$$y = c_0 + c_1 x_1 + c_2 x_2 + \cdots + c_n x_n + \varepsilon$$

where y is the response variable, $x_1, x_2, ..., x_n$ are the predictor variables (or predictors), $c_1, c_2, ..., c_n$ are the *coefficients* for the predictor variables, c_0 is the *intercept*, and ε is the *error term*. (For more details on the assumptions of the least squares model, see "Assumptions of Least Squares Regression" on page 384.) The predictors can be simple variables or even nonlinear functions of variables.

Suppose that you have a matrix of observed predictor variables X and a vector of response variables Y. (In this sentence, I'm using the terms "matrix" and "vector" in the mathematical sense.) We have assumed a linear model, so given a set of coefficients c, we can calculate a set of estimates $\hat{y}$ for the input data X by calculating $\hat{y} = cX$. The differences between the estimates $\hat{y}$ and the actual values Y are called the *residuals*. You can think of the residuals as a measure of the prediction error; small residuals mean that the predicted values are close to the actual values. We assume that the expected difference between the actual response values and the residual values (the error term in the model) is 0. This is important to remember: at best, a model is probabilistic.[†]

Our goal is to find the set of coefficients c that does the best job of estimating Y given X; we'd like the estimates $\hat{y}$ to be as close as possible to Y. In a classical linear regression model, we find coefficients c that minimize the sum of squared differences between the estimates $\hat{y}$ and the observed values Y. Specifically, we want to find values for c that minimize:

$$\mathrm{RSS}(c) = \sum_{i=1}^{N} (y_i - \hat{y}_i)^2$$

This is called the least squares method for regression. You can use the lm function in R to estimate the coefficients in a linear model:[‡]

```
lm(formula, data, subset, weights, na.action,
    method = "qr", model = TRUE, x = FALSE, y = FALSE, qr = TRUE,
    singular.ok = TRUE, contrasts = NULL, offset, ...)
```

Arguments to lm include the following.

Argument	Description	Default
formula	A formula object that specifies the form of the model to fit.	
data	A data frame, list, or environment (or an object that can be coerced to a data frame) in which the variables in formula can be evaluated.	
subset	A vector specifying the observations in data to include in the model.	
weights	A numeric vector containing weights for each observation in data.	NULL
na.action	A function that specifies what lm should do if there are NA values in the data. If NULL, lm uses na.omit.	getOption("na.action"), which defaults to na.fail
method	The method to use for fitting. Only method="qr" fits a model, though you can specify method="model.frame" to return a model frame.	"qr"
model	A logical value specifying whether the "model frame" should be returned.	TRUE
x	Logical values specifying whether the "model matrix" should be returned.	FALSE
y	A logical value specifying whether the response vector should be returned.	FALSE
qr	A logical value specifying whether the QR-decomposition should be returned.	TRUE
singular.ok	A logical value that specifies whether a singular fit results is an error.	TRUE

[†] By the way, the estimate returned by a model is not an exact prediction. It is, instead, the expected value of the response variable given the predictor variables. To be precise, the estimate $\hat{y}$ means:
$$\hat{y} = E[y|x_1, x_2, \ldots, x_n]$$
This observation is important when we talk about generalized linear models later.

[‡] To efficiently calculate the coefficients, R uses several matrix calculations. R uses a method called QR-decomposition to transform X into an orthogonal matrix Q and an upper triangular matrix R, where $X = QR$, and then calculates the coefficients as $c = R^{-1}Q^T Y$.

Argument	Description	Default
contrasts	A list of contrasts for factors in the model, specifying one contrast for each factor in the model. For example, for formula y~a+b, to specify a Helmert contrast for a and a treatment contrast for b, you would use the argument `contrasts=(a="contr.helmert", b="contr.treatment")`. Some options in R are `"contr.helmert"` for Helmert contrasts, `"contr.sum"` for sum-to-zero contrasts, `"contr.treatment"` to contrast each level with the baseline level, and `"contr.poly"` for contrasts based on orthogonal polynomials. See [Venables 2002] for an explanation of why contrasts are important and how they are used.	When contrasts=NULL (the default), lm uses the value from options("contrasts")
offset	A vector of offsets to use when building the model. (An offset is a linear term that is included in the model without fitting.)	
...	Additional arguments passed to lower-level functions such as `lm.fit` (for unweighted models) or `lm.wfit` (for weighted models).	

Model-fitting functions in R return model objects. A model object contains a lot of information about the fitted model (and the fitting operation). Different model objects contain slightly different information.

You may notice that most modeling functions share a few common variables: `formula`, `data`, `na.action`, `subset`, `weights`. These arguments mean the same thing for most modeling functions.

If you are working with a very large data set, you may want to consider using the `biglm` function instead of `lm`. This function uses only p^2 memory for p variables, which is much less than the memory required for `lm`.

Assumptions of Least Squares Regression

Linear models fit with the least squares method are one of the oldest statistical methods, dating back to the age of slide rules. Even today, when computers are ubiquitous, high-quality statistical software is free, and statisticians have developed thousands of new estimation methods, they are still popular. One reason why linear regression is still popular is because linear models are easy to understand. Another reason is that the least squares method has the smallest variance among all unbiased linear estimates (proven by the Gauss-Markov theorem).

Technically, linear regression is not always appropriate. Ordinary least squares (OLS) regression (implemented through `lm`) is only guaranteed to work when certain properties of the training data are true. Here are the key assumptions:

1. Linearity. We assume that the response variable y is a linear function of the predictor variables $x_1, x_2, ..., c_n$.
2. Full rank. There is no linear relationship between any pair of predictor variables. (Equivalently, the predictor matrix is not singular.) Technically, $\forall\ x_i, x_j, \nexists\ c$ such that $x_i = cx_j$.
3. Exogenicity of the predictor variables. The expected value of the error term ε is 0 for all possible values of the predictor variables.

4. Homoscedasticity. The variance of the error term ε is constant and is not correlated with the predictor variables.

5. Nonautocorrelation. In a sequence of observations, the values of y are not correlated with each other.

6. Exogenously generated data. The predictor variables $x_1, x_2, ..., x_n$ are generated independently of the process that generates the error term ε.

7. The error term ε is normally distributed with standard deviation σ and mean 0.

In practice, OLS models often make accurate predictions even when one (or more) of these assumptions are violated.

By the way, it's perfectly OK for there to be a *nonlinear* relationship between some of the predictor variables. Suppose that one of the variables is age. You could add age^2, log(age), or other nonlinear mathematical expressions using age to the model and not violate the assumptions above. You are effectively defining a set of new predictor variables: $w_1 = age$, $w_2 = age^2$, $w_3 = \log(age)$. This doesn't violate the linearity assumption (because the model is still a linear function of the predictor variables) or the full rank assumption (as long as the relationship between the new variables is not linear).

If you want to be careful, you can use test functions to check if the OLS assumptions apply:

- You can test for heteroscedasticity using the function ncv.test in the car (Companion to Applied Regression) package, which implements the Breusch-Pagan test. (Alternatively, you could use the bptest function in the lmtest library, which implements the same test. The lmtest library includes a number of other functions for testing for heteroscedasticity; see the documentation for more details.)

- You can test for autocorrelation in a model using the function durbin.watson in the car package, which implements the Durbin-Watson test. You can also use the function dwtest in the library lmtest by specifying a formula and a data set. (Alternatively, you could use the function bgtest in the lmtest package, which implements the Breusch-Godfrey test. This functions also tests for higher-order disturbances.)

- You can check that the predictor matrix is not singular by using the singular.ok=FALSE argument in lm.

Incidentally, the example used in "Example: A Simple Linear Model" on page 373 is not heteroscedastic:

```
> ncv.test(runs.mdl)
Non-constant Variance Score Test
Variance formula: ~ fitted.values
Chisquare = 1.411893    Df = 1    p = 0.2347424
```

Nor is there a problem with autocorrelation:

```
> durbin.watson(runs.mdl)
 lag Autocorrelation D-W Statistic p-value
   1     0.003318923      1.983938   0.884
 Alternative hypothesis: rho != 0
```

Or with singularity:

```
> runs.mdl <- lm(
+     formula=runs~singles+doubles+triples+homeruns+
+                   walks+hitbypitch+sacrificeflies+
+                   stolenbases+caughtstealing,
+     data=team.batting.00to08,singular.ok=FALSE)
```

If the model has problems with heteroscedasticity or outliers, consider using a re-sistant or robust regression function, as described in "Robust and Resistant Regres-sion" on page 386. If the data is homoscedastic and not autocorrelated, but the error form is not normal, a good choice is ridge regression, which is described in "Ridge Regression" on page 389. If the predictors are closely correlated (and nearly collinear), a good choice is principal components regression, as described in "Prin-cipal Components Regression and Partial Least Squares Regression" on page 391.

Robust and Resistant Regression

Often, ordinary least squares regression works well even with imperfect data. How-ever, it's better in many situations to use regression techniques that are less sensitive to outliers and heteroscedasticity. With R, there are alternative options for fitting linear models.

Resistant regression

If you would like to fit a linear regression model to data with outliers, consider using resistant regression. Using the least median squares (LMS) and least trimmed squares (LTS) estimators:

```
library(MASS)
## S3 method for class 'formula':
lqs(formula, data, ...,
    method = c("lts", "lqs", "lms", "S", "model.frame"),
    subset, na.action, model = TRUE,
    x.ret = FALSE, y.ret = FALSE, contrasts = NULL)

## Default S3 method:
lqs(x, y, intercept = TRUE, method = c("lts", "lqs", "lms", "S"),
    quantile, control = lqs.control(...), k0 = 1.548, seed, ...)
```

Robust regression

Robust regression methods can be useful when there are problems with heteroscedasticity and outliers in the data. The function rlm in the MASS package fits a model using *MM*-estimation:

```
## S3 method for class 'formula':
rlm(formula, data, weights, ..., subset, na.action,
```

```
    method = c("M", "MM", "model.frame"),
    wt.method = c("inv.var", "case"),
    model = TRUE, x.ret = TRUE, y.ret = FALSE, contrasts = NULL)

## Default S3 method:
rlm(x, y, weights, ..., w = rep(1, nrow(x)),
    init = "ls", psi = psi.huber,
    scale.est = c("MAD", "Huber", "proposal 2"), k2 = 1.345,
    method = c("M", "MM"), wt.method = c("inv.var", "case"),
    maxit = 20, acc = 1e-4, test.vec = "resid", lqs.control = NULL)
```

You may also want to try the function lmRob in the robust package, which fits a model using *MS*- and *S*-estimation:

```
library(robust)
lmRob(formula, data, weights, subset, na.action, model = TRUE, x = FALSE,
      y = FALSE, contrasts = NULL, nrep = NULL,
      control = lmRob.control(...), genetic.control = NULL, ...)
```

Comparing lm, lqs, and rlm

As a quick exercise, we'll look at how lm, lqs, and rlm perform on some particularly ugly data: U.S. housing prices. We'll use Robert Schiller's home price index as an example, looking at home prices between 1890 and 2009.[§] First, we'll load the data and fit the data using an ordinary linear regression model, a robust regression model, and a resistant regression model:

```
> library(nutshell)
> data(schiller.index)
> hpi.lm <- lm(Index~Year,data=schiller.index)
> hpi.rlm <- rlm(Index~Year,data=schiller.index)
> hpi.lqs <- lqs(Index~Year,data=schiller.index)
```

Now, we'll plot the data to compare how each method worked. We'll plot the models using the abline function because it allows you to specify a model as an argument (as long as the model function has a coefficient function):

```
> plot(hpi,pch=19,cex=0.3)
> abline(reg=hpi.lm,lty=1)
> abline(reg=hpi.rlm,lty=2)
> abline(reg=hpi.lqs,lty=3)
> legend(x=1900,y=200,legend=c("lm","rlm","lqs"),lty=c(1,2,3))
```

As you can see from Figure 20-2, the standard linear model is influenced by big peaks (such as the growth between 2001 and 2006) and big valleys (such as the dip between 1920 and 1940). The robust regression method is less sensitive to peaks and valleys in this data, and the resistant regression method is the least sensitive.

Subset Selection and Shrinkage Methods

Modeling functions like lm will include every variable specified in the formula, calculating a coefficient for each one. Unfortunately, this means that lm may calculate

---

[§] The data is available from *http://www.irrationalexuberance.com/*.

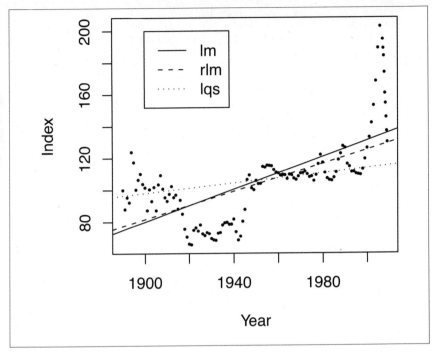

Figure 20-2. Home prices and lm, rlm, and lqs models

coefficients for variables that aren't needed. You can manually tune a model using diagnostics like `summary` and `lm.influence`. However, you can also use some other statistical techniques to reduce the effect of insignificant variables or remove them from a model altogether.

Stepwise Variable Selection

A simple technique for selecting the most important variables is stepwise variable selection. The stepwise algorithm works by repeatedly adding or removing variables from the model, trying to "improve" the model at each step. When the algorithm can no longer improve the model by adding or subtracting variables, it stops and returns the new (and usually smaller) model.

Note that "improvement" does not just mean reducing the residual sum of squares (RSS) for the fitted model. Adding an additional variable to a model will not increase the RSS (see a statistics book for an explanation of why), but it does increase model complexity. Typically, AIC (Akaike's information criterion) is used to measure the value of each additional variable. The AIC is defined as AIC = $-2 * \log(L) + k * \text{edf}$, where L is the likelihood and edf is the equivalent degrees of freedom.

In R, you perform stepwise selection through the `step` function:

```
step(object, scope, scale = 0,
    direction = c("both", "backward", "forward"),
    trace = 1, keep = NULL, steps = 1000, k = 2, ...)
```

Here is a description of the arguments to step.

Argument	Description	Default
object	An object representing a model, such as the objects returned by lm, glm, or aov.	
scope	An argument specifying a set of variables that you want in the final model and a list of all variables that you want to consider including in the model. The first set is called the *lower bound*, and the second is called the *upper bound*. If a single formula is specified, it is interpreted as the upper bound. To specify both an upper and a lower bound, pass a list with two formulas labeled as upper and lower.	
scale	A value used in the definition of AIC for lm and aov models. See the help file for extractAIC for more information.	0
direction	Specifies whether variables should be only added to the model (direction="forward"), removed from the model (direction="backward"), or both (direction="both").	c("both", "backward", "forward")
trace	A numeric value that specifies whether to print out details of the fitting process. Specify trace=0 (or a negative number) to suppress printing, trace=1 for normal detail, and higher numbers for even more detail.	1
keep	A function used to select a subset of arguments to keep from an object. The function accepts a fitted model object and an AIC statistic.	NULL
steps	A numeric value that specifies the maximum number of steps to take before the function halts.	1000
k	The multiple of the number of degrees of freedom to be used in the penalty calculation (extractAIC).	2
...	Additional arguments for extractAIC.	

There is an alternative implementation of stepwise selection in the MASS library: the stepAIC function. This function works similarly to step, but operates on a wider range of model objects.

Ridge Regression

Stepwise variable selection simply fits a model using lm, but limits the number of variables in the model. In contrast, ridge regression places constraints on the size of the coefficients and fits a model using different computations.

Ridge regression can be used to mitigate problems when there are several highly correlated variables in the underlying data. This condition (called *multicollinearity*) causes high variance in the results. Reducing the number, or impact, of regressors in the data can help reduce these problems.‖

In "Details About the lm Function" on page 382, we described how ordinary linear regression finds the coefficients that minimize the residual sum of squares. Ridge regression does something similar. Ridge regression attempts to minimize the sum of squared residuals plus a penalty for the coefficient sizes. The penalty is a constant

‖ For example, see [Greene2007].

λ times the sum of squared coefficients. Specifically, ridge regression tries to minimize the following quantity:

$$\text{RSS}_{\text{ridge}}(c) = \sum_{i=1}^{N} (y_i - \hat{y}_i)^2 + \lambda \sum_{j=1}^{m} c_i^2$$

To estimate a model using ridge regression, you can use the `lm.ridge` function from the `MASS` package:

```
library(MASS)
lm.ridge(formula, data, subset, na.action, lambda = 0, model = FALSE,
         x = FALSE, y = FALSE, contrasts = NULL, ...)
```

Arguments to `lm.ridge` are the following.

Argument	Description	Default
formula	A formula object that specifies the form of the model to fit.	
data	A data frame, list, or environment (or an object that can be coerced to a data frame) in which the variables in `formula` can be evaluated.	
subset	A vector specifying the observations in `data` to include in the model.	
na.action	A function that specifies what `lm` should do if there are NA values in the data. If NULL, `lm` uses `na.omit`.	
lambda	A scalar or vector of ridge constants.	0
model	A logical value specifying whether the "model frame" should be returned.	FALSE
x	Logical values specifying whether the "model matrix" should be returned.	FALSE
y	A logical value specifying whether the response vector should be returned.	FALSE
contrasts	A list of contrasts for factors in the model.	NULL
...	Additional arguments to `lm.fit`.	

Lasso and Least Angle Regression

Another technique for reducing the size of the coefficients (and thus reducing their impact on the final model) is the lasso. Like ridge regression, lasso regression puts a penalty on the size of the coefficients. However, the lasso algorithm uses a different penalty: instead of a sum of squared coefficients, the lasso sums the absolute value of the coefficients. (In math terms, ridge uses L^2-norms, while lasso uses L^1-norms.) Specifically, the lasso algorithm tries to minimize the following value:

$$\text{RSS}_{\text{lasso}}(c) = \sum_{i=1}^{N} (y_i - \hat{y}_i)^2 + \lambda \sum_{j=1}^{m} |c_i|$$

The best way to compute lasso regression in R is through the `lars` function:

```
library(lars)
lars(x, y, type = c("lasso", "lar", "forward.stagewise", "stepwise"),
```

```
trace = FALSE, normalize = TRUE, intercept = TRUE, Gram,
eps = .Machine$double.eps, max.steps, use.Gram = TRUE)
```

The `lars` function computes the entire lasso path at once. Specifically, it begins with a model with no variables. It then computes the lambda values for which each variable enters the model and shows the resulting coefficients. Finally, the `lars` algorithm computes a model with all the coefficients present, which is the same as an ordinary linear regression fit.

This function actually implements a more general algorithm called *least angle regression*; you have the option to choose least angle regression, forward stagewise regression, or stepwise regression instead of lasso. Here are the arguments to the `lars` function.

Argument	Description	Default
x	A matrix of predictor variables.	
y	A numeric vector containing the response variable.	
type	The type of model to fit. Use `type="lasso"` for lasso, `type="lar"` for least angle regression, `type="forward.stagewise"` for infinitesimal forward stagewise, and `type="stepwise"` for stepwise.	c("lasso", "lar", "forward.stagewise", "stepwise")
trace	A logical value specifying whether to print details as the function is running.	FALSE
normalize	A logical value specifying whether each variable will be standardized to have an L^2-norm of 1.	TRUE
intercept	A logical value indicating whether an intercept should be included in the model.	TRUE
Gram	The $X'X$ matrix used in the calculations. To rerun lars with slightly different parameters, but the same underlying data, you may reuse the Gram matrix from a prior run to increase efficiency.	
eps	An effective 0.	.Machine$double.eps
max.steps	A limit on the number of steps taken by the `lars` function.	
use.Gram	A logical value specifying whether `lars` should precompute the Gram matrix. (For large N, this can be time consuming.)	TRUE

Principal Components Regression and Partial Least Squares Regression

Ordinary least squares regression doesn't always work well with closely correlated variables. A useful technique for modeling effects in this form of data is principal components regression. Principal components regression works by first transforming the predictor variables using principal components analysis. Next, a linear regression is performed on the transformed variables.

A closely related technique is partial least squares regression. In partial least squares regression, both the predictor and the response variables are transformed before fitting a linear regression. In R, principal components regression is available through the function `pcr` in the `pls` package:

```
library(pls)
pcr(..., method = pls.options()$pcralg)
```

Partial least squares is available through the function `plsr` in the same package:

```
plsr(..., method = pls.options()$plsralg)
```

Both functions are actually aliases to the function `mvr`:

```
mvr(formula, ncomp, data, subset, na.action,
    method = pls.options()$mvralg,
    scale = FALSE, validation = c("none", "CV", "LOO"),
    model = TRUE, x = FALSE, y = FALSE, ...)
```

Nonlinear Models

The regression models shown above all produced linear models. In this section, we'll look at some algorithms for fitting nonlinear models when you know the general form of the model.

Generalized Linear Models

Generalized linear modeling is a technique developed by John Nelder and Robert Wedderburn to compute many common types of models using a single framework. You can use generalized linear models (GLMs) to fit linear regression models, logistic regression models, Poisson regression models, and other types of models.

As the name implies, GLMs are a generalization of linear models. Like linear models, there is a response variable y and a set of predictor variables x_1, x_2, ..., x_n. GLMs introduce a new quantity called the *linear predictor*. The linear predictor takes the following form:

$$\eta = c_1 x_1 + c_2 x_2 + \cdots + c_n x_n$$

In a general linear model, the predicted value is a function of the linear predictor. The relationship between the response and predictor variables does not have to be linear. However, the relationship between the predictor variables and the linear predictor must be linear. Additionally, the only way that the predictor variables influence the predicted value is through the linear predictor.

In "Example: A Simple Linear Model" on page 373, we noted that a good way to interpret the predicted value of a model is as the expected value (or mean) of the response variable, given a set of predictor variables. This is also true in GLMs, and the relationships between that mean and the linear predictor is what makes GLMs so flexible. To be precise, there must be a smooth, invertible function m such that:

$$\mu = m(\eta), \eta = m^{-1}(\mu) = l(\mu)$$

The inverse of m (denoted by l above) is called the *link function*. You can use many different function families with a GLM, each of which lets you predict a different form of model. For GLMs, the underlying probability distribution needs to be part of the exponential family of probability distributions. More precisely, distributions that can be modeled by GLMs have the following form:

$$f_y(y;\mu;\varphi)=\exp\left(\frac{A}{\varphi}\left(y\lambda(\mu)-y\lambda(\mu)\right)+\tau(y,\varphi)\right)$$

As a simple example, if you use the identity function for m and assume a normal distribution for the error term, then $\eta = \mu$ and we just have an ordinary linear regression model. However, you can specify some much more interesting forms of models with GLMs. You can model functions with Gaussian, binomial, Poisson, gamma, and other distributions, and use a variety of link functions, including identity, logit, probit, inverse, log, and other functions.

In R, you can model all of these different types of models using the `glm` function:

```
glm(formula, family = gaussian, data, weights, subset,
    na.action, start = NULL, etastart, mustart,
    offset, control = glm.control(...), model = TRUE,
    method = "glm.fit", x = FALSE, y = TRUE, contrasts = NULL,
    ...)
```

Here are the arguments to `glm`.

Argument	Description	Default
formula	A formula object that specifies the form of the model to fit.	
family	Describes the probability distribution of the disturbance term and the link function for the model. (See below for information on different families.)	gaussian
data	A data frame, list, or environment (or an object that can be coerced to a data frame) in which the variables in `formula` can be evaluated.	
weights	A numeric vector containing weights for each observation in `data`.	
subset	A vector specifying the observations in `data` to include in the model.	
na.action	A function that specifies what `lm` should do if there are NA values in the data. If `NULL`, `lm` uses `na.omit`.	getOption("na.action"), which defaults to na.fail
start	A numeric vector containing starting values for parameters in the linear predictor.	NULL
etastart	A numeric vector containing starting values for the linear predictor.	
mustart	A numeric vector containing starting values for the vector of means.	
offset	A set of terms that are added to the linear term with a constant coefficient of 1. (You can use an offset to force a variable, or a set of variables, into the model.)	
control	A list of parameters for controlling the fitting process. Parameters include `epsilon` (which specifies the convergence tolerance), `maxit` (which specifies the maximum number of iterations), and `trace` (which specifies whether to output information on each iteration). See `glm.control` for more information.	glm.control(...), which, in turn, has defaults epsilon=1e-8, maxit=25, trace=FALSE
model	A logical value specifying whether the "model frame" should be returned.	TRUE

Argument	Description	Default
method	The method to use for fitting. Only method="glm.fit" fits a model, though you can specify method="model.frame" to return a model frame.	"glm.fit"
x	Logical values specifying whether the "model matrix" should be returned.	FALSE
y	A logical value specifying whether the "response vector" should be returned.	TRUE
contrasts	A list of contrasts for factors in the model.	NULL
...	Additional arguments passed to glm.control.	

GLM fits a model using iteratively reweighted least squares (IWLS).

As noted above, you can model many different types of functions using GLM. The following function families are available in R:

```
binomial(link = "logit")
gaussian(link = "identity")
Gamma(link = "inverse")
inverse.gaussian(link = "1/mu^2")
poisson(link = "log")
quasi(link = "identity", variance = "constant")
quasibinomial(link = "logit")
quasipoisson(link = "log")
```

You may specify an alternative link function for most of these function families. Here is a list of the possible link functions for each family.

Family function	Allowed link functions	Default link function
binomial	"logit", "probit", "cauchit", "log", and "cloglog"	"logit"
gaussian	"identity", "log", and "inverse"	"identity"
Gamma	"inverse", "identity", and "log"	"inverse"
inverse.gaussian	"1/mu^2", "inverse", "identity", and "log"	"1/mu^2"
poisson	"log", "identity", and "sqrt"	"log"
quasi	"logit", "probit", "cloglog", "identity", "inverse", "log", "1/mu^2", and "sqrt", or use the power function to create a power link function	"identity"
quasibinomial		"logit"
quasipoisson		"log"

The quasi function also takes a variance argument (with default constant); see the help file for quasi for more information.

If you are working with a large data set and have limited memory, you may want to consider using the bigglm function in the biglm package.

As an example, let's use the glm function to fit the same model that we used for lm. By default, glm assumes a Gaussian error distribution, so we expect the fitted model to be identical to the one fitted above:

```
> runs.glm <- glm(
+     formula=runs~singles+doubles+triples+homeruns+
```

```
+                   walks+hitbypitch+sacrificeflies+
+                       stolenbases+caughtstealing,
+       data=team.batting.00to08)
> runs.glm

Call:  glm(formula = runs ~ singles + doubles + triples + homeruns +
           walks + hitbypitch + sacrificeflies + stolenbases + caughtstealing,
           data = team.batting.00to08)

Coefficients:
    (Intercept)          singles           doubles           triples
      -507.16020          0.56705           0.69110           1.15836
        homeruns            walks        hitbypitch    sacrificeflies
         1.47439          0.30118           0.37750           0.87218
     stolenbases   caughtstealing
         0.04369         -0.01533

Degrees of Freedom: 269 Total (i.e. Null);   260 Residual
Null Deviance:     1637000
Residual Deviance: 140100  AIC: 2476
```

As expected, the fitted model is identical to the model from lm. (Typically, it's better to use lm rather than glm when fitting an ordinary linear regression model because lm is more efficient.) Notice that glm provides slightly different information through the print statement, such as the degrees of freedom, null deviance, residual deviance, and AIC. We'll revisit glm when talking about logistic regression models for classification; see "Logistic Regression" on page 435.

Nonlinear Least Squares

Sometimes, you know the form of a model, even if the model is extremely nonlinear.

To fit nonlinear models (minimizing least squares error), you can use the nls function:

```
nls(formula, data, start, control, algorithm,
    trace, subset, weights, na.action, model,
    lower, upper, ...)
```

Here is a description of the arguments to the nls function.

Argument	Description
formula	A formula object that specifies the form of the model to fit.
data	A data frame in which formula can be evaluated.
start	A named list or named vector with starting estimates for the fit.
control	A list of arguments to pass to control the fitting process (see the help file for nls.control for more information).
algorithm	The algorithm to use for fitting the model. Use algorithm="plinear" for the Golub-Pereyra algorithm for partially linear least squares models and algorithm="port" for the 'nl2sol' algorithm from the Port library.
trace	A logical value specifying whether to print the progress of the algorithm while nls is running.

Argument	Description
subset	An optional vector specifying the set of rows to include.
weights	An optional vector specifying weights for observations.
na.action	A function that specifies how to treat NA values in the data.
model	A logical value specifying whether to include the model frame as part of the model object.
lower	An optional vector specifying lower bounds for the parameters of the model.
upper	An optional vector specifying upper bounds for the parameters of the model.
...	Additional arguments (not currently used).

The nls function is actually a wrapper for the nlm function. The nlm function is similar to nls, but takes an R function (not a formula) and list of starting parameters as arguments. It's usually easier to use nls because nls allows you to specify models using formulas and data frames, like other R modeling functions. For more information about nlm, see the help file.

By the way, you can actually use nlm to fit a linear model. It will work, but it will be slow and inefficient.

Survival Models

Survival analysis is concerned with looking at the amount of time that elapses before an event occurs. An obvious application is to look at mortality statistics (predicting how long people live), but it can also be applied to mechanical systems (the time before a failure occurs), marketing (the amount of time before a consumer cancels an account), or other areas.

In R, there are a variety of functions in the **survival** library for modeling survival data.

To estimate a survival curve for censored data, you can use the **survfit** function:

```
survfit(formula, data, weights, subset, na.action,
        etype, id, ...)
```

This function accepts the following arguments.

Argument	Description
formula	Describes the relationship between the response value and the predictors. The response value should be a Surv object.
data	The data frame in which to evaluate formula.
weights	Weights for observations.
subset	Subset of observation to use in fitting the model.
na.action	Function to deal with missing values.
etype	The variable giving the type of event.
id	The variable that identifies individual subjects.

Argument	Description
type	Specifies the type of survival curve. Options include "kaplan-meier", "fleming-harrington", and "fh2".
error	Specifies the type of error. Possible values are "greenwood" for the Greenwood formula or "tsiatis" for the Tsiatis formula.
conf.type	Confidence interval type. One of "none", "plain", "log" (the default), or "log-log".
conf.lower	A character string to specify modified lower limits to the curve, the upper limit remains unchanged. Possible values are "usual" (unmodified), "peto", and "modified".
start.time	Numeric value specifying a time to start calculating survival information.
conf.int	The level for a two-sided confidence interval on the survival curve(s).
se.fit	A logical value indicating whether standard errors should be computed.
...	Additional variables passed to internal functions.

As an example, let's fit a survival curve for the GSE2034 data set. This data comes from the Gene Expression Omnibus of the National Center for Biotechnology Information (NCBI), which is accessible from *http://www.ncbi.nlm.nih.gov/geo/*. The experiment examined how the expression of certain genes affected breast cancer relapse-free survival time. In particular, it tested estrogen receptor binding sites. (We'll revisit this example in Chapter 24.)

First, we need to create a Surv object within the data frame. A Surv object is an R object for representing survival information, in particular, censored data. Censored data occurs when the outcome of the experiment is not known for all observations. In this case, the data is censored. There are three possible outcomes for each observation: the subject had a recurrence of the disease, the subject died without having a recurrence of the disease, or the subject was still alive without a recurrence at the time the data was reported. The last outcome—the subject was still alive without a recurrence—results in the censored values:

```
> library(survival)
> GSE2034.Surv <- transform(GSE2034,
+   surv=Surv(
+     time=GSE2034$months.to.relapse.or.last.followup,
+     event=GSE2034$relapse,
+     type="right"
+   )
+ )
# show the first 26 observations:
> GSE2034.Surv$surv[1:26,]
 [1] 101+ 118+   9  106+  37  125+ 109+  14   99+ 137+  34   32  128+
[14]  14  130+  30  155+  25   30   84+   7  100+  30    7  133+  43
```

Now, let's calculate the survival model. We'll just make it a function of the ER.status flag (which stands for "estrogen receptor"):

```
> GSE2034.survfit <- survfit(
+   formula=surv~ER.Status,
```

```
+    data=GSE2034.Surv,
+ )
```

The easiest way to view a **survfit** object is graphically. Let's plot the model:

```
> plot(GSE2034.survfit,lty=1:2,log=T)
> legend(135,1,c("ER+","ER-"),lty=1:2,cex=0.5)
```

The plot is shown in Figure 20-3. Note the different curve shape for each cohort.

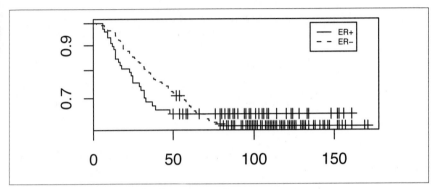

Figure 20-3. Survival curves for the GSE2034 data

To fit a parametric survival model, you can use the **survreg** function in the **survival** package:

```
survreg(formula, data, weights, subset,
        na.action, dist="weibull", init=NULL, scale=0,
        control,parms=NULL,model=FALSE, x=FALSE,
        y=TRUE, robust=FALSE, score=FALSE, ...)
```

Here is a description of the arguments to **survreg**.

Argument	Description	Default
formula	A formula that describes the form of the model; the response is usually a Surv object (created by the Surv function).	
data	A data frame containing the training data for the model.	
weights	A vector of weights for observations in data.	
subset	An expression describing a subset of observations in data to use for fitting the model.	
na.action	A function that describes how to treat NA values.	options()$na.action
dist	A character value describing the form of the y variable (either "weibull", "exponential","gaussian","logistic","lognormal",or "loglogistic") or a distribution like the ones in survreg.distributions.	"weibull"
init	Optional vector of initial parameters.	NULL
scale	Value specifying the scale of the estimates. Estimated if scale <= 0.	0
control	A list of control values, usually produced by survreg.control.	
parms	A list of fixed parameters for the distribution function.	NULL

Argument	Description	Default
model, x, y	Logical values indicating whether to return the model frame, X matrix, or Y vector (respectively) with the results.	FALSE
robust	A logical value indicating whether to use "robust sandwich standard methods."	FALSE
score	A logical value indicating whether to return the score vector.	FALSE
...	Other arguments passed to `survreg.control`.	

You can compute the expected survival for a set of subjects (or individual expectations for each subject) with the function `survexp`:

```
library(survival)
survexp(formula, data, weights, subset, na.action, times, cohort=TRUE,
        conditional=FALSE, ratetable=survexp.us, scale=1, npoints,
        se.fit, model=FALSE, x=FALSE, y=FALSE)
```

Here is a description of the arguments to `survexp`.

Argument	Description	Default
formula	A formula object describing the form of the model. The (optional) response should contain a vector of follow-up times, and the predictors should contain grouping variables separated by + operators.	
data	A data frame containing source data on which to predict values.	
weights	A vector of weights for the cases.	
subset	An expression indicating which observations in data should be included in the prediction.	
na.action	A function specifying how to deal with missing (NA) values in the data.	options()$na.action
times	A vector of follow-up times at which the resulting survival curve is evaluated. (This may also be included in the formula; see above.)	
cohort	A logical value indicating whether to calculate the survival of the whole cohort (`cohort=TRUE`) or individual observations (`cohort=FALSE`).	TRUE
conditional	A logical value indicating whether to calculate conditional expected survival. Specify `conditional=TRUE` if the follow-up times are times of death, and `conditional=FALSE` if the follow-up times are potential censoring times.	FALSE
ratetable	A fitted Cox model (from `coxph`) or a table of survival times.	survexp.us
scale	A numeric value specifying how to scale the results.	1
npoints	A numeric value indicating the number of points at which to calculate individual results.	
se.fit	A logical value indicating whether to include the standard error of the predicted survival.	
model, x, y	Specifies whether to return the model frame, the X matrix, or the Y vector (respectively) in the results.	FALSE for all three

The Cox proportional hazard model is a nonparametric method for fitting survival models. It is available in R through the `coxph` function in the `survival` library:

```
coxph(formula, data=, weights, subset,
      na.action, init, control,
      method=c("efron","breslow","exact"),
```

```
        singular.ok=TRUE, robust=FALSE,
        model=FALSE, x=FALSE, y=TRUE, ...)
```

Here is a description of the arguments to coxph.

Argument	Description	Default
formula	A formula that describes the form of the model; the response must be a Surv object (created by the Surv function).	
data	A data frame containing source data on which to predict values.	
weights	A vector of weights for the cases.	
subset	An expression indicating which observations in data should be fit.	
na.action	A function specifying how to deal with missing (NA) values in the data.	
init	A vector of initial parameter values for the fitting process.	0 for all variables
control	Object of class coxph.control specifying the iteration limit and other control options.	coxph.control(...)
method	A character value specifying the method for handling ties. Choices include "efron", "breslow", and "exact".	"efron"
singular.ok	A logical value indicating whether to stop with an error if the X matrix is singular or to simply skip variables that are linear combinations of other variables.	TRUE
robust	A logical value indicating whether to return a robust variance estimate.	FALSE
model	A logical value specifying whether to return the model frame.	FALSE
x	A logical value specifying whether to return the X matrix.	FALSE
y	A logical value specifying whether to return the Y vector.	TRUE
...	Additional arguments passed to coxph.control.	

As an example, let's fit a Cox proportional hazard model to the GSE2034 data:

```
> GSE2034.coxph <- coxph(
+     formula=surv~ER.Status,
+     data=GSE2034.Surv,
+ )
> GSE2034.coxph
Call:
coxph(formula = surv ~ ER.Status, data = GSE2034.Surv)


                  coef exp(coef) se(coef)       z    p
ER.StatusER+ -0.00378     0.996    0.223 -0.0170 0.99

Likelihood ratio test=0  on 1 df, p=0.986  n= 286
```

The summary method for coxph objects provides additional information about the fit:

```
> summary(GSE2034.coxph)
Call:
coxph(formula = surv ~ ER.Status, data = GSE2034.Surv)

  n= 286
```

```
                 coef exp(coef) se(coef)       z Pr(>|z|)
ER.StatusER+ -0.00378   0.99623  0.22260 -0.017    0.986


             exp(coef) exp(-coef) lower .95 upper .95
ER.StatusER+    0.9962      1.004     0.644     1.541


Rsquare= 0   (max possible= 0.983 )
Likelihood ratio test= 0  on 1 df,   p=0.9865
Wald test            = 0  on 1 df,   p=0.9865
Score (logrank) test = 0  on 1 df,   p=0.9865
```

Another useful function is `cox.zph`, which tests the proportional hazards assumption for a Cox regression model fit:

```
> cox.zph(GSE2034.coxph)
             rho chisq       p
ER.StatusER+ 0.33  11.6 0.000655
```

There are additional methods available for viewing information about `coxph` fits, including `residuals`, `predict`, and `survfit`; see the help file for `coxph.object` for more information.

There are other functions in the `survival` package for fitting survival models, such as `cch` which fits proportional hazard models to case-cohort data. See the help files for more information.

Smoothing

This section describes a number of functions for fitting piecewise smooth curves to data. Functions in this section are particularly useful for plotting charts; there are even convenience functions for using these functions to show fitted values in some graphics packages.

Splines

One method for fitting a function to source data is with splines. With a linear model, a single line is fitted to all the data. With spline methods, a set of different polynomials is fitted to different sections of the data.

You can compute simple cubic splines with the `spline` function in the `stats` package:

```
spline(x, y = NULL, n = 3*length(x), method = "fmm",
       xmin = min(x), xmax = max(x), xout, ties = mean)
```

Here is a description of the arguments to `smooth.spline`.

Argument	Description	Default
x	A vector specifying the predictor variable, or a two-column matrix specifying both the predictor and the response variables.	
y	If x is a vector, then y is a vector containing the response variable.	NULL
n	If xout is not specified, interpolation is done at n equally spaced points between xmin and xmax.	3*length(x)

Argument	Description	Default
method	Specifies the type of spline. Allowed values include `"fmm"`, `"natural"`, `"periodic"`, and `"monoH.FC"`.	"fmm"
xmin	Lowest x value for interpolations.	min(x)
xmax	Highest x value for interpolations.	max(x)
xout	An optional vector of values specifying where interpolation should be done.	
ties	A method for handling ties. Either the string `"ordered"` or a function that returns a single numeric value.	mean

To return a function instead of a list of parameters, use the function `splinefun`:

```
splinefun(x, y = NULL, method = c("fmm", "periodic", "natural", "monoH.FC"),
          ties = mean)
```

To fit a cubic smoothing spline model to supplied data, use the `smooth.spline` function:

```
smooth.spline(x, y = NULL, w = NULL, df, spar = NULL,
              cv = FALSE, all.knots = FALSE, nknots = NULL,
              keep.data = TRUE, df.offset = 0, penalty = 1,
              control.spar = list())
```

Here is a description of the arguments to `smooth.spline`.

Argument	Description	Default
x	A vector specifying the predictor variable, or a two-column matrix specifying both the predictor and the response variables.	
y	If x is a vector, then y is a vector containing the response variable.	NULL
w	An (optional) numeric vector containing weights for the input data.	NULL
df	Degrees of freedom.	
spar	Numeric value specifying the smoothing parameter.	NULL
cv	A logical value specifying whether to use ordinary cross-validation (cv=TRUE) or generalized cross-validation (cv=FALSE).	FALSE
all.knots	A logical value specifying whether to use all values in x as knots.	FALSE
nknots	An integer value specifying the number of knots to use when `all.knots=FALSE`.	NULL
keep.data	A logical value indicating whether the input data should be kept in the result.	TRUE
df.offset	A numeric value specifying how much to allow the df to be increased in cross-validation.	0
penalty	The penalty for degrees of freedom during cross-validation.	1
control.spar	A list of parameters describing how to compute spar (when not explicitly specified). See the help file for more information.	list()

For example, we can calculate a smoothing spline on the Schiller home price index. This data set contains one annual measurement through 2006, but then has fractional measurements after 2006, making it slightly difficult to align with other data:

```
> schiller.index[schiller.index$Year>2006,]
        Year Real.Home.Price.Index
118 2007.125            194.6713
119 2007.375            188.9270
120 2007.625            184.1683
121 2007.875            173.8622
122 2008.125            160.7639
123 2008.375            154.4993
124 2008.625            145.6642
125 2008.875            137.0083
126 2009.125            130.0611
```

We can use smoothing splines to find values for 2007 and 2008:

```
> library(nutshell)
> data(schiller.index)
> schiller.index.spl <- smooth.spline(schiller.index$Year,
+    schiller.index$Real.Home.Price.Index)
> predict(schiller.index.spl,x=c(2007,2008))
$x
[1] 2007 2008

$y
[1] 195.6682 168.8219
```

Fitting Polynomial Surfaces

You can fit a polynomial surface to data (by local fitting) using the loess function. (This function is used in many graphics functions; for example, panel.loess uses loess to fit a curve to data and plot the curve.)

```
loess(formula, data, weights, subset, na.action, model = FALSE,
      span = 0.75, enp.target, degree = 2,
      parametric = FALSE, drop.square = FALSE, normalize = TRUE,
      family = c("gaussian", "symmetric"),
      method = c("loess", "model.frame"),
      control = loess.control(...), ...)
```

Here is a description of the arguments to loess.

Argument	Description	Default
formula	A formula specifying the relationship between the response and the predictor variables.	
data	A data frame, list, or environment specifying the training data for the model fit. (If none is specified, formula is evaluated in the calling environment.)	
weights	A vector of weights for the cases in the training data.	
subset	An optional expression specifying a subset of cases to include in the model.	
na.action	A function specifying how to treat missing values.	getOption("na.action")
model	A logical value indicating whether to return the model frame.	FALSE
span	A numeric value specifying the parameter α, which controls the degree of smoothing.	0.75

Argument	Description	Default
enp.target	A numeric value specifying the equivalent number of parameters to be used (replaced span).	
degree	The degree of polynomials used.	2
parametric	A vector specifying any terms that should be fit globally rather than locally. (May be specified by name, number, or as a logical vector.)	FALSE
drop.square	Specifies whether to drop the quadratic term for some predictors.	FALSE
normalize	A logical value specifying whether to normalize predictors to a common scale.	TRUE
family	Specifies how fitting is done. Specify `family="gaussian"` to fit by least squares, and `family="symmetric"` to fit with Tukey's biweight function.	"gaussian"
method	Specifies whether to fit the model or just return the model frame.	"loess"
control	Control parameters for loess, typically generated by a call to `loess.control`.	loess.control(...)
...	Additional arguments are passed to `loess.control`.	

Using the same example as above:

```
> schiller.index.loess <- loess(Real.Home.Price.Index~Year,data=schiller.index)
> predict(schiller.index.loess, newdata=data.frame(Year=c(2007,2008)))
[1] 156.5490 158.8857
```

Kernel Smoothing

To estimate a probability density function, regression function, or their derivatives using polynomials, try the function `locpoly` in the library `KernSmooth`:

```
library(KernSmooth)
locpoly(x, y, drv = 0L, degree, kernel = "normal",
        bandwidth, gridsize = 401L, bwdisc = 25,
        range.x, binned = FALSE, truncate = TRUE)
```

Here is a description of the arguments to `locpoly`.

Argument	Description	Default
x	A vector of x values (with no missing values).	
y	A vector of y values (with no missing values).	
drv	Order of derivative to estimate.	0L
degree	Degree of local polynomials.	drv + 1
kernel	Kernel function to use. Currently ignored ("normal" is used).	"normal"
bandwidth	A single value or an array of length `gridsize` that specifies the kernel bandwidth smoothing parameter.	
gridsize	Specifies the number of equally spaced points over which the function is estimated.	401L
bwdisc	Number of (logarithmically equally spaced) values on which `bandwidth` is discretized.	25
range.x	A vector containing the minimum and maximum values of x on which to compute the estimate.	

Argument	Description	Default
binned	A logical value specifying whether to interpret x and y as grid counts (as opposed to raw data).	FALSE
truncate	A logical value specifying whether to ignore x values outside range.x.	TRUE

R also includes an implementation of local regression through the `locfit` function in the `locfit` library:

```
library(locfit)
locfit(formula, data=sys.frame(sys.parent()), weights=1, cens=0, base=0,
    subset, geth=FALSE, ..., lfproc=locfit.raw)
```

Machine Learning Algorithms for Regression

Most of the models above assumed that you knew the basic form of the model equation and error function. In each of these cases, our goal was to find the coefficients of variables in a known function. However, sometimes you are presented with data where there are many predictive variables, and the relationships between the predictors and response are very complicated.

Statisticians have developed a variety of different techniques to help model more complex relationships in data sets and to predict values for large, complicated data sets. This section describes a variety of techniques for finding not only the coefficients of a model function but also the function itself.

In this section, I use the San Francisco home sales data set described in "More About the San Francisco Real Estate Prices Data Set" on page 290. This is a pretty ugly data set, with lots of nonlinear relationships. Real estate is all about location, and we have several different variables in the data set that represent location. (The relationships between these variables is not linear, in case you were worried.)

Before modeling, we'll split the data set into training and testing data sets. Splitting data into training and testing data sets (and, often, validation data sets as well) is a standard practice when fitting models. Statistical models have a tendency to "overfit" the training data; they do a better job predicting trends in the training data than in other data.

I chose this approach because it works with all of the modeling functions in this section. There are other statistical techniques available for making sure that a model doesn't overfit the data, including cross-validation and bootstrapping. Functions for cross-validation are available for some models (for example, `xpred.rpart` for `rpart` trees); look at the detailed help files for a package (in this case, with the command `help(package="rpart")`) to see if these functions are available for a specific modeling tool. Bootstrap resampling is available through the `boot` library.

Because this section presents many different types of models, I decided to use a simple, standard approach for evaluating model fits. For each model, I estimated the root mean square (RMS) error for the training and validation data sets. Don't interpret the results as authoritative: I didn't try too hard to tune each model's parameters and know that the models that worked best for this data set do not work best for all data sets. However, I thought I'd include the results because I was interested in them (in good fun) and thought readers would be as well.

Anyway, I wrote the following function to evaluate the performance of each function:

```
calculate_rms_error <- function(mdl, train, test, yval) {
  train.yhat <- predict(object=mdl,newdata=train)
  test.yhat  <- predict(object=mdl,newdata=test)
  train.y    <- with(train,get(yval))
  test.y     <- with(test,get(yval))
  train.err  <- sqrt(mean((train.yhat - train.y)^2))
  test.err   <- sqrt(mean((test.yhat - test.y)^2))
  c(train.err=train.err,test.err=test.err)
}
```

To create a random sample, I used the `sample` function to pick 70% of values for the training data. I saved the sample indices to a vector for later reuse (so that I could derive the same sample later, and allow you to use the same sample as well). I also saved the sample indices to make it easy to define the testing data set.

```
> nrow(sanfrancisco.home.sales) * .7
[1] 2296.7
> sanfrancisco.home.sales.training.indices <-
+   sample(1:nrow(sanfrancisco.home.sales),2296)
> sanfrancisco.home.sales.testing.indices <-
+   setdiff(rownames(sanfrancisco.home.sales),
+           sanfrancisco.home.sales.training.indices)
> sanfrancisco.home.sales.training <-
    sanfrancisco.home.sales[sanfrancisco.home.sales.training.indices,]
> sanfrancisco.home.sales.testing <-
+   sanfrancisco.home.sales[sanfrancisco.home.sales.testing.indices,]
> save(sanfrancisco.home.sales.training.indices,
+   sanfrancisco.home.sales.testing.indices,
+   sanfrancisco.home.sales,
+   file="~/Documents/book/current/data/sanfrancisco.home.sales.RData")
```

Note that the sampling is random, so you will get a different subset each time you run this code. The vectors sanfrancisco.home.sales.training.indices and sanfrancisco.home.sales.testing.indices that I used in this section are included in the nutshell package. (Use the command data(sanfrancisco.home.sales) to access them. The data sets sanfrancisco.home.sales.training and sanfrancisco.home.sales.testing are not included.) You can use the same training and testing sets to re-create the results in this section, or you can pick your own subsets.

Regression Tree Models

Most of the models that we have seen in this chapter are in the form of a single equation. You can use the model to predict values by plugging new data values into a single equation.

Tree models have a slightly different form. Instead of a single, compact equation, tree models represent data by a set of binary decision rules. Instead of plugging numbers into an equation, you follow the rules in a tree to determine the predicted value. Tree models are very easy to interpret, but don't usually predict values as accurately as other types of models. Tree models are particularly popular in medicine and biology, perhaps because they resemble the process that doctors use to make decisions. In this section, we'll show how to use some popular tree methods for regression in R.

Recursive partitioning trees

One of the most popular algorithms for building tree models is classification and regression trees, or CART. CART uses a greedy algorithm to build a tree from the training data. Here's an explanation of how CART works:

1. Grow the tree using the following (recursive) method:
   A. Start with a single set containing all the training data.
   B. If the number of observations is less than the minimum required for a split, stop splitting the tree. Output the average of all the y-values in the training data as the predicted value for the terminal node.
   C. Find a variable x_j and value s that minimizes the RMS error when you split the data into two sets.
   D. Repeat the splitting process (starting at step B) on each of the two sets.
2. Prune the tree using the following (iterative) method:
   A. Stop if there is only one node in the tree.
   B. Measure the cost/complexity of the overall tree. (The cost/complexity measurement is a measurement that takes into account the number of observations in each node, the RMS prediction error, and the number of nodes in the tree.)
   C. Try collapsing each internal node on the tree and measure which subtree has the best cost/complexity.
   D. Repeat the process (starting at step A) on the subtree with the best cost/complexity.
3. Output the tree with the lowest cost/complexity.

R includes an implementation of classification and regression trees in the rpart package. To fit a model, use the rpart function:

```
library(rpart)
rpart(formula, data, weights, subset, na.action = na.rpart, method,
      model = FALSE, x = FALSE, y = TRUE, parms, control, cost, ...)
```

Here are the arguments to rpart.

Argument	Description	Default
formula	A formula describing the relationship between the response and the predictor variables.	
data	A data frame to use for fitting the model.	
weights	An optional vector of weights to use for the training data.	

Argument	Description	Default
subset	An optional expression specifying which observations to use in fitting the model.	
na.action	The function to call for missing values.	na.rpart
method	A character value that specifies the fitting method. Must be one of "exp", "poisson", "class", or "anova".	If y is a survival object, then method="exp", if y has two columns then method="poisson", or y if a factor then method="class", otherwise method="anova"
model	A logical value specifying whether to keep the model frame in the results.	FALSE
x	A logical value specifying whether to return the x matrix in the results.	FALSE
y	A logical value specifying whether to return the y matrix in the results.	TRUE
parms	A list of parameters passed to the fitting function.	
control	Options that control details of the rpart algorithm; see `rpart.control` for more information.	
cost	A numeric vector of costs, one for each variable in the model.	1 for all variables
...	Additional argument passed to `rpart.control`.	

The CART algorithm handles missing values differently from many other modeling algorithms. With an algorithm like linear regression, missing values need to be filtered out in order for the math to work. However, CART takes advantage of the rule-based model structure to handle missing values differently. When a value is missing for an observation at a split, CART can instead split values using a *surrogate* variable. See the help files for **rpart** for more information on how to control the process of finding and using surrogates.

As an example, let's build a regression tree on the San Francisco home sales data set. We'll start off naively, adding some redundant information and fields that could lead to a model that overfits the data:

```
> library(rpart)
> sf.price.model.rpart <- rpart(
+     price~bedrooms+squarefeet+lotsize+latitude+
+     longitude+neighborhood+month,
+     data=sanfrancisco.home.sales.training)
```

Let's take a look at the model returned by this call to **rpart**. The simplest way to examine the object is to use **print.rpart** to print it on the console. The output below has been modified slightly to fit in this book:

```
> sf.price.model.rpart
n= 2296

node), split, n, deviance, yval
      * denotes terminal node

 1) root 2296 8.058726e+14  902088.0
```

```
    2) neighborhood=Bayview,Bernal Heights,Chinatown,Crocker Amazon,
        Diamond Heights,Downtown,Excelsior,Inner Sunset,Lakeshore,
        Mission,Nob Hill,Ocean View,Outer Mission,Outer Richmond,
        Outer Sunset,Parkside,Potrero Hill,South Of Market,
        Visitacion Valley,Western Addition 1524 1.850806e+14  723301.8
      4) squarefeet< 1772 1282 1.124418e+14  675471.1
        8) neighborhood=Bayview,Chinatown,Crocker Amazon,
            Diamond Heights,Downtown,Excelsior,Lakeshore,Ocean View,
            Outer Mission,Visitacion Valley 444 1.408221e+13  539813.1 *
        9) neighborhood=Bernal Heights,Inner Sunset,Mission,Nob Hill,
            Outer Richmond,Outer Sunset,Parkside,Potrero Hill,
            South Of Market,Western Addition 838 8.585934e+13  747347.3 *
      5) squarefeet>=1772 242 5.416861e+13  976686.0 *
  3) neighborhood=Castro-Upper Market,Financial District,Glen Park,
      Haight-Ashbury,Inner Richmond,Marina,Noe Valley,North Beach,
      Pacific Heights,Presidio Heights,Russian Hill,Seacliff,
      Twin Peaks,West Of Twin Peaks 772 4.759124e+14 1255028.0
    6) squarefeet< 2119 591 1.962903e+14 1103036.0
      12) neighborhood=Castro-Upper Market,Glen Park,Haight-Ashbury,
          Inner Richmond,Noe Valley,North Beach,Pacific Heights,
          Russian Hill,Twin Peaks,
          West Of Twin Peaks 479 1.185669e+14 1032675.0
        24) month=2008-02-01,2008-03-01,2008-06-01,2008-07-01,
            2008-08-01,2008-09-01,2008-10-01,2008-11-01,2008-12-01,
            2009-01-01,2009-02-01,2009-03-01,2009-04-01,2009-05-01,
            2009-06-01,2009-07-01 389 5.941085e+13  980348.3 *
        25) month=2008-04-01,2008-05-01 90 5.348720e+13 1258844.0
          50) longitude< -122.4142 81 1.550328e+13 1136562.0 *
          51) longitude>=-122.4142 9 2.587193e+13 2359389.0 *
      13) neighborhood=Financial District,Marina,Presidio Heights,
          Seacliff 112 6.521045e+13 1403951.0 *
    7) squarefeet>=2119 181 2.213886e+14 1751315.0
      14) neighborhood=Castro-Upper Market,Glen Park,Haight-Ashbury,
          Inner Richmond,Marina,Noe Valley,North Beach,Russian Hill,
          Twin Peaks,West Of Twin Peaks 159 1.032114e+14 1574642.0
        28) month=2008-04-01,2008-06-01,2008-07-01,2008-10-01,
            2009-02-01,2009-03-01,2009-04-01,2009-05-01,
            2009-06-01,2009-07-01 77 2.070744e+13 1310922.0 *
        29) month=2008-02-01,2008-03-01,2008-05-01,2008-08-01,
            2008-09-01,2008-11-01,2008-12-01,
            2009-01-01 82 7.212013e+13 1822280.0
          58) lotsize< 3305.5 62 3.077240e+13 1598774.0 *
          59) lotsize>=3305.5 20 2.864915e+13 2515150.0
            118) neighborhood=Glen Park,Inner Richmond,Twin Peaks,
                West Of Twin Peaks 13 1.254738e+13 1962769.0 *
            119) neighborhood=Castro-Upper Market,Marina,
                Russian Hill 7 4.768574e+12 3541000.0 *
      15) neighborhood=Financial District,Pacific Heights,
          Presidio Heights,Seacliff 22 7.734568e+13 3028182.0
        30) lotsize< 3473 12 7.263123e+12 2299500.0 *
        31) lotsize>=3473 10 5.606476e+13 3902600.0 *
```

Notice the key on the second line of the output. (Each line contains the node number, description of the split, number of observations under that node in the tree, deviance, and predicted value.) This tree model tells us some obvious things, like location and size are good predictors of price. Reading a textual description of an **rpart** object is somewhat confusing. The method `plot.rpart` will draw the tree structure in an **rpart** object:

```
plot(x, uniform=FALSE, branch=1, compress=FALSE, nspace,
    margin=0, minbranch=.3, ...)
```

You can label the tree using `text.rpart`:

```
text(x, splits=TRUE, label, FUN=text, all=FALSE,
    pretty=NULL, digits=getOption("digits") - 3, use.n=FALSE,
    fancy=FALSE, fwidth=0.8, fheight=0.8, ...)
```

For both functions, the argument x specifies the **rpart** object; the other options control the way the output looks. See the help file for more information about these parameters. As an example, let's plot the tree we just created above:

```
> plot(sf.price.model.rpart,uniform=TRUE,compress=TRUE,lty=3,branch=0.7)
> text(sf.price.model.rpart,all=TRUE,digits=7,use.n=TRUE,cex=0.4,xpd=TRUE,)
```

As you can see from Figure 20-4, it's difficult to read a small picture of a big tree. To keep the tree somewhat readable, we have abbreviated neighborhood names to single letters (corresponding to their order in the factor). Sometimes, the function `draw.tree` in the package **maptree** can produce prettier diagrams. See "Classification Tree Models" on page 446 for more details.

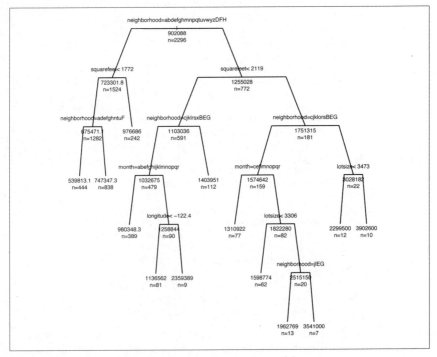

Figure 20-4. rpart tree for the San Francisco home sales model

To predict a value with a tree model, you would start at the top of the tree and follow the tree down, depending on the rules for a specific observation. For example, suppose that we had a property in Pacific Heights with 2,500 square feet of living space and a lot size of 5,000 square feet. We would traverse the tree starting at node 1, then go to node 3, then node 7, then node 15, and, finally, land on node 31. The estimated price of this property would be $3,902,600.

There are a number of other functions available in the **rpart** package for viewing (or manipulating) tree objects. To view the approximate r-square and relative error at each split, use the function **rsq.rpart**. The graphical output is shown in Figure 20-5; here is the output on the R console:

```
> rsq.rpart(sf.price.model.rpart)

Regression tree:
rpart(formula = price ~ bedrooms + squarefeet + lotsize + latitude +
    longitude + neighborhood + month, data = sanfrancisco.home.sales.training)

Variables actually used in tree construction:
[1] longitude   lotsize     month        neighborhood squarefeet

Root node error: 8.0587e+14/2296 = 3.5099e+11

n= 2296

          CP nsplit rel error  xerror     xstd
1  0.179780      0   1.00000 1.00038 0.117779
2  0.072261      1   0.82022 0.83652 0.105103
3  0.050667      2   0.74796 0.83211 0.096150
4  0.022919      3   0.69729 0.80729 0.094461
5  0.017395      4   0.67437 0.80907 0.096560
6  0.015527      5   0.65698 0.82365 0.097687
7  0.015511      6   0.64145 0.81720 0.097579
8  0.014321      7   0.62594 0.81461 0.097575
9  0.014063      9   0.59730 0.81204 0.097598
10 0.011032     10   0.58323 0.81559 0.097691
11 0.010000     12   0.56117 0.80271 0.096216
```

As you can probably tell, the initial tree was a bit complicated. You can remove nodes where the cost/complexity trade-off isn't great by using the **prune** function:

```
prune(tree, cp, ...)
```

The argument **cp** is a complexity parameter that controls how much to trim the tree. To help choose a complexity parameter, try the function **plotcp**:

```
plotcp(x, minline = TRUE, lty = 3, col = 1,
    upper = c("size", "splits", "none"), ...)
```

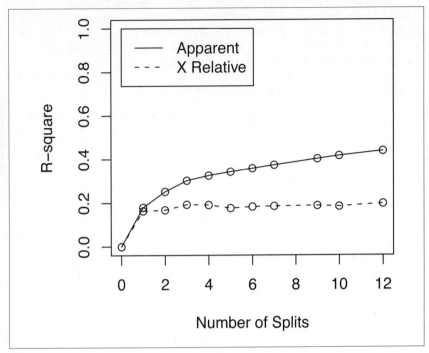

Figure 20-5. Plot from rsq.rpart(sf.price.model.rpart)

The `plotcp` function plots tree sizes and relative errors for different parameters of the complexity parameter. For the example above, it looks like a value of .011 is a good balance between complexity and performance. Here is the pruned model (see also Figure 20-6):

```
> prune(sf.price.model.rpart,cp=0.11)
n= 2296

node), split, n, deviance, yval
      * denotes terminal node

1) root 2296 8.058726e+14  902088.0
2) neighborhood=Bayview,Bernal Heights,Chinatown,Crocker Amazon,
      Diamond Heights,Downtown,Excelsior,Inner Sunset,Lakeshore,Mission,
      Nob Hill,Ocean View,Outer Mission,Outer Richmond,Outer Sunset,
      Parkside,Potrero Hill,South Of Market,Visitacion Valley,
      Western Addition 1524 1.850806e+14  723301.8 *
3) neighborhood=Castro-Upper Market,Financial District,Glen Park,
      Haight-Ashbury, Inner Richmond,Marina,Noe Valley,North Beach,
      Pacific Heights,Presidio Heights,Russian Hill,Seacliff,Twin Peaks,
      West Of Twin Peaks 772 4.759124e+14 1255028.0 *
```

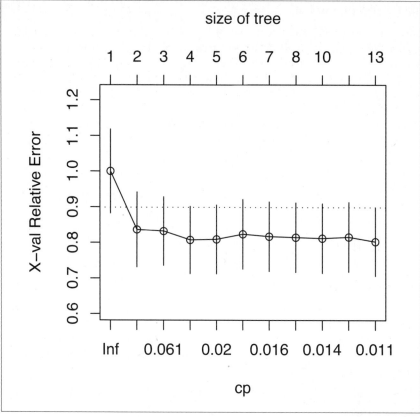

Figure 20-6. Output of plotcp for the sf.prices.rpart model

And if you're curious, here is the error of this model on the training and test populations:

```
> calculate_rms_error(sf.price.model.rpart,
+    sanfrancisco.home.sales.training,
+    sanfrancisco.home.sales.testing,
+    "price")
train.err   test.err
 443806.8   564986.8
```

The units, incidentally, are dollars.

There is an alternative implementation of CART trees available with R through the **tree** package. It was written by W. N. Venables, one of the authors of [Venables2002]. He notes that **tree** can give more explicit output while running, but recommends **rpart** for most users.

Patient rule induction method

Another technique for building rule-based models is the patient rule induction method (PRIM) algorithm. PRIM doesn't actually build trees. Instead, it partitions the data into a set of "boxes" (in p dimensions). The algorithm starts with a box containing all the data and then shrinks the box one side at a time, trying to maximize the average value in the box. After reaching a minimum number of observations in the box, the algorithm tries expanding the box again, as long as it can increase the average value in the box. When the algorithm finds the best initial box, it then repeats the process on the remaining observations, until there are no observations left. The algorithm leads to a set of rules that can be used to predict values.

To try out PRIM in R, there are functions in the library `prim`:

```
prim.box(x, y, box.init=NULL, peel.alpha=0.05, paste.alpha=0.01,
    mass.min=0.05, threshold, pasting=TRUE, verbose=FALSE,
    threshold.type=0)

prim.hdr(prim, threshold, threshold.type)
prim.combine(prim1, prim2)
```

Bagging for regression

Bagging (or bootstrap aggregation) is a technique for building predictive models based on other models (most commonly trees). The idea of bagging is to use bootstrapping to build a number of different models and then average the results. The weaker models essentially form a committee to vote for a result, which leads to more accurate predictions.

To build regression bagging models in R, you can use the function `bagging` in the `ipred` library:

```
library(ipred)
bagging(formula, data, subset, na.action=na.rpart, ...)
```

The `formula`, `data`, `subset`, and `na.action` arguments work the same way as in most modeling functions. The additional arguments are passed on to the function `ipredbagg`, which does all the work (but doesn't have a method for formulas):

```
ipredbagg(y, X=NULL, nbagg=25, control=rpart.control(xval=0),
          comb=NULL, coob=FALSE, ns=length(y), keepX = TRUE, ...)
```

You can specify the number of trees to build by `nbagg`, control parameters for `rpart` through `control`, a list of models to use for double-bagging through `comb`, `coob` to indicate if an out-of-bag error rate should be computed, and `ns` to specify the number of observations to draw from the learning sample.

Let's try building a model on the pricing data using bagging. We'll pick 100 `rpart` trees (for fun):

```
> sf.price.model.bagging <- bagging(
+    price~bedrooms+squarefeet+lotsize+latitude+
+    longitude+neighborhood+month,
+    data=sanfrancisco.home.sales.training, nbagg=100)
> summary(sf.price.model.bagging)
```

```
        Length Class         Mode
y         1034  -none-        numeric
X            7  data.frame    list
mtrees     100  -none-        list
OOB          1  -none-        logical
comb         1  -none-        logical
call         4  -none-        call
```

Let's take a quick look at how bagging worked on this data set:

```
> calculate_rms_error(sf.price.model.bagging,
+      sanfrancisco.home.sales.training,
+      sanfrancisco.home.sales.testing,
+    "price")
train.err   test.err
 491003.8   582056.5
```

Boosting for regression

Boosting is a technique that's closely related to bagging. Unlike bagging, the individual models don't all have equal votes. Better models are given stronger votes.

You can find a variety of tools for computing boosting models in R in the package mboost. The function blackboost builds boosting models from regression trees, glmboost from general linear models, and gamboost for boosting based on additive models. Here, we'll just build a model using regression trees:

```
> library(mboost)
Loading required package: modeltools
Loading required package: stats4
Loading required package: party
Loading required package: grid
Loading required package: coin
Loading required package: mvtnorm
Loading required package: zoo
> sf.price.model.blackboost <- blackboost(
+    price~bedrooms+squarefeet+lotsize+latitude+
+    longitude+neighborhood+month,
+    data=sanfrancisco.home.sales.training)
```

Here is a summary of the model object:

```
> summary(sf.price.model.blackboost)
                Length Class                 Mode
ensemble         100   -none-                list
fit             2296   -none-                numeric
offset             1   -none-                numeric
ustart          2296   -none-                numeric
risk             100   -none-                numeric
control            8   boost_control         list
family             1   boost_family          S4
response        2296   -none-                numeric
weights         2296   -none-                numeric
update             1   -none-                function
tree_controls      1   TreeControl           S4
data               1   LearningSampleFormula S4
```

```
predict          1   -none-          function
call             3   -none-          call
```

And here is a quick evaluation of the performance of this model:

```
> calculate_rms_error(sf.price.model.blackboost,
+       sanfrancisco.home.sales.training,
+       sanfrancisco.home.sales.testing,
+       "price")
train.err  test.err
  1080520   1075810
```

Random forests for regression

Random forests are another technique for building predictive models using trees. Like boosting and bagging, random forests work by combining a set of other tree models. Unlike boosting and bagging, which use an existing algorithm like CART to build a series of trees from a random sample of the observations in the test data, random forests build trees from a random sample of the columns in the test data.

Here's a description of how the random forest algorithm creates the underlying trees (using variable names from the R implementation):

1. Take a sample of size `sampsize` from the training data.
2. Begin with a single node.
3. Run the following algorithm, starting with the starting node:
   A. Stop if the number of observations is less than `nodesize`.
   B. Select `mtry` variables (at random).
   C. Find the variable and value that does the "best" job splitting the observations. (Specifically, the algorithm uses MSE [mean square error] to measure regression error, and Gini to measure classification error.)
   D. Split the observations into two nodes.
   E. Call step A on each of these nodes.

Unlike trees generated by CART, trees generated by random forest aren't pruned; they're just grown to a very deep level.

For regression problems, the estimated value is calculated by averaging the prediction of all the trees in the forest. For classification problems, the prediction is made by predicting the class using each tree in the forest and then outputting the choice that received the most votes.

To build random forest models in R, use the `randomForest` function in the `random Forest` package:

```
library(randomForest)
## S3 method for class 'formula':
randomForest(formula, data=NULL, ..., subset, na.action=na.fail)
## Default S3 method:
randomForest(x, y=NULL,  xtest=NULL, ytest=NULL, ntree=500,
             mtry=if (!is.null(y) && !is.factor(y))
             max(floor(ncol(x)/3), 1) else floor(sqrt(ncol(x))),
```

```
        replace=TRUE, classwt=NULL, cutoff, strata,
        sampsize = if (replace) nrow(x) else ceiling(.632*nrow(x)),
        nodesize = if (!is.null(y) && !is.factor(y)) 5 else 1,
        importance=FALSE, localImp=FALSE, nPerm=1,
        proximity, oob.prox=proximity,
        norm.votes=TRUE, do.trace=FALSE,
        keep.forest=!is.null(y) && is.null(xtest), corr.bias=FALSE,
        keep.inbag=FALSE, ...)
```

Unlike some other functions we've seen so far, randomForest will fail if called on data with missing observations. So, we'll set na.action=na.omit to omit NA values. Additionally, randomForest cannot handle categorical predictors with more than 32 levels, so we will cut out the neighborhood variable:

```
> sf.price.model.randomforest <- randomForest(
+    price~bedrooms+squarefeet+lotsize+latitude+
+    longitude+month,
+    data=sanfrancisco.home.sales.training,
+    na.action=na.omit)
```

The print method for randomForest objects returns some useful information about the fit:

```
> sf.price.model.randomforest

Call:
 randomForest(formula = price ~ bedrooms + squarefeet + lotsize +
             latitude + longitude + month,
             data = sanfrancisco.home.sales.training,
             na.action = na.omit)
               Type of random forest: regression
                     Number of trees: 500
No. of variables tried at each split: 2

        Mean of squared residuals: 258521431697
                  % Var explained: 39.78
```

Here is how the model performed:

```
> calculate_rms_error(sf.price.model.randomforest,
+      na.omit(sanfrancisco.home.sales.training),
+      na.omit(sanfrancisco.home.sales.testing),
+      "price")
train.err  test.err
 241885.2  559461.0
```

As a point of comparison, here are the results of the rpart model, also with NA values omitted:

```
> calculate_rms_error(sf.price.model.rpart,
+      na.omit(sanfrancisco.home.sales.training),
+      na.omit(sanfrancisco.home.sales.testing),
+      "price")
train.err  test.err
 442839.6  589583.1
```

MARS

Another popular algorithm for machine learning is multivariate adaptive regression splines, or MARS. MARS works by splitting input variables into multiple *basis functions* and then fitting a linear regression model to those basis functions. The basis functions used by MARS come in pairs: $f(x) = \{x - t$ if $x > t$, 0 otherwise$\}$ and $g(x) = \{t - x$ if $x < t$, 0 otherwise$\}$. These functions are *piecewise linear* functions. The value t is called a *knot*.

MARS is closely related to CART. Like CART, it begins by building a large model and then prunes back unneeded terms until the best model is found. The MARS algorithm works by gradually building up a model out of basis functions (or products of basis functions) until it reaches a predetermined depth. This results in an over-fitted, overly complex model. Then the algorithm deletes terms from the model, one by one, until it has pared back everything but a constant term. At each stage, the algorithm uses generalized cross-validation (GCV) to measure how well each model fits. Finally, the algorithm returns the model with the best cost/benefit ratio.

To fit a model using MARS in R, use the function `earth` in the package `earth`:

```
library(earth)
earth(formula = stop("no 'formula' arg"),
    data, weights = NULL, wp = NULL, scale.y = (NCOL(y)==1), subset = NULL,
    na.action = na.fail, glm = NULL, trace = 0,
    keepxy = FALSE, nfold=0, stratify=TRUE, ...)
```

Arguments to `earth` include the following.

Argument	Description	Default
formula	A formula describing the relationship between the response and the predictor variables.	stop("no 'formula' arg")
data	A data frame containing the training data.	
weights	An optional vector of weights to use for the fitting data. (It is especially optional, because it is not supported as of `earth` version 2.3-2.)	NULL
wp	A numeric vector of response weights. Must include a value for each column of y.	NULL
scale.y	A numeric value specifying whether to scale y in the forward pass. (See the help file for more information.)	(NCOL(y)==1)
subset	A logical vector specifying which observations from `data` to include.	NULL
na.action	A function specifying how to treat missing values. Only `na.fail` is currently supported.	na.fail
glm	A list of arguments to `glm`.	NULL
trace	A numeric value specifying whether to print a "trace" of the algorithm execution.	0
keepxy	A logical value specifying whether to keep x and y (or data), subset, and weights in the model object. (Useful if you plan to use `update` to modify the model at a later time.)	FALSE
nfold	A numeric value specifying the number of cross-validation folds.	0
stratify	A logical value specifying whether to stratify the cross-validation folds.	TRUE

Argument	Description	Default
...	Additional options are passed to earth.fit. There are many, many options available to tune the fitting process. See the help file for earth for more information.	

The **earth** function is very flexible. By default, lm is used to fit models. Note that glm can be used instead to allow finer control of the model. The function **earth** can't cope directly with missing values in the data set. To deal with NA values, you need to explicitly deal with them in the input data. You could, for example, impute median values or model imputed values. In the example below, I picked the easy solution and just used the na.omit function to filter them out.

Let's build an earth model on the San Francisco home sales data set. We'll add the **trace=1** option to show some details of the computation:

```
> sf.price.model.earth <- earth(
+     price~bedrooms+squarefeet+latitude+
+     longitude+neighborhood+month,
+     data=na.omit(sanfrancisco.home.sales.training), trace=1)
x is a 957 by 54 matrix: 1=bedrooms, 2=squarefeet, 3=latitude,
   4=longitude, 5=neighborhoodBernalHeights, 6=neighborhoodCastro-UpperMarket,
   7=neighborhoodChinatown, 8=neighborhoodCrockerAmazon,
   9=neighborhoodDiamondHeights, 10=neighborhoodDowntown,
   11=neighborhoodExcelsior, 12=neighborhoodFinancialDistrict,
   13=neighborhoodGlenPark, 14=neighborhoodHaight-Ashbury,
   15=neighborhoodInnerRichmond, 16=neighborhoodInnerSunset,
   17=neighborhoodLakeshore, 18=neighborhoodMarina,
   19=neighborhoodMission, 20=neighborhoodNobHill,
   21=neighborhoodNoeValley, 22=neighborhoodNorthBeach,
   23=neighborhoodOceanView, 24=neighborhoodOuterMission,
   25=neighborhoodOuterRichmond, 26=neighborhoodOuterSunset,
   27=neighborhoodPacificHeights, 28=neighborhoodParkside,
   29=neighborhoodPotreroHill, 30=neighborhoodPresidioHeights,
   31=neighborhoodRussianHill, 32=neighborhoodSeacliff,
   33=neighborhoodSouthOfMarket, 34=neighborhoodTwinPeaks,
   35=neighborhoodVisitacionValley, 36=neighborhoodWestOfTwinPeaks,
   37=neighborhoodWesternAddition, 38=month2008-03-01, 39=month2008-04-01,
   40=month2008-05-01, 41=month2008-06-01, 42=month2008-07-01,
   43=month2008-08-01, 44=month2008-09-01, 45=month2008-10-01,
   46=month2008-11-01, 47=month2008-12-01, 48=month2009-01-01,
   49=month2009-02-01, 50=month2009-03-01, 51=month2009-04-01,
   52=month2009-05-01, 53=month2009-06-01, 54=month2009-07-01
y is a 957 by 1 matrix: 1=price
Forward pass term 1, 2, 4, 6, 8, 10, 12, 14, 16, 18, 20, 22, 24, 26, 28,
      30, 32, 34, 36, 38, 40, 42, 44, 46, 48, 50, 52, 54, 56, 58,
      60, 62, 64, 66, 68, 70, 72, 74, 76, 78, 80
Reached delta RSq threshold (DeltaRSq 0.000861741 < 0.001)
After forward pass GRSq 0.4918 RSq 0.581
Prune method "backward" penalty 2 nprune 44: selected 36 of 44 terms, and 26
   of 54 predictors
After backward pass GRSq 0.5021 RSq 0.5724
```

The **earth** object has an informative **print** method, showing the function call and statistics about the model fit:

```
> sf.price.model.earth
Selected 31 of 41 terms, and 22 of 55 predictors
Importance: squarefeet, neighborhoodPresidioHeights,
  latitude, neighborhoodSeacliff, neighborhoodNoeValley,
  neighborhoodCastro-UpperMarket, neighborhoodNobHill,
  lotsize, month2008-07-01, neighborhoodWesternAddition, ...
Number of terms at each degree of interaction: 1 30 (additive model)
GCV 216647913449    RSS 1.817434e+14    GRSq 0.5162424    RSq 0.5750596
```

The **summary** method will show the basis functions for the fitted model in addition to information about the fit:

```
> summary(sf.price.model.earth)
Call: earth(formula=price~bedrooms+squarefeet+lotsize+latitude+
       longitude+neighborhood+month,
       data=na.omit(sanfrancisco.home.sales.training))
```

	coefficients
(Intercept)	1452882
h(bedrooms-3)	130018
h(bedrooms-5)	-186130
h(squarefeet-2690)	81
h(2690-squarefeet)	-178
h(lotsize-2495)	183
h(lotsize-3672)	-141
h(latitude-37.7775)	-112301793
h(37.7775-latitude)	-7931270
h(latitude-37.7827)	420380414
h(latitude-37.7888)	-188726623
h(latitude-37.8015)	-356738902
h(longitude- -122.464)	-6056771
h(-122.438-longitude)	-6536227
neighborhoodCastro-UpperMarket	338549
neighborhoodChinatown	-1121365
neighborhoodInnerSunset	-188192
neighborhoodMarina	-2000574
neighborhoodNobHill	-2176350
neighborhoodNoeValley	368772
neighborhoodNorthBeach	-2395955
neighborhoodPacificHeights	-1108284
neighborhoodPresidioHeights	1146964
neighborhoodRussianHill	-1857710
neighborhoodSeacliff	2422127
neighborhoodWesternAddition	-442262
month2008-03-01	181640
month2008-04-01	297754
month2008-05-01	187684
month2008-07-01	-322801
month2008-10-01	115435

```
Selected 31 of 41 terms, and 22 of 55 predictors
Importance: squarefeet, neighborhoodPresidioHeights, latitude,
  neighborhoodSeacliff, neighborhoodNoeValley,
```

```
  neighborhoodCastro-UpperMarket, neighborhoodNobHill,
    lotsize, month2008-07-01, neighborhoodWesternAddition, ...
  Number of terms at each degree of interaction: 1 30 (additive model)
  GCV 216647913449    RSS 1.817434e+14    GRSq 0.5162424    RSq 0.5750596
```

The output of summary includes a short synopsis of variable importance in the model.
You can use the function evimp to return a matrix showing the relative importance
of variables in the model:

```
evimp(obj, trim=TRUE, sqrt.=FALSE)
```

The argument obj specifies an earth object, trim specifies whether to delete rows in
the matrix for variables that don't appear in the fitted model, and sqrt specifies
whether to take the square root of the GCV and RSS importances before normalizing
them. For the example above, here is the output:

```
> evimp(sf.price.model.earth)
                                  col used nsubsets         gcv
squarefeet                          2    1       30 100.00000000 1
neighborhoodPresidioHeights        31    1       29  62.71464260 1
latitude                            4    1       28  45.85760472 1
neighborhoodSeacliff               33    1       27  33.94468291 1
neighborhoodNoeValley              22    1       25  22.55538880 1
neighborhoodCastro-UpperMarket      7    1       24  18.84206296 1
neighborhoodNobHill                21    1       23  14.79044745 1
lotsize                             3    1       21  10.94876414 1
month2008-07-01                    43    1       20   9.54292889 1
neighborhoodWesternAddition        38    1       19   7.47060804 1
longitude                           5    1       18   6.37068263 1
neighborhoodNorthBeach             23    1       16   4.64098864 1
neighborhoodPacificHeights         28    1       14   3.21207679 1
neighborhoodMarina                 19    1       13   3.25260354 0
neighborhoodRussianHill            32    1       12   3.02881439 1
month2008-04-01                    40    1       10   2.22407575 1
bedrooms                            1    1        8   1.20894174 1
neighborhoodInnerSunset            17    1        5   0.54773450 1
month2008-03-01                    39    1        4   0.38402626 1
neighborhoodChinatown               8    1        3   0.24940165 1
month2008-10-01                    46    1        2   0.15317304 1
month2008-05-01                    41    1        1   0.09138073 1
                                              rss
squarefeet                          100.0000000 1
neighborhoodPresidioHeights          65.9412651 1
latitude                             50.3490370 1
neighborhoodSeacliff                 39.2669043 1
neighborhoodNoeValley                28.3043535 1
neighborhoodCastro-UpperMarket       24.6223129 1
neighborhoodNobHill                  20.6738425 1
lotsize                              16.5523065 1
month2008-07-01                      14.9572215 1
neighborhoodWesternAddition          12.8021914 1
longitude                            11.4928253 1
neighborhoodNorthBeach                9.2983004 1
neighborhoodPacificHeights            7.3843377 1
neighborhoodMarina                    7.0666997 1
neighborhoodRussianHill               6.5297824 1
```

month2008-04-01	5.1687163	1
bedrooms	3.6503604	1
neighborhoodInnerSunset	2.1002700	1
month2008-03-01	1.6337090	1
neighborhoodChinatown	1.1922930	1
month2008-10-01	0.7831185	1
month2008-05-01	0.4026390	1

The function `plot.earth` will plot model selection, cumulative distribution of residuals, residuals versus fitted values, and the residual Q-Q plot for an **earth** object:

```
> plot(sf.price.model.earth)
```

The output of this call is shown in Figure 20-7. There are many options for this function that control the output; see the help file for more information. Another useful function for looking at **earth** objects is `plotmo`:

```
> plotmo(sf.price.model.earth)
```

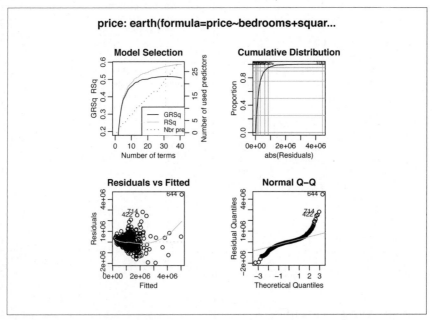

Figure 20-7. Output of plot.earth

The `plotmo` function plots the predicted model response when varying one or two predictors while holding other predictors constant. The output of `plotmo` for the San Francisco home sales data set is shown in Figure 20-8.

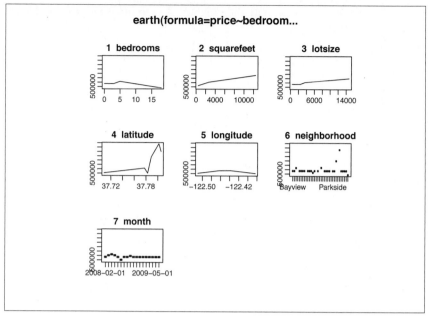

Figure 20-8. Output of plotmo

For the fun of it, let's look at the predictions from `earth`:

```
> calculate_rms_error(sf.price.model.earth,
+       na.omit(sanfrancisco.home.sales.training),
+       na.omit(sanfrancisco.home.sales.testing),
+       "price")
train.err  test.err
 435786.1  535941.5
```

Neural Networks

Neural networks are a very popular type of statistical model. Neural networks were originally designed to approximate how neurons work in the human brain; much of the original research on neural networks came from artificial intelligence researchers. Neural networks are very flexible and can be used to model a large number of different problems. By changing the structure of neural networks, it's possible to model some very complicated nonlinear relationships. Neural networks are so popular that there are entire academic journals devoted to them (such as *Neural Networks*, published by Elsevier).

The base distribution of R includes an implementation of one of the simplest types of neural networks: single-hidden-layer neural networks. Even this simple form of neural network can be used to model some very complicated relationships in data sets. Figure 20-9 is a graphical representation of what these neural networks look like. As you can see, each input value feeds into each "hidden layer" node. The output of each hidden-layer node feeds into each output node. What the modeling function actually does is to estimate the weights for each input into each hidden node and output node.

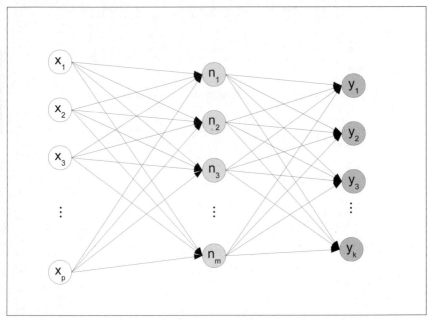

Figure 20-9. Single-hidden-layer, feed-forward neural network

The diagram omits two things: bias units and skip layer connectors. A *bias unit* is just a constant input term; it lets a constant term be mixed into each unit. *Skip layer connections* allow values from the inputs to be mixed into the outputs, skipping over the hidden layer. Both of these additions are included in the R implementation.

In equation form, here is the formula for neural network models:

$$y_k = g_0\left(\alpha_k + \sum_{i=1}^{i=m} w_{i,k}, g_i\left(\alpha_i + \sum_{j=1}^{j=p} w_{j,i} x_j\right)\right)$$

The function g_i used for the hidden nodes is the sigmoid function: $\sigma(x) = e^x/(1 + e^x)$. The function used for the output nodes is usually the identity function for regression, and the softmax function for classification. (We'll discuss the softmax function in "Neural Networks" on page 450.) For classification models, there are k outputs corresponding to the different levels. For regression models, there is only one output node.

To fit neural network models, use the function nnet in the package nnet:

```
library(nnet)
## S3 method for class 'formula':
nnet(formula, data, weights, ...,
     subset, na.action, contrasts = NULL)

## Default S3 method:
nnet(x, y, weights, size, Wts, mask,
     linout = FALSE, entropy = FALSE, softmax = FALSE,
     censored = FALSE, skip = FALSE, rang = 0.7, decay = 0,
     maxit = 100, Hess = FALSE, trace = TRUE, MaxNWts = 1000,
     abstol = 1.0e-4, reltol = 1.0e-8, ...)
```

Arguments to nnet include the following.

Argument	Description	Default
formula	A formula describing the relationship between the response and the predictor variables.	
data	A data frame containing the training data.	
weights	An optional vector of weights to use for the training data.	
...	Additional arguments passed to other functions (such as the nnet.default if using the nnet method, or optim).	
subset	An optional vector specifying the subset of observations to use in fitting the model.	
na.action	A function specifying how to treat missing values.	
contrasts	A list of factors to use for factors that appear in the model.	NULL
size	Number of units in the hidden layer.	
Wts	Initial parameter vector.	Randomly chosen, if not specified
mask	A logical vector indicating which parameters should be optimized.	All parameters
linout	Use linout=FALSE for logistic output units, linout=TRUE for linear units.	FALSE
entropy	A logical value specifying whether to use entropy/maximum conditional likelihood fitting.	FALSE
softmax	A logical value specifying whether to use a softmax/log-linear model and maximum conditional likelihood fitting.	FALSE
censored	A logical value specifying whether to treat the input data as censored data. (By default, a response variable value of c(1, 0, 1) means "both classes 1 and 3." If we treat the data as censored, then c(1, 0, 1) is interpreted to mean "not 2, but possibly 1 or 3."	FALSE
skip	A logical value specifying whether to add skip-layer connections from input to output.	FALSE
rang	A numeric value specifying the range for initial random weights. Weights are chosen between -rang and rang.	0.7
decay	A numeric parameter for weight decay.	0

Argument	Description	Default
maxit	Maximum number of iterations.	100
Hess	A logical value specifying whether to return the Hessian of fit.	FALSE
trace	A logical value specifying whether to print out a "trace" as nnet is running.	TRUE
maxNWts	A numeric value specifying the maximum number of weights.	1000
abstol	A numeric value specifying absolute tolerance. (Fitting process halts if the fit criterion falls below abstol.)	1.0e-4
reltol	A numeric value specifying relative tolerance. (Fitting process halts if the algorithm can't reduce the error by reltol in each step.)	1.0e-8

There is no simple, closed-form solution for finding the optimal weights for a neural network model. So, the nnet function uses the Broyden-Fletcher-Goldfarb-Shanno (BFGS) optimization method of the optim function to fit the model.

Let's try nnet on the San Francisco home sales data set. I had to play with the parameters a little bit to get a decent fit. I settled on 12 hidden units, linear outputs (which is appropriate for regression), skip connections, and a decay of 0.025:

```
> sf.price.model.nnet <- nnet(
+    price~bedrooms+squarefeet+lotsize+latitude+
+    longitude+neighborhood+month,
+    data=sanfrancisco.home.sales.training, size=12,
+    skip=TRUE, linout=TRUE, decay=0.025, na.action=na.omit)
# weights:  740
initial  value 1387941951981143.500000
iter  10 value 292963198488371.437500
iter  20 value 235738652534232.968750
iter  30 value 215547308140618.656250
iter  40 value 212019186628667.375000
iter  50 value 210632523063203.562500
iter  60 value 208381505485842.656250
iter  70 value 207265136422489.750000
iter  80 value 207023188781434.906250
iter  90 value 206897724524820.937500
iter 100 value 206849625163830.156250
final  value 206849625163830.156250
stopped after 100 iterations
```

To view the model, you can use the print or summary methods. Neither is particularly informative, though the summary method will show weights for all the units. Here is a small portion of the output for summary (the omitted portion is replaced with an ellipsis):

```
> summary(sf.price.model.nnet)
a 55-12-1 network with 740 weights
options were - skip-layer connections  linear output units  decay=0.025
     b->h1      i1->h1     i2->h1     i3->h1     i4->h1     i5->h1
     12.59       9.83    21398.35   29597.88     478.93   -1553.28
     i6->h1     i7->h1     i8->h1     i9->h1    i10->h1    i11->h1
     -0.15      -0.27       0.34      -0.05      -0.31       0.16
...
```

Here's how this model performed:

```
> calculate_rms_error(sf.price.model.nnet,
+       na.omit(sanfrancisco.home.sales.training),
+       na.omit(sanfrancisco.home.sales.testing),
+     "price")
train.err   test.err
 447567.2  566056.4
```

For more complex neural networks (such as networks with multiple hidden layers), see the packages AMORE, neural, and neuralnet.

Project Pursuit Regression

Projection pursuit regression is another very general model for representing non-linear relationships. Projection pursuit models have the form:

$$f(X) = \sum_{m=1}^{M} g_m(w_m^T X)$$

The functions g_m are called *ridge functions*. The project pursuit algorithm tries to optimize parameters for the parameters w_m by trying to minimize the sum of the residuals. In equation form:

$$\min_{g_m, w_m} \left(\sum_{i=1}^{N} \left[y_i - \sum_{m-1}^{M} g_m(w_m^T x_i) \right] \right)$$

Project pursuit regression is closely related to the neural network models that we saw above. (Note the similar form of the equations.) If we were to use the sigmoid function for the ridge functions g_m, projection pursuit would be identical to a neural network. In practice, projection pursuit regression is usually used with some type of smoothing method for the ridge functions. The default in R is to use Friedman's supersmoother function. (This function is actually pretty complicated and chooses the best of three relationships to pick the best smoothing function. See the help file for supsmu for more details. Note that this function finds the best smoother for the input data, not the smoother that leads to the best model.)

To use projection pursuit regression in R, use the function ppr:

```
## S3 method for class 'formula':
ppr(formula, data, weights, subset, na.action,
      contrasts = NULL, ..., model = FALSE)

## Default S3 method:
ppr(x, y, weights = rep(1,n),
      ww = rep(1,q), nterms, max.terms = nterms, optlevel = 2,
      sm.method = c("supsmu", "spline", "gcvspline"),
      bass = 0, span = 0, df = 5, gcvpen = 1, ...)
```

Arguments to ppr include the following.

Argument	Description	Default
formula/data/subset/ na.action, x/y	Specifies the data to use for modeling, depending on the form of the function.	
weights	A vector of weights for each case.	
contrasts	A list specifying the contrasts to use for factors.	NULL
model	A logical value indicating whether to return the model frame.	FALSE
ww	A vector of weights for each response.	rep(1, q)
nterms	Number of terms to include in the final model.	
max.terms	Maximum number of terms to choose from when building the model.	nterms
optlevel	An integer value between 0 and 3, which determines how optimization is done. See the help file for more information.	2
sm.method	A character value specifying the method used for smoothing the ridge functions. Specify sm.method="supsmu" for Friedman's supersmoother, sm.method="spline" to use the code from smooth.spline, or sm.method="gcvspline" to choose the smoothing method with gcv.	"supsmu"
bass	When sm.method="supsmu", a numeric value specifying the "bass" tone control for the supersmoother algorithm.	0
span	When sm.method="supsmu", a numeric value specifying the "span" control for the supersmoother.	0
df	When sm.method="spline", specifies the degrees of freedom for the spline function.	5
gcvpen	When sm.method="gcvspline", a numeric value specifying the penalty for each degree of freedom.	1
...		

Let's try projection pursuit regression on the home sales data:

```
> sf.price.model.ppr <- ppr(
+    price~bedrooms+squarefeet+lotsize+latitude+
+    longitude+neighborhood+month,
+    data=sanfrancisco.home.sales.training, nterms=20)
> sf.price.model.ppr
Call:
ppr(formula = price ~ bedrooms + squarefeet + lotsize + latitude +
        longitude + neighborhood + month,
    data = sanfrancisco.home.sales.training,
    nterms = 20)

Goodness of fit:
    20 terms
1.532615e+13
```

The summary function for ppr models prints out an enormous amount of information, including the function call, goodness-of-fit measurement, projection pursuit vectors, and coefficients of ridge terms; I have omitted the output from the book to save space.

You can plot the ridge functions from a `ppr` model using the `plot` function. To plot them all at the same time, I used the graphical parameter `mfcol=c(4, 4)` to plot them on a 4 × 4 grid. (I also narrowed the margins to make them easier to read.)

```
par(mfcol=c(4,4), mar=c(2.5,2.5,1.5,1.5))
plot(sf.price.model.ppr)
```

The ridge functions are shown in Figure 20-10. I picked 12 explanatory variables, which seemed to do best on the validation data (though not on the training data):

```
> calculate_rms_error(sf.price.model.ppr,
+       na.omit(sanfrancisco.home.sales.training),
+       na.omit(sanfrancisco.home.sales.testing),
+       "price")
train.err  test.err
194884.8  585613.9
```

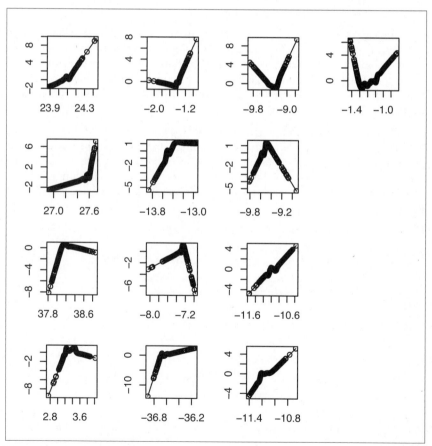

Figure 20-10. Ridge functions from projection pursuit model

Generalized Additive Models

Generalized additive models are another regression model technique for modeling complicated relationships in high-dimensionality data sets. Generalized additive models have the following form:

$$Y = \alpha + \sum_{j=1}^{p} f_j(X_j) + \varepsilon$$

Notice that each predictor variable x_j is first processed by a function f_j and is then used in a linear model. The generalized additive model algorithm finds the form of the functions f. These functions are often called *basis functions*.

The simplest way to fit generalized additive models in R is through the function gam in the library gam:

```
gam(formula, family = gaussian, data, weights, subset, na.action,
        start, etastart, mustart, control = gam.control(...),
        model=FALSE, method, x=FALSE, y=TRUE, ...)
```

This implementation is similar to the version from S and includes support for both local linear regression and smoothing spline basis functions. The gam package currently includes two different types of basis functions: smoothing splines and local regression. The gam function uses a back-fitting method to estimate parameters for the basis functions, and also estimates weights for the different terms in the model using penalized residual sum of squares.

When using the gam function to specify a model, you need to specify which type of basis function to use for which term. For example, suppose that you wanted to fit a model where the response variable was y, and the predictors were u, v, w, and x. To specify a model with smoothing functions for u and v, a local regression term for w, and an identity basis functions for x, you would specify the formula as y~s(u)+s(v)+lo(w)+x.

Here is a detailed description of the arguments to gam.

Argument	Description	Default
formula	A GAM formula specifying the form of the model. (See the help files for s and lo for more information on how to specify options for the basis functions.)	
family	A family object specifying the distribution and link function. See "Generalized Linear Models" on page 392 for a list of families.	gaussian()
data	A data frame containing the data to use for fitting.	list
weights	An (optional) numeric vector of weights for the input data.	NULL
subset	An optional vector specifying the subset of observations to use in fitting the model.	NULL
na.action	A function that indicates how to deal with missing values.	options("na.action"), which is na.omit by default

Argument	Description	Default
offset (through gam.fit)	A numeric value specifying an a priori known component to include in the additive predictor during fitting.	NULL
start	Starting values for the parameters in the additive predictors.	
etastart	Starting values for the additive predictors.	
mustart	Starting values for the vector of means.	
control	A list of parameters for controlling the fitting process. Use the function gam.control to generate a suitable list (and see the help file for that function to get the tuning parameters).	gam.control()
model	A logical value indicating whether the model frame should be included in the returned object.	FALSE
method	A character value specifying the method that should be used to fit the parametric part of the model. The only allowed values are method="glm.fit" (which uses iteratively reweighted least squares) or method="model.frame" (which does nothing except return the model frame).	NULL
x	A logical value specifying whether to return the X matrix (the predictors) with the model frame.	FALSE
y	A logical value specifying whether to return the Y vector (the response) with the model frame.	TRUE
...	Additional parameters passed to other methods (particularly, gam.fit).	

In R, there is an alternative implementation of generalized additive models available through the function gam in the package mgcv:

```
library(mgcv)
gam(formula,family=gaussian(),data=list(),weights=NULL,subset=NULL,
    na.action,offset=NULL,method="GCV.Cp",
    optimizer=c("outer","newton"),control=gam.control(),scale=0,
    select=FALSE,knots=NULL,sp=NULL,min.sp=NULL,H=NULL,gamma=1,
    fit=TRUE,paraPen=NULL,G=NULL,in.out,...)
```

This function allows a variety of different basis functions to be used: thin-plate regression splines (the default), cubic regression splines, and p-splines. The alternative gam function will estimate parameters for the basis functions as part of the fitting process using penalized likelihood maximization. The gam function in the mgcv package has many more options than the gam function in the gam package, but it is also a lot more complicated. See the help files in the mgcv package for more on the technical differences between the two packages.

Support Vector Machines

Support vector machines (SVMs) are a fairly recent algorithm for nonlinear models. They are a lot more difficult to explain to nonmathematicians than most statistical modeling algorithms. Explaining how SVMs work in detail is beyond the scope of this book, but here's a quick synopsis:

- SVMs don't rely on all of the underlying data to train the model. Only some observations (called the *support vectors*) are used. This makes SVMs somewhat resistant to outliers (like robust regression techniques) when used for regression. (It's also possible to use SVMs in the opposite way: to detect anomalies in the data.) You can control the range of values considered through the insensitive-loss function parameter `epsilon`.

- SVMs use a nonlinear transformation of the input data (like the basis functions in additive models or kernels in kernel methods). You can control the type of kernel used in SVMs through the parameter `kernel`.

- The final SVM model is fitted using a standard regression, with maximum likelihood estimates.

SVMs are black-box models; it's difficult to learn anything about a problem by looking at the parameters from a fitted SVM model. However, SVMs have become very popular, and many people have found that SVMs perform well in real-world situations. (An interesting side note is that SVMs are included as part of the Oracle Data Mining software, while many other algorithms are not.)

In R, SVMs are available in the library `e1071`,[#] through the function `svm`:

```
library(e1071)
## S3 method for class 'formula':
svm(formula, data = NULL, ..., subset, na.action =
na.omit, scale = TRUE)
## Default S3 method:
svm(x, y = NULL, scale = TRUE, type = NULL, kernel =
"radial", degree = 3, gamma = if (is.vector(x)) 1 else 1 / ncol(x),
coef0 = 0, cost = 1, nu = 0.5,
class.weights = NULL, cachesize = 40, tolerance = 0.001, epsilon = 0.1,
shrinking = TRUE, cross = 0, probability = FALSE, fitted = TRUE,
..., subset, na.action = na.omit)
```

Other implementations are available through the `ksvm` and `lssvm` functions in the `kernlab` library, `svmlight` in the `klaR` library, and `svmpath` in the `svmpath` library.

Let's try building an `svm` model for the home sales data:

```
> sf.price.model.svm <- svm(
+   price~bedrooms+squarefeet+lotsize+latitude+
+   longitude+neighborhood+month,
+   data=sanfrancisco.home.sales.training)
```

[#]Incidentally, this is, by far, the worst named package available on CRAN. It's named for a class given by the Department of Statistics, TU Wien. The package contains a number of very useful functions: SVM classifiers, algorithms for tuning other modeling functions, naive Bayes classifiers, and some other useful functions. It really should be called something like "veryusefulstatisticalfunctions."

Here is how the model performed:

```
> calculate_rms_error(sf.price.model.svm,
+       na.omit(sanfrancisco.home.sales.training),
+       na.omit(sanfrancisco.home.sales.testing),
+    "price")
train.err  test.err
 518647.9  641039.5
```

21

Classification Models

In Chapter 20, I provided an overview of R's statistical modeling software for regression problems. However, not all problems can be solved by predicting a continuous numerical quantity like a drug dose, or a person's wage, or the value of a customer. Often, an analyst's goal is to classify an item into a category or maybe to estimate the probability that an item belongs to a certain category. Models that describe this relationship are called *classification models*.

This chapter gives an overview of R's statistical modeling software for linear classification models.

Linear Classification Models

In this section, we'll look at a few popular linear classification models.

Logistic Regression

Suppose that you were trying to estimate the probability of a certain outcome (which we'll call A) for a categorical variable with two values. You could try to predict the probability of A as a linear function of the predictor variables, assuming $y = c_0 + c_1 x_1 + x_2 x_2 + \ldots + c_n x_n = \Pr(A)$. The problem with this approach is that the value of y is unconstrained; probabilities are only valid for values between 0 and 1. A good approach for dealing with this problem is to pick a function for y that varies between 0 and 1 for all possible predictor values. If we were to use that function as a link function in a general linear model, then we could build a model that estimates the probability of different outcomes. That is the idea behind logistic regression.

In a logistic regression, the relationship between the predictor variables and the probability that an observation is a member of a given class is given by the logistic function:

$$\Pr(A) = \frac{1}{1 + e^{-\eta}}$$

The logit function (which is used as the link function) is:

$$\text{logit}\left(\Pr(A)\right)=\ln\left(\frac{\Pr(A)}{1-\Pr(A)}\right)=\eta$$

Let's take a look at a specific example of logistic regression. In particular, let's look at the field goal data set. Each time a kicker attempts a field goal, there is a chance that the goal will be successful, and a chance that it will fail. The probability varies according to distance; attempts closer to the end zone are more likely to be successful. To model this relationship, we'll try to use a logistic regression. To begin, let's load the data and create a new binary variable for field goals that are either good or bad:

```
> library(nutshell)
> data(field.goals)
> field.goals.forlr <- transform(field.goals,
+   good=as.factor(ifelse(play.type=="FG good","good","bad")))
```

Let's take a quick look at the percentage of good field goals by distance. We'll start by tabulating the results with the table function:

```
> field.goals.table <- table(field.goals.forlr$good,
+                             field.goals.forlr$yards)
> field.goals.table
```

	18	19	20	21	22	23	24	25	26	27	28	29	30	31	32	33	34	35	36	37	38
bad	0	0	1	1	1	1	0	0	0	3	5	5	2	6	7	5	3	0	4	3	11
good	1	12	24	28	24	29	30	18	27	22	26	32	22	21	30	31	21	25	20	23	29

	39	40	41	42	43	44	45	46	47	48	49	50	51	52	53	54	55	56	57	58	59
bad	6	7	5	6	11	5	9	12	11	10	9	5	8	11	10	3	1	2	1	1	1
good	35	27	32	21	15	24	16	15	26	18	14	11	9	12	10	2	1	3	0	1	0

	60	61	62
bad	1	1	1
good	0	0	0

We'll also plot the results (as percentages):

```
> plot(colnames(field.goals.table),
+       field.goals.table["good",]/
+         (field.goals.table["bad",] +
+           field.goals.table["good",]),
+   xlab="Distance (Yards)", ylab="Percent Good")
```

The resulting plot is shown in Figure 21-1. As you can see, field goal percentage tapers off linearly between about 25 and 55 yards (with a few outliers at the end).

Each individual field goal attempt corresponds to a Bernoulli trial; the number of successful field goals at each position on the field will be given by a binomial distribution. So, we specify family="binomial" when calling glm. To model the probability of a successful field goal using a logistic regression, we would make the following call to glm:

```
> field.goals.mdl <- glm(formula=good~yards,
+    data=field.goals.forlr,
+    family="binomial")
```

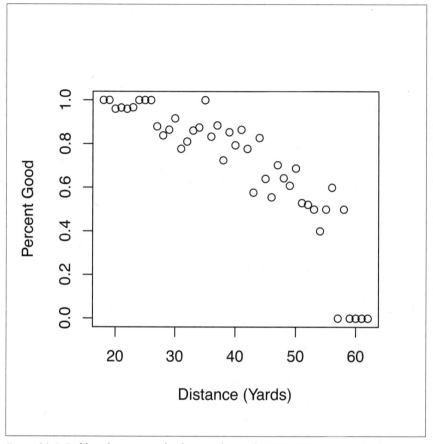

Figure 21-1. Field goal percentage by distance during the 2005 NFL season

Just like lm, the glm function returns no results by default. The print method will show some details about the model fit:

```
> field.goals.mdl

Call:  glm(formula = good ~ yards, family = "binomial",
           data = field.goals.forlr)

Coefficients:
(Intercept)        yards
    5.17886      -0.09726

Degrees of Freedom: 981 Total (i.e. Null);  980 Residual
Null Deviance:      978.9
Residual Deviance: 861.2 AIC: 865.2
```

And, as with lm, you can get more detailed results about the model object with the summary method:

```
> summary(field.goals.mdl)

Call:
glm(formula = good ~ yards, family = "binomial", data = field.goals.forlr)

Deviance Residuals:
    Min       1Q   Median       3Q      Max
-2.5582   0.2916   0.4664   0.6979   1.3790

Coefficients:
             Estimate Std. Error z value Pr(>|z|)
(Intercept)  5.178856   0.416201  12.443   <2e-16 ***
yards       -0.097261   0.009892  -9.832   <2e-16 ***
---
Signif. codes:  0 '***' 0.001 '**' 0.01 '*' 0.05 '.' 0.1 ' ' 1

(Dispersion parameter for binomial family taken to be 1)

    Null deviance: 978.90  on 981  degrees of freedom
Residual deviance: 861.22  on 980  degrees of freedom
AIC: 865.22

Number of Fisher Scoring iterations: 5
```

Let's take a quick look at how well this model fits the data. First, let's start by plotting the field goals from 2005 as above:

```
> plot(colnames(field.goals.table),
+      field.goals.table["good",]/
+        (field.goals.table["bad",] +
+          field.goals.table["good",]),
+    xlab="Distance (Yards)", ylab="Percent Good")
```

Next, we'll add a line to this chart showing the estimated probability of success at each point. We'll create a function to calculate the probability and then use that function to plot the curve:

```
> fg.prob <- function(y) {
+   eta <- 5.178856 + -0.097261 * y;
+   1 / (1 + exp(-eta))
+ }
> lines(15:65,fg.prob(15:65),new=TRUE)
```

The chart is shown in Figure 21-2. As expected from the statistics above, the model look like it fits the data reasonably well.

For more than two (unordered) categories, you need to use a different method to predict probabilities. One method is to use the multinom function:

```
multinom(formula, data, weights, subset, na.action,
         contrasts = NULL, Hess = FALSE, summ = 0, censored = FALSE,
         model = FALSE, ...)
```

This actually fits multinomial log-linear models using neural networks.

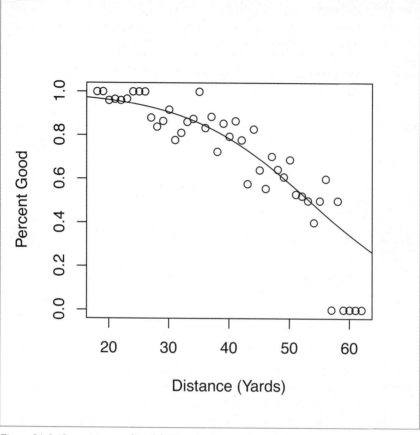

Figure 21-2. Comparing predicted field goal success to actual success

Here is a description of the arguments to the multinom function.

Argument	Description	Default
formula	A formula specifying the form of the model to fit.	
data	A data frame to use for the training data for the model.	
weights	An optional vector of weights for the training data.	
subset	An (optional) expression describing the set of observations to use for fitting the model.	
na.action	A function specifying how to treat missing values.	
contrast	A list of contrasts to use for factors appearing as variables in formula.	NULL
Hess	A logical value specifying whether the Hessian (observed observation matrix) should be returned.	FALSE
summ	An integer value describing the method used to summarize data. Use summ=0 not to summarize, summ=1 or summ=2 to replace duplicate observations with a single observation (and appropriately adjusting the weights), and summ=3 to also combine rows with the same predictor variables but different response variables.	0

Argument	Description	Default
censored	If the response variable is a matrix with more than two columns, changes how the values are interpreted. If censored=FALSE, values are interpreted as counts; if censored=TRUE, values of 1 are interpreted as possible values, and values of 0 as impossible.	FALSE
model	A logical value specifying whether the model matrix should be returned.	FALSE
...	Additional arguments passed to nnet.	

For more than two ordered categories, you can also use proportional odds linear regression. To do this in R, you can use the polr function in the MASS package:

```
polr(formula, data, weights, start, ..., subset, na.action,
     contrasts = NULL, Hess = FALSE, model = TRUE,
     method = c("logistic", "probit", "cloglog", "cauchit"))
```

Here is a description of the arguments to the polr function.

Argument	Description	Default
formula	A formula specifying the form of the model to fit.	
data	A data frame to use for the training data for the model. (If omitted, variables from the current environment are used instead.)	
weights	An optional vector of weights for the training data.	1
start	A vector of initial values for the parameters in the form c(coefficients, zeta).	
...	Additional arguments passed to the function optim.	
subset	An (optional) expression describing the set of observations to use for fitting the model.	
na.action	A function specifying how to treat missing values.	
contrasts	A list of contrasts to use for factors appearing as variables in formula.	NULL
Hess	A logical value specifying whether the Hessian (observed observation matrix) should be returned.	FALSE
model	A logical value specifying whether the model matrix should be returned.	TRUE
method	Specifies the form of the model. Use method="logistic" for logistic, method="probit" for probit, method="cloglog" for complementary log-log, and method="cauchit" for a Cauchy latent variable.	"logistic"

Linear Discriminant Analysis

Linear discriminant analysis (LDA) is a statistical technique for finding the linear combination of features that best separate observations into different classes. LDA assumes that the data in each class is normally distributed and that there is a unique covariance matrix for each class. To use linear discriminant analysis in R, use the function lda:

```
library(MASS)
## S3 method for class 'formula':
lda(formula, data, ..., subset, na.action)

## Default S3 method:
```

```
lda(x, grouping, prior = proportions, tol = 1.0e-4,
    method, CV = FALSE, nu, ...)

## S3 method for class 'data.frame':
lda(x, ...)

## S3 method for class 'matrix':
lda(x, grouping, ..., subset, na.action)
```

Here is a description of the arguments to the lda function.

Argument	Description	Default
formula	A formula specifying the form of the model to fit.	
data	If a formula is given, specifies a data frame for the training.	
x	Specifies a matrix or data frame for the fitting data (when no formula is provided).	
grouping	A factor specifying the response variable (when no formula is provided).	
prior	A vector of prior probabilities for class membership (in the same order as the levels of grouping).	proportions
tol	A numeric value specifying a tolerance for testing if the input data is a singular matrix; if the variance of any variable is less than tol^2, it will be rejected.	1.0e-4
subset	A vector specifying the set of observations in data to include.	
na.action	A function specifying how to deal with missing values.	
method	The method for fitting. Use method="moment" for standard estimators, method="mle" for MLEs, method="mve" to use cov.mve, or method="t" for estimates based on the *t*-distribution.	
CV	A logical value specifying whether to use "leave-one-out" cross-validation. (See the help file for more information.)	FALSE
nu	A numeric value specifying degrees of freedom when method="t".	
...	Arguments passed to other methods.	

A closely related function for classification is quadratic discriminant analysis (QDA), available through the function qda. QDA looks for a quadratic combination of features that best separate observations into different classes:

```
library(MASS)
qda(x, ...)

## S3 method for class 'formula':
qda(formula, data, ..., subset, na.action)

## Default S3 method:
qda(x, grouping, prior = proportions,
    method, CV = FALSE, nu, ...)

## S3 method for class 'data.frame':
qda(x, ...)

## S3 method for class 'matrix':
qda(x, grouping, ..., subset, na.action)
```

The arguments to `qda` are the same as the arguments to `lda`.

For the remainder of this chapter, I'll rely on a single data set for examples: the Spambase data set. The Spambase data set was created by Mark Hopkins, Erik Reeber, George Forman, and Jaap Suermondt at Hewlett-Packard Labs. It includes 4,601 observations corresponding to email messages, 1,813 of which are spam. From the original email messages, 58 different attributes were computed. This data set is really nice to use in examples because it's already been cleaned and preprocessed.

Here is how I loaded the raw data into R:

```
# code to load it in
spambase <- read.csv(
  file="~/Documents/book/data/spam/spambase.data.txt", header=FALSE)
names(spambase) <-
c("word_freq_make", "word_freq_address", "word_freq_all", "word_freq_3d",
  "word_freq_our", "word_freq_over", "word_freq_remove",
  "word_freq_internet", "word_freq_order", "word_freq_mail",
  "word_freq_receive", "word_freq_will", "word_freq_people",
  "word_freq_report", "word_freq_addresses", "word_freq_free",
  "word_freq_business", "word_freq_email", "word_freq_you",
  "word_freq_credit", "word_freq_your", "word_freq_font",
  "word_freq_000", "word_freq_money", "word_freq_hp", "word_freq_hpl",
  "word_freq_george", "word_freq_650", "word_freq_lab", "word_freq_labs",
  "word_freq_telnet", "word_freq_857", "word_freq_data", "word_freq_415",
  "word_freq_85", "word_freq_technology", "word_freq_1999",
  "word_freq_parts", "word_freq_pm", "word_freq_direct", "word_freq_cs",
  "word_freq_meeting", "word_freq_original", "word_freq_project",
  "word_freq_re", "word_freq_edu", "word_freq_table",
  "word_freq_conference", "char_freq_semicolon", "char_freq_left_paren",
  "char_freq_left_bracket", "char_freq_exclamation", "char_freq_dollar",
  "char_freq_pound", "capital_run_length_average",
  "capital_run_length_longest", "capital_run_length_total", "is_spam")
> spambase <- transform(spambase, is_spam=as.factor(is_spam))
```

I've included a copy with the `nutshell` package, so you can load this data set with the commands:

```
> library(nutshell)
> data(spambase)
```

To use this data set for our examples, we'll split it into training and validation data sets. We'll split the data set into 70% and 30% samples, stratified by the `is_spam` factor. To do this, we'll use the function `strata` in the `sampling` library to do the sampling:

```
> library(sampling)
> table(spambase$is_spam)

   0    1
2788 1813
> spambase.strata <- strata(spambase,
+   stratanames=c("is_spam"), size=c(1269, 1951), method="srswor")
```

This function returns a data frame that describes the set of values in the sample:

```
> names(spambase.strata)
[1] "is_spam" "ID_unit" "Prob"     "Stratum"
```

The variable ID_unit tells us the row numbers in the sample. To create training (and validation) data sets, we'll extract observations that match (or don't match) ID_unit values in the stratified sample:

```
> spambase.training <- spambase[
+   rownames(spambase) %in% spambase.strata$ID_unit,]
> spambase.validation <- spambase[
+   !(rownames(spambase) %in% spambase.strata$ID_unit),]
> nrow(spambase.training)
[1] 3220
> nrow(spambase.validation)
[1] 1381
```

Let's try quadratic discriminant analysis with the Spambase data set:

```
> spam.qda <- qda(formula=is_spam~., data=spambase.training)
> summary(spam.qda)
         Length Class  Mode
prior       2   -none- numeric
counts      2   -none- numeric
means     114   -none- numeric
scaling  6498   -none- numeric
ldet        2   -none- numeric
lev         2   -none- character
N           1   -none- numeric
call        3   -none- call
terms       3   terms  call
xlevels     0   -none- list
> # check with training
> table(actual=spambase.training$is_spam,
    predicted=predict(spam.qda,newdata=spambase.training)$class)
      predicted
actual    0    1
     0 1481  470
     1   56 1213
> # check with validation
> table(actual=spambase.validation$is_spam,
    predicted=predict(spam.qda,newdata=spambase.validation)$class)
      predicted
actual   0   1
     0 625 212
     1  28 516
```

Flexible discriminant analysis (FDA) is another technique related to LDA. This algorithm is based on the observation that LDA essentially fits a model by linear regression, so FDA substitutes a nonparametric regression for the linear regression. To compute flexible discriminant analysis:

```
libary(mda)
fda(formula, data, weights, theta, dimension, eps, method,
    keep.fitted, ...)
```

Repeating the example from above:

```
> spam.fda <- fda(formula=is_spam~., data=spambase.training)
> table(actual=spambase.validation$is_spam,
+    predicted=predict(spam.fda,newdata=spambase.validation,type="class"))
       predicted
actual   0   1
     0 800  37
     1 120 424
```

Another related technique is mixture discriminant analysis (MDA). MDA represents each class with a Gaussian mixture. This is available in R from the `mda` function in the `mda` library:

```
library(mda)
mda(formula, data, subclasses, sub.df, tot.df, dimension, eps,
    iter, weights, method, keep.fitted, trace, ...)
```

Here is an example using the Spambase data set:

```
> spam.mda <- mda(formula=is_spam~., data=spambase.training)
> table(actual=spambase.validation$is_spam,
+    predicted=predict(spam.mda,newdata=spambase.validation))
       predicted
actual   0   1
     0 800  37
     1 109 435
```

Log-Linear Models

There are several ways to fit log-linear models in R. One of the simplest is to use the function `loglin`:

```
loglin(table, margin, start = rep(1, length(table)), fit = FALSE,
       eps = 0.1, iter = 20, param = FALSE, print = TRUE)
```

The `loglin` function fits models using iterative proportional fitting (IPF). Here is a description of the arguments to the `loglin` function.

Argument	Description	Default
table	A contingency table to be fit	
margin	A list of vectors with the marginal totals to be fit	
start	A starting estimate for the fitted table	rep(1, length(table))
fit	A logical value specifying whether to return the fitted values	FALSE
eps	A numeric value specifying the maximum deviation allowed between observed and fitted margins	0.1
iter	A numeric value specifying the maximum number of iterations	20
param	A logical value specifying whether to return the parameter values	FALSE
print	A logical value specifying whether to print the number of iterations and the final deviation	TRUE

A more user friendly version is `loglm`:

```
library(MASS)
loglm(formula, data, subset, na.action, ...)
```

By using `loglm`, you can specify a data frame, a model formula, a subset of observations, and an action for NA variables, just like the `lm` function. (Other arguments are passed to `loglin`.)

An alternative method for fitting log-linear models is to use generalized linear models. See "Generalized Linear Models" on page 392 for more details.

Machine Learning Algorithms for Classification

Much like regression, there are problems where linear methods don't work well for classification. This section describes some machine learning algorithms for classification problems.

k Nearest Neighbors

One of the simplest techniques for classification problems is *k* nearest neighbors. Here's how the algorithm works:

1. The analyst specifies a "training" data set.
2. To predict the class of a new value, the algorithm looks for the *k* observations in the training set that are closest to the new value.
3. The prediction for the new value is the class of the "majority" of the *k* nearest neighbors.

To use *k* nearest neighbors in R, use the function `knn` in the `class` package:

```
libary(class)
knn(train, test, cl, k = 1, l = 0, prob = FALSE, use.all = TRUE)
```

Here is the description of the arguments to the `knn` function.

Argument	Description	Default
train	A matrix or data frame containing the training data.	
test	A matrix or data frame containing the test data.	
cl	A factor specifying the classification of observations in the training set.	
k	A numeric value specifying the number of neighbors to consider.	1
l	When $k > 0$, specifies the minimum vote for a decision. (If there aren't enough votes, the value doubt is returned.)	0
prob	If prob=TRUE, then the proportion of votes for the winning class is returned as attribute prob.	FALSE
use.all	Controls the handling of ties when selecting nearest neighbors. If use.all=TRUE, then all distances equal to the *k*th largest are included. If use.all=FALSE, then a random selection of observations is used to select *k* neighbors.	TRUE

Let's use knn to classify email messages as spam (or not spam) within the Spambase data set. Unlike some other model types in R, *k* nearest neighbors doesn't create a model object. Instead, you provide both the training and the test data as arguments to knn:

```
> spambase.knn <- knn(train=spambase.training,
+    test=spambase.validation,
+    cl=spambase.training$is_spam)
> summary(spambase.knn)
  0   1
861 520
```

The knn function returns an index of classes for each row in the test data. Let's compare the results returned by knn to the correct classification results in the original data:

```
> table(predicted=spambase.knn,actual=spambase.validation$is_spam)
         actual
predicted   0   1
        0 740 121
        1  97 423
```

As you can see, using *k* nearest neighbors with the default parameters correctly classifies 423 out of 544 messages as spam, but incorrectly classifies 97 out of 837 legitimate messages as spam.

As an alternative, suppose that we examined the five nearest neighbors, instead of just the nearest neighbor. To do this, we would set the argument k=5:

```
> spambase.knn5 <- knn(train=spambase.training,
+    test=spambase.validation,
+    cl=spambase.training$is_spam, k=5)
> summary(spambase.knn5)
  0   1
865 516
> table(predicted=spambase.knn5,actual=spambase.validation$is_spam)
         actual
predicted   0   1
        0 724 141
        1 113 403
```

Classification Tree Models

We introduced regression trees in "Regression Tree Models" on page 406. Classification trees work almost the same way. There are two key differences. First, CART uses a different error function to measure how well different splits divide the training data (or to measure cost/complexity trade-offs). Typically, Gini is used to measure cost/complexity. Second, CART uses a different method to choose predicted values. The predicted value at each terminal node is chosen by taking the most common value among the response values in the test data.

As an example of how to use recursive partitioning trees for classification, let's build a quick tree model on the Spambase data set (output modified slightly to fit on page):

```
> spam.tree <- rpart(is_spam~., data=spambase.training)
> spam.tree
n= 3220

node), split, n, loss, yval, (yprob)
      * denotes terminal node

 1) root 3220 1269 0 (0.60590062 0.39409938)
   2) char_freq_dollar< 0.0395 2361   529 0 (0.77594240 0.22405760)
     4) word_freq_remove< 0.065 2148   333 0 (0.84497207 0.15502793)
       8) char_freq_exclamation< 0.3905 1874   178 0 (0.905016 0.094984) *
       9) char_freq_exclamation>=0.3905 274   119 1 (0.43430657 0.56569343)
        18) capital_run_length_total< 65.5 141    42 0 (0.7021277 0.2978723)
          36) word_freq_free< 0.77 126    28 0 (0.77777778 0.22222222) *
          37) word_freq_free>=0.77 15     1 1 (0.06666667 0.93333333) *
        19) capital_run_length_total>=65.5 133    20 1 (0.150376 0.849624) *
     5) word_freq_remove>=0.065 213    17 1 (0.07981221 0.92018779) *
   3) char_freq_dollar>=0.0395 859   119 1 (0.13853318 0.86146682)
     6) word_freq_hp>=0.385 69     7 0 (0.89855072 0.10144928) *
     7) word_freq_hp< 0.385 790    57 1 (0.07215190 0.92784810) *
```

You can get much more detail about the tree object (and the process used to build it) by calling the summary method. I've omitted the results because they are quite lengthy.

You can use the printcp function to show the cp table for the fitted object:

```
> printcp(spam.tree)

Classification tree:
rpart(formula = is_spam ~ ., data = spambase.training)

Variables actually used in tree construction:
[1] capital_run_length_total char_freq_dollar
[3] char_freq_exclamation    word_freq_free
[5] word_freq_hp             word_freq_remove

Root node error: 1269/3220 = 0.3941

n= 3220

          CP nsplit rel error  xerror     xstd
1 0.489362      0   1.00000 1.00000 0.021851
2 0.141056      1   0.51064 0.51931 0.018041
3 0.043341      2   0.36958 0.37431 0.015857
4 0.036643      3   0.32624 0.34358 0.015300
5 0.010244      5   0.25296 0.28526 0.014125
6 0.010000      6   0.24271 0.27344 0.013866
```

Let's take a look at the generated tree:

```
> plot(spam.tree,uniform=TRUE)
> text(spam.tree,all=TRUE,cex=0.75,splits=TRUE,use.n=TRUE,xpd=TRUE)
```

The results are shown in Figure 21-3. The library maptree contains an alternative function for plotting classification trees. In many contexts, this function is more readable and easier to use. Here is an example for this tree (see Figure 21-4):

```
> library(maptree)
> draw.tree(spam.tree,cex=0.5,nodeinfo=TRUE,col=gray(0:8 / 8))
```

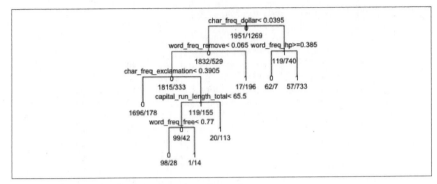

Figure 21-3. rpart tree for Spambase data, from plot.tree

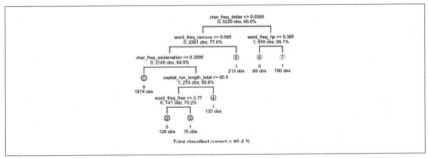

Figure 21-4. rpart tree for Spambase data, plotted by draw.tree

Let's take a look at how well the **rpart** model works:

```
> table(actual=spambase.validation$is_spam,
+   predicted=predict(spam.tree,newdata=spambase.validation,type="class"))
      predicted
actual   0   1
     0 795  42
     1  96 448
```

Bagging

To use bagging models in R for classification problems, you can use the function **bagging** in the package **adabag** (this function only works for classification, not regression):

```
library(adabag)
bagging(formula, data, mfinal = 100, minsplit = 5, cp = 0.01,
        maxdepth = nlevels(vardep))
```

Here are the results for the Spambase data set:

```
> spam.bag <- bagging(formula=is_spam~., data=spambase.training)
> summary(spam.bag)
           Length Class   Mode
formula         3 formula call
trees         100 -none-  list
votes        6440 -none-  numeric
class        3220 -none-  character
samples    322000 -none-  numeric
importance     57 -none-  numeric
> table(actual=spambase.training$is_spam,
+     predicted=predict(spam.bag,newdata=spambase.training)$class)
       predicted
actual    0    1
     0 1878   73
     1  344  925
> table(actual=spambase.validation$is_spam,
+     predicted=predict(spam.bag,newdata=spambase.validation)$class)
       predicted
actual    0    1
     0  804   33
     1  162  382
```

You can also try the function bagging in the ipred library, which we used in "Bagging for regression" on page 414.

Boosting

You can build boosting models for classification with the function ada in the package ada (this function does not work for regression problems):

```
## Default S3 method:
ada(x, y,test.x,test.y=NULL, loss=c("exponential","logistic"),
    type=c("discrete","real","gentle"),iter=50, nu=0.1, bag.frac=0.5,
    model.coef=TRUE,bag.shift=FALSE,max.iter=20,delta=10^(-10),
    verbose=FALSE,na.action=na.rpart,...)

## S3 method for class 'formula':
ada(formula, data, ..., subset, na.action=na.rpart)
```

Let's use ada to build a boosting model for the Spambase data set:

```
> spam.ada <- ada(formula=is_spam~., data=spambase.training,loss="logistic")
> spam.ada
Call:
ada(is_spam ~ ., data = spambase.training, loss = "logistic")

Loss: logistic Method: discrete   Iteration: 50

Final Confusion Matrix for Data:
          Final Prediction
True value    0    1
        0 1922   29
        1   48 1221

Train Error: 0.024
```

```
Out-Of-Bag Error:  0.038  iteration= 50

Additional Estimates of number of iterations:

train.err1 train.kap1
       48         48
```

Here is how ada performed on this problem:

```
> table(actual=spambase.training$is_spam,
+   predicted=predict(spam.ada,newdata=spambase.training))
      predicted
actual    0    1
     0 1922   29
     1   48 1221
> table(actual=spambase.validation$is_spam,
+   predicted=predict(spam.ada,newdata=spambase.validation))
      predicted
actual    0    1
     0  803   34
     1   36  508
```

As you can see, we achieved a very low error rate with boosting (4% false positive and 6.6% false negative), comparable with the results in the original study.

Additional implementations of boosting are available in the library mboost, which we introduced in "Boosting for regression" on page 415.

Neural Networks

We introduced neural network models in "Neural Networks" on page 423; see that section for a description of the arguments to nnet. As an example of how neural network models can be used for classification problems, we'll build a neural network model to classify messages as "spam" or "not spam" in the Spambase data set:

```
> spam.nnet <- nnet(is_spam~.,data=spambase.training,size=10,decay=0.1)
# weights:  591
initial  value 2840.007029
iter  10 value 1902.105150
iter  20 value 1086.933253
iter  30 value 724.134231
iter  40 value 682.122500
iter  50 value 607.033261
iter  60 value 550.845571
iter  70 value 520.489178
iter  80 value 483.315802
iter  90 value 449.411087
iter 100 value 438.685285
final  value 438.685285
stopped after 100 iterations
```

Let's take a look at how the neural network model performed:

```
> table(actual=spambase.training$is_spam,
+   predicted=predict(spam.nnet,type="class"))
      predicted
```

```
actual    0    1
     0 1889   62
     1   82 1187
> table(actual=spambase.validation$is_spam,
+    predicted=predict(spam.nnet,
+      newdata=spambase.validation,
+      type="class"))
       predicted
actual   0    1
     0 796   41
     1  39  505
```

Note that neural network algorithms are nondeterministic (they use some random values), so you might get different results even if you use the same code.

SVMs

Like neural networks, support vector machine models can also be used for either regression or classification. As an example of how to use SVMs for classification, we'll also use the Spambase data set:

```
> library(e1071)
> spam.svm <- svm(is_spam~.,data=spambase.training)
> spam.svm

Call:
svm(formula = is_spam ~ ., data = spambase.training)


Parameters:
   SVM-Type:  C-classification
 SVM-Kernel:  radial
       cost:  1
      gamma:  0.01754386

Number of Support Vectors:  975

> table(actual=spambase.validation$is_spam,
+    predicted=predict(spam.svm,
+      newdata=spambase.validation,
+      type="class"))
       predicted
actual   0    1
     0 807   30
     1  65  479
```

Random Forests

Random forests are another algorithm that can be used for either regression or classification problems. Here is how random forests can be used with the Spambase data set:

```
> library(randomForest)
randomForest 4.5-30
> spam.rf <- randomForest(is_spam~.,data=spambase.training)
```

```
> spam.rf

Call:
 randomForest(formula = is_spam ~ ., data = spambase.training)
               Type of random forest: classification
                     Number of trees: 500
No. of variables tried at each split: 7

        OOB estimate of  error rate: 5.16%
Confusion matrix:
      0    1 class.error
0 1890   61  0.03126602
1  105 1164  0.08274232
```

Notice the confusion matrix, showing how well the random forest performed on the training data. Let's take a look at how it did on the validation data:

```
> table(actual=spambase.validation$is_spam,
+   predicted=predict(spam.rf,
+     newdata=spambase.validation,
+     type="class"))
     predicted
actual   0   1
     0 812  25
     1  40 504
```

22

Machine Learning

This chapter covers machine learning algorithms that were not included in Chapter 20. In Chapters 20 and 21, we showed techniques for predicting values when you were interested in a specific value. This chapter shows methods for finding patterns in data when you aren't quite sure what you're looking for.

The techniques in this chapter are often called *data mining*. Data mining means something very simple: looking for patterns in data. Unfortunately, the term "data mining" now has negative connotations, much like the term "hacking" has negative connotations. When properly used, data mining algorithms can be a good technique when you are looking for patterns in large, unstructured data sources. R provides implementations of several popular data mining algorithms.

Market Basket Analysis

Association rules are a popular technique for data mining. The association rule algorithm was developed initially by Rakesh Agrawal, Tomasz Imielinski, and Arun Swami at the IBM Almaden Research Center.[*] It was originally designed as an efficient algorithm for finding interesting relationships in large databases of customer transactions. The algorithm finds sets of associations, items that are frequently associated with each other. For example, when analyzing supermarket data, you might find that consumers often purchase eggs and milk together. The algorithm was designed to run efficiently on large databases, especially databases that don't fit into a computer's memory.

R includes several algorithms implementing association rules. One of the most popular is the a priori algorithm. To try it in R, use the `apriori` function in the `arules` package:

```
library(arules)
apriori(data, parameter = NULL, appearance = NULL, control = NULL)
```

[*] You can read their paper here: *http://rakesh.agrawal-family.com/papers/sigmod93assoc.pdf*.

Here is a description of the arguments to `apriori`.

Argument	Description	Default
data	An object of class `transactions` (or a matrix or data frame that can be coerced into that form) in which associations are to be found.	
parameter	An object of class `ASParameter` (or a list with named components) that is used to specify mining parameters. Parameters include support level, minimum rule length, maximum rule length, and types of rules (see the help file for `ASParameter` for more information).	NULL
appearance	An object of class `APappearance` (or a list with named components) that is used to specify restrictions on the associations found by the algorithm.	NULL
control	An object of class `APcontrol` (or a list with named components) that is used to control the performance of the algorithm.	NULL

The `apriori` implementation is well engineered and thought out: it makes ample use of S4 classes to define data types (including a `transactions` class for data and classes to control parameters), and prints useful information when it is run. However, it currently requires data sets to be loaded completely into memory.

As an example, we will look at a set of transactions from Audioscrobbler. Audioscrobbler was an online service that tracked the listening habits of users. The company is now part of Last.fm and still provides application programming interfaces (APIs) for looking at music preferences. However, in 2005, the company released a database of information on music preferences under a Creative Commons license. The database consists of a set of records showing how many times each user listened to different artists. For our purposes, we'll ignore the count and just look at users and artists. For this example, I used a random sample of 20,000 user IDs from the database. Specifically, we will try to look for patterns in the artists that users listen to.

I loaded the data into R using the `read.transactions` function (in the `arules` package):

```
> library(arules)
> audioscrobbler <- read.transactions(
+   file="~/Documents/book/data/profiledata_06-May-2005/transactions.csv",
+   format="single",
+   sep=",",
+   cols=c(1,2))
```

You can find the data in the `nutshell` package:

```
> library(nutshell)
> data(audioscrobbler)
```

To find some results, I needed to change the default settings. I looked for associations at a 6.45% support level, which I specified through the `parameter` argument. (Why 6.45%? Because that returned exactly 10 rules on the test data, and 10 rules seemed like the right length for an example.)

```
> audioscrobbler.apriori <- apriori(
+   data=audioscrobbler,
+   parameter=new("APparameter",support=0.0645)
  )
```

```
parameter specification:
 confidence minval smax arem  aval originalSupport support minlen
       0.8    0.1    1 none FALSE            TRUE  0.0645      1
 maxlen target    ext
      5  rules FALSE

algorithmic control:
 filter tree heap memopt load sort verbose
    0.1 TRUE TRUE  FALSE TRUE    2    TRUE

apriori - find association rules with the apriori algorithm
version 4.21 (2004.05.09)        (c) 1996-2004   Christian Borgelt
set item appearances ...[0 item(s)] done [0.00s].
set transactions ...[429033 item(s), 20001 transaction(s)] done [2.36s].
sorting and recoding items ... [287 item(s)] done [0.16s].
creating transaction tree ... done [0.03s].
checking subsets of size 1 2 3 4 done [0.25s].
writing ... [10 rule(s)] done [0.00s].
creating S4 object  ... done [0.17s].
```

As you can see, the `apriori` function includes some information on what it is doing while running. After it finishes, you can inspect the returned object to learn more. The returned object consists of association rules (and is an object of class `arules`). Like most modeling algorithms, the object has an informative `summary` function that tells you about the rules:

```
> summary(audioscrobbler.apriori)
set of 10 rules

rule length distribution (lhs + rhs):sizes
 3
10

   Min. 1st Qu.  Median    Mean 3rd Qu.    Max.
      3       3       3       3       3       3

summary of quality measures:
    support              confidence              lift
 Min.   :0.06475   Min.   :0.8008   Min.   :2.613
 1st Qu.:0.06536   1st Qu.:0.8027   1st Qu.:2.619
 Median :0.06642   Median :0.8076   Median :2.651
 Mean   :0.06640   Mean   :0.8128   Mean   :2.696
 3rd Qu.:0.06707   3rd Qu.:0.8178   3rd Qu.:2.761
 Max.   :0.06870   Max.   :0.8399   Max.   :2.888

mining info:
          data ntransactions support confidence
  audioscrobbler         20001  0.0645        0.8
```

You can view the returned rules with the `inspect` function:

```
> inspect(audioscrobbler.apriori)
  lhs                        rhs          support confidence      lift
1 {Jimmy Eat World,
   blink-182}            => {Green Day} 0.06524674  0.8085502 2.780095
```

```
 2 {The Strokes,
    Coldplay}              => {Radiohead} 0.06619669   0.8019382 2.616996
 3 {Interpol,
    Beck}                  => {Radiohead} 0.06474676   0.8180670 2.669629
 4 {Interpol,
    Coldplay}              => {Radiohead} 0.06774661   0.8008274 2.613371
 5 {The Beatles,
    Interpol}              => {Radiohead} 0.06719664   0.8047904 2.626303
 6 {The Offspring,
    blink-182}             => {Green Day} 0.06664667   0.8399496 2.888058
 7 {Foo Fighters,
    blink-182}             => {Green Day} 0.06669667   0.8169014 2.808810
 8 {Pixies,
    Beck}                  => {Radiohead} 0.06569672   0.8066298 2.632306
 9 {The Smashing Pumpkins,
    Beck}                  => {Radiohead} 0.06869657   0.8287093 2.704359
10 {The Smashing Pumpkins,
    Pink Floyd}            => {Radiohead} 0.06514674   0.8018462 2.616695
```

The lefthand side of the rules (lhs) forms the predicate of the rule; the righthand side (rhs) forms the conclusion. For example, consider rule 1. This rule means "if the user has listened to Jimmy Eat World and Blink-182, then for 6.524675% of the time, he or she also listened to Green Day." You can draw your own conclusions about whether these results mean anything, other than that Audioscrobbler's users were fans of alternative and classic rock.

The arules package also includes an implementation of the Eclat algorithm, which finds frequent item sets. To find item sets using the Eclat algorithm, try the function eclat:

```
eclat(data, parameter = NULL, control = NULL
```

The eclat function accepts similar arguments as apriori (some of the parameters within the arguments are slightly different). I tightened up the support level for the eclat function in order to keep the number of results low. If you keep the default parameters, then the algorithm will return item sets with only one item, which is not very interesting. So, I set the minimum length to 2, and the support level to 12.9%. Here is an example of running eclat on the Audioscrobbler data:

```
> audioscrobbler.eclat <- eclat(
+     data=audioscrobbler,
+     parameter=new("ECparameter",support=0.129,minlen=2)
+ )

parameter specification:
 tidLists support minlen maxlen          target   ext
    FALSE   0.129      2      5 frequent itemsets FALSE

algorithmic control:
 sparse sort verbose
      7   -2    TRUE

eclat - find frequent item sets with the eclat algorithm
version 2.6 (2004.08.16)        (c) 2002-2004   Christian Borgelt
create itemset ...
```

```
set transactions ...[429033 item(s), 20001 transaction(s)] done [2.44s].
sorting and recoding items ... [74 item(s)] done [0.14s].
creating bit matrix ... [74 row(s), 20001 column(s)] done [0.01s].
writing  ... [10 set(s)] done [0.01s].
Creating S4 object  ... done [0.02s].
```

You can view information about the results with the **summary** function:

```
> summary(audioscrobbler.eclat)
set of 10 itemsets

most frequent items:
           Green Day              Radiohead Red Hot Chili Peppers
                   5                      5                      3
             Nirvana          The Beatles              (Other)
                   3                      2                      2

element (itemset/transaction) length distribution:sizes
 2
10

  Min. 1st Qu.  Median   Mean 3rd Qu.   Max.
     2       2       2      2       2      2

summary of quality measures:
    support
 Min.   :0.1291
 1st Qu.:0.1303
 Median :0.1360
 Mean   :0.1382
 3rd Qu.:0.1394
 Max.   :0.1567

includes transaction ID lists: FALSE

mining info:
          data ntransactions support
 audioscrobbler        20001   0.129
```

You can also view the item sets with the **inspect** function:

```
> inspect(audioscrobbler.eclat)
   items                      support
1  {Red Hot Chili Peppers,
    Radiohead}                0.1290935
2  {Red Hot Chili Peppers,
    Green Day}                0.1397430
3  {Red Hot Chili Peppers,
    Nirvana}                  0.1336433
4  {Nirvana,
    Radiohead}                0.1384931
5  {Green Day,
    Nirvana}                  0.1382931
6  {Coldplay,
    Radiohead}                0.1538423
7  {Coldplay,
    Green Day}                0.1292435
```

```
 8 {Green Day,
    Radiohead}          0.1335433
 9 {The Beatles,
    Green Day}          0.1290935
10 {The Beatles,
    Radiohead}          0.1566922
```

As above, you can draw your own conclusions about whether the results are interesting.

Clustering

Another important data mining technique is clustering. Clustering is a way to find similar sets of observations in a data set; groups of similar observations are called *clusters*. There are several functions available for clustering in R.

Distance Measures

To effectively use clustering algorithms, you need to begin by measuring the distance between observations. A convenient way to do this in R is through the function `dist` in the `stats` package:

```
dist(x, method = "euclidean", diag = FALSE, upper = FALSE, p = 2)
```

The `dist` function computes the distance between pairs of objects in another object, such as as matrix or a data frame. It returns a distance matrix (an object of type "dist") containing the computed distances. Here is a description of the arguments to `dist`.

Argument	Description	Default
x	The object on which to compute distances. Must be a data frame, matrix, or "dist" object.	
method	The method for computing distances. Specify method="euclidean" for Euclidean distances (2-norm), method="maximum" for the maximum distance between observations (supremum norm), method="manhattan" for the absolute distance between two vectors (1-norm), method="canberra" for Canberra distances (see the help file), method="binary" to regard nonzero values as 1 and zeros as 0, or method="minkowski" to use the p-norm (the pth root of the sum of the pth powers of the differences of the components).	"eucli- dean"
diag	A logical value specifying whether the diagonal of the distance matrix should be printed by print.dist.	FALSE
upper	A logical value specifying whether the upper triangle of the distance matrix should be printed.	FALSE
p	The power of the Minkowski distance (when method="minkowski").	2

An alternative method for computing distances between points is the `daisy` function in the `cluster` package:

```
daisy(x, metric = c("euclidean", "manhattan", "gower"),
      stand = FALSE, type = list())
```

The `daisy` function computes the pairwise dissimilarities between observations in a data set. Here is a description of the arguments to `daisy`.

Argument	Description	Default
x	A numeric matrix or data frame on which to compute distances.	
metric	A character value specifying the distance metric to use. Specify `metric="euclidean"` for Euclidean distances, `metric="manhattan"` for Manhattan distances (like walking around blocks in Manhattan), or `metric="gower"` to use Gower's distance.	"euclidean"
stand	A logical flag indicating whether to standardize measurements before computing distances.	FALSE
type	A list of values specifying the types of variables in x. Use `"ordratio"` for ratio-scaled variables to be treated as ordinal variables, `"logratio"` for ratio-scaled variables that must be logarithmically transformed, `"assym"` for asymmetric binary, and `"symm"` for symmetric binary.	

Clustering Algorithms

k-means clustering is one of the simplest clustering algorithms. To use k-means clustering, use the function `kmeans` from the `stats` package:

```
kmeans(x, centers, iter.max = 10, nstart = 1,
       algorithm = c("Hartigan-Wong", "Lloyd", "Forgy",
                     "MacQueen")
```

Here is a description of the arguments to `kmeans`.

Argument	Description	Default
x	A numeric matrix (or an object that can be coerced to a matrix) on which to cluster.	
centers	If a numeric value, specifies the number of clusters. If a numeric vector, specifies the initial cluster centers.	
iter.max	A numeric value specifying the maximum number of iterations.	10
nstart	Specifies the number of random sets to choose (if centers is a number).	1
algorithm	A character value specifying the clustering algorithm to use. Legal values include `algorithm="Hartigan-Wong"`, `algorithm="Lloyd"`, `algorithm="Forgy"`, `algorithm="MacQueen"`.	"Hartigan-Wong"

As an example, let's try building clusters on the San Francisco home sales data set. First, we need to create a distance matrix from the data frame. To do this, we'll need to only include a subset of variables:

```
> sf.dist <- daisy(
+   na.omit(sanfrancisco.home.sales[,
+     c("price", "bedrooms", "squarefeet", "lotsize",
+       "year", "latitude", "longitude")]),
+   metric="euclidean",
+   stand=TRUE
+ )
> summary(sf.dist)
973710 dissimilarities, summarized :
```

```
      Min.   1st Qu.    Median      Mean   3rd Qu.      Max.
  0.015086   3.167900  4.186900  4.617100  5.432400 25.519000
Metric :   euclidean
Number of objects :  1396
```

Next, we'll try *k*-means clustering. After some experimentation with different numbers of clusters, I found that six clusters gave some interesting results:

```
> sf.price.model.kmeans <- kmeans(sf.dist,centers=6)
> sf.price.model.kmeans$size
[1] 502    4 324 130  42 394
> sf.price.model.kmeans$withinss
[1] 346742.69  26377.99 446048.17 254858.23 211858.99 280531.60
```

Let's label the original data set with the clusters so that we can show summary statistics for each cluster:

```
> sanfrancisco.home.sales$cluster <- NA
> for (i in names(sf.price.model.kmeans$cluster)) {
+   sanfrancisco.home.sales[i,"cluster"] <-
+     sf.price.model.kmeans$cluster[i]
+ }
```

Here are the mean values for each cluster:

```
> by(sanfrancisco.home.sales[ ,
+     c("price", "bedrooms", "squarefeet",
+       "lotsize", "year", "latitude", "longitude") ],
+   INDICES=sanfrancisco.home.sales$cluster,
+   FUN=mean)

sanfrancisco.home.sales$cluster: 1
        price       bedrooms     squarefeet       lotsize         year
  620227.091633      1.123506   1219.633466   2375.193227  1933.109562
     latitude      longitude
    37.729114    -122.428059
------------------------------------------------------------
sanfrancisco.home.sales$cluster: 2
        price       bedrooms     squarefeet       lotsize         year
 7258750.00000        7.25000   7634.75000    5410.25000   1926.75000
     latitude      longitude
    37.79023      -122.44317
------------------------------------------------------------
sanfrancisco.home.sales$cluster: 3
        price       bedrooms     squarefeet       lotsize         year
  1.151657e+06   2.040123e+00  2.150068e+03  3.003188e+03  1.931238e+03
     latitude      longitude
  3.776289e+01  -1.224434e+02
------------------------------------------------------------
sanfrancisco.home.sales$cluster: 4
        price       bedrooms     squarefeet       lotsize         year
  1.571292e+06   2.907692e+00  2.718185e+03  4.677015e+03  1.934446e+03
     latitude      longitude
  3.777158e+01  -1.224429e+02
------------------------------------------------------------
sanfrancisco.home.sales$cluster: 5
        price       bedrooms     squarefeet       lotsize         year
```

```
  2.297417e+06  2.928571e+00  4.213286e+03  6.734905e+03  1.924929e+03
       latitude      longitude
  3.777424e+01 -1.224362e+02
------------------------------------------------------------
sanfrancisco.home.sales$cluster: 6
         price       bedrooms     squarefeet        lotsize           year
  886409.898477      1.284264   1518.230964    2857.159898    1931.637056
       latitude      longitude
      37.752869    -122.474225
```

As an alternative, you may want to try partitioning around medoids, which is a more robust version of *k*-means clustering. To use this algorithm in R, try the pam function in the cluster library:

```
libary(cluster)
pam(x, k, diss = inherits(x, "dist"), metric = "euclidean",
    medoids = NULL, stand = FALSE, cluster.only = FALSE,
    do.swap = TRUE,
    keep.diss = !diss && !cluster.only && n < 100,
    keep.data = !diss && !cluster.only, trace.lev = 0)
```

Let's try pam on the San Francisco home sales data set:

```
> sf.price.model.pam <- pam(sf.dist,k=6)
```

There is a plot method for partition objects (like the object returned by pam), which will display some useful information about the clusters:

```
> plot(sf.price.model.pam)
```

The results of this call are shown in Figure 22-1. The call produces two different plots: a cluster plot and a silhouette plot.

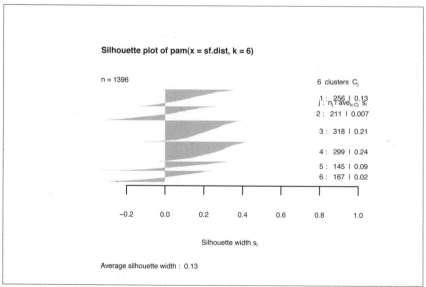

Figure 22-1. Information about San Francisco house price pam model

Many other clustering algorithms are available in R:

- Agglomerative clustering is available through the `agnes` function in the `cluster` package.
- Divisive hierarchical clustering is available through the `diana` function in the `cluster` package or through `mona` (if only binary variables are used).
- Fuzzy clustering is available through the `fanny` function in the `cluster` package.
- Self-organizing maps are available through the `batchSOM` and `SOM` functions in the `class` package.

23

Time Series Analysis

Time series are a little different from other types of data. Time series data often has long-term trends or periodic patterns that traditional summary statistics don't capture. To find these patterns, you need to use different types of analyses. As an example of a time series, we will revisit the turkey price data that we first saw in "Time Series" on page 89.

Autocorrelation Functions

One important property of a time series is the autocorrelation function. You can estimate the autocorrelation function for time series using R's `acf` function:

```
acf(x, lag.max = NULL,
    type = c("correlation", "covariance", "partial"),
    plot = TRUE, na.action = na.fail, demean = TRUE, ...)
```

The function `pacf` is an alias for `acf`, except with the default type of `"partial"`:

```
pacf(x, lag.max, plot, na.action, ...)
```

By default, this function plots the results. (An example plot is shown in "Plotting Time Series" on page 218.) As an example, let's show the autocorrelation function of the turkey price data:

```
> library(nutshell)
> data(turkey.price.ts)
> acf(turkey.price.ts,plot=FALSE)

Autocorrelations of series 'turkey.price.ts', by lag

0.0000 0.0833 0.1667 0.2500 0.3333 0.4167 0.5000 0.5833 0.6667 0.7500
 1.000  0.465 -0.019 -0.165 -0.145 -0.219 -0.215 -0.122 -0.136 -0.200
0.8333 0.9167 1.0000 1.0833 1.1667 1.2500 1.3333 1.4167 1.5000 1.5833
-0.016  0.368  0.723  0.403 -0.013 -0.187 -0.141 -0.180 -0.226 -0.130

> pacf(turkey.price.ts,plot=FALSE)
```

```
Partial autocorrelations of series 'turkey.price.ts', by lag

0.0833 0.1667 0.2500 0.3333 0.4167 0.5000 0.5833 0.6667 0.7500 0.8333
 0.465 -0.300 -0.020 -0.060 -0.218 -0.054 -0.061 -0.211 -0.180  0.098
0.9167 1.0000 1.0833 1.1667 1.2500 1.3333 1.4167 1.5000 1.5833
 0.299  0.571 -0.122 -0.077 -0.075  0.119  0.064 -0.149 -0.061
```

The function ccf plots the cross-correlation function for two time series:

```
ccf(x, y, lag.max = NULL, type = c("correlation", "covariance"),
    plot = TRUE, na.action = na.fail, ...)
```

By default, this function will plot the results. You can suppress the plot (to just view the function) with the argument plot=FALSE.

As an example of cross-correlations, we can use average ham prices in the United States. These are included in the nutshell package as ham.price.ts:

```
> library(nutshell)
> data(ham.price.ts)
> ccf(turkey.price.ts, ham.price.ts, plot=FALSE)

Autocorrelations of series 'X', by lag

-1.0833 -1.0000 -0.9167 -0.8333 -0.7500 -0.6667 -0.5833 -0.5000 -0.4167
  0.147   0.168  -0.188  -0.259  -0.234  -0.098  -0.004   0.010   0.231
-0.3333 -0.2500 -0.1667 -0.0833  0.0000  0.0833  0.1667  0.2500  0.3333
  0.228   0.059  -0.038   0.379   0.124  -0.207  -0.315  -0.160  -0.084
 0.4167  0.5000  0.5833  0.6667  0.7500  0.8333  0.9167  1.0000  1.0833
 -0.047  -0.005   0.229   0.223  -0.056  -0.099   0.189   0.039  -0.108
```

You can apply filters to a time series with the filter function or convolutions (using fast Fourier transforms [FFTs]) with the convolve function.

Time Series Models

Time series models are a little different from other models that we've seen in R. With most other models, the goal is to predict a value (the response variable) from a set of other variables (the predictor variables). Usually, we explicitly assume that there is no autocorrelation: that the sequence of observations does not matter.

With time series, we assume the opposite: we assume that previous observations help predict future observations (see Figure 23-1).

To fit an autoregressive model to a time series, use the function ar:

```
ar(x, aic = TRUE, order.max = NULL,
   method=c("yule-walker", "burg", "ols", "mle", "yw"),
   na.action, series, ...)
```

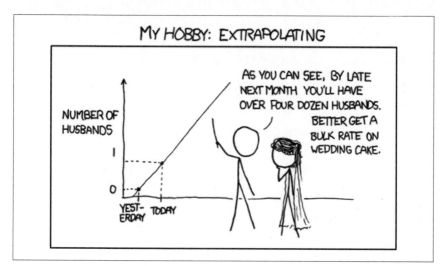

MY HOBBY: EXTRAPOLATING

NUMBER OF
HUSBANDS

AS YOU CAN SEE, BY LATE
NEXT MONTH YOU'LL HAVE
OVER FOUR DOZEN HUSBANDS.
BETTER GET A
BULK RATE ON
WEDDING CAKE.

YEST-
ERDAY TODAY

Figure 23-1. Extrapolating times series

Here is a description of the arguments to `ar`.

Argument	Description	
x	A time series.	
aic	A logical value that specifies whether the Akaike information criterion is used to choose the order of the model.	TRUE
order.max	A numeric value specifying the maximum order of the model to fit.	NULL
method	A character value that specifies the method to use for fitting the model. Specify `method="yw"` (or `method="yule-walker"`) for the Yule-Walker method, `method="burg"` for the Burg method, `method="ols"` for ordinary least squares, or `method="mle"` for maximum likelihood estimation.	c("yule-walker", "burg", "ols", "mle", "yw")
na.action	A function that specifies how to handle missing values.	
series	A character vector of names for the series.	
demean	A logical value specifying if a mean should be estimated during fitting.	
var.method	Specifies the method used to estimate the innovations variance when `method="ar.burg"`.	
...	Additional arguments, depending on method.	

The `ar` function actually calls one of four other functions, depending on the fit method chosen: `ar.yw`, `ar.burg`, `ar.ols`, or `ar.mle`. As an example, let's fit an autoregressive model to the turkey price data:

```
> library(nutshell)
> data(turkey.price.ts)
> turkey.price.ts.ar <- ar(turkey.price.ts)
> turkey.price.ts.ar

Call:
ar(x = turkey.price.ts)
```

```
Coefficients:
      1        2        3        4        5        6        7
 0.3353  -0.1868  -0.0024   0.0571  -0.1554  -0.0208   0.0914
      8        9       10       11       12
-0.0658  -0.0952   0.0649   0.0099   0.5714

Order selected 12  sigma^2 estimated as  0.05182
```

You can use the model to predict future values. To do this, use the **predict** function. Here is the method for **ar** objects:

```
predict(object, newdata, n.ahead = 1, se.fit = TRUE, ...)
```

The argument **object** specifies the model object to use for prediction. You can use **newdata** to specify new data to use for prediction, or **n.ahead** to specify a number of periods ahead to predict. The argument **se.fit** specifies whether to return standard errors of the prediction error.

Here is a forecast for the next 12 months for turkey prices:

```
> predict(turkey.price.ts.ar,n.ahead=12)
$pred
           Jan       Feb       Mar       Apr       May       Jun
2008                                           1.8827277 1.7209182
2009 1.5439290 1.6971933 1.5849406 1.7800358
           Jul       Aug       Sep       Oct       Nov       Dec
2008 1.7715016 1.9416776 1.7791961 1.4822070 0.9894343 1.1588863
2009

$se
           Jan       Feb       Mar       Apr       May       Jun
2008                                           0.2276439 0.2400967
2009 0.2450732 0.2470678 0.2470864 0.2480176
           Jul       Aug       Sep       Oct       Nov       Dec
2008 0.2406938 0.2415644 0.2417360 0.2429339 0.2444610 0.2449850
2009
```

To take a look at a forecast from an autoregressive model, you can use the function **ts.plot**. This function plots multiple time series on a single chart, even if the times are not overlapping. You can specify colors, line types, or other characteristics of each series as vectors; the ith place in the vector determines the property for the ith series.

Here is how to plot the turkey price time series as a solid line, and a projection 24 months into the future as a dashed line:

```
ts.plot(turkey.price.ts,
        predict(turkey.price.ts.ar,n.ahead=24)$pred,
        lty=c(1:2))
```

The plot is shown in Figure 23-2. You can also fit autoregressive integrated moving average (ARIMA) models in R using the **arima** function:

```
arima(x, order = c(0, 0, 0),
      seasonal = list(order = c(0, 0, 0), period = NA),
      xreg = NULL, include.mean = TRUE,
```

```
transform.pars = TRUE,
fixed = NULL, init = NULL,
method = c("CSS-ML", "ML", "CSS"),
n.cond, optim.method = "BFGS",
optim.control = list(), kappa = 1e6)
```

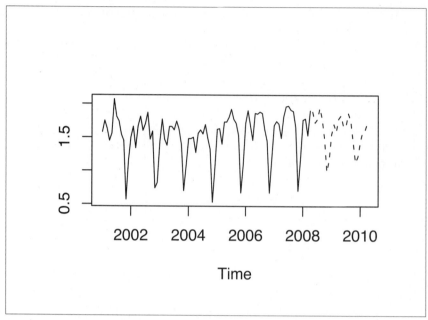

Figure 23-2. Forecast of turkey prices using an autoregressive model

Here is a description of the arguments to `arima`.

Argument	Description	Default
x	A time series.	
order	A numeric vector (p, d, q), where p is the AR order, d is the degree of differencing, and q is the MA order.	c(0, 0, 0)
seasonal	A list specifying the seasonal part of the model. The list contains two parts: the order and the period.	list(order = c(0, 0, 0), period = NA)
xreg	An (optional) vector or matrix of external regressors (with the same number of rows as x).	NULL
include.mean	A logical value specifying whether the model should include a mean/intercept term.	TRUE
tranform.pars	A logical value specifying whether the AR parameters should be transformed to ensure that they remain in the region of stationarity.	TRUE
fixed	An optional numeric vector specifying fixed values for parameters. (Only NA values are varied.)	NULL
init	A numeric vector of initial parameter values.	NULL

Argument	Description	Default
method	A character value specifying the fitting method to use. The default setting, method="CSS-ML", uses conditional sum of squares to find starting values, then maximum likelihood. Specify method="ML" for maximum likelihood only, or method="CSS" for conditional sum of squares only.	c("CSS-ML", "ML", "CSS")
n.cond	A numeric value indicating the number of initial values to ignore (only used for conditional sum of squares).	
optim.method	A character value that is passed to optim as method.	"BFGS"
optim.control	A list of values that is passed to optim as control.	list()
kappa	The prior variance for the past observations in a differenced model. See the help file for more information.	1e-6

The arima function uses the optim function to fit models. You can use the result of an ARIMA model to smooth a time series with the tsSmooth function. For more information, see the help file for tsSmooth.

24

Bioconductor

Most of this book is applicable across multiple areas of study, but this chapter focuses on a single field: bioinformatics. In particular, we're going to focus on the Bioconductor project. Bioconductor is an open source software project for analyzing genomic data in R. Initially, it focused on just gene expression data, but now includes tools for analyzing other types of data such as serial analysis of gene expression (SAGE), proteomic, single-nucleotide polymorphism (SNP), and gene sequence data.

Biological data isn't much different from other types of data we've seen in the book: data is stored in vectors, arrays, and data frames. You can process and analyze this data using the same tools that R provides for other types of data, including data access tools, statistical models, and graphics.

Bioconductor provides tools for each step of the analysis process: loading, cleaning, and analyzing data. Depending on the type of data that you are working with, you might need to use other software in conjunction with Bioconductor. For example, if you are working with Affymetrix GeneChip arrays, you will need to use the Affymetrix GeneChip Command Console software to scan the arrays and produce probe cell intensity data (CEL files) that can be loaded into R. You can then load the probe cell intensity files into Bioconductor for futher processing.

This chapter provides a very brief overview of Bioconductor. In this chapter, we'll first look at an example, using publically available gene expression data. Next, I'll describe some popular packages in Bioconductor. After that, I will describe some of the key data structures in Bioconductor. Finally, I'll provide some pointers for additional information.

An Example

In this chapter, we will load a data set from NCBI's Gene Expression Omnibus (GEO) website (*http://www.ncbi.nlm.nih.gov/geo/*). GEO is a public repository that archives and freely distributes microarray, next-generation sequencing, and other forms of high-throughput functional genomic data submitted by the scientific community. It is one of many resources available through the National Center for Biotechnology Information (NCBI), an organization that is part of the National Library of Medicine, and, in turn, part of the U.S. National Institutes of Health (NIH). This is a very useful resource when learning to use Bioconductor, because you can find not only raw data but also references to papers that analyzed that data.

As an example, we'll use the data files from GSE2034 (*http://www.ncbi.nlm.nih.gov/ projects/geo/query/acc.cgi?acc=GSE2034*), a study that looked for predictors of relapse-free breast cancer survival. (I used data from the same study as an example in "Survival Models" on page 396.) My goal was not to re-create the results shown in the original papers (which I did not do), but instead to show how Bioconductor tools could be used to load and inspect this data.

Loading Raw Expression Data

Let's start with an example of loading raw data into R. We'll show how to load Affymetrix CEL files, which are output from Affymetrix's scanner software. If you would like to try this yourself, you can download the raw CEL files from *ftp://ftp .ncbi.nih.gov/pub/geo/DATA/supplementary/series/GSE2034/GSE2034_RAW.tar*.

 The CEL files are immense: almost 1 GB compressed. See "Loading Data from GEO" on page 474 for instructions on how to get pre-processed expression files for this experiment.

Affymetrix is a leading provider of tools for genetic analysis, including high-density arrays, scanners, and analysis software. For this study, the authors used Affymetrix GeneChip Human Genome U133 Arrays,[*] which are used to measure the expression level of over 22,000 probe sets that translate to 14,500 human genes. These arrays work by measuring the amount of thousands of different RNA fragments using thousands of different probes. Each probe is 25 bases long. The CEL files contain scanner data for each probe for each sample. Data processing software (like Bioconductor) is used to translate combinations of probes to probe sets, which can, in turn, be mapped to genes. A probe set is composed of a set of perfect-match (PM) probes (for which all 25 bases match) and mismatch (MM) probes (for which the 13th base is reversed); the software measures the actual expression level of genes by comparing the two types of probes. Typically, each probe set comprises 11 to 20 different probes. Data for each sample is stored in a separate CEL file.

You can load these files into R as a single batch using ReadAffy. The ReadAffy function will load all files in the current working directory by default. If you are using a machine with a *lot* of memory and have placed the files in the directory ~/GSE2034/ CEL, you could load the data with the following commands:

```
> library(affy)
> # assuming the files are in ~/GSE2034/CEL
```

[*] See *http://www.affymetrix.com/products_services/arrays/specific/hgu133av2.affx* for more information on this platform.

```
> setwd("~/GSE2034/CEL")
> GSE2034 <- ReadAffy()
```

I have 4 GB on my computer, which wasn't enough to read all the raw files into memory. So, I took a subset of the CEL files for a random sample of subjects.

To pick the stratified sample, I used several R functions from outside Bioconductor. I used a stratified sample, selecting 50 subjects with no relapse and 50 with relapse. To select the set of filenames to load, I used the `strata` function from the `sampling` package to pick a set of GEO accession numbers to load. (These are the identifiers for each subject.) Next, I pasted the prefix ".CEL" on the end of each number to generate filenames. Finally, I passed this vector as an argument to `ReadAffy`.

Here is the code I used to read in the data:

```
> library(nutshell)
> data(GSE2034)
> library(sampling)
> setwd("~/Documents/book/data/GSE2034/CEL")
> GSE2034.fromcel.smpl <-
+   ReadAffy(filenames=paste(
+                             GSE2034[strata(GSE2034,
+                                            stratanames="relapse",
+                                            size=c(50,50),
+                                            method="srswor"
+                                           )$ID_unit,
+                                     ]$GEO.asscession.number,
+                             ".CEL",
+                             sep=""))
```

The `ReadAffy` function returns an `AffyBatch` object, containing unprocessed gene expression data:

```
> GSE2034.fromcel.smpl
AffyBatch object
size of arrays=712x712 features (16 kb)
cdf=HG-U133A (22283 affyids)
number of samples=100
number of genes=22283
annotation=hgu133a
notes=
```

Before we can analyze this data, we need to attach phenotype (patient) data, normalize the data, summarize by probe, and associate the expression data with annotation (gene symbol) data.

First, the sample names in the `AffyBatch` object match the filenames, not the identifiers (GEO accession numbers) in the patient data table:

```
> sampleNames(GSE2034.fromcel.smpl)[1:10]
 [1] "GSM36796.CEL" "GSM36834.CEL" "GSM36873.CEL" "GSM36917.CEL"
 [5] "GSM36919.CEL" "GSM36938.CEL" "GSM36944.CEL" "GSM36965.CEL"
 [9] "GSM36991.CEL" "GSM36993.CEL"
```

Let's clean up the sample names in the `AffyBatch` object, so that we can match them to names in the patient data table. (We'll do that in "Matching Phenotype Data" on page 476.)

```
> sampleNames(GSE2034.fromcel.smpl) <-
+   sub("\\.CEL$","",sampleNames(GSE2034.fromcel.smpl))
> sampleNames(GSE2034.fromcel.smpl)[1:10]
 [1] "GSM36796" "GSM36834" "GSM36873" "GSM36917" "GSM36919" "GSM36938"
 [7] "GSM36944" "GSM36965" "GSM36991" "GSM36993"
```

An important step in data processing is quality control (QC). You want to make sure that no errors occurred in handling the experimental data or scanning the arrays. You can use the `qc` function in the `simpleaffy` package for quality control. This function calculates a set of QC metrics (recommended by Affymetrix) to check that arrays have hybridized correctly and that sample quality is acceptable. It returns an object of class `QCStats` that you can plot to check for problematic samples. As an example, we'll calculate QC metrics on the first 20 samples that we loaded into R. (I picked 20 so the plot would be readable in print.)

```
> myqc <- qc(GSE2034.fromcel.smpl[,1:20])
> plot(myqc,cex=0.7)
```

The results are shown in Figure 24-1. Each line represents a separate sample. The vertical solid line in the middle of the diagram corresponds to zero fold change, the dotted line to the left and right to three fold downregulation and three fold upregulation change, respectively. The lines plotted on each row show which scale factors are acceptable. Good values are blue, suspicious are red, when viewed on screen. In this example, all the bars are acceptable. For more information on how to read this diagram, see the help file for `plot.qc.stats`.

Before analyzing the microarray data, it needs additional preprocessing. First, the raw data needs to be background corrected and normalized between arrays. You can do this with the Bioconductor `vsn` package, using the `vsn2` function. The `vsn2` function returns a `vsn` object containing background-corrected and normalized probe intensity data.

Next, the data needs to be log transformed, summarized by probe set, and transformed into an `ExpressionSet` that can be used in further analysis. As we noted above, CEL files include information on all probes; these need to be grouped into probe sets and adjusted for mismatches. Raw expression data values are exponentially distributed; a log transformation makes the distribution normal. You can do this through the `rma` function in the `affy` package.

If you don't plan to tweak parameters, you can execute both steps at once through the `vsnrma` function in the `vsn` package. (The `vsn` function requires a lot of memory to process large `AffyBatch` objects. My computer couldn't handle all 100 arrays at once, so I took a subset of 50 observations.)

```
> library(affy)
> GSE2034.fromcel.smpl.vsnrma <- vsnrma(GSE2034.fromcel.smpl[,1:50])
vsn2: 506944 x 50 matrix (1 stratum).
Please use 'meanSdPlot' to verify the fit.
Calculating Expression
```

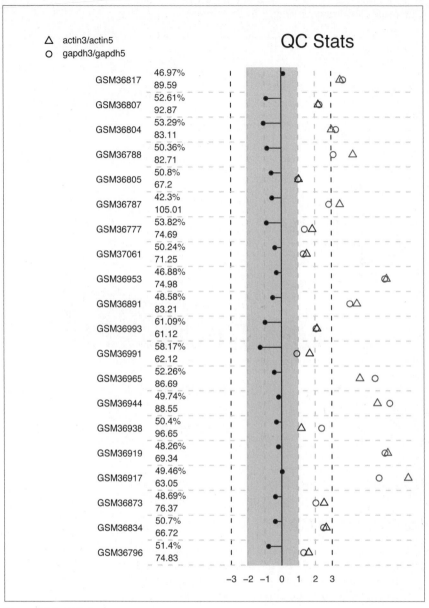

Figure 24-1. QC plot

Following the recommendation above (in the output of `vsn2`), let's use `meanSdPlot` to plot the row standard deviation versus row means for the output:

```
> meanSdPlot(GSE2034.fromcel.smpl.vsnrma)
```

The results are shown in Figure 24-2.

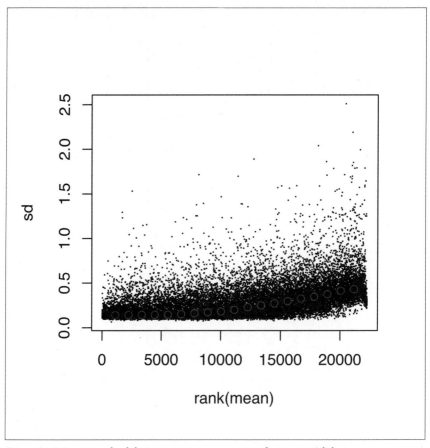

Figure 24-2. Row standard deviation versus row means, from meanSdPlot

Loading Data from GEO

In this specific case, we can cheat. This example uses a data set that was shared through GEO, so we can use the getGEO function in the GEOquery package to download preprocessed expression sets directly into R. (Clearly this won't work with data that isn't available on GEO, but it does make this step simpler.)

```
> library(GEOquery)
Loading required package: Biobase

Welcome to Bioconductor

    Vignettes contain introductory material. To view, type
    'openVignette()'. To cite Bioconductor, see
    'citation("Biobase")' and for packages 'citation(pkgname)'.
```

```
Loading required package: RCurl
Loading required package: bitops
> GSE2034.geo <- getGEO("GSE2034")
Found 2 file(s)
GSE2034_series_matrix-1.txt.gz
trying URL 'ftp://ftp.ncbi.nih.gov/pub/geo/DATA/
  SeriesMatrix/GSE2034/GSE2034_series_matrix-1.txt.gz'
ftp data connection made, file length 12800217 bytes
opened URL
==================================================
downloaded 12.2 Mb

trying URL 'http://www.ncbi.nlm.nih.gov/geo/query/
  acc.cgi?targ=self&acc=GPL96&form=text&view=full'
Content type 'geo/text' length unknown
opened URL
.......... .......... .......... .......... ..........

downloaded 27.9 Mb

File stored at:
/var/folders/gj/gj60srEiEVq4hTWB5lvMak+++TM/-Tmp-//RtmpnO9uT5/GPL96.soft
GSE2034_series_matrix-2.txt.gz
trying URL 'ftp://ftp.ncbi.nih.gov/pub/geo/DATA/
  SeriesMatrix/GSE2034/GSE2034_series_matrix-2.txt.gz'
ftp data connection made, file length 1662337 bytes
opened URL
==================================================
downloaded 1.6 Mb

trying URL 'http://www.ncbi.nlm.nih.gov/geo/query/
  acc.cgi?targ=self&acc=GPL96&form=text&view=full'
Content type 'geo/text' length unknown
opened URL
.......... .......... .......... .......... ..........

downloaded 27.9 Mb

File stored at:
/var/folders/gj/gj60srEiEVq4hTWB5lvMak+++TM/-Tmp-//RtmpnO9uT5/GPL96.soft
```

In this case, the object is a list of two ExpressionSet objects:

```
> class(GSE2034.geo)
[1] "list"
> class(GSE2034.geo[[1]])
[1] "ExpressionSet"
attr(,"package")
[1] "Biobase"
> GSE2034.geo
$`GSE2034_series_matrix-1.txt.gz`
ExpressionSet (storageMode: lockedEnvironment)
assayData: 22283 features, 255 samples
  element names: exprs
phenoData
  sampleNames: GSM36777, GSM36778, ..., GSM37031  (255 total)
```

```
      varLabels and varMetadata description:
        title: NA
        geo_accession: NA
        ...: ...
        data_row_count: NA
        (23 total)
    featureData
      featureNames: 1007_s_at, 1053_at, ..., AFFX-TrpnX-M_at  (22283 total)
      fvarLabels and fvarMetadata description:
        ID: NA
        GB_ACC: NA
        ...: ...
        Gene.Ontology.Molecular.Function: NA
        (16 total)
      additional fvarMetadata: Column, Description
    experimentData: use 'experimentData(object)'
    Annotation: GPL96

    $`GSE2034_series_matrix-2.txt.gz`
    ExpressionSet (storageMode: lockedEnvironment)
    assayData: 22283 features, 31 samples
      element names: exprs
    phenoData
      sampleNames: GSM37032, GSM37033, ..., GSM37062  (31 total)
      varLabels and varMetadata description:
        title: NA
        geo_accession: NA
        ...: ...
        data_row_count: NA
        (23 total)
    featureData
      featureNames: 1007_s_at, 1053_at, ..., AFFX-TrpnX-M_at  (22283 total)
      fvarLabels and fvarMetadata description:
        ID: NA
        GB_ACC: NA
        ...: ...
        Gene.Ontology.Molecular.Function: NA
        (16 total)
      additional fvarMetadata: Column, Description
    experimentData: use 'experimentData(object)'
    Annotation: GPL96
```

In the rest of this chapter, I'll focus on the first object in the list:

```
> GSE2034.geo1 <- GSE2034.geo[[1]]
```

Matching Phenotype Data

Neither the CEL files nor the series matrix files from GEO contain clinical information. In "Survival Models" on page 396, we used a file from GEO containing the experimental outcomes for this experiment, including an indicator of which patients experienced a relapse and the time until relapse or last checkup. We'll add this information to the AffyBatch file by matching observations in the GSE2034 data set to the expression data. We can add data to the files created from the CEL files using the following code:

```
> matches <- match(
+    subset(GSE2034,
+      GSE2034$GEO.asscession.number %in%
+      sampleNames(GSE2034.fromcel.smpl))$GEO.asscession.number,
+    sampleNames(GSE2034.fromcel.smpl))
> phenoData(GSE2034.fromcel.smpl) <- new(
+    "AnnotatedDataFrame",
+    data=subset(GSE2034,
+      GSE2034$GEO.asscession.number %in%
+      sampleNames(GSE2034.fromcel.smpl))[matches,])
```

To add patient information to the matrix files from GEO, we'll use a slightly different strategy. Loading the matrix files created ExpressionSet objects that were already tagged with phenotype information. Instead of replacing this information, we'll just add more patient information. (Again, notice that I'm using R code to create the new data frame of phenotype data and the data frame with variable metadata. There's nothing fancy about Bioconductor; it's just a set of R functions for dealing with a certain type of data.)

```
# matching in new version
> matches <- match(
+    subset(GSE2034,
+        GSE2034$GEO.asscession.number %in%
+        sampleNames(GSE2034.geo1))$GEO.asscession.number,
+    sampleNames(GSE2034.geo1))
> names(GSE2034) <- c("PID", "geo_accession","lymph.node.status",
+    "months.to.relapse.or.last.followup", "relapse", "ER.Status",
+    "Brain.relapses")
> GSE2034.pdata <- merge(pData(GSE2034.geo1),GSE2034[,2:7])
> GSE2034.varMetadata <- rbind(varMetadata(GSE2034.geo1),
+ data.frame(row.names=names(GSE2034)[3:7],labelDescription=rep(NA,5)))
> pData(GSE2034.geo1) <- GSE2034.pdata
> varMetadata(GSE2034.geo1) <- GSE2034.varMetadata
```

Analyzing Expression Data

As an analysis example, I'll use the file downloaded in "Loading Data from GEO" on page 474. The expression set we are examining contains 22,283 features on 255 subjects. Fitting a model to a data set this large could take a long time, so we'll start the analysis by filtering out some probes. Specifically, we'll filter out probes with low variance using the nsFilter function in the genefilter package:

```
> annotation(GSE2034.geo1)
[1] "GPL96"
# there is no GPL96 annotation package that is easily available,
# though this is the same as affy hgu133a, so use that instead
> annotation(GSE2034.geo1) <- "hgu133a"
> library(genefilter)
> GSE2034.geo1.f <- nsFilter(GSE2034.geo1, var.cutoff=0.5)
```

The filtered expression set contains only 6,534 features, which is much more manageable. Let's start by drawing a "volcano plot" using the expression data. To draw the plot, we'll start by calculating a *t*-test on each row, segmenting observations based on relapse status:

```
> tt <- rowttests(GSE2034.geo1.f$eset,"relapse")
```

Next, we'll plot the log of the *p*-value (from the *t*-test) for each probe versus the difference in group means. (Both are included in the output of the `rowttests` function.)

```
> plot(tt$dm,-log10(tt$p.value),pch=".",xlab="log-ratio",
+    ylab=expression(-log[10]~p))
```

The plot is shown in Figure 24-3. As you can see, there are a few values far to the right of the plot.

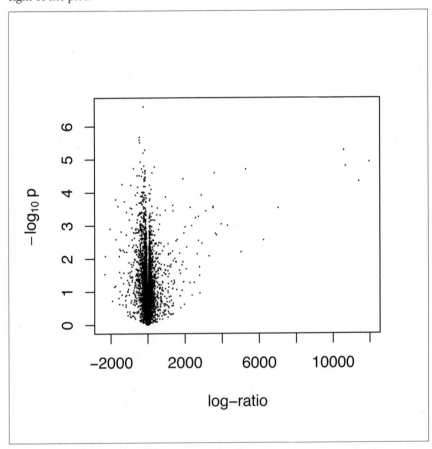

Figure 24-3. Volcano plot of filtered GSE2034 data

In the original paper on this study, the authors fit a Cox proportional hazard model using the expression data. We can do the same thing, using the `rbsurv` package from Bioconductor, which stands for "robust survival" analysis. The `rbsurv` function allows you to fit a model to an expression object, choosing predictive variables using *n*-fold cross-validation. I chose to restrict the model to 75 genes, and the fitting to six iterations of threefold validation to keep the running time manageable (though this function still required an hour to fit the model):

```
> library(rbsurv)
Loading required package: survival
Loading required package: splines
> GSE2034.rbsurv <- rbsurv(
+     time=pData(GSE2034.geo1.f$eset)$months.to.relapse.or.last.followup,
+     status=pData(GSE2034.geo1.f$eset)$relapse,
+     x=assayData(GSE2034.geo1.f$eset)$exprs,
+     gene.ID=row.names(assayData(GSE2034.geo1.f$eset)$exprs),
+     max.n.genes=75,
+     n.fold=3,
+     n.iter=6)
Please wait... Done.
```

This function uses bootstrap resampling to generate the robust estimate, so you may get different results, depending on the state of your random number generator. This function returns a list containing a number of different objects:

```
> typeof(GSE2034.rbsurv)
[1] "list"
> names(GSE2034.rbsurv)
[1] "n.genes"    "n.samples"  "method"     "n.iter"     "n.fold"
[6] "covariates" "model"      "gene.list"
```

We can take a look at the coefficients in the fitted model to see which probes are significant. Not surprisingly, with over 6,000 predictors, there are a lot of genes that are highly correlated with relapse-free survival time. The fitted model contained 62 probes; here are the first 20:

```
> GSE2034.rbsurv$model[1:20]
     Seq Order        Gene nloglik    AIC Selected
0      1     0           0  495.16 990.32
110    1     1 221286_s_at  487.39 976.79 *
2      1     2   209096_at  481.33 966.65 *
3      1     3   201817_at  475.60 957.20 *
4      1     4 214459_x_at  471.54 951.08 *
5      1     5   207165_at  468.15 946.30 *
6      1     6 211430_s_at  465.70 943.40 *
7      1     7   203218_at  458.58 931.16 *
8      1     8   209539_at  455.17 926.33 *
9      1     9 202666_s_at  452.74 923.48 *
10     1    10 222201_s_at  449.92 919.83 *
11     1    11   212898_at  445.67 913.34 *
12     1    12 216598_s_at  441.15 906.29 *
13     1    13 203530_s_at  440.51 907.01 *
14     1    14   201178_at  437.32 902.63 *
15     1    15   203764_at  435.25 900.49 *
16     1    16 202324_s_at  435.14 902.27 *
17     1    17 220757_s_at  434.71 903.41 *
18     1    18 201010_s_at  433.90 903.80 *
19     1    19   218919_at  432.35 902.70 *
```

The gene.list element contains the Affymetrix probe names from a Human Genome U133A Array:

```
> annotation(GSE2034.geo1)
[1] "hgu133a"
```

To show a list of gene symbols corresponding to these probes, we can use the getSYMBOL function from the annotate package:

```
> library(annotate)
> getSYMBOL(GSE2034.rbsurv$gene.list,"hgu133a")
    221286_s_at      209096_at      201817_at    214459_x_at
    "MGC29506"       "UBE2V2"       "UBE3C"       "HLA-C"
     207165_at     211430_s_at      203218_at      209539_at
       "HMMR"        "IGHG3"        "MAPK9"       "ARHGEF6"
   202666_s_at     222201_s_at      212898_at    216598_s_at
      "ACTL6A"      "CASP8AP2"     "KIAA0406"       "CCL2"
   203530_s_at       201178_at      203764_at    202324_s_at
       "STX4"         "FBX07"       "DLGAP5"       "ACBD3"
   220757_s_at     201010_s_at      218919_at      200726_at
      "UBXN6"        "TXNIP"        "ZFAND1"       "PPP1CC"
   221432_s_at     215088_s_at    215379_x_at    219215_s_at
     "SLC25A28"    "hCG_1776980"     "CKAP2"      "SLC39A4"
     204252_at       212900_at    209380_s_at      209619_at
       "CDK2"        "SEC24A"        "ABCC5"        "CD74"
   208843_s_at     203524_s_at    209312_x_at    222077_s_at
     "GORASP2"        "MPST"       "HLA-DRB1"     "RACGAP1"
   202824_s_at       212687_at    221500_s_at    217258_x_at
      "TCEB1"        "LIMS1"        "STX16"         "IVD"
     205034_at       201849_at      201664_at    215946_x_at
      "CCNE2"        "BNIP3"         "SMC4"        "IGLL3"
     219494_at       208757_at    221671_x_at      212149_at
      "RAD54B"       "TMED9"         "IGKC"        "EFR3A"
     202969_at     209831_x_at      204641_at    217378_x_at
       "DYRK2"       "DNASE2"        "NEK2"    "LOC100130100"
   204670_x_at     211761_s_at    205812_s_at    216401_x_at
     "HLA-DRB5"      "CACYBP"       "SULT1C4"     "LOC652493"
   217816_s_at       201368_at      209422_at      213391_at
       "PCNP"        "ZFP36L2"       "PHF20"       "DPY19L4"
   208306_x_at       201288_at      206102_at
     "HLA-DRB4"      "ARHGDIB"       "GINS1"
```

To get more information on these probes, we can use functions from the anaffy package to annotate the results. This package can provide a lot of information on each probe; the function aaf.handler shows the available fields:

```
> aaf.handler()
 [1] "Probe"         "Symbol"                "Description"
 [4] "Chromosome"    "Chromosome Location"   "GenBank"
 [7] "Gene"          "Cytoband"              "UniGene"
[10] "PubMed"        "Gene Ontology"         "Pathway"
```

Let's include the Probe, Symbol, Description, PubMed ID, Gene Ontology, and Pathway for each probe. To do this, we first create an aafTable object with the annotation information and then save it as an HTML file so we can view it:

```
> anntable <- aafTableAnn(probeid=GSE2034.rbsurv$gene.list,
+                     chip="hgu133a.db",
+                     colnames=c("Probe", "Symbol", "Description",
+                                "PubMed", "Gene Ontology", "Pathway"))
> saveHTML(anntable,filename="~/results.html")
```

Figure 24-4 shows a screen shot of the results. As you can see, the annotation package can provide a lot of supplemental information about each probe, hopefully allowing you to learn something interesting from the experiment.

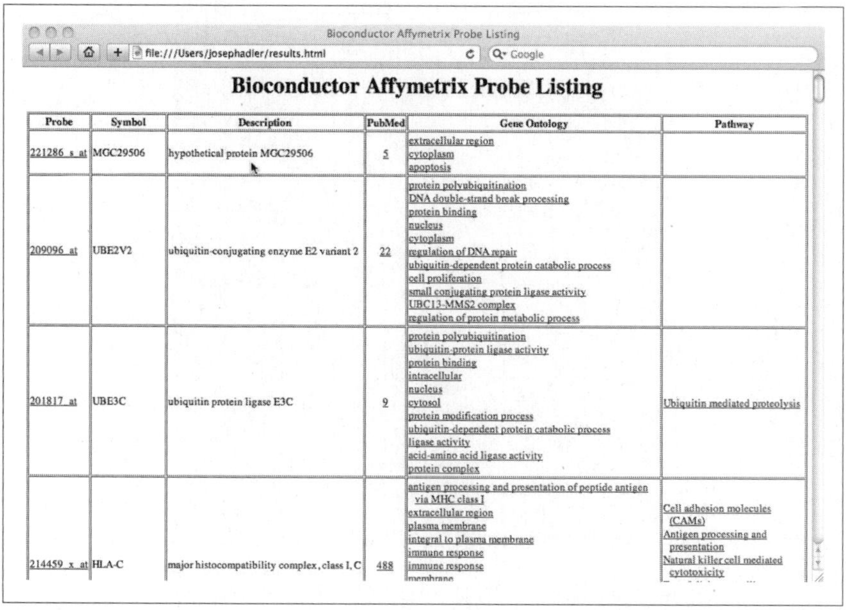

Figure 24-4. Screen shot of Safari showing annotated results from the GSE2034.rbsurv model

Finally, you can use R to visualize the expression levels using a heat map. Heat maps are like image plots or level plots, but automatically reorder observations using clustering to show hot or cold spots. To make the diagram legible, we'll pick 50 subjects: 25 with relapse, 25 without:

```
> relapse.df <- data.frame(row.names=GSE2034.geo1$geo_accession,
+                      relapse=GSE2034.geo1$relapse)
> library(sampling)
> smpl <- strata(relapse.df,c("relapse"), size=c(25,25), method="srswor")
```

Now, let's plot the heat map using R's heatmap function. By default, R uses hierarchical clustering to group similar observations together. Dendrograms are plotted in the margins showing the clustering. Heat maps are plotted with colors ranging from yellow to red on screen, though you can use the col parameter to control the color palette. Here is the code that I used to generate the heat map shown in Figure 24-5:

```
> heatmap(assayData(GSE2034.geo1.f$eset)$exprs[
+          GSE2034.rbsurv$gene.list,smpl$ID_unit],
+          Colv=smpl$relapse, cexRow=0.45, cexCol=0.45)
```

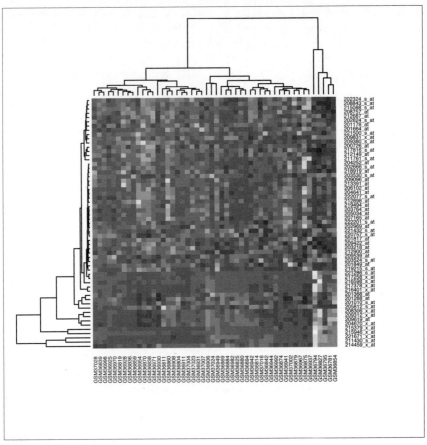

Figure 24-5. Heat map showing expression level for 50 subjects

Key Bioconductor Packages

The Bioconductor repository contains over 300 packages for working with genetic data. Below is a list of some popular packages, with short descriptions of the classes and functions that they contain.

Category	Package	Description
Loading, preprocessing	aCGH	Classes and functions for array comparative genomic hybridization data. Functions for reading aCGH data from image analysis output files and clone information files and for creating aCGH S3 objects for storing these data. Basic methods for accessing/replacing, subsetting, printing, and plotting aCGH objects.
	affy	Methods for Affymetrix oligonucleotide arrays. Includes class definitions for representing microarray data. Also includes methods for importing data, quality control, and normalization.
	affyQCReport	A package to generate QC reports for Affymetrix array data.

Category	Package	Description
	arrayQuality	Functions for performing print-run and array-level quality assessment.
	gcrma	Background adjustment using sequence information. The main function gcrma converts background-adjusted probe intensities to expression measures using the same normalization and summarization methods as RMA (robust multiarray average).
	limma	Limma is an R package for the analysis of gene expression microarray data, especially the use of linear models for analyzing designed experiments and the assessment of differential expression.
	lumi	Functions to preprocess Illumina microarray (BeadArray) data. It includes functions of Illumina data input, quality control, variance stabilization, normalization, and gene annotation.
	marray	Diagnostic plots and normalization of cDNA microarray data.
	oligo	The oligo package includes tools for preprocessing data from oligonucleotide arrays. It supports all microarray designs provided by Affymetrix and NimbleGen: expression, tiling, SNP, and exon arrays.
	prada	Tools for analyzing and navigating data from high-throughput phenotyping experiments based on cellular assays and fluorescent detection (flow cytometry [FACS], high-content screening microscopy).
	PROcess	The PROcess package contains a collection of functions for processing spectra (particularly, Ciphergen SELDI-TOF spectra for proteomic data) to remove baseline drifts, if any, detect peaks, and align them to a set of protobiomarkers.
	Ringo	Tools for working with two-color oligoarrays (particularly NimlbeGen arrays). Stands for R Investigation of NimbleGen Oligoarrays.
	simpleaffy	Provides high-level functions for reading Affy .CEL files and phenotypic data and then computing simple things with it, such as t-tests, fold changes, and the like. Also has some basic scatter plot functions and mechanisms for generating high-resolution journal figures.
	vsn	Variance stabilization and calibration for microarray data. The package implements a method for normalizing microarray intensities, both between colors within an array and between arrays.
Annotation	annotate	The basic purpose of annotate is to supply interface routines that support user actions that rely on the different metadata packages provided through the Bioconductor Project.
	annaffy	This package is designed to help interface between Affymetrix analysis results and web-based databases. It provides classes and functions for accessing those resources both interactively and through statically generated HTML pages.
	annBuilder	AnnBuilder constructs annotation data packages for given sets of genes with known mappings to GenBank accession numbers, UniGene identifiers, Image identifiers, or Entrez Gene identifiers.
	biomaRt	Interface to BioMart databases (e.g., Ensembl, Wormbase, and Gramene).
	GOstats	A set of tools for interacting with Gene Ontology (GO) and microarray data. A variety of basic manipulation tools for graphs, hypothesis testing, and other simple calculations.

Category	Package	Description
Analysis	affypdnn	Probe-dependent nearest neighbors for affy probes.
	affyPLM	Methods for fitting probe-level models.
	bioDist	A collection of software tools for calculating distance measures.
	factDesign	This package provides a set of tools for analyzing data from a factorial designed microarray experiment or any microarray experiment for which a linear model is appropriate. The functions can be used to evaluate tests of contrast of biological interest and perform single outlier detection.
	genefilter	Methods for filtering genes from microarray experiments.
	GSEABase	This package provides classes and methods to support Gene Set Enrichment Analysis (GSEA).
	hopach	Hierarchical Ordered Partitioning and Collapsing Hybrid.
	MLInterfaces	Uniform interfaces to machine learning code for data in Bioconductor containers. Includes clustering, classification, and regression algorithms.
	limma	Limma is an R package for the analysis of gene expression microarray data, especially the use of linear models for analyzing designed experiments and the assessment of differential expression.
	marray	Diagnostic plots and normalization of cDNA microarray data.
	multtest	The multtest package contains a collection of functions for multiple hypothesis testing. These functions can be used to identify differentially expressed genes in microarray experiments (i.e., genes whose expression levels are associated with a response or covariate of interest).
	ROC	Functions for calculating and plotting receiver operating characteristic (ROC) curves with microarray data.
	simpleaffy	Provides high-level functions for reading Affy .CEL files and phenotypic data and then computing simple things with it, such as t-tests, fold changes, and the like. Also has some basic scatter plot functions and mechanisms for generating high-resolution journal figures.
Visualization	affycomp	Graphical tools for assessing Affymetrix expression measures. These tools rely on two studies: a dilution study and a spike-in study.
	geneplotter	Graphics-related functions for Bioconductor.
	graph	The graph package provides an implementation of graphs (the kind with nodes and edges) in R.
	RBGL	Provides an interface to graph algorithms (such as shortest path, connectivity, etc.).
	Rgraphviz	Provides graph-rendering functionality. Different layout algorithms are provided, and parameters like node plotting, line type, and color can be controlled by the user.
	SNPchip	This package defines classes and functions for plotting copy number and genotype in high-throughput SNP platforms such as Affymetrix and Illumina. In particular, SNPchip is a useful add-on to the oligo package for visualizing SNP-level estimates after preprocessing.

Category	Package	Description
Utilities	Biobase	Biobase contains standardized data structures to represent genomic data.
	Biostrings	Memory-efficient string containers, string-matching algorithms, and other utilities, for fast manipulation of large biological sequences or set of sequences.
	BSgenome	Infrastructure shared by all the Biostrings-based genome data packages.
	convert	Tools to convert between limma, marray, and Biobase data objects.

Data Structures

One of the best features of Bioconductor is the use of structured data to represent biological concepts. This section presents a few important classes that are used through Bioconductor.

Bioconductor classes are implemented using formal class methods; see Chapter 10 for more details. Most of these classes inherit from the basic classes in the Biobase package, so you can use the same methods to work with different types of objects. For example, you can use the same method to read phenotype data for expression data from different vendors (such as Affymetrix arrays and Illumina arrays). You could also use the same method to read phenotype data for expression data from completely different types of data (such as gene expression data and proteomic data).

Objects in Bioconductor contain many different types of information about an experiment: the experimental platform, information about the samples, information about the phenotypes, the experimental results, and almost anything else that is relevant for describing an experiment or the results of the experiment. Classes defined in the Biobase package provide a general framework that fits many different types of experimental data. Classes defined in other packages can be used to represent data from specific types of microarrays, often for specific products from specific vendors. This section contains descriptions of a few key classes defined in Biobase.

eSet

eSet is a virtual class that is used by many Bioconductor functions. Objects based on eSet package together all the relevant information about a high-throughput experiment: expression data, metadata describing the experiment, annotation about the chip or technology used, and a description of the experiment itself.

Many other classes inherit from eSet: In Biobase, the classes ExpressionSet (for high-throughput expression-level assays), MultiSet (also for high-throughput expression-level assays), SnpSet (for high-throughput SNP assays), and NChannelSet (for multiple-channel arrays) are children of eSet. In the affy package, the class AffyBatch (used to represent Affymetrix GeneChip probe-level data) inherits from eSet. In the lumi package, LumiBatch (used to represent Illumina microarray data) is based on eSet. In oligoClasses, the classes SnpLevelSet, SnpCallSet, SnpCopyNumberSet, oligoSnpSet, and SnpCallSetPlus all inherit from eSet.

An eSet object has the following slots:

assayData
> An assayData object containing the expression data. (The expression data must contain matrices with equal dimensions and with column numbers equal to nrow(phenoData).)

phenoData
> An AnnotatedDataFrame object describing the sample phenotypes.

featureData
> An AnnotatedDataFrame object describing the features or probes (corresponding to columns in assayData) for this experiment.

experimentData
> A MIAME object containing detailed information on the experimental method(s).

annotation
> A character value describing the annotation package used for the experiment.

There are included methods for getting or setting the object in each of these slots directly. (For example, assayData(x) <- y will set the assayData slot in eSet x to y.) Additionally, methods are defined for directly accessing commonly used slots within each of these objects:

sampleNames, sampleNames<-
> Get or set sample names in assayData and phenoData.

featureNames, featureNames<-
> Get or set feature names in assayData.

dims
> Gets the dimensions for the expression data in assayData.

pData
> Gets or sets sample data (pData slot in phenoData).

fData
> Gets or sets feature data information (pData slot in featureData).

varMetadata
> Gets or sets metadata describing variables in pData.

varLabels
> Gets or sets variable labels in phenoData.

fvarMetadata
> Gets or sets metadata describing features in fData.

fVarLabels
> Gets or sets variable labels in featureData.

description
> Alias for experimentData.

pMedIds
> Gets or sets PubMed Identifiers (PMIDs) from experimentData.

abstract
> Gets abstract from `experimentData`.

preproc, preproc<-
> Get or set preprocessing information in `experimentData`.

storageMode, storageMode<-
> Get or set storage mode for `assayData`.

assayDataElement
> Gets or sets an element in an `AssayData` object.

notes
> Used to add free-form notes to an `AssayData` object.

There are methods to coerce `eSet` objects to `ExpressionSet` and `MultiSet` objects. See the help file for `eSet` for more details.

AssayData

`AssayData` objects hold expression data. You can access the contents of an `Assay Data` object with the following methods:

featureNames, featureNames<-
> Get or set the feature names (or probe names) for an object.

sampleNames, sampleNames<-
> Get or set the sample names for an object.

storageMode, storageMode<-
> Get or set the storage mode for an `AssayData` object.[†]

assayDataElement
> Gets or sets a specific element in an `AssayData` object.

`AssayData` objects are used in `eSet` objects to hold expression data.

AnnotatedDataFrame

`AnnotatedDataFrame` objects are what they sound like: a data frame plus annotation. Typically, they are used to include a data frame containing some experimental data, plus information about each column/variable in the data frame. `AnnotatedData Frame` objects contain two slots:

data
> A data frame. Rows represent samples; columns represent variables.

varMetaData
> A data frame with one row corresponding to each column in `data`. This data frame must include a column called `labelDescription`, but may contain additional information.

---

† `AssayData` objects can hold the expression data in a list, environments, or "locked" environments; see the help file for more information.

The `Biobase` package defines a few useful methods for accessing information in `AnnotatedDataFrame` objects:

pData, pData<-
    Get or set the data stored in the object.

varMetaData, varMetaData<-
    Get or set the metadata.

sampleNames, sampleNames<-
    Get or set the sample names.

featureNames, featureNames<-
    Alias for `sampleNames, sampleNames<-`.

varLabels, varLabels<-
    Get or set the variable labels.

dimLabels, dimLabels<-
    Get or set the dimension labels (`rowNames, columnNames`).

`AnnotatedDataFrame` objects are used to hold information about samples in `eSet` objects.

MIAME

MIAME stands for Minimum Information About a Microarray Experiment.[‡] `MIAME` objects are used to contain information about an experiment. Slots in `MIAME` objects include:

name
    Experimenter name

lab
    Lab where the experiment was conducted

contact
    Contact information for the experimenter

title
    Single-sentence description of the experiment

abstract
    An abstract describing the experiment

url
    A URL reference with information about the experiment

samples
    Information about the samples

hybridization
    Information about the hybridizations

‡ MIAME is a standard developed by the MGED Society. See *http://www.mged.org/Workgroups/ MIAME/miame.html* for more information.

normControls
> Information about the controls

preprocessing
> Information about preprocessing steps performed on raw data from the experiment

pubMedIds
> PubMed Identifiers of papers relevant for this data

other
> Other information about the experiment that doesn't fit elsewhere

MIAME objects are used in eSet objects to describe an experiment.

Other Classes Used by Bioconductor Packages

There are a variety of other classes used in different Bioconductor packages and functions:

AssayData
> A container class defined as a class union of list and environment. Designed to contain one or more matrices of the same dimension.

ProbeSet
> A simple class that contains the raw probe data (PM and MM data) for a probe set from one or more samples.

RGList
> A class used to store raw intensities as they are read in from an image analysis output file.

MAList
> A simple list-based class for storing M-values and A-values for a batch of spotted microarrays.

Elist
> A simple list-based class for storing expression values (E-values) for a set of one-channel microarrays.

Elistraw
> A simple list-based class for storing expression values (E-values) for a set of one-channel microarrays (in raw form).

MArray-LM
> A list-based class for storing the results of fitting gene-wise linear models to a batch of microarrays.

TestResults
> A matrix-based class for storing the results of simultaneous tests.

DBPDInfo
> A class for Platform Design Information objects, stored using a database approach.

QuantificationSet
> A virtual class to store summarized measures.

FeatureSet
> A class to store data from expression/exon/SNP/tiling arrays at the feature level.

See the help files for more information on these classes.

Where to Go Next

This chapter just scratches the surface of the tools available through Bioconductor; there are dozens of packages available on Bioconductor for doing different types of analysis. The best place to start is the Bioconductor website: *http://www.bioconductor.org*.

Here are some suggestions for learning more about this project and how to use the Bioconductor tools.

Resources Outside Bioconductor

If you are working with genetic data, there are a variety of R packages outside Bioconductor that you might find useful. See *http://cran.r-project.org/web/views/Genetics.html* for more information.

Vignettes

In "Getting Help" on page 32, I introduced vignettes. There is at least one vignette for every package in Bioconductor. For example, let's attach the `affy` package and look at the available vignettes:

```
> library(affy)
> vignette(all=FALSE)
```

This shows the following list of available vignettes (from `affy` and `Biobase`):

```
Vignettes in package 'affy':

affy                   1. Primer (source, pdf)
builtinMethods         2. Built-in Processing Methods (source,
                          pdf)
customMethods          3. Custom Processing Methods (source, pdf)
vim                    4. Import Methods (source, pdf)


Vignettes in package 'Biobase':

BiobaseDevelopment     Notes for eSet developers (source, pdf)
Bioconductor           Bioconductor Overview (source, pdf)
ExpressionSetIntroduction
                       An introduction to Biobase and
                       ExpressionSets (source, pdf)
HowTo                  Notes for writing introductory 'how to'
                       documents (source, pdf)
Qviews                 quick views of eSet instances (source, pdf)
esApply                esApply Introduction (source, pdf)
```

If you are not familiar with a package, but think it could be useful for your work, try reading the included vignettes. In many cases, the vignettes will guide you through the whole analysis process: loading, cleaning, and analyzing data.

Courses

The Bioconductor project offers classes on Bioconductor. See *http://www.biocon ductor.org/workshops* for a list of past course materials and upcoming events.

Books

The developers of Bioconductor have published several books; I found these very helpful when learning Bioconductor. If you are not familiar with the methods of modern biology or Bioconductor, then [Gentleman2005] is a very good choice. If you are familiar with modern biology and just want to see more examples, try [Hahne2008]. [Foulkes2009] provides a good introduction to statistical genetics using a number of tools outside Bioconductor (such as the `genetics` package). Finally, [Ewens2005] is a good book on statistical genetics, though it does not specifically discuss R.

R Reference

base

This package contains the basic functions that let R function as a language: arithmetic, input/output, basic programming support, and so on. Its contents are available through inheritance from any environment.

Functions

Function	Description
!	Not operator.
!=	Not equal operator.
$, $<-	Select or set named element from a list.
%%	Modulo operator.
%*%	Binary operator to multiply two matrices, if they are conformable.
%/%	Integer division operator.
%in%	Binary operator that returns a logical vector indicating if there is a match or not for its left operand.
%o%	Operator to calculate the outer product of two arrays.
%x%	Operator to calculate the Kronecker product of two arrays.
&	Operator that performs elementwise logical AND.
&&	Operator that performs logical AND, evaluating expressions from left to right until the result is determined.
*	Multiplication operator.
+	Addition operator.
-	Unary negation or binary subtraction operator.
/	Binary division operator.
:	Generates regular sequences.

Function	Description
::	Accesses an exported variable in a namespace.
:::	Accesses an internal variable in a namespace.
<	Less-than operator.
>	Greater-than operator.
>=	Greater-than-or-equal-to operator.
<=	Less-than-or-equal-to operator.
==	Equality operator.
@	Extracts the contents of an slot in an object with a formal (S4) class structure.
Arg	Returns argument of a complex value.
Conj	Returns conjugate of a complex value.
Cstack_info	Reports information on the C stack size and usage (if available).
Encoding, Encoding<-	Read or set the declared encodings for a character vector.
Filter	Extracts the elements of a vector for which a predicate (logical) function gives true.
Find	Returns the first or last element in a vector for which a condition is true.
I	Changes the class of an object to indicate that it should be treated "as is."
ISOdate, ISOdatetime	Functions to convert between character representations and objects of classes `"POSIXlt"` and `"POSIXct"` representing calendar dates and times.
Im	Extracts the imaginary part of a complex value.
La.svd	Computes the singular-value decomposition of a rectangular matrix.
Map	Applies a function to the corresponding elements of given vectors.
Mod	Returns the modulus of a complex number.
NCOL, NROW	`nrow` and `ncol` return the number of rows or columns present in x. NCOL and NROW do the same, treating a vector as one-column matrix.
Negate	Creates the negation of a given function.
NextMethod	For S3 generic functions, dispatches to the method for the next class in the object's class vector.
Position	Gives the position of an element in a matrix for which a predicate (logical) function is true.
R.Version	Provides detailed information about the version of R running.
R.home	Returns the R home directory.
RNGkind	Allows you to query or set the kind of random number generator (RNG) in use.
RNGversion	Can be used to set the random number generators as they were in an earlier R version (for reproducibility).
R_system_version	Simple S3 class for representing numeric versions, including package versions, and associated methods.
Re	Returns the real part of a complex number.
Recall	Used as a placeholder for the name of the function in which it is called. It allows the definition of recursive functions that still work after being renamed.

Function	Description
Reduce	Uses a binary function to successively combine the elements of a given vector and a possibly given initial value.
Sys.Date, Sys.time	Return the system's idea of the current date with and without time.
Sys.chmod	Provides a low-level interface to the computer's file system.
Sys.getenv	Obtains the values of the environment variables.
Sys.getlocale	Gets details of or sets aspects of the locale for the R process.
Sys.getpid	Gets the process ID of the R session.
Sys.glob	Performs wildcard expansion (also known as "globbing") on file paths.
Sys.info	Reports system and user information.
Sys.localeconv	Gets details of the numerical and monetary representations in the current locale.
Sys.setenv	Sets environment variables (for other processes called from within R or future calls to Sys.getenv from this R process).
Sys.setlocale	Gets details of or sets aspects of the locale for the R process.
Sys.sleep	Suspends execution of R expressions for a given number of seconds.
Sys.timezone	Returns the current time zone.
Sys.umask	Provides a low-level interface to the computer's file system.
Sys.unsetenv	Removes environment variables.
Sys.which	Interface to the system command which.
UseMethod	Dispatches to the appropriate method for an S3 generic function.
Vectorize	Returns a new function that acts as if mapply was called.
^	Exponentiation operator.
abbreviate	Abbreviates strings to at least minlength characters, such that they remain *unique* (if they were), unless strict=TRUE.
abs	Absolute value.
acos	Computes the arccosine.
acosh	Computes the hyperbolic arccosine.
addNA	Modifies a factor by turning NA into an extra level (so that NA values are counted in tables, for instance).
addTaskCallback	Registers an R function that is to be called each time a top-level task is completed.
agrep	Searches for approximate matches to pattern (the first argument) within the string x (the second argument) using the Levenshtein edit distance.
alist	Function to construct, coerce, and check for both kinds of R lists.
all	Given a set of logical vectors, are all of the values true?
all.equal	all.equal(x, y) is a utility to compare R objects x and y testing "near equality." If they are different, comparison is still made to some extent, and a report of the differences is returned.
all.names, all.vars	Return a character vector containing all the names that occur in an expression or call.
any	Given a set of logical vectors, is at least one of the values true?

Function	Description
anyDuplicated	Determines which elements of a vector or data frame are duplicates of elements with smaller subscripts and returns a logical vector indicating which elements (rows) are duplicates.
aperm	Transposes an array by permuting its dimensions and optionally resizing it.
append	Adds elements to a vector.
apply	Returns a vector or array or list of values obtained by applying a function to margins of an array.
args	Displays the argument names and corresponding default values of a function or primitive.
array	Creates arrays.
as.Date	Function to convert between character representations and objects of class "Date" representing calendar dates.
as.POSIXct, as.POSIXlt	Functions to manipulate objects of classes "POSIXlt" and "POSIXct" representing calendar dates and times.
as.array	Coerces to arrays.
as.call	Coerces to "call" objects.
as.character	Coerces to "character" objects.
as.complex	Coerces to "complex" objects.
as.data.frame	Coerces to "data.frame" objects.
as.difftime	Coerces to "difftime" objects.
as.double	Coerces to "double" objects.
as.environment	Converts a number or a character string to the corresponding environment on the search path.
as.expression	Coerces to "expression" objects.
as.factor	Coerces to "factor".
as.function	Coerces to "function".
as.hexmode	Coerces to "hexmode".
as.integer	Creates or tests for objects of type "integer".
as.list	Coerces to "list".
as.logical	Coerces to "logical" objects.
as.matrix	Coerces to "matrix" objects.
as.name	Coerces to "name" objects.
as.null	Ignores its argument and returns the value NULL.
as.numeric	Coerces to "numeric".
as.numeric_version	Coerces to "numeric_version".
as.octmode	Coerces to "octmode".
as.ordered	Coerces to ordered factors.
as.package_version	Coerces to "package_version" object.
as.pairlist	Coerces to "pairlist" object.

Function	Description
as.raw	Coerces to type "raw".
as.real	Coerces to type "real".
as.symbol	Coerces to "symbol".
as.table	Coerces to "table".
as.vector	Coerces to "vector".
asS4	Tests whether the object is an instance of an S4 class.
asin	Computes the arcsine.
asinh	Computes the hyperbolic arcsine.
assign	Assigns a value to a name in an environment.
atan	Computes the arctangent.
atan2	Computes the two-argument arctangent.
atanh	Computes the hyperbolic arctangent.
attach	Attaches a database (usually a list, data frame, or environment) to the R search path. This means that the database is searched by R when evaluating a variable, so objects in the database can be accessed by simply giving their names.
attachNamespace	Function to load and unload namespaces.
attr, attr<-	Get or set specific attributes of an object.
attributes, attributes<-	Access an object's attributes.
autoload, autoloader	`autoload` creates a promise-to-evaluate `autoloader` and stores it with name name in the `.AutoloadEnv` environment.
backsolve	Solves a system of linear equations where the coefficient matrix is upper or lower triangular.
baseenv	Gets, sets, tests for, and creates environments.
basename	Removes all of the path up to the last path separator (if any).
besselI, besselJ, besselK, besselY	Bessel functions of integer and fractional order, of first and second kind, $J(v)$ and $Y(v)$, and modified Bessel functions (of first and third kind), $I(v)$ and $K(v)$.
beta	Special mathematical function related to the beta and gamma functions.
bindingIsActive, bindingIsLocked	These functions represent an experimental interface for adjustments to environments and bindings within environments. They allow for locking environments as well as individual bindings and for linking a variable to a function.
bindtextdomain	If Native Language Support was enabled in this build of R, attempts to translate character vectors or sets where the translations are to be found.
body, body<-	Get or set the body of a function.
bquote	Analog of the LISP backquote, macro. `bquote` quotes its argument except that terms wrapped in `.()` are evaluated in the specified `where` environment.
break	Basic control-flow constructs of the R language. They function in much the same way as control statements in any Algol-like language. They are all reserved words.
browser	Interrupts the execution of an expression and allows the inspection of the environment where `browser` was called from.

Function	Description
builtins	Returns the names of all the built-in objects. These are fetched directly from the symbol table of the R interpreter.
by	An object-oriented wrapper for `tapply` applied to data frames.
bzfile	Function to create, open, and close connections.
c	Combines its arguments.
call	Creates or tests for objects of mode "call".
callCC	Downward-only version of Scheme's call with current continuation.
capabilities	Reports on the optional features that have been compiled into this build of R.
casefold	Translates characters in character vectors, in particular, from upper- to lowercase or vice versa.
cat	Outputs the objects, concatenating the representations. `cat` performs much less conversion than `print`.
cbind	Takes a sequence of vector, matrix, or data frame arguments and combines by columns.
ceiling	Takes a single numeric argument x and returns a numeric vector containing the smallest integers not less than the corresponding elements of x.
char.expand	Seeks a unique match of its first argument among the elements of its second. If successful, it returns this element; otherwise, it performs an action specified by the third argument.
charToRaw	Conversion and manipulation of objects of type "raw".
character	Creates or tests for objects of type "character".
charmatch	Seeks matches for the elements of its first argument among those of its second.
chartr	Translates characters in character vectors, in particular, from upper- to lowercase or vice versa.
check_tzones	Description of the classes `"POSIXlt"` and `"POSIXct"` representing calendar dates and times (to the nearest second).
chol	Computes the Choleski factorization of a real, symmetric, positive-definite square matrix.
chol2inv	Inverts a symmetric, positive-definite square matrix from its Choleski decomposition. Equivalently, computes $(X'X)^{-1}$ from the (R part) of the QR-decomposition of X.
choose	Special mathematical function related to the beta and gamma functions.
class, class<-	The function `class` prints the vector of names of classes an object inherits from. Correspondingly, `class<-` sets the classes an object inherits from.
close, close.connection	Close connections.
closeAllConnections	Displays aspects of connections.
col	Returns a matrix of integers, indicating their column number in a matrix-like object or a factor of column labels.
colMeans	Forms row and column means for numeric arrays.
colSums	Forms row and column sums for numeric arrays.
colnames, colnames<-	Retrieve or set the row or column names of a matrix-like object.

Function	Description
commandArgs	Provides access to a copy of the command-line arguments supplied when this R session was invoked.
comment, comment<-	Set and query a *comment* attribute for any R objects.
computeRestarts	Provides a mechanism for handling unusual conditions, including errors and warnings.
conditionCall, conditionCall.condition, conditionMessage, conditionMessage.condition	Provide a mechanism for handling unusual conditions, including errors and warnings.
conflicts	Reports on objects that exist with the same name in two or more places on the search path, usually because an object in the user's workspace or a package is masking a system object of the same name. This helps discover unintentional masking.
contributors	The R Who's Who, describing who made significant contributions to the development of R.
crossprod	Given matrices x and y as arguments, returns a matrix cross-product. This is formally equivalent to (but usually slightly faster than) the call t(x) \%*\% y.
cummax, cummin, cumprod, cum-sum	Returns a vector whose elements are the cumulative sums, products, minima, or maxima of the elements of the argument.
cut	Divides the range of x into intervals and codes the values in x according to which interval they fall. The leftmost interval corresponds to level 1, the next leftmost to level 2 and so on.
cut.Date, cut.POSIXt	Method for cut applied to date-time objects.
dQuote	Single- or double-quote text by combining with appropriate single or double left and right quotation marks.
data.class	Determines the class of an arbitrary R object.
data.frame	Creates data frames.
data.matrix	Returns the matrix obtained by converting all the variables in a data frame to numeric mode and then binding them together as the columns of a matrix. Factors and ordered factors are replaced by their internal codes.
date	Returns a character string of the current system date and time.
debug	Sets, unsets, or queries the debugging flag on a function.
default.stringsAsFactors	Creates data frames, tightly coupled collections of variables that share many of the properties of matrices and lists, used as the fundamental data structure by most of R's modeling software.
delayedAssign	Creates a *promise* to evaluate a given expression if its value is requested. This provides direct access to the *lazy evaluation* mechanism used by R for the evaluation of (interpreted) functions.
deparse	Turns unevaluated expressions into character strings.
det, determinant	det calculates the determinant of a matrix. determinant is a generic function that returns separately the modulus of the determinant, optionally on the logarithm scale, and the sign of the determinant.

Function	Description
detach	Detaches a database, i.e., removes it from the `search()` path of available R objects. Usually, this is either a `data.frame` that has been `attached` or a package that was required previously.
dget	Writes an ASCII (American Standard Code for Information Interchange) text representation of an R object to a file or connection or uses one to re-create the object.
diag, diag<-	Extract or replace the diagonal of a matrix or construct a diagonal matrix.
diff, diff.Date, diff.POSIXt, diff.default	Return suitably lagged and iterated differences.
difftime	Creates, prints, and rounds time intervals.
digamma	Special mathematical function related to the beta and gamma functions.
dim, dim.data.frame, dim<-, dimnames, dimnames.data.frame, dimnames<-, dimnames<-.data.frame	Retrieve or set the dimension of an object.
dir, dir.create	Produce a character vector of the names of files in the named directory.
dirname	Returns the part of the `path` up to (but excluding) the last path separator, or "." if there is no path separator.
do.call	Constructs and executes a function call from a name or a function and a list of arguments to be passed to it.
double	Creates, coerces to, or tests for a double-precision vector.
dput	Writes an ASCII text representation of an R object to a file or connection or uses one to re-create the object.
drop	Deletes the dimensions of an array that has only one level.
dump	Takes a vector of names of R objects and produces text representations of the objects on a file or connection. A dump file can usually be `sourced` into another R (or S) session.
duplicated	Determines which elements of a vector or data frame are duplicates of elements with smaller subscripts and returns a logical vector indicating which elements (rows) are duplicates.
dyn.load, dyn.unload	Load or unload DLLs (also known as shared objects) and test whether a C function or FORTRAN subroutine is available.
eapply	Applies FUN to the named values from an environment and returns the results as a list.
eigen	Computes eigenvalues and eigenvectors of real or complex matrices.
emptyenv	Gets, sets, tests for, and creates environments.
encodeString	Escapes the strings in a character vector in the same way `print.default` does and optionally fits the encoded strings within a field width.
env.profile	This function is intended to assess the performance of hashed environments.
environment, environment<-	Gets or sets the environment associated with a function or formula.
environmentIsLocked	Returns a logical environment indicating if an environment is locked.
environmentName	Returns the name of an environment.

Function	Description
eval, eval.parent, evalq	Evaluate an R expression in a specified environment.
exists	Looks for an R object of a given name.
exp	Computes the exponential function.
expand.grid	Creates a data frame from all combinations of the supplied vectors or factors.
expm1	Computes $e^x - 1$ accurately for $x \ll 1$.
expression	Creates objects of mode `"expression"`.
factor	Used to encode a vector as a factor.
factorial	Special mathematical function related to the beta and gamma functions.
fifo	Creates a FIFO connection.
file	Creates a file connection.
file.access	Utility function to access information about files on the user's file systems.
file.append	Provides a low-level interface to the computer's file system.
file.choose	Chooses a file interactively.
file.copy, file.create, file.exists	Provide a low-level interface to the computer's file system.
file.info	Utility function to extract information about files on the user's file systems.
file.path	Constructs the path to a file from components in a platform-independent way.
file.remove, file.rename	Provides a low-level interface to the computer's file system.
file.show	Displays one or more files.
file.symlink	Provides a low-level interface to the computer's file system.
findInterval	Finds the indices of x in vec, where vec must be sorted (nondecreasingly).
findRestart	Provides a mechanism for handling unusual conditions, including errors and warnings.
floor	Takes a single numeric argument x and returns a numeric vector containing the largest integers not greater than the corresponding elements of x.
flush, flush.connection	Functions to create, open, and close connections.
force	Forces the evaluation of a function argument.
formals, formals<-	Get or set the formal arguments of a function.
format, format.AsIs	Format an R object for pretty printing.
formatC	Formats numbers individually and flexibly, using C-style format specifications.
formatDL	Formats vectors of items and their descriptions as two-column tables or LaTeX-style description lists.
forwardsolve	Solves a system of linear equations where the coefficient matrix is upper or lower triangular.
function	Provides the base mechanisms for defining new functions in the R language.
gamma	Special mathematical function related to the beta and gamma functions.
gc	Causes garbage collection to take place.
gc.time	Reports the time spent in garbage collection so far in the R session while GC timing was enabled.

Function	Description
gcinfo	Sets a flag so that automatic collection is either silent (`verbose=FALSE`) or prints memory usage statistics (`verbose=TRUE`).
gctorture	Provokes garbage collection on (nearly) every memory allocation. Intended to ferret out memory protection bugs. Also makes R run *very* slowly, unfortunately.
get	Searches for an R object with a given name and returns it.
getAllConnections	Displays aspects of connections.
getCConverterDescriptions, getCConverterStatus	Provide facilities to manage the extensible list of converters used to translate R objects into C pointers for use in `.C` calls. The number and a description of each element in the list can be retrieved. One can also query and set the activity status of individual elements, temporarily ignoring them. And one can remove individual elements.
getConnection	Displays aspects of connections.
getDLLRegisteredRoutines, getDLLRegistered Routines.DLLInfo, getDLLRegistered Routines.character	These functions allow us to query the set of routines in a DLL that are registered with R to enhance dynamic lookup, error handling when calling native routines, and potentially security in the future. These functions provide a description of each of the registered routines in the DLL for the different interfaces, i.e., `.C`, `.Call`, `.Fortran`, and `.External`.
getExportedValue	Function to support reflection on namespace objects.
getHook	Allows users to set actions to be taken before packages are attached/detached and namespaces are (un)loaded.
getLoadedDLLs	Provides a way to get a list of all the DLLs (see `dyn.load`) that are currently loaded in the R session.
getNamespace, getNamespaceExports, getNamespaceImports, getNamespaceName, getNamespaceUsers, getNamespaceVersion	Functions to support reflection on namespace objects.
getNativeSymbolInfo	Finds and returns as comprehensive a description of one or more dynamically loaded or "exported" built-in native symbols.
getNumCConverters	Used to manage the extensible list of converters used to translate R objects into C pointers for use in `.C` calls. Returns an integer giving the number of elements in a specified list, both active and inactive.
getOption	Allows the user to set and examine a variety of global *options* that affect the way in which R computes and displays its results.
getRversion	A simple S3 class for representing numeric versions, including package versions, and associated methods.
getSrcLines	This function is for working with source files.
getTaskCallbackNames	Provides a way to get the names (or identifiers) for the currently registered task callbacks that are invoked at the conclusion of each top-level task. These identifiers can be used to remove a callback.
geterrmessage	Gives the last error message.
gettext	If Native Language Support was enabled in this build of R, attempts to translate character vectors or set where the translations are to be found.

Function	Description
gettextf	A wrapper for the C function `sprintf` that returns a character vector containing a formatted combination of text and variable values.
getwd	Returns an absolute filename representing the current working directory of the R process.
gl	Generates factors by specifying the pattern of their levels.
globalenv	Gets, sets, tests for, and creates environments.
gregexpr, grep, grepl, gsub	`grep` searches for matches to `pattern` (its first argument) within the character vector x (second argument). `grepl` is an alternative way to return the results. `regexpr` and `gregexpr` return results, too, but they return more detail in a different format.
gsub	`sub` and `gsub` perform replacement of matches determined by regular expression matching.
gzcon	Provides a modified connection that wraps an existing connection and decompresses reads or compresses writes through that connection. Standard `gzip` headers are assumed.
gzfile	Function to create, open, and close connections.
iconv, iconvlist	These use system facilities to convert a character vector between encodings: the "i" stands for "internationalization."
icuSetCollate	Controls the way collation is done by ICU (an optional part of the R build).
identical	The safe and reliable way to test two objects for being *exactly* equal. It returns TRUE in this case, FALSE in every other case.
identity	A trivial identity function returning its argument.
ifelse	Returns a value with the same shape as `test` that is filled with elements selected from either yes or no, depending on whether the element of `test` is TRUE or FALSE.
inherits	Indicates whether its first argument inherits from any of the classes specified in the what argument.
intToBits	Conversion and manipulation of objects of type `"raw"`.
intToUtf8	Conversion of UTF-8 encoded character vectors to and from integer vectors.
integer	Creates or tests for objects of type `"integer"`.
interaction	Computes a factor that represents the interaction of the given factors. The result of `interaction` is always unordered.
interactive	Returns TRUE when R is being used interactively and FALSE otherwise.
intersect	Performs *set* union, intersection, (asymmetric!) difference, equality, and membership on two vectors.
inverse.rle	Computes the lengths and values of runs of equal values in a vector—or the reverse operation.
invisible	Returns a (temporarily) invisible copy of an object.
invokeRestart, invokeRestartInteractively	Provide a mechanism for handling unusual conditions, including errors and warnings.
is.R	Tests if running under R.

Function	Description
is.array	Creates or tests for arrays.
is.atomic	Returns TRUE if x is an atomic vector (or NULL) and FALSE otherwise.
is.call	Tests for objects of mode "call".
is.character	Tests for objects of type "character".
is.complex	Tests for objects of type "complex".
is.data.frame	Tests if an object is a data frame.
is.double	Tests for a double-precision vector.
is.element	Tests if an element is a member of a set.
is.environment	Tests if an object is an environment.
is.expression	Tests if an object is an "expression".
is.factor	Returns a logical value indicating if an object is a factor.
is.finite, is.infinite	Return a vector of the same length as x, indicating which elements are finite (not infinite and not missing).
is.function	Checks whether its argument is a (primitive) function.
is.integer	Creates or tests for objects of type "integer".
is.language	Returns TRUE if x is a variable name, a call, or an expression.
is.list	Tests if an object is a list.
is.loaded	Tests whether a C function or FORTRAN subroutine is available.
is.logical	Tests for objects of type "logical".
is.matrix	Tests if its argument is a (strict) matrix.
is.na, is.na<-	The generic function is.na indicates which elements in an object are missing. The generic function is.na<- sets elements to NA.
is.name	Returns TRUE or FALSE, depending on whether the argument is a name or not.
is.nan	Tests if an object is a NaN (meaning "not a number").
is.null	Returns TRUE if its argument is NULL and FALSE otherwise.
is.numeric, is.numeric.Date, is.numeric.POSIXt	A general test of an object being interpretable as numbers.
is.numeric_version	A simple S3 class for representing numeric versions, including package versions, and associated methods.
is.ordered	Tests if an object is an ordered factor.
is.package_version	Tests for a package_version object.
is.pairlist	Tests for a pairlist object.
is.primitive	Checks whether its argument is a (primitive) function.
is.qr	Tests whether an object is the QR-decomposition of a matrix (created by the qr function).
is.raw	Tests for objects of type "raw".
is.recursive	Returns TRUE if x has a recursive (listlike) structure and FALSE otherwise.

Function	Description
is.symbol	is.symbol (and the identical is.name) returns TRUE or FALSE, depending on whether the argument is a name or not.
is.table	table uses the cross-classifying factors to build a contingency table of the counts at each combination of factor levels.
is.unsorted	Tests if an object is not sorted, without the cost of sorting it.
is.vector	Returns TRUE if x is a vector (of mode logical, integer, real, complex, character, raw or a list if not specified) or expression and FALSE otherwise.
isIncomplete, isOpen	Functions to create, open, and close connections.
isRestart	Provides a mechanism for handling unusual conditions, including errors and warnings.
isS4	Tests whether the object is an instance of an S4 class.
isSeekable	Function to reposition connections.
isSymmetric, isSymmetric.matrix	Generic functions to test if object is symmetric or not. Currently, only a matrix method is implemented.
isTRUE	This operator acts on logical vectors.
isdebugged	Sets, unsets, or queries the debugging flag on a function.
jitter	Adds a small amount of noise to a numeric vector.
julian, julian.Date, julian.POSIXt	Extract the weekday, month, or quarter, or the Julian time (days since some origin). These are generic functions: the methods for the internal date-time classes are documented here.
kappa, kappa.defaultm, kappa.lm, kappa.qr, kappa.tri	The condition number of a regular (square) matrix is the product of the norm of the matrix and the norm of its inverse (or pseudoinverse) and hence depends on the kind of matrix norm. kappa() computes an estimate of the 2-norm condition number of a matrix or of the R matrix of a QR-decomposition, perhaps of a linear fit. The 2-norm condition number can be shown to be the ratio of the largest to the smallest *nonzero* singular value of the matrix.
kronecker	Computes the generalized Kronecker product of two arrays, X and Y. \%x\% is an alias for kronecker (where FUN is hardwired to "*").
l10n_info	Reports on localization information.
labels	Finds a suitable set of labels from an object for use in printing or plotting, for example.
lapply	Returns a list of the same length as X, each element of which is the result of applying FUN to the corresponding element of X.
lazyLoad	Lazy loads a database of R objects.
lbeta, lchoose	Special mathematical functions related to the beta and gamma functions.
length, length<-, length<-.factor	Get or set the length of vectors (including lists) and factors and of any other R object for which a method has been defined.
levels, levels.default, levels<-, levels<-.factor	Provide access to the levels attribute of a variable.
lfactorial, lgamma	Special mathematical functions related to the beta and gamma functions.
library	library and require load add-on packages.

Function	Description
library.dynam	Loads a specified file of compiled code if it has not been loaded already, or unloads it.
library.dynam.unload	Loads a specified file of compiled code if it has not been loaded already, or unloads it.
licence	The license terms under which R is distributed.
list	Function to construct, coerce, and check for both kinds of R lists.
list.files	Produces a character vector of the names of files in the named directory.
load	Reloads data sets written with the function `save`.
loadNamespace	Loads the specified namespace and registers it in an internal database.
loadedNamespaces	Returns a character vector of the names of the loaded namespaces.
loadingNamespaceInfo	Returns a list of the arguments that would be passed to .onLoad when a namespace is being loaded.
local	Evaluates an R expression in a specified environment.
lockBinding	Locks individual bindings in a specified environment.
lockEnvironment	Locks its environment argument, which must be a normal environment (not base).
log	Computes logarithms, by default natural logarithms.
log10	Computes common (i.e., base 10) logarithms.
log1p	Computes $\log(1 + x)$ accurately for $|x| \ll 1$ (and less accurately when $x \approx -1$).
log2	Computes binary (i.e., base 2) logarithms.
logical	Creates or tests for objects of type `"logical"` and the basic logical constants.
lower.tri	Returns a matrix of logicals the same size of a given matrix with entries TRUE in the lower or upper triangle.
ls	`ls` and `objects` return a vector of character strings giving the names of the objects in a specified environment.
make.names	Makes syntactically valid names out of character vectors.
make.unique	Makes the elements of a character vector unique by appending sequence numbers to duplicates.
makeActiveBinding	Installs fun so that getting the value of sym calls fun with no arguments, and assigning to sym calls fun with one argument, the value to be assigned.
mapply	A multivariate version of `sapply`. mapply applies FUN to the first elements of each ... argument, the second elements, the third elements, and so on.
margin.table	For a contingency table in array form, computes the sum of table entries for a given index.
mat.or.vec	Creates an nr by nc zero matrix if nc is greater than 1, and a zero vector of length nr if nc equals 1.
match	Returns a vector of the positions of (first) matches of its first argument in its second.
match.arg	Matches arg against a table of candidate values as specified by `choices`, where NULL means to take the first one.
match.call	Returns a call in which all of the specified arguments are specified by their full names.

Function	Description
match.fun	When called inside functions that take a function as argument, extracts the desired function object while avoiding undesired matching to objects of other types.
matrix	Creates a matrix from a given set of values. `as.matrix` attempts to turn its argument into a matrix. `is.matrix` tests if its argument is a (strict) matrix.
max	Returns the (parallel) maxima and minima of the input values.
max.col	Finds the maximum position for each row of a matrix, breaking ties at random.
mean	Generic function for the (trimmed) arithmetic mean.
memory.profile	Lists the usage of the cons cells by SEXPREC type.
merge, merge.data.frame, merge.default	Merge two data frames by common columns or row names or perform other versions of database *join* operations.
message	Generates a diagnostic message from its arguments.
mget	Searches for an R object with a given name and returns it.
min	Returns the (parallel) maxima and minima of the input values.
missing	Tests whether a value was specified as an argument to a function.
mode, mode<-	Get or set the type of storage mode of an object.
months	Extracts the months from an object.
mostattributes<-	The mostattributes assignment takes special care of the dim, names, and dimnames attributes and assigns them only when valid, whereas an attributes assignment would give an error if any were not.
names, names<-	Functions to get or set the names of an object.
nargs	When used inside a function body, returns the number of arguments supplied to that function, *including* positional arguments left blank.
nchar	Takes a character vector as an argument and returns a vector whose elements contain the sizes of the corresponding elements of x.
ncol, nrow	Return the number of rows or columns present in x.
new.env	Gets, sets, tests for, and creates environments.
ngettext	If Native Language Support was enabled in this build of R, attempts to translate character vectors or set where the translations are to be found.
nlevels	Returns the number of levels that its argument has.
noquote	Prints character strings without quotes.
numeric	Creates or coerces objects of type `"numeric"`. `is.numeric` is a more general test of an object being interpretable as numbers.
numeric_version	A simple S3 class for representing numeric versions, including package versions, and associated methods.
nzchar	A fast way to find out if elements of a character vector are nonempty strings.
objects	`ls` and `objects` return a vector of character strings giving the names of the objects in a specified environment.
oldClass, oldClass<-	Get and set the `class` attribute.
on.exit	Records the expression given as its argument as needing to be executed when the current function exits (either naturally or as the result of an error).

Function	Description
open, open.connection	Functions to create, open, and close connections.
open.srcfile, open.srcfilecopy	These functions are for working with source files.
options	Allows the user to set and examine a variety of global *options* that affect the way in which R computes and displays its results.
order	Returns a permutation that rearranges its first argument into ascending or descending order, breaking ties by further arguments.
ordered	Used to create ordered factors.
outer	The outer product of the arrays X and Y is the array A with dimension `c(dim(X)`, `dim(Y))`, where element `A[c(arrayindex.x, arrayindex.y)]` = `FUN(X[arrayindex.x], Y[arrayindex.y], ...)`.
packBits	Conversion and manipulation of objects of type `"raw"`.
packageEvent	`setHook` provides a general mechanism for users to register hooks, a list of functions to be called from system (or user) functions. The initial set of hooks is associated with events on packages/name spaces: these hooks are named via calls to `package Event`.
packageStartupMessage	Generates a diagnostic message from its arguments.
package_version	Creates a package_version object (a simple S3 class for representing numeric versions, including package versions, and associated methods).
pairlist	Function to construct, coerce, and check for both kinds of R lists.
parent.env, parent.env<-	Get, set, test for, and create environments.
parent.frame	Provides access to `environments` ("frames" in S terminology) associated with functions farther up the calling stack.
parse	Returns the parsed but unevaluated expressions in a list.
parseNamespaceFile	Internal namespace support function. Not intended to be called directly.
paste	Concatenates vectors after converting to character.
path.expand	Expands a path name, for example, by replacing a leading tilde by the user's home directory (if defined on that platform).
pipe	Function to create, open, and close connections.
pmatch	Seeks matches for the elements of its first argument among those of its second.
pmax, pmax.int, pmin, pmin.int	Return the (parallel) maxima and minima of the input values.
polyroot	Finds zeros of a real or complex polynomial.
pos.to.env	Returns the environment at a specified position in the search path.
pretty	Computes a sequence of about n+1 equally spaced "round" values that cover the range of the values in x. The values are chosen so that they are 1, 2, or 5 times a power of 10.
prettyNum	Formats numbers individually and flexibly, using C-style format specifications.
print	Prints its argument and returns it *invisibly* (via `invisible(x)`).
prmatrix	An earlier method for printing matrices, provided for S compatibility.
proc.time	Determines how much real and CPU time (in seconds) the currently running R process has already taken.

Function	Description
prod	Returns the product of all the values present in its arguments.
prop.table	This is really `sweep(x, margin, margin.table(x, margin), "/")` for newbies, except that if `margin` has length 0, then one gets `x/sum(x)`.
psigamma	Special mathematical function related to the beta and gamma functions.
pushBack, pushBackLength	Functions to push back text lines onto a connection and to inquire about how many lines are currently pushed back.
q	Alias for quit.
qr	Computes the *QR*-decomposition of a matrix. It provides an interface to the techniques used in the LINPACK routine DQRDC or the LAPACK routines DGEQP3 and (for complex matrices) ZGEQP3.
qr.coef	Returns the coefficients obtained when fitting y to the matrix with *QR*-decomposition qr.
qr.qy	Returns Q %*% y, where Q is the (complete) Q matrix.
qr.qty	Returns t(Q) %*% y, where Q is the (complete) Q matrix.
qr.resid	Returns the residuals obtained when fitting y to the matrix with *QR*-decomposition qr.
qr.solve	Solves systems of equations via the *QR*-decomposition: if *a* is a *QR*-decomposition, it is the same as solve.qr, but if *a* is a rectangular matrix, the *QR*-decomposition is computed first.
qr.fitted	Returns the fitted values obtained when fitting y to the matrix with *QR*-decomposition qr.
qr.Q, qr.R, qr.X	Returns the original matrix from which the object was constructed or the components of the decomposition.
quarters	Extracts the quarter from an object.
quit	The function `quit` or its alias q terminates the current R session.
quote	Simply returns its argument. The argument is not evaluated and can be any R expression.
range	Returns a vector containing the minimum and maximum of all the given arguments.
rank	Returns the sample ranks of the values in a vector. Ties (i.e., equal values) and missing values can be handled in several ways.
rapply	A recursive version of `lapply`.
raw	Creates or tests for objects of type `"raw"`.
rawConnection, rawConnectionValue	Input and output raw connections.
rawShift, rawToBits, rawToChar	Conversion and manipulation of objects of type `"raw"`.
rbind	Takes a sequence of vector, matrix, or data frame argument and combines by rows, respectively.
rcond	Computes the 1- and inf-norm condition numbers for a matrix, also for complex matrices, using standard LAPACK routines.
read.dcf	Reads or writes an R object from/to a file in Debian control file format.
readBin	Reads binary data from a connection or writes binary data to a connection.

Function	Description
readChar	Transfers character strings to and from connections, without assuming they are null terminated on the connection.
readLines	Reads some or all text lines from a connection.
readline	Reads a line from the terminal.
real	This function is the same as its `double` equivalents and is provided for backward compatibility only.
reg.finalizer	Registers an R function to be called upon garbage collection of objects or (optionally) at the end of an R session.
regexpr	Searches for matches to `pattern` (its first argument) within the character vector x (second argument), and returns detailed results.
remove, rm	Used to remove objects.
removeCConverter	Returns TRUE if an element in the converter list was identified and removed. (This function provides facilities to manage the extensible list of converters used to translate R objects into C pointers for use in `.C` calls.)
removeTaskCallback	Un-registers a function that was registered earlier via `addTaskCallback`.
rep, rep.int	`rep` replicates the values in x. It is a generic function, and the (internal) default method is described here. `rep.int` is a faster simplified version for the most common case.
replace	Replaces the values in x with indices given in `list` by those given in `values`. If necessary, the values in `values` are recycled.
replicate	A wrapper for the common use of `sapply` for repeated evaluation of an expression (which will usually involve random number generation).
require	`library` and `require` load add-on packages.
retracemem	Marks an object so that a message is printed whenever the internal function `duplicate` is called. This happens when two objects share the same memory and one of them is modified. It is a major cause of hard-to-predict memory use in R.
return	Provides the base mechanisms for defining new functions in the R language.
rev	Provides a reversed version of its argument. It is generic function with a default method for vectors and one for `dendrograms`.
rle	Computes the lengths and values of runs of equal values in a vector—or the reverse operation.
round	Rounds the values in its first argument to the specified number of decimal places (default 0).
round.POSIXt	Rounds or truncates date-time objects.
row	Returns a matrix of integers indicating their row number in a matrix-like object or a factor indicating the row labels.
row.names, row.names<-	Get or set the row names attribute from an object (such as a data frame).
rowMeans, rowSums	Form row and column sums and means for numeric arrays.
rownames, rownames<-	Retrieve or set the row or column names of a matrix-like object.
rowsum	Computes column sums across rows of a matrix-like object for each level of a grouping variable.

Function	Description
sQuote	Single- or double-quote text by combining with appropriate single or double left and right quotation marks.
sample	Takes a sample of specified size from the elements of x using either with or without replacement.
sapply	A user-friendly version of `lapply` by default returning a vector or matrix if appropriate.
save	Writes an external representation of R objects to a specified file. The objects can be read back from the file at a later date by using the function `load` (or `data` in some cases).
save.image	Just a shortcut for "save my current workspace," i.e., `save(list = ls(all=TRUE), file = ".RData")`. It is also what happens with `q("yes")`.
saveNamespaceImage	Low-level namespace support function.
scale	A generic function whose default method centers and/or scales the columns of a numeric matrix.
scan	Reads data into a vector or list from the console or file.
search	Gives a list of `attached` *packages* (see `library`), and R objects, usually `data.frames`.
searchpaths	Gives a similar character vector to `search`, with the entries for packages being the path to the package used to load the code.
seek, seek.connection	Functions to reposition connections.
seq, seq.int, seq_along, seq_len	Generate regular sequences. `seq` is a standard generic with a default method. `seq.int` is an internal generic that can be much faster but has a few restrictions. `seq_along` and `seq_len` are very fast primitives for two common cases.
sequence	For each element of nvec, the sequence `seq_len(nvec[i])` is created. These are concatenated and the result returned.
serialize	A simple low-level interface for serializing to connections.
set.seed	The recommended way to specify seeds for randon number generation.
setCConverterStatus	Provides facilities to manage the extensible list of converters used to translate R objects into C pointers for use in `.C` calls. The number and a description of each element in the list can be retrieved. One can also query and set the activity status of individual elements, temporarily ignoring them. And one can remove individual elements.
setHook	Allows users to set actions to be taken before packages are attached/detached and namespaces are (un)loaded.
setSessionTimeLimit, setTimeLimit	Functions to set CPU and/or elapsed time limits for top-level computations or the current session.
setdiff, setequal	Perform *set* union, intersection, (asymmetric!) difference, equality, and membership on two vectors.
setwd	Used to set the working directory to `dir`.
shQuote	Quotes a string to be passed to an operating system shell.
showConnections	Displays aspects of connections.

Function	Description
sign	Returns a vector with the signs of the corresponding elements of x (the sign of a real number is 1, 0, or −1 if the number is positive, zero, or negative, respectively).
signalCondition	Provides a mechanism for handling unusual conditions, including errors and warnings.
signif	Rounds the values in its first argument to the specified number of significant digits.
simpleCondition, simpleError, simpleMessage, simpleWarning	Provide a mechanism for handling unusual conditions, including errors and warnings.
sin	Computes the sine.
sinh	Computes the hyperbolic sine.
sink, sink.number	`sink` diverts R output to a connection. `sink.number()` reports how many diversions are in use. `sink.number(type = "message")` reports the number of the connection currently being used for error messages.
slice.index	Returns a matrix of integers indicating the number of their slice in a given array.
socketConnection	Function to create, open, and close connections.
socketSelect	Waits for the first of several socket connections to become available.
solve	This generic function solves the equation a `\%*\%` x = b for x, where b can be either a vector or a matrix.
solve.qr	The method of solve for qr objects.
sort	Sorts (or *orders*) a vector or factor (partially) into ascending (or descending) order.
source	Causes R to accept its input from the named file or URL (the name must be quoted) or connection.
split, split<-	`split` divides the data in the vector x into the groups defined by f. The replacement forms replace values corresponding to such a division.
sprintf	A wrapper for the C function `sprintf` that returns a character vector containing a formatted combination of text and variable values.
sqrt	Computes miscellaneous mathematical functions. The naming follows the standard for computer languages such as C or FORTRAN.
srcfile	This function is for working with source files.
srcfilecopy	This function is for working with source files.
srcref	This function is for working with source files.
standardGeneric	Dispatches the method defined for a generic function f, using the actual arguments in the frame from which it is called.
stderr	Displays aspects of connections.
stdin	Displays aspects of connections.
stdout	Displays aspects of connections.
stop	Stops execution of the current expression and executes an error action.
stopifnot	If any of the expressions in `...` are not `all` TRUE, `stop` is called, producing an error message indicating the *first* of the elements of `...` which were not true.
storage.mode, storage.mode<-	Get or set the type of storage mode of an object.

Function	Description
strftime, strptime	Functions to convert between character representations and objects of classes "POSIXlt" and "POSIXct" representing calendar dates and times.
strsplit	Splits the elements of a character vector x into substrings according to the presence of substring `split` within them.
strtrim	Trims character strings to specified display widths.
structure	Returns a given object with further attributes set.
strwrap	Wraps character strings to format paragraphs.
sub	Performs replacement of matches determined by regular expression matching.
subset	Returns subsets of vectors, matrices, or data frames that meet conditions.
substitute	Returns the parse tree for the (unevaluated) expression `expr`, substituting any variables bound in `env`.
substr, substr<-, substring, substring<-	Extract or replace substrings in a character vector.
sum	Returns the sum of all the values present in its arguments.
summary	A generic function used to produce summaries of the results of various model fitting.
suppressMessages, suppressPackageStartupMessages	Generate a diagnostic message from their arguments.
suppressWarnings	Generates a warning message that corresponds to its argument(s) and (optionally) the expression or function from which it was called.
svd	Computes the singular-value decomposition of a rectangular matrix.
sweep	Returns an array obtained from an input array by sweeping out a summary statistic.
switch	Evaluates EXPR and accordingly chooses one of the additional arguments (in `. . .`).
sys.call, sys.calls, sys.frame, sys.frames, sys.function, sys.nframe, sys.on.exit, sys.parent, sys.parents	Provide access to `environments` ("frames" in S terminology) associated with functions farther up the calling stack.
sys.source	Parses expressions in a given file and then successively evaluates them in the specified environment.
sys.status	Provides access to `environments` ("frames" in S terminology) associated with functions farther up the calling stack.
system	Invokes the OS command specified by `command`.
system.file	Finds the full filenames of files in packages, etc.
system.time	Returns CPU (and other) times that `expr` used.
t	Given a matrix or `data.frame` x, t returns the transpose of x. Methods include `t.data.frame` and `t.default`.
table	Uses the cross-classifying factors to build a contingency table of the counts at each combination of factor levels.
tabulate	Takes the integer-valued vector `bin` and counts the number of times each integer occurs in it.
tan	Computes the tangent.

Function	Description
tanh	Computes the hyperbolic tangent.
tapply	Applies a function to each cell of a ragged array, i.e., to each (nonempty) group of values given by a unique combination of the levels of certain factors.
taskCallbackManager	Provides an entirely S-language mechanism for managing callbacks or actions that are invoked at the conclusion of each top-level task. Essentially, we register a single R function from this manager with the underlying native task-callback mechanism, and this function handles invoking the other R callbacks under the control of the manager. The manager consists of a collection of functions that access shared variables to manage the list of user-level callbacks.
tcrossprod	Given matrices x and y as arguments, returns a matrix cross-product. This is formally equivalent to (but usually slightly faster than) the call x \%*\% t(y) (tcros sprod).
tempdir, tempfile	tempfile returns a vector of character strings that can be used as names for temporary files.
textConnection, textConnection-Value	Input and output text connections.
toString	This is a helper function for format to produce a single-character string describing an R object.
tolower	Translates characters in character vectors, in particular, from upper- to lowercase or vice versa.
topenv	Finds the top-level environment.
toupper	Translates characters in character vectors, in particular, from upper-to lower case or vice versa.
trace	A call to trace allows you to insert debugging code (e.g., a call to browser or recover) at chosen places in any function. A call to untrace cancels the tracing.
traceback	By default, traceback() prints the call stack of the last uncaught error, i.e., the sequence of calls that led to the error.
tracemem	Marks an object so that a message is printed whenever the internal function duplicate is called.
tracingState	Tracing can be temporarily turned on or off globally by calling tracingState.
transform	Returns a new data frame by applying a set of transformations to an existing data frame.
trigamma	Special mathematical functions related to the beta and gamma functions.
trunc	Takes a single numeric argument x and returns a numeric vector containing the integers formed by truncating the values in x toward 0.
truncate, truncate.connection	Functions to reposition connections.
try	A wrapper to run an expression that might fail and allow the user's code to handle error recovery.
tryCatch	Provides a mechanism for handling unusual conditions, including errors and warnings.
typeof	Determines the (R internal) type of storage mode of any object.
unclass	Returns (a copy of) its argument with its class attribute removed.

Function	Description
undebug	Sets, unsets, or queries the debugging flag on a function.
union	Performs *set* union, intersection, (asymmetric!) difference, equality, and membership on two vectors.
unique	Returns a vector, data frame, or array like x but with duplicate elements/rows removed.
units, units<-	Extracts units from a difftime object.
unix.time	Returns CPU (and other) times that expr used.
unlink	Deletes the file(s) or directories specified by x.
unlist	Given a list structure x, simplifies it to produce a vector that contains all the atomic components that occur in x.
unloadNamespace	Function to load and unload namespaces.
unlockBinding	Unlocks individual bindings in a specified environment.
unname	Removes the names or dimnames attribute of an R object.
unserialize	A simple low-level interface for serializing to connections.
unsplit	Reverses the effect of split.
untrace	A call to trace allows you to insert debugging code (e.g., a call to browser or recover) at chosen places in any function. A call to untrace cancels the tracing.
untracemem	Undoes a call to tracemem.
unz	Function to create, open, and close connections.
upper.tri	Returns a matrix of logicals the same size of a given matrix with entries TRUE in the lower or upper triangle.
url	Function to create, open, and close connections.
utf8ToInt	Conversion of UTF-8 encoded character vectors to and from integer vectors.
vector	Produces a vector of a given length and mode.
warning	Generates a warning message that corresponds to its argument(s) and (optionally) the expression or function from which it was called.
warnings	warnings and its print method print the variable last.warning in a pleasing form.
weekdays	Extracts the weekdays from an object.
which	Gives the TRUE indices of a logical object, allowing for array indices.
which.max, which.min	Determines the location, i.e., index of the (first) minimum or maximum of a numeric vector.
with	Evaluates an R expression in an environment constructed from data, possibly modifying the original data.
withCallingHandlers	Calling handlers are established by withCallingHandlers.
withRestarts	Restarts are used for establishing recovery protocols. They can be established using withRestarts.
withVisible	Evaluates an expression, returning it in a two-element list containing its value and a flag showing whether it would automatically print.

Function	Description
within	Evaluates an R expression in an environment constructed from data, possibly modifying the original data.
write	The data (usually a matrix) x is written to file file. If the object is a two-dimensional matrix, you need to transpose it to get the columns in file the same as those in the internal representation.
write.dcf	Reads or writes an R object from/to a file in Debian control file format.
writeBin	Reads binary data from a connection or writes binary data to a connection.
writeChar	Transfers character strings to and from connections, without assuming they are null terminated on the connection.
writeLines	Writes text lines to a connection.
xor	This operator acts on logical vectors.
xpdrows.data.frame	Auxiliary function for use with data frames.
xtfrm	A generic auxiliary function that produces a numeric vector that will sort in the same order as x.
zapsmall	Determines a digits argument dr for calling round(x, digits = dr) such that values close to 0 (compared with the maximal absolute value) are "zapped," i.e., treated as 0.
~	Tilde is used to separate the left- and righthand sides in model formula.

Data Sets

Dataset	Class	Description
F	logical	Alias for FALSE.
LETTERS	character	Constants built into R.
R.version, version	simple.list	R.Version() provides detailed information about the version of R running. R.version is a variable (a list) holding this information (and version is a copy of it for S compatibility).
R.version.string	character	R.version.string is a copy of R.version$version.string.
T	logical	Alias for TRUE.
letters, month.abb, month.name	character	Vectors of constants built into R.
pi	numeric	Alias for the constant pi.

boot

This package provides functions for bootstrap resampling.

Functions

Function	Description
EEF.profile	Calculates the log-likelihood for a mean using an empirical exponential family likelihood.
EL.profile	Calculates the log-likelihood for a mean using an empirical likelihood.
abc.ci	Calculates equitailed two-sided nonparametric approximate bootstrap confidence intervals for a parameter, given a set of data and an estimator of the parameter, using numerical differentiation.
boot	Generates R bootstrap replicates of a statistic applied to data.
boot.array	Takes a bootstrap object calculated by one of the functions `boot`, `censboot`, or `tilt.boot` and returns the frequency (or index) array for the bootstrap resamples.
boot.ci	Generates five different types of equitailed two-sided nonparametric confidence intervals. These are the first-order normal approximation, the basic bootstrap interval, the Studentized bootstrap interval, the bootstrap percentile interval, and the adjusted bootstrap percentile (BCa) interval. All or a subset of these intervals can be generated.
censboot	Applies types of bootstrap resampling that have been suggested to deal with right-censored data. It can also perform model-based resampling using a Cox regression model.
control	Finds control variate estimates from a bootstrap output object.
corr	Calculates the weighted correlation given a data set and a set of weights.
cum3	Calculates an estimate of the third cumulant, or skewness, of a vector. Also, if more than one vector is specified, a product-moment of order 3 is estimated.
cv.glm	Calculates the estimated K-fold cross-validation prediction error for generalized linear models.
empinf	Calculates the empirical influence values for a statistic applied to a data set.
envelope	Calculates overall and pointwise confidence envelopes for a curve based on bootstrap replicates of the curve evaluated at a number of fixed points.
exp.tilt	Calculates exponentially tilted multinomial distributions such that the resampling distributions of the linear approximation to a statistic have the required means.
freq.array	Takes a matrix of indices for nonparametric bootstrap resamples and returns the frequencies of the original observations in each resample.
glm.diag	Calculates jackknife deviance residuals, standardized deviance residuals, standardized Pearson residuals, approximate Cook statistic, leverage, and estimated dispersion.
glm.diag.plots	Makes plot of jackknife deviance residuals against linear predictor, normal scores plots of standardized deviance residuals, plot of approximate Cook statistics against leverage/(1 − leverage), and case plot of Cook statistic.
imp.moments, imp.prob, imp.quantile	Central moment, tail probability, and quantile estimates for a statistic under importance resampling.
imp.weights	Calculates the importance sampling weight required to correct for simulation from a distribution with probabilities p when estimates are required assuming that simulation was from an alternative distribution with probabilities q.

Function	Description
inv.logit	Given a numeric object, returns the inverse logit of the values.
jack.after.boot	Calculates the jackknife influence values from a bootstrap output object and plots the corresponding jackknife-after-bootstrap plot.
k3.linear	Estimates the skewness of a statistic from its empirical influence values.
lik.CI	Function for use with the practicals in Davison and Hinkley (1997), *Bootstrap Methods and Their Applications*, Cambridge Series in Statistical and Probabilistic Mathematics, No. 1.
linear.approx	Takes a bootstrap object and, for each bootstrap replicate, calculates the linear approximation to the statistic of interest for that bootstrap sample.
logit	Calculates the logit of proportions.
nested.corr	Function for use with the practicals in Davison and Hinkley (1997), *Bootstrap Methods and Their Applications*, Cambridge Series in Statistical and Probabilistic Mathematics, No. 1.
norm.ci	Using the normal approximation to a statistic, calculates equitailed two-sided confidence intervals.
saddle	Calculates a saddlepoint approximation to the distribution of a linear combination of W at a particular point u, where W is a vector of random variables.
saddle.distn	Approximates an entire distribution using saddlepoint methods.
simplex	This function will optimize the linear function a\%*\%x subject to the constraints A1\%*\%x <= b1, A2\%*\%x >= b2, A3\%*\%x = b3, and x >= 0. Either maximization or minimization is possible but the default is minimization.
smooth.f	Uses the method of frequency smoothing to find a distribution on a data set that has a required value, theta, of the statistic of interest.
tilt.boot	This function will run an initial bootstrap with equal resampling probabilities (if required) and will use the output of the initial run to find resampling probabilities that put the value of the statistic at required values. It then runs an importance resampling bootstrap using the calculated probabilities as the resampling distribution.
tsboot	Generates R bootstrap replicates of a statistic applied to a time series. The replicate time series can be generated using fixed or random block lengths or can be model-based replicates.
var.linear	Estimates the variance of a statistic from its empirical influence values.

Data Sets

Data Set	Class	Description
acme	data.frame	The acme data frame has 60 rows and 3 columns. The excess returns for the Acme Cleveland Corporation, along with those for all stocks listed on the New York and American Stock Exchanges, were recorded over a 5-year period. These excess returns are relative to the return on a riskless investment such as U.S. Treasury bills.

Data Set	Class	Description
aids	data.frame	The aids data frame has 570 rows and 6 columns. Although all cases of AIDS in England and Wales must be reported to the Communicable Disease Surveillance Centre, there is often a considerable delay between the time of diagnosis and the time that it is reported. In estimating the prevalence of AIDS, account must be taken of the unknown number of cases that have been diagnosed but not reported. The data set here records the reported cases of AIDS diagnosed from July 1983 until the end of 1992. The data is cross-classified by the date of diagnosis and the time delay in the reporting of the cases.
aircondit	data.frame	Proschan reported on the times between failures of the air-conditioning equipment in 10 Boeing 720 aircraft. The aircondit data frame contains the intervals for the ninth aircraft, while aircondit7 contains those for the seventh aircraft. Both data frames have just one column. Note that the data has been sorted into increasing order.
aircondit7	data.frame	Proschan reported on the times between failures of the air-conditioning equipment in 10 Boeing 720 aircraft. The aircondit data frame contains the intervals for the ninth aircraft, while aircondit7 contains those for the seventh aircraft. Both data frames have just one column. Note that the data has been sorted into increasing order.
amis	data.frame	The amis data frame has 8,437 rows and 4 columns. In a study into the effect that warning signs have on speeding patterns, Cambridgeshire County Council considered 14 pairs of locations. The locations were paired to account for factors such as traffic volume and type of road. One site in each pair had a sign erected warning of the dangers of speeding and asking drivers to slow down. No action was taken at the second site. Three sets of measurements were taken at each site. Each set of measurements was nominally of the speeds of 100 cars, but not all sites have exactly 100 measurements. These speed measurements were taken before the erection of the sign, shortly after the erection of the sign, and again after the sign had been in place for some time.
aml	data.frame	The aml data frame has 23 rows and 3 columns. A clinical trial to evaluate the efficacy of maintenance chemotherapy for acute myelogenous leukemia was conducted by Embury et al. at Stanford University. After reaching a stage of remission through treatment by chemotherapy, patients were randomized into two groups. The first group received maintenance chemotherapy, and the second group did not. The aim of the study was to see if maintenance chemotherapy increased the length of the remission. The data here formed a preliminary analysis that was conducted in October 1974.
beaver	ts	The beaver data frame has 100 rows and 4 columns. It is a multivariate time series of class "ts" and also inherits from class "data.frame". This data set is part of a long study into body temperature regulation in beavers. Four adult female beavers were live-trapped and had a temperature-sensitive radio transmitter surgically implanted. Readings were taken every 10 minutes. The location of the beaver was also recorded, and her activity level was dichotomized by whether she was in the retreat or outside of it, since high-intensity activities only occur outside of the retreat. The data in this data frame comes from those readings for one of the beavers on a day in autumn.

Data Set	Class	Description
bigcity	data.frame	The bigcity data frame has 49 rows and 2 columns. The city data frame has 10 rows and 2 columns. The measurements are the populations (in 1000s) of 49 U.S. cities in 1920 and 1930. The 49 cities are a random sample taken from the 196 largest cities in 1920. The city data frame consists of the first 10 observations in bigcity.
brambles	data.frame	The brambles data frame has 823 rows and 3 columns. The location of living bramble canes in a 9-m square plot was recorded. We take 9 m to be the unit of distance so that the plot can be thought of as a unit square. The bramble canes were also classified by their age.
breslow	data.frame	The breslow data frame has 10 rows and 5 columns. In 1961, Doll and Hill sent out a questionnaire to all men on the British Medical Register inquiring about their smoking habits. Almost 70% of the men replied. Death certificates were obtained for medical practitioners, and causes of death were assigned on the basis of these certificates. The breslow data set contains the person-years of observations and deaths from coronary artery disease accumulated during the first 10 years of the study.
calcium	data.frame	The calcium data frame has 27 rows and 2 columns. Howard Grimes of the Botany Department, North Carolina State University, conducted an experiment for bio-chemical analysis of intracellular storage and transport of calcium across plasma membrane. Cells were suspended in a solution of radioactive calcium for a certain length of time, and then the amount of radioactive calcium that was absorbed by the cells was measured. The experiment was repeated independently with nine different times of suspension each replicated three times.
cane	data.frame	The cane data frame has 180 rows and 5 columns. The data frame represents a randomized block design with 45 varieties of sugarcane and 4 blocks. The aim of the experiment was to classify the varieties into resistant, intermediate, and sus-ceptible to a disease called "coal of sugarcane" (*carvao da cana-de-acucar*). This is a disease that is common in sugar-cane plantations in certain areas of Brazil. For each plot, 50 pieces of sugarcane stem were put in a solution containing the disease agent, and then some were planted in the plot. After a fixed period of time, the total number of shoots and the number of diseased shoots were recorded.
capability	data.frame	The capability data frame has 75 rows and 1 column. The data consists of simulated successive observations from a process in equilibrium. The process is assumed to have specification limits (5.49, 5.79).
catsM	data.frame	The catsM data frame has 97 rows and 3 columns. One hundred and forty-four adult (over 2 kg in weight) cats used for experiments with the drug digitalis had their heart and body weight recorded. Forty-seven of the cats were female, and 97 were male. The catsM data frame consists of the data for the male cats. The full data can be found in data set \link[MASS]{cats} in package MASS.
cav	data.frame	The cav data frame has 138 rows and 2 columns. The data gives the positions of the individual caveolae in a square region with sides of length 500 units. This grid was originally on a 2.65µm square of muscle fiber. The data consist of those points falling in the lower-left quarter of the region used for the data set caveolae.dat.
cd4	data.frame	The cd4 data frame has 20 rows and 2 columns. CD4 cells are carried in the blood as part of the human immune system. One of the effects of the human immuno-deficiency virus (HIV) is that these cells die. The count of CD4 cells is used in deter-mining the onset of full-blown AIDS in a patient. In this study of the effectiveness of a new antiviral drug on HIV, 20 HIV-positive patients had their CD4 counts recorded

		and then were put on a course of treatment with this drug. After using the drug for 1 year, their CD4 counts were again recorded. The aim of the experiment was to show that patients taking the drug had increased CD4 counts, which is not generally seen in HIV-positive patients.
cd4.nested	boot	This is an example of a nested bootstrap for the correlation coefficient of the cd4 data frame.
channing	data.frame	The channing data frame has 462 rows and 5 columns. Channing House is a retirement center in Palo Alto, California. The data was collected between the opening of the house in 1964 until July 1, 1975. During that time, 97 men and 365 women passed through the center. For each of these, their age on entry and also on leaving or death was recorded. A large number of the observations were censored mainly due to the resident being alive on July 1, 1975, when the data was collected. Over the course of the study, 130 women and 46 men died at Channing House. Differences between the survival of the sexes, taking age into account, was one of the primary concerns of this study.
city	data.frame	The bigcity data frame has 49 rows and 2 columns. The city data frame has 10 rows and 2 columns. The measurements are the populations (in 1000s) of 49 U.S. cities in 1920 and 1930. The 49 cities are a random sample taken from the 196 largest cities in 1920. The city data frame consists of the first 10 observations in bigcity.
claridge	data.frame	The claridge data frame has 37 rows and 2 columns. The data comes from an experiment that was designed to look for a relationship between a certain genetic characteristic and handedness. The 37 subjects were women who had a son with mental retardation due to inheriting a defective X-chromosome. For each such mother, a genetic measurement of her DNA was made. Larger values of this measurement are known to be linked to the defective gene, and it was hypothesized that larger values might also be linked to a progressive shift away from right-handedness. Each woman also filled in a questionnaire regarding which hand she used for various tasks. From these questionnaires, a measure of hand preference was found for each mother. The scale of this measure goes from 1, indicating women who always favor their right hand, to 8, indicating women who always favor their left hand. Between these two extremes are women who favor one hand for some tasks and the other for other tasks.
cloth	data.frame	The cloth data frame has 32 rows and 2 columns.
co.transfer	data.frame	The co.transfer data frame has 7 rows and 2 columns. Seven smokers with chickenpox had their levels of carbon monoxide transfer measured upon being admitted to the hospital and then again after 1 week. The main question was whether 1 week of hospitalization had changed the carbon monoxide transfer factor.
coal	data.frame	The coal data frame has 191 rows and 1 column. This data frame gives the dates of 191 explosions in coal mines that resulted in 10 or more fatalities. The time span of the data is from March 15, 1851, until March 22, 1962.
darwin	data.frame	The darwin data frame has 15 rows and 1 column. Charles Darwin conducted an experiment to examine the superiority of cross-fertilized plants over self-fertilized plants. Fifteen pairs of plants were used. Each pair consisted of one cross-fertilized plant and one self-fertilized plant that germinated at the same time and grew in the same pot. The plants were measured at a fixed time after planting, and the differences in heights between the cross- and self-fertilized plants were recorded in eighths of an inch.

Data Set	Class	Description
dogs	data.frame	The dogs data frame has 7 rows and 2 columns. Data on the cardiac oxygen consumption and left ventricular pressure was gathered on seven domestic dogs.
downs.bc	data.frame	The downs.bc data frame has 30 rows and 3 columns. Down's syndrome is a genetic disorder caused by an extra chromosome 21 or a part of chromosome 21 being translocated to another chromosome. The incidence of Down's syndrome is highly dependent on the mother's age and rises sharply after age 30. In the 1960s, a large-scale study of the effect of maternal age on the incidence of Down's syndrome was conducted at the British Columbia Health Surveillance Registry. This data frame consists of the data that was collected in that study. Mothers were classified by age. Most groups correspond to the age in years, but the first group comprises all mothers aged 15–17 and the last is those aged 46–49. No data for mothers over 50 or below 15 was collected.
ducks	data.frame	The ducks data frame has 11 rows and 2 columns. Each row of the data frame represents a male duck that is a second-generation cross between a mallard and a pintail. For 11 such ducks, a behavioral index and plumage index were calculated. These were measured on scales devised for this experiment, which was to examine whether there was any link between which species the ducks resembled physically and which they resembled in behavior. The scale for physical appearance ranged from 0 (identical in appearance to a mallard) to 20 (identical to a pintail). The behavioral traits of the ducks were on a scale of 0 to 15, with lower numbers indicating more mallard-like behavior.
fir	data.frame	The fir data frame has 50 rows and 3 columns. The number of balsam-fir seedlings in each quadrant of a grid of 50 five-foot-square quadrants were counted. The grid consisted of 5 rows of 10 quadrants in each row.
frets	data.frame	The frets data frame has 25 rows and 4 columns. The data consists of measurements of the length and breadth of the heads of pairs of adult brothers in 25 randomly sampled families. All measurements are expressed in millimeters.
grav	data.frame	The gravity data frame has 81 rows and 2 columns. The grav data set has 26 rows and 2 columns. Between May 1934 and July 1935, the U.S. National Bureau of Standards conducted a series of experiments to estimate the acceleration due to gravity, g, at Washington, DC. Each experiment produced a number of replicate estimates of g using the same methodology. Although the basic method remained the same for all experiments, that of the reversible pendulum, there were changes in configuration. The gravity data frame contains the data from all eight experiments. The grav data frame contains the data from experiments 7 and 8. The data is expressed as deviations from 980.000 in centimeters per second squared.
gravity	data.frame	The gravity data frame has 81 rows and 2 columns. The grav data set has 26 rows and 2 columns. Between May 1934 and July 1935, the U.S. National Bureau of Standards conducted a series of experiments to estimate the acceleration due to gravity, g, at Washington, DC. Each experiment produced a number of replicate estimates of g using the same methodology. Although the basic method remained the same for all experiments, that of the reversible pendulum, there were changes in configuration. The gravity data frame contains the data from all eight experiments. The grav data frame contains the data from experiments 7 and 8. The data is expressed as deviations from 980.000 in centimeters per second squared.
hirose	data.frame	The hirose data frame has 44 rows and 3 columns. PET film is used in electrical insulation. In this accelerated life test, the failure times for 44 samples in gas-insulated transformers were estimated. Four different voltage levels were used.

Data Set	Class	Description
islay	data.frame	The `islay` data frame has 18 rows and 1 column. Measurements were taken of paleocurrent azimuths from the Jura Quartzite on the Scottish island of Islay.
manaus	ts	The `manaus` time series is of class `"ts"` and has 1,080 observations on one variable. The data values are monthly averages of the daily stages (heights) of the Rio Negro at Manaus. Manaus is 18 km upstream from the confluence of the Rio Negro with the Amazon but because of the tiny slope of the water surface and the lower courses of its flatland affluents, they may be regarded as a good approximation of the water level in the Amazon at the confluence. The data here covers 90 years from January 1903 until December 1992. The Manaus gauge is tied in with an arbitrary benchmark of 100m set in the steps of the Municipal Prefecture; gauge readings are usually referred to sea level, on the basis of a mark on the steps leading to the Parish Church (Matriz), which is assumed to lie at an altitude of 35.874 m according to observations made many years ago under the direction of Samuel Pereira, an engineer in charge of the Manaus Sanitation Committee Whereas such an altitude cannot, by any means, be considered to be a precise datum point, observations have been provisionally referred to it. The measurements are in meters.
melanoma	data.frame	The `melanoma` data frame has 205 rows and 7 columns. The data consists of measurements made on patients with malignant melanoma. Each patient had his or her tumor surgically removed at the Department of Plastic Surgery, University Hospital of Odense, Denmark, during the period 1962–1977. The surgery consisted of complete removal of the tumor together with about 2.5 cm of the surrounding skin. Among the measurements taken were the thickness of the tumor and whether it was ulcerated or not. These are thought to be important prognostic variables in that patients with a thick and/or ulcerated tumor have an increased chance of death from melanoma. Patients were followed until the end of 1977.
motor	data.frame	The `motor` data frame has 94 rows and 4 columns. The rows were obtained by removing replicate values of `time` from the data set `mcycle`. Two extra columns were added to allow for strata with a different residual variance in each stratum.
neuro	matrix	`neuro` is a matrix containing times of observed firing of a neuron in windows of 250 ms either side of the application of a stimulus to a human subject. Each row of the matrix is a replication of the experiment, and there are a total of 469 replicates.
nitrofen	data.frame	The `nitrofen` data frame has 50 rows and 5 columns. Nitrofen is a herbicide that was used extensively for the control of broad-leaved and grass weeds in cereals and rice. Although it is relatively nontoxic to adult mammals, nitrofen is a significant teratogen and mutagen. It is also acutely toxic and reproductively toxic to cladoceran zooplankton. Nitrofen is no longer in commercial use in the United States, having been the first pesticide to be withdrawn due to teratogenic effects. The data here comes from an experiment to measure the reproductive toxicity of nitrofen on a species of zooplankton (*Ceriodaphnia dubia*). Fifty animals were randomized into batches of 10, and each batch was put in a solution with a measured concentration of nitrofen. Then the number of live offspring in each of the three broods of each animal was recorded.
nodal	data.frame	The `nodal` data frame has 53 rows and 7 columns. The treatment strategy for a patient diagnosed with prostate cancer depends highly on whether the cancer has spread to the surrounding lymph nodes. It is common to operate on the patient to get samples from the nodes, which can then be analyzed under a microscope, but clearly it would be preferable if an accurate assessment of nodal involvement could be made without surgery. For a sample of 53 prostate cancer patients, a number of

Data Set	Class	Description
		possible predictor variables were measured before surgery. The patients then had surgery to determine nodal involvement. The point of the study was to see if nodal involvement could be accurately predicted from the predictor variables and which ones were most important.
nuclear	data.frame	The `nuclear` data frame has 32 rows and 11 columns. The data relates to the construction of 32 light-water reactor (LWR) plants constructed in the United States in the late 1960s and early 1970s. The data was collected with the aim of predicting the cost of construction of additional LWR plants. Six of the power plants had partial turnkey guarantees, and it is possible that, for these plants, some manufacturers' subsidies may be hidden in the quoted capital costs.
paulsen	data.frame	The `paulsen` data frame has 346 rows and 1 column. Sections were prepared from the brain of adult guinea pigs. Spontaneous currents that flowed into individual brain cells were then recorded and the peak amplitude of each current measured. The aim of the experiment was to see if the current flow was quantal in nature (i.e., that it is not a single burst but instead is built up of many smaller bursts of current). If the current was indeed quantal, then it would be expected that the distribution of the current amplitude would be multimodal with modes at regular intervals. The modes would be expected to decrease in magnitude for higher current amplitudes.
poisons	data.frame	The `poisons` data frame has 48 rows and 3 columns. The data form a 3×4 factorial experiment, the factors being three poisons and four treatments. Each combination of the two factors was used on four animals, the allocation to animals having been completely randomized.
polar	data.frame	The `polar` data frame has 50 rows and 2 columns. The data consists of the pole positions from a paleomagnetic study of New Caledonian laterites.
remission	data.frame	The `remission` data frame has 27 rows and 3 columns.
salinity	data.frame	The `salinity` data frame has 28 rows and 4 columns. Biweekly averages of the water salinity and river discharge in Pamlico Sound, North Carolina, were recorded between the years 1972 and 1977. The data in this set consists only of those measurements in March, April, and May.
survival	data.frame	The `survival` data frame has 14 rows and 2 columns. The data measured the survival percentages of batches of rats who were given varying doses of radiation. At each of six doses there were two or three replications of the experiment.
tau	data.frame	The `tau` data frame has 60 rows and 2 columns. The tau particle is a heavy electron-like particle discovered in the 1970s by Martin Perl at the Stanford Linear Accelerator Center. Soon after its production, the tau particle decays into various collections of more stable particles. About 86% of the time, the decay involves just one charged particle. This rate has been measured independently 13 times. The one-charged-particle event is made up of four major modes of decay as well as a collection of other events. The four main types of decay are denoted rho, pi, e, and mu. These rates have been measured independently 6, 7, 14, and 19 times, respectively. Due to physical constraints, each experiment can only estimate the composite one-charged-particle decay rate or the rate of one of the major modes of decay. Each experiment consists of a major research project involving many years' work. One of the goals of the experiments was to estimate the rate of decay due to events other than the four main modes of decay. These are uncertain events and so cannot themselves be observed directly.

Data Set	Class	Description
tuna	data.frame	The tuna data frame has 64 rows and 1 column. The data comes from an aerial line transect survey of southern bluefin tuna in the Great Australian Bight. An aircraft with two spotters on board flew randomly allocated line transects. Each school of tuna sighted was counted and its perpendicular distance from the transect measured. The survey was conducted in summer when tuna tend to stay on the surface.
urine	data.frame	The urine data frame has 79 rows and 7 columns. Seventy-nine urine specimens were analyzed in an effort to determine if certain physical characteristics of the urine might be related to the formation of calcium oxalate crystals.
wool	ts	wool is a time series of class "ts" and contains 309 observations. Each week that the market was open, the Australian Wool Corporation set a floor price that determined its policy on intervention and was therefore a reflection of the overall price of wool for the week in question. Actual prices paid varied considerably about the floor price. The series here is the log of the ratio between the price for fine-grade wool and the floor price, each market week between July 1976 and June 1984.

class

This package provides functions for classification.

Functions

Function	Description
SOM, batchSOM	Kohonen's self-organizing maps (SOMs) are a crude form of multidimensional scaling.
condense	Condenses training set for k-nearest-neighbor (k-NN) classifier.
knn	k-nearest-neighbor classification for test set from training set. For each row of the test set, the k-nearest (in Euclidean distance) training set vectors are found, and the classification is decided by majority vote, with ties broken at random. If there are ties for the kth nearest vector, all candidates are included in the vote.
knn.cv	k-nearest-neighbor cross-validatory classification from training set.
knn1	Nearest-neighbor classification for test set from training set. For each row of the test set, the nearest neighbor (by Euclidean distance) training set vector is found, and its classification used. If there is more than one nearest neighbor, a majority vote is used, with ties broken at random.
lvq1, lvq2, lvq3	Moves examples in a codebook to better represent the training set.
lvqinit	Constructs an initial codebook for learning vector quantization (LVQ) methods.
lvqtest	Classifies a test set by 1-NN from a specified LVQ codebook.
multiedit	Multiedit for k-NN classifier.
olvq1	Moves examples in a codebook to better represent the training set.
reduce.nn	Reduces training set for a k-NN classifier. Used after condense.
somgrid	Plotting functions for SOM results.

cluster

This package provides functions for cluster analysis.

Functions

Function	Description
agnes	Computes agglomerative hierarchical clustering of the data set.
bannerplot	Draws a "banner," i.e., basically a horizontal `barplot` visualizing the (agglomerative or divisive) hierarchical clustering or an other binary dendrogram structure.
clara	Computes a "`clara`" object, a list representing a clustering of the data into k clusters.
clusplot	Draws a two-dimensional (2D) "clusplot" on the current graphics device.
coef.hclust	Computes the "agglomerative coefficient," measuring the clustering structure of the data set.
daisy	Computes all the pairwise dissimilarities (distances) between observations in the data set.
diana	Computes a divisive hierarchical clustering of the data set, returning an object of class `diana`.
ellipsoidPoints	Computes points on the ellipsoid boundary, mostly for drawing.
ellipsoidhull	Computes the "ellipsoid hull" or "spanning ellipsoid," i.e., the ellipsoid of minimal volume ("area" in 2D) such that all given points lie just inside or on the boundary of the ellipsoid.
fanny	Computes a fuzzy clustering of the data into k clusters.
lower.to.upper.tri.inds	Computes index vectors for extracting or reordering of lower or upper triangular matrices that are stored as contiguous vectors.
mona	Returns a list representing a divisive hierarchical clustering of a data set with binary variables only.
pam	Partitioning (clustering) of the data into k clusters "around medoids," a more robust version of k-means clustering.
pltree	Generic function drawing a clustering tree ("dendrogram") on the current graphics device. There is a `twins` method; see `pltree.twins` for usage and examples.
predict.ellipsoid	Computes points on the ellipsoid boundary, mostly for drawing.
silhouette	Computes silhouette information according to a given clustering in k clusters.
sizeDiss	Returns the number of observations (*sample size*) corresponding to a dissimilarity-like object or, equivalently, the number of rows or columns of a matrix when only the lower or upper triangular part (without diagonal) is given. It is nothing else but the inverse function of $f(n) = n(n-1)/2$.
sortSilhouette	Computes silhouette information according to a given clustering in k clusters.
upper.to.lower.tri.inds	Computes index vectors for extracting or reordering of lower or upper triangular matrices that are stored as contiguous vectors.

Function	Description
volume	Computes the volume of a planar object. This is a generic function and a method for `ellipsoid` objects.

Data Sets

Data Set	Class	Description
agriculture	data.frame	Gross national product (GNP) per capita and percentage of the population working in agriculture for each country belonging to the European Union in 1993.
animals	data.frame	This data set considers 6 binary attributes for 20 animals.
chorSub	matrix	This is a small rounded subset of the C-horizon data.
flower	data.frame	This data set consists of 8 characteristics for 18 popular flowers.
plant-Traits	data.frame	This data set constitutes a description of 136 plant species according to biological attributes (morphological or reproductive).
pluton	data.frame	The `pluton` data frame has 45 rows and 4 columns, containing percentages of isotopic composition of 45 plutonium batches.
ruspini	data.frame	The Ruspini data set, consisting of 75 points in 4 groups, is popular for illustrating clustering techniques.
votes.repub	data.frame	A data frame with the percents of votes given to the Republican candidates in presidential elections from 1856 to 1976. Rows represent the 50 states, and columns the 31 elections.
xclara	data.frame	An artificial data set consisting of 3,000 points in 3 well-separated clusters of size 1,000 each.

codetools

This package provides tools for analyzing R code. It is mainly intended to support the other tools in this package and byte code compilation. See the help file for more information.

foreign

This package provides functions for reading data stored by Minitab, S, SAS, SPSS, Stata, Systat, dBase, and so forth.

Functions

Function	Description
data.restore	Reads binary data files or `data.dump` files that were produced in S version 3.
lookup.xport	Scans a file as a SAS XPORT format library and returns a list containing information about the SAS library.
read.S	Reads binary data files or `data.dump` files that were produced in S version 3.
read.arff	Reads data from Weka Attribute-Relation File Format (ARFF) files.

Function	Description
read.dbf	Reads a DBF file into a data frame, converting character fields to factors and trying to respect NULL fields.
read.dta	Reads a file in Stata version 5–10 binary format into a data frame.
read.epiinfo	Reads data files in the .REC format used by Epi Info versions 6 and earlier and by EpiData. Epi Info is a public-domain database and statistics package produced by the U.S. Centers for Disease Control and Prevention, and EpiData is a freely available data entry and validation system.
read.mtp	Returns a list with the data stored in a file as a Minitab Portable Worksheet.
read.octave	Reads a file in Octave text data format into a list.
read.spss	Reads a file stored by the SPSS save or export commands.
read.ssd	Generates a SAS program to convert the ssd contents to SAS transport format and then uses read.xport to obtain a data frame.
read.systat	Reads a rectangular data file stored by the Systat SAVE command as (legacy) *.sys or, more recently, *.syd files.
read.xport	Reads a file as a SAS XPORT format library and returns a list of data.frames.
write.arff	Writes data into Weka Attribute-Relation File Format (ARFF) files.
write.dbf	Tries to write a data frame to a DBF file.
write.dta	Writes the data frame to file in the Stata binary format. Does not write array variables unless they can be drop-ed to a vector.
write.foreign	Exports simple data frames to other statistical packages by writing the data as free-format text and writing a separate file of instructions for the other package to read the data.

grDevices

This package provides functions for graphics devices and support for base and grid graphics.

Functions

Function	Description
CIDFont	Used to define the translation of an R graphics font family name to a Type 1 or CID font description, used by both the postscript and the pdf graphics devices.
Type1Font	Used to define the translation of an R graphics font family name to a Type 1 or CID font description, used by both the postscript and the pdf graphics devices.
X11	Starts a graphics device driver for the X Window System (version 11). This can only be done on machines/accounts that have access to an X server.
X11.options	Sets options for an X11 device.
X11Font, X11Fonts	Handle the translation of a device-independent R graphics font family name to an X11 font description.
as.graphicsAnnot	Coerces an R object into a form suitable for graphics annotation.

Function	Description
bitmap	Generates a graphics file. `dev2bitmap` copies the current graphics device to a file in a graphics format.
bmp	Graphics device for generating BMP(bitmap) files.
boxplot.stats	This function is typically called by another function to gather the statistics necessary for producing box plots, but may be invoked separately.
cairo_pdf	A Cairo-based graphics device for generating PDF files.
cairo_ps	A Cairo-based graphics device for generating PostScript files.
check.options	Utility function for setting options with some consistency checks. The `attrib utes` of the new settings in `new` are checked for consistency with the *model* (often default) list in `name.opt`.
chull	Computes the subset of points that lie on the convex hull of the set of points specified.
cm	Translates from inches to centimeters (cm).
cm.colors	Creates a vector of *n* contiguous colors.
col2rgb	R color to RGB (red/green/blue) conversion.
colorConverter	Specifies color spaces for use in `convertColor`.
colorRamp, colorRampPalette	These functions return functions that interpolate a set of given colors to create new color palettes (like `topo.colors`) and color ramps, functions that map the interval [0, 1] to colors (like `gray`).
colors, colours	Returns the built-in color names that R knows about.
contourLines	Calculates contour lines for a given set of data.
convertColor	Converts colors between standard color space representations. This function is experimental.
densCols	Produces a vector containing colors that encode the local densities at each point in a scatter plot.
dev.control	Allows the user to control the recording of graphics operations in a device.
dev.copy	Copies the graphics contents of the current device to the device specified by `which` or to a new device that has been created by the function specified by `device` (it is an error to specify both `which` and `device`).
dev.copy2eps	Copies the graphics contents of the current device to an Encapsulated PostScript Format (EPSF) output file in portrait orientation (`horizontal = FALSE`).
dev.copy2pdf	Copies the graphics contents of the current device to a PDF output file in portrait orientation (`horizontal = FALSE`).
dev.cur	Returns a named integer vector of length 1, giving the number and name of the active device, or 1, the null device, if none is active.
dev.interactive	Tests if the current graphics device (or that which would be opened) is interactive.
dev.list	Returns the numbers of all open devices, except device 1, the null device. This is a numeric vector with a names attribute giving the device names, or `NULL` if there is no open device.
dev.new	Opens a new graphics device.
dev.next	Returns the number and name of the next device in the list of devices.
dev.off	Shuts down the specified (by default the current) graphics device.

Function	Description
dev.prev	Returns the number and name of the previous device in the list of devices.
dev.print	Copies the graphics contents of the current device to a new device that has been created by the function specified by `device` and then shuts the new device.
dev.set	Makes the specified graphics device the active device.
dev.size	Finds the dimensions of the device surface of the current device.
dev2bitmap	`bitmap` generates a graphics file. `dev2bitmap` copies the current graphics device to a file in a graphics format.
devAskNewPage	Used to control (for the current device) whether the user is prompted before starting a new page of output.
deviceIsInteractive	Tests if the current graphics device (or that which would be opened) is interactive.
embedFonts	Runs Ghostscript to process a PDF or PostScript file and embed all fonts in the file.
extendrange	Extends a numeric range by a small percentage, i.e., fraction, *on both sides*.
getGraphicsEvent	Waits for input from a graphics window in the form of a mouse or keyboard event.
graphics.off	Provides control over multiple graphics devices.
gray	Creates a vector of colors from a vector of gray levels.
gray.colors	Creates a vector of n gamma-corrected gray colors.
grey	Creates a vector of colors from a vector of gray levels.
grey.colors	Creates a vector of n gamma-corrected gray colors.
hcl	Creates a vector of colors from vectors specifying hue, chroma, and luminance.
heat.colors	Creates a vector of n contiguous colors.
hsv	Creates a vector of colors from vectors specifying hue, saturation, and value.
jpeg	Creates a graphics device for generating JPEG format files.
make.rgb	Specifies color spaces for use in `convertColor`.
n2mfrow	Easy setup for plotting multiple figures (in a rectangular layout) on one page. This computes a sensible default for `par(mfrow)`.
nclass.FD	Computes the number of classes for a histogram using the Freedman-Diaconis choice based on the interquartile range (IQR), unless that's 0, where it reverts to `mad(x, constant = 2)`, and when that is 0 as well, returns 1.
nclass.Sturges	Computes the number of classes for a histogram using Sturges's formula, implicitly basing bin sizes on the range of the data.
nclass.scott	Computes the number of classes for a histogram using Scott's choice for a normal distribution based on the estimate of the standard error, unless that is 0 where it returns 1.
palette	Views or manipulates the color palette that is used when a `col=` has a numeric index.
pdf	Starts the graphics device driver for producing PDF graphics.
pdf.options	The auxiliary function `pdf.options` can be used to set or view (if called without arguments) the default values for some of the arguments to `pdf`.
pdfFonts	Lists existing mapping for PDF fonts or creates new mappings.
pictex	Produces graphics suitable for inclusion in TeX and LaTeX documents.

Function	Description
png	Creates a new graphics device for producing Portable Network Graphics (PNG) files.
postscript	Starts the graphics device driver for producing PostScript graphics.
postscriptFonts	Lists existing mapping for PostScript fonts or creates new mappings.
ps.options	The auxiliary function ps.options can be used to set or view (if called without arguments) the default values for some of the arguments to postscript.
quartz	Starts a graphics device driver for the Mac OS X system.
quartz.options	Sets options for a quartz device.
quartzFont	Translates from a device-independent R graphics font family name to a quartz font description.
quartzFonts	Lists existing mappings of device-independent R graphics to a quartz font description, or defines new mappings.
rainbow	Creates a vector of n contiguous colors.
recordGraphics	Records arbitrary code on the graphics engine display list. Useful for encapsulating calculations with graphical output that depends on the calculations. Intended *only* for expert use.
recordPlot, replayPlot	Functions to save the current plot in an R variable and to replay it.
rgb	Creates colors corresponding to the given intensities (between 0 and max) of the red, green, and blue primaries.
rgb2hsv	Transforms colors from RGB space (red/green/blue) into HSV space (hue/saturation/value).
savePlot	Saves the current page of a Cairo X11() device to a file.
setEPS	A wrapper to ps.options that sets defaults appropriate for figures for inclusion in documents (the default size is 7 inches square unless width or height is supplied).
setPS	A wrapper to ps.options to set defaults appropriate for figures for spooling to a PostScript printer.
svg	Creates a new graphics device for outputting graphics in Scalable Vector Graphics (SVG) format.
terrain.colors	Creates a vector of n contiguous colors.
tiff	Creates a new graphics device for outputting graphics in Tagged Image File Format (TIFF) format.
topo.colors	Creates a vector of n contiguous colors.
trans3d	Projection of three-dimensional to two-dimensional points using a 4×4 viewing transformation matrix.
x11	A synonym for X11 (which opens a new X11 device for plotting graphics).
xfig	Starts the graphics device driver for producing XFig (version 3.2) graphics.
xy.coords	Used by many functions to obtain x and y coordinates for plotting. The use of this common mechanism across all relevant R functions produces a measure of consistency.
xyTable	Given (x, y) points, determines their multiplicity—checking for equality only up to some (crude kind of) noise. Note that this is a special kind of 2D binning.

Function	Description
xyz.coords	Utility for obtaining consistent x, y, and z coordinates and labels for three-dimensional (3D) plots.

Data Sets

Data Set	Class	Description
Hershey	list	If the `family` graphical parameter (see `par`) has been set to one of the Hershey fonts, Hershey vector fonts are used to render text. When using the `text` and `contour` functions, Hershey fonts may be selected via the `vfont` argument, which is a character vector of length 2. This allows Cyrillic to be selected, which is not available via the font families.
blues9	character	`densCols` produces a vector containing colors that encode the local densities at each point in a scatter plot.
colorspaces	list	Converts colors between standard color space representations. This function is experimental.

graphics

This package contains functions for base graphics. Base graphics are traditional S graphics, as opposed to the newer grid graphics.

Functions

Function	Description
Axis	Generic function to add a suitable axis to the current plot.
abline	Adds one or more straight lines through the current plot.
arrows	Draws arrows between pairs of points.
assocplot	Produces a Cohen-Friendly association plot indicating deviations from independence of rows and columns in a two-dimensional contingency table.
axTicks	Computes pretty tick mark locations, the same way as R does internally. This is only nontrivial when *log* coordinates are active. By default, gives the `at` values that `axis(side)` would use.
axis	Adds an axis to the current plot, allowing the specification of the side, position, labels, and other options.
barplot	Creates a bar plot with vertical or horizontal bars.
box	Draws a box around the current plot in a given color and line type. The `bty` parameter determines the type of box drawn. See `par` for details.
boxplot	Produces box-and-whisker plot(s) of the given (grouped) values.
boxplot.matrix	Interprets the columns (or rows) of a matrix as different groups and draws a box plot for each.
bxp	Draws box plots based on the given summaries in z. It is usually called from within `boxplot`, but can be invoked directly.

Function	Description
cdplot	Computes and plots conditional densities describing how the conditional distribution of a categorical variable y changes over a numeric variable x.
clip	Sets clipping region in user coordinates.
close.screen	Removes the specified screen definition(s) created by `split.screen`.
co.intervals	Produces two variants of the *conditioning* plots.
contour	Creates a contour plot or adds contour lines to an existing plot. Methods include `contour.default`.
coplot	Produces two variants of the *conditioning* plots.
curve	Draws a curve corresponding to a given function or, for `curve()`, also an expression (in x) over the interval `[from,to]`.
dotchart	Draws a Cleveland dot plot.
erase.screen	Used to clear a single screen (when using `split.screen`), which it does by filling with the background color.
filled.contour	Produces a contour plot with the areas between the contours filled in solid color (Cleveland calls this a level plot).
fourfoldplot	Creates a fourfold display of a 2-by-2-by-k contingency table on the current graphics device, allowing for the visual inspection of the association between two dichotomous variables in one or several populations (strata).
frame	This function (`frame` is an alias for `plot.new`) causes the completion of plotting in the current plot (if there is one) and an advance to a new graphics frame.
grconvertX, grconvertY	Convert between graphics coordinate systems.
grid	Adds an nx-by-ny rectangular grid to an existing plot.
hist	The generic function `hist` computes a histogram of the given data values. If `plot=TRUE`, the resulting object of `\link[base]{class "histogram"}` is plotted by `plot.histogram`, before it is returned. Methods include `hist.default`.
identify	Reads the position of the graphics pointer when the (first) mouse button is pressed. It then searches the coordinates given in x and y for the point closest to the pointer. If this point is close enough to the pointer, its index will be returned as part of the value of the call.
image	Creates a grid of colored or grayscale rectangles with colors corresponding to the values in z.
layout, layout.show	`layout` divides the device up into as many rows and columns as there are in matrix mat, with the column widths and the row heights specified in the respective arguments.
lcm	`layout` divides the device up into as many rows and columns as there are in matrix mat, with the column widths and the row heights specified in the respective arguments.
legend	Used to add legends to plots. Note that a call to the function `locator(1)` can be used in place of the x and y arguments.
lines	A generic function taking coordinates given in various ways and joining the corresponding points with line segments. Methods include `lines.default` and `lines.ts`.

Function	Description
locator	Reads the position of the graphics cursor when the (first) mouse button is pressed.
matlines, matplot, matpoints	Plot the columns of one matrix against the columns of another.
mosaicplot	Plots a mosaic on the current graphics device.
mtext	Text is written in one of the four margins of the current figure region or one of the outer margins of the device region.
pairs	A matrix of scatter plots is produced.
panel.smooth	An example of a simple useful `panel` function to be used as an argument in, e.g., `coplot` or `pairs`.
par	Used to set or query graphical parameters.
persp	Draws perspective plots of surfaces over the x–y plane.
pie	Draws a pie chart.
plot	Generic function for plotting R objects.
plot.design	Plots univariate effects of one or more `factors`, typically for a designed experiment as analyzed by `aov()`.
plot.new	This function (`plot.new` is an alias for `frame`) causes the completion of plotting in the current plot (if there is one) and an advance to a new graphics frame.
plot.window	Sets up the world coordinate system for a graphics window. It is called by higher-level functions such as `plot.default` (*after* `plot.new`).
plot.xy	This is *the* internal function that does the basic plotting of points and lines. Usually, one should rather use the higher-level functions instead and refer to their help pages for explanation of the arguments.
points	A generic function to draw a sequence of points at the specified coordinates. The specified character(s) are plotted, centered at the coordinates. Methods include `points.default`.
polygon	Draws the polygons whose vertices are given in x and y.
rect	Draws a rectangle (or sequence of rectangles) with the given coordinates, fill, and border colors.
rug	Adds a *rug* representation (1D plot) of the data to the plot.
screen	Used to select which screen to draw in (when using split.screen).
segments	Draws line segments between pairs of points.
smoothScatter	Produces a smoothed color density representation of the scatter plot, obtained through a kernel density estimate.
spineplot	Spine plots are a special case of mosaic plots and can be seen as a generalization of stacked (or highlighted) bar plots. Analogously, spinograms are an extension of histograms.
split.screen	Defines a number of regions within the current device that can, to some extent, be treated as separate graphics devices. It is useful for generating multiple plots on a single device.
stars	Draws star plots or segment diagrams of a multivariate data set. With one single location, also draws "spider" (or "radar") plots.
stem	Produces a stem-and-leaf plot of the values in x.

Function	Description
strheight	Computes the height of the given strings or mathematical expressions s[i] on the current plotting device in *user* coordinates, *inches*, or as a fraction of the figure width par("fin").
stripchart	Produces one-dimensional scatter plots (or dot plots) of the given data. These plots are a good alternative to box plots when sample sizes are small.
strwidth	Computes the width of the given strings or mathematical expressions s[i] on the current plotting device in *user* coordinates, *inches*, or as a fraction of the figure width par("fin").
sunflowerplot	Multiple points are plotted as "sunflowers" with multiple leaves ("petals") such that overplotting is visualized instead of accidental and invisible.
symbols	Draws symbols on a plot. One of six symbols, *circles*, *squares*, *rectangles*, *stars*, *thermometers*, and *boxplots*, can be plotted at a specified set of x and y coordinates.
text	Draws the strings given in the vector labels at the coordinates given by x and y. y may be missing since xy.coords(x,y) is used for construction of the coordinates.
title	Used to add labels to a plot.
xinch, xyinch, yincch	xinch and yinch convert the specified number of inches given as their arguments into the correct units for plotting with graphics functions. Usually, this only makes sense when normal coordinates are used, i.e., *no* log scale (see the log argument to par). xyinch does the same for a pair of numbers xy, simultaneously.
xspline	Draws an X-spline, a curve drawn relative to control points.

grid

This package is a low-level graphics system that provides a great deal of control and flexibility in the appearance and arrangement of graphical output. It does not provide high-level functions that create complete plots. What it does provide is a basis for developing such high-level functions (e.g., the lattice package), the facilities for customizing and manipulating lattice output, the ability to produce high-level plots or non-statistical images from scratch, and the ability to add sophisticated annotations to the output from base graphics functions (see the gridBase package). For more information, see the help files for grid.

KernSmooth

This package provides functions for kernel smoothing.

Functions

Function	Description
bkde	Returns *x* and *y* coordinates of the binned kernel density estimate of the probability density of the data.

Function	Description
bkde2D	Returns the set of grid points in each coordinate direction, and the matrix of density estimates over the mesh induced by the grid points. The kernel is the standard bivariate normal density.
bkfe	Returns an estimate of a binned approximation to the kernel estimate of the specified density function. The kernel is the standard normal density.
dpih	Uses direct plug-in methodology to select the bin width of a histogram.
dpik	Uses direct plug-in methodology to select the bandwidth of a kernel density estimate.
dpill	Uses direct plug-in methodology to select the bandwidth of a local linear Gaussian kernel regression estimate.
locpoly	Estimates a probability density function, regression function, or their derivatives using local polynomials. A fast binned implementation over an equally spaced grid is used.

lattice

Trellis graphics is a framework for data visualization developed at Bell Labs by Richard Becker, William Cleveland, et al., extending ideas presented in Bill Cleveland's 1993 book *Visualizing Data*.

Lattice is best thought of as an implementation of Trellis graphics for R. It is built upon the grid graphics engine and requires the grid add-on package. It is not (readily) compatible with traditional R graphics tools. The public interface is based on the implementation in S-PLUS, but features several extensions, in addition to incompatibilities introduced through the use of grid. To the extent possible, care has been taken to ensure that existing Trellis code written for S-PLUS works unchanged (or with minimal change) in lattice. If you are having problems porting S-PLUS code, read the entry for panel in the documentation for xyplot. Most high-level Trellis functions in S-PLUS are implemented, with the exception of piechart.

Functions

Function	Description
Rows	Convenience function to extract a subset of a list. Usually used in creating keys.
as.shingle, as.factorOrShingle	Functions to handle shingles.
axis.default	Default function for drawing axes in lattice plots.
banking	Calculates banking slope.
barchart	Draws bar charts.
bwplot	Draws box plots.
canonical.theme	Initialization of a display device with appropriate graphical parameters.
cloud	Generic function to draw 3D scatter plots and surfaces. The "formula" methods do most of the actual work.
col.whitebg	Initialization of a display device with appropriate graphical parameters.

Function	Description
contourplot	Draws level plots and contour plots.
current.column	Returns an integer index specifying which column in the layout is currently active.
current.panel.limits	Used to retrieve a panel's x and y limits.
current.row	Returns an integer index specifying which row in the layout is currently active.
densityplot	Draws histograms and kernel density plots, possibly conditioned on other variables.
diag.panel.splom	This is the default superpanel function for `splom`.
do.breaks	Draws histograms and kernel density plots, possibly conditioned on other variables.
dotplot	Draws Cleveland dot plots.
draw.colorkey	Produces (and possibly draws) a grid frame grob, which is a color key that can be placed in other grid plots. Used in levelplot.
draw.key	Produces (and possibly draws) a grid frame grob, which is a legend (aka key) that can be placed in other grid plots.
equal.count	Function to handle shingles.
histogram	Draws histograms and kernel density plots, possibly conditioned on other variables.
is.shingle	Function to handle shingles.
larrows, llines, lplot.xy, lpoints, lpolygon, lrect, lsegments, ltext, panel.points, panel.polygon, panel.rect, panel.segments, panel.text	These functions are intended to replace common low-level traditional graphics functions, primarily for use in panel functions. The originals cannot be used (at least not easily) because lattice panel functions need to use grid graphics. Low-level drawing functions in grid can be used directly as well and are often more flexible. These functions are provided for convenience and portability.
lattice.getOption, lattice.options	Functions to handle settings used by lattice. Their main purpose is to make code maintenance easier, and users normally should not need to use these functions. However, fine control at this level may be useful in certain cases.
latticeParseFormula	Used by high-level lattice functions like `xyplot` to parse the formula argument and evaluate various components of the data.
level.colors	Calculates false colors from a numeric variable (including factors, using their numeric codes) given a color scheme and break points.
levelplot	Draws level plots and contour plots.
ltransform3dMatrix, ltransform3dto3d	These are (related to) the default panel functions for `cloud` and `wireframe`.
make.groups	Combines two or more vectors, possibly of different lengths, producing a data frame with a second column indicating which of these vectors that row came from. This is mostly useful for getting data into a form suitable for use in high-level lattice functions.
oneway	Fits a one-way model to univariate data grouped by a factor, the result often being displayed using `rfs`.
packet.number	A function that identifies which packet each observation in the data is part of.
packet.panel.default	Default function in lattice to determine, given the column, row, page, and other relevant information, the packet (if any) that should be used in a panel.
panel.3dscatter, panel	Default panel functions controlling `cloud` and `wireframe` displays.
panel.abline	Adds a line of the form $y = a + bx$ or vertical and/or horizontal lines.

Function	Description
panel.average	Treats one of *x* and *y* as a factor (according to the value of horizontal), calculates fun applied to the subsets of the other variable determined by each unique value of the factor, and joins them by a line.
panel.arrows	Draws arrows in a panel.
panel.axis	The function used by lattice to draw axes. It is typically not used by users, except those wishing to create advanced annotation. Keep in mind issues of clipping when trying to use it as part of the panel function. `current.panel.limits` can be used to retrieve a panel's *x* and *y* limits.
panel.barchart	Default panel function for `barchart`.
panel.brush.splom	`panel.link.splom` is meant for use with splom and requires a panel to be chosen using `trellis.focus` before it is called. Clicking on a point causes that and the corresponding projections in other pairwise scatter plots to be highlighted.
panel.bwplot	Default panel function for `bwplot`.
panel.cloud	Default panel function controlling `cloud` and `wireframe` displays.
panel.contourplot	Default panel function for `levelplot`.
panel.curve	Adds a curve, similar to what curve does with add=TRUE. Graphical parameters for the line are obtained from the `add.line` setting.
panel.densityplot	Default panel function for `densityplot`.
panel.dotplot	Default panel function for `dotplot`.
panel.error	Default handler used when an error occurs while executing a panel function.
panel.fill	Fills the panel with a specified color.
panel.grid	Draws a reference grid.
panel.histogram	Default panel function for `histogram`.
panel.identify	Similar to identify. When called, it waits for the user to identify points (in the panel being drawn) via mouse clicks.
panel.levelplot	Default panel function for `levelplot`.
panel.linejoin	`panel.linejoin` is an alias for `panel.average` that was retained for back-compatibility and may go away in the future.
panel.lines	Plots lines in a panel.
panel.link.splom	The classic Trellis paradigm is to plot the whole object at once, without the possibility of interacting with it afterward. However, by keeping track of the grid viewports where the panels and strips are drawn, it is possible to go back to them afterward and enhance them one panel at a time. This function provides convenient interfaces to help in this. Note that this is still experimental and the exact details may change in the future.
panel.lmline	`panel.lmline(x, y)` is equivalent to `panel.abline(lm(y~x))`.
panel.loess	Adds a smooth curve (fitted by `loess`).
panel.mathdensity	Plots a (usually theoretical) probability density function.
panel.number	Returns an integer counting which panel is being drawn (starting from 1 for the first panel, aka the panel order).
panel.pairs	Default superpanel function for `splom`.

Function	Description
panel.parallel	Default panel function for `parallel`.
panel.qq	Default panel function for `qq`.
panel.qqmath	Default panel function for `qqmath`.
panel.qqmathline	Useful panel function with `qqmath`. Draws a line passing through the points (usually) determined by the .25 and .75 quantiles of the sample and the theoretical distribution.
panel.refline	Similar to `panel.abline`, but uses the "reference.line" settings for the defaults.
panel.rug	Adds a rug representation of the (marginal) data to the panel.
panel.smoothScatter	Allows the user to place `smoothScatter` plots in lattice graphics.
panel.splom	Default panel function for `splom`.
panel.stripplot	Default panel function for `stripplot`. Also see `panel.superpose`.
panel.superpose, panel.superpose.2	These are panel functions for Trellis displays, which are useful when a grouping variable is specified for use within panels. The x (and y where appropriate) variables are plotted with different graphical parameters for each distinct value of the grouping variable.
panel.tmd.default, panel.tmd.qqmath	Default panel functions for tmd.
panel.violin	This is a panel function that can create a violin plot. It is typically used in a high-level call to `bwplot`.
panel.wireframe	Default panel functions controlling `cloud` and `wireframe` displays.
panel.xyplot	Default panel function for `xyplot`.
parallel	Draws conditional scatter plot matrices and parallel coordinate plots.
prepanel.default.bwplot, prepanel.default.cloud, prepanel.default.densityplot, prepanel.default.histogram, prepanel.default.levelplot, prepanel.default.parallel, prepanel.default.qq, prepanel.default.qqmath, prepanel.default.splom, prepanel.default.xyplot	These prepanel functions are used as fallback defaults in various high-level plot functions in lattice. These are rarely useful to normal users, but may be helpful in developing new displays.
prepanel.lmline, prepanel.loess, prepanel.qqmathline	These are predefined prepanel functions available in lattice.
prepanel.tmd.default, prepanel.tmd.qqmath	tmd creates Tukey mean-difference plots from a `trellis` object returned by `xyplot`, `qq`, or `qqmath`. The prepanel and panel functions are used as appropriate. The `formula` method for `tmd` is provided for convenience and simply calls `tmd` on the object created by calling `xyplot` on that formula.
qq	Quantile-quantile plots for comparing two distributions.
qqmath	Quantile-quantile plot of a sample and a theoretical distribution.
rfs	Plots fitted values and residuals (via `qqmath`) on a common scale for any object that has methods for fitted values and residuals.

Function	Description
shingle	Function to handle shingle.
show.settings	Function used to query, display, and modify graphical parameters for fine control of Trellis displays. Modifications are made to the settings for the currently active device only.
simpleKey	Simple interface to generate a list appropriate for `draw.key`.
simpleTheme	Simple interface to generate a list appropriate as a theme, typically used as the `par.settings` argument in a high-level call.
splom	Draws conditional scatter plot matrices and parallel coordinate plots.
standard.theme	Initialization of a display device with appropriate graphical parameters.
strip.custom	Provides a convenient way to obtain new strip functions that differ from `strip.default` only in the default values of certain arguments.
strip.default	Function that draws the strips by default in Trellis plots. Users can write their own strip functions, but most commonly this involves calling `strip.default` with slightly different arguments.
stripplot	Draws strip plots in lattice.
tmd	tmd creates Tukey mean-difference plots from a `trellis` object returned by `xyplot`, `qq`, or `qqmath`. The `formula` method for tmd is provided for convenience and simply calls tmd on the object created by calling `xyplot` on that formula.
trellis.currentLayout	Returns a matrix with as many rows and columns as in the layout of panels in the current plot.
trellis.device	Initialization of a display device with appropriate graphical parameters.
trellis.focus, trellis.grobname	`trellis.focus` can be used to move to a particular panel or strip, identified by its position in the array of panels.
trellis.last.object	Updates method for objects of class `"trellis"` and is a way to retrieve the last printed `trellis` object (that was saved).
trellis.panelArgs	Once a panel or strip is in focus (e.g., by using `trellis.switchFocus`), `trellis.panelArgs` can be used to retrieve the arguments that were available to the `panel` function at that position.
trellis.par.get, trellis.par.set	Functions used to query, display, and modify graphical parameters for fine control of Trellis displays. Modifications are made to the settings for the currently active device only.
trellis.switchFocus	A convenience function to switch from one viewport to another, while preserving the current row and column.
trellis.unfocus	Unsets the focus and makes the top-level viewport the current viewport.
trellis.vpname	Returns the name of a viewport.
which.packet	Returns the combination of levels of the conditioning variables in the form of a numeric vector as long as the number of conditioning variables, with each element an integer indexing the levels of the corresponding variable.
wireframe	Generic function to draw 3D scatter plots and surfaces. The `"formula"` methods do most of the actual work.

Function	Description
xscale.components.default, yscale.components.default	Return a list of the form suitable as the components argument of `axis.default`.
xyplot	Produces conditional scatter plots.

Data Sets

Data Set	Class	Description
barley	data.frame	Total yield in bushels per acre for 10 varieties at 6 sites in each of 2 years.
environmental	data.frame	Daily measurements of ozone concentration, wind speed, temperature, and solar radiation in New York City from May to September of 1973.
ethanol	data.frame	Ethanol fuel was burned in a single-cylinder engine. For various settings of the engine compression and equivalence ratio, the emissions of nitrogen oxides were recorded.
melanoma	data.frame	This data from the Connecticut Tumor Registry presents age-adjusted numbers of melanoma skin cancer incidences per 100,000 people in Connecticut for the years 1936–1972.
singer	data.frame	Heights, in inches, of the singers in the New York Choral Society in 1979. The data is grouped according to voice part. The vocal range for each voice part increases in pitch according to the following order: Bass 2, Bass 1, Tenor 2, Tenor 1, Alto 2, Alto 1, Soprano 2, Soprano 1.

MASS

This is the main package of Venables and Ripley's MASS.

Functions

Function	Description
Null	Given a matrix M, finds a matrix N giving a basis for the null space. That is, `t(N) \ %*\% M` is the 0, and N has the maximum number of linearly independent columns.
Shepard	One form of nonmetric multidimensional scaling.
addterm	Tries fitting all models that differ from the current model by adding a single term from those supplied, maintaining marginality.
area	Integrates a function of one variable over a finite range using a recursive adaptive method. This function is mainly for demonstration purposes.
as.fractions	Finds rational approximations to the components of a real numeric object using a standard continued fraction method.
bandwidth.nrd	A well-supported rule of thumb for choosing the bandwidth of a Gaussian kernel density estimator.
bcv	Uses biased cross-validation to select the bandwidth of a Gaussian kernel density estimator.

Function	Description
boxcox	Computes and optionally plots profile log-likelihoods for the parameter of the Box-Cox power transformation.
con2tr	Converts lists to data frames for use by lattice.
contr.sdif	A coding for unordered factors based on successive differences.
corresp	Finds the principal canonical correlation and corresponding row and column scores from a correspondence analysis of a two-way contingency table.
cov.mcd, cov.mve, cov.rob	Compute a multivariate location and scale estimate with a high breakdown point. This can be thought of as estimating the mean and covariance of the good part of the data. `cov.mve` and `cov.mcd` are compatibility wrappers.
cov.trob	Estimates a covariance or correlation matrix assuming the data came from a multivariate t-distribution: this provides some degree of robustness to outliers without giving a high breakdown point.
denumerate	`loglm` allows dimension numbers to be used in place of names in the formula. `denumerate` modifies such a formula into one that `terms` can process.
dose.p	Calibrates binomial assays, generalizing the calculation of LD50.
dropterm	Tries fitting all models that differ from the current model by dropping a single term, maintaining marginality.
eqscplot	Version of a scatter plot with scales chosen to be equal on both axes, i.e., 1 cm represents the same units on each.
fitdistr	Maximum likelihood fitting of univariate distributions, allowing parameters to be held fixed, if desired.
fractions	Finds rational approximations to the components of a real numeric object using a standard continued fraction method.
gamma.dispersion	A frontend to `gamma.shape` for convenience. Finds the reciprocal of the estimate of the shape parameter only.
gamma.shape	Finds the maximum likelihood estimate of the shape parameter of the gamma distribution after fitting a Gamma generalized linear model.
ginv	Calculates the Moore-Penrose generalized inverse of a matrix X.
glm.convert	Modifies an output object from `glm.nb()` to one that looks like the output from `glm()` with a negative binomial family. This allows it to be updated keeping the theta parameter fixed.
glm.nb	A modification of the system function `glm()` to include estimation of the additional parameter, `theta`, for a negative binomial generalized linear model.
glmmPQL	Fits a generalized linear mixed model (GLMM) with multivariate normal random effects, using penalized quasi-likelihood.
hist.FD	Plots a histogram with automatic bin width selection, using the Scott or Freedman-Diaconis formula.
hist.scott	Plots a histogram with automatic bin width selection, using the Scott or Freedman-Diaconis formula.
huber	Finds the Huber M-estimator of location with the median absolute deviation (MAD) scale.

Function	Description
hubers	Finds the Huber M-estimator for location with scale specified, scale with location specified, or both if neither is specified.
is.fractions	Finds rational approximations to the components of a real numeric object using a standard continued fraction method.
isoMDS	One form of nonmetric multidimensional scaling.
kde2d	Two-dimensional kernel density estimation with an axis-aligned bivariate normal kernel, evaluated on a square grid.
lda	Linear discriminant analysis.
ldahist	Plots histograms or density plots of data on a single Fisher linear discriminant.
lm.gls	Fits linear models by generalized least squares.
lm.ridge	Fits a linear model by ridge regression.
lmsreg	Fits a regression to the *good* points in the data set, thereby achieving a regression estimator with a high breakdown point. (`lmsreg` is a compatibility wrapper for `lqs`.)
lmwork	The standardized residuals. These are normalized to unit variance, fitted including the current data point.
loglm	Provides a frontend to the standard function, `loglin`, to allow log-linear models to be specified and fitted in a manner similar to that of other fitting functions, such as `glm`.
logtrans	Finds and optionally plots the marginal (profile) likelihood for alpha for a transformation model of the form `log(y + alpha) ~ x1 + x2 + ...`.
lqs, lqs.formula	Fit a regression to the *good* points in the data set, thereby achieving a regression estimator with a high breakdown point. `lmsreg` and `ltsreg` are compatibility wrappers.
ltsreg	A compatibility wrapper for `lqs`.
mca	Computes a multiple-correspondence analysis of a set of factors.
mvrnorm	Produces one or more samples from the specified multivariate normal distribution.
negative.binomial	Specifies the information required to fit a negative binomial generalized linear model, with known `theta` parameter, using `glm()`.
parcoord	Parallel coordinates plot.
polr	Fits a logistic or probit regression model to an ordered factor response. The default logistic case is *proportional odds logistic regression*, after which the function is named.
psi.bisquare, psi.hampel, psi.huber	Psi functions for `rlm`.
qda	Fits quadratic discriminant analysis models.
rational	Finds rational approximations to the components of a real numeric object using a standard continued fraction method.
renumerate	`denumerate` converts a formula written using the conventions of `loglm` into one that `terms` is able to process. `renumerate` converts it back again to a form like the original.
rlm	Fits a linear model by robust regression using an *M*-estimator.

Function	Description
rms.curv	Calculates the root mean square parameter effects and intrinsic relative curvatures, c^{θ} and c^i, for a fitted nonlinear regression.
rnegbin	Generates random outcomes from a negative binomial distribution, with mean mu and variance `mu + mu^2/theta`.
sammon	One form of nonmetric multidimensional scaling.
select	Fits a linear model by ridge regression.
stdres	The standardized residuals. These are normalized to unit variance, fitted including the current data point.
stepAIC	Performs stepwise model selection by AIC.
studres	Extracts the Studentized residuals from a linear model.
theta.md, theta.ml, theta.mm	Given the estimated mean vector, estimate `theta` of the negative binomial distribution.
truehist	Creates a histogram on the current graphics device.
ucv	Uses unbiased cross-validation to select the bandwidth of a Gaussian kernel density estimator.
width.SJ	Uses the method of Sheather and Jones to select the bandwidth of a Gaussian kernel density estimator.
write.matrix	Writes a matrix or data frame to a file or the console, using column labels and a layout respecting columns.

Data Sets

Data Set	Class	Description
Aids2	data.frame	Data on patients diagnosed with AIDS in Australia before July 1, 1991.
Animals	data.frame	Average brain and body weights for 28 species of land animals.
Boston	data.frame	The Boston data frame has 506 rows and 14 columns.
Cars93	data.frame	The Cars93 data frame has 93 rows and 27 columns.
Cushings	data.frame	Cushing's syndrome is a hypertensive disorder associated with oversecretion of cortisol by the adrenal gland. The observations are urinary excretion rates of two steroid metabolites.
DDT	numeric	A numeric vector of 15 measurements by different laboratories of the pesticide DDT in kale, in ppm (parts per million), using the multiple pesticide residue measurement.
GAGurine	data.frame	Data was collected on the concentration of the chemical glycosaminoglycan (GAG) in the urine of 314 children aged 0 to 17 years. The aim of the study was to produce a chart to help a pediatrican to assess if a child's GAG concentration is "normal."
Insurance	data.frame	The data given in data frame Insurance consists of the numbers of policyholders of an insurance company who were exposed to risk, and the numbers of car insurance claims made by those policyholders in the third quarter of 1973.
Melanoma	data.frame	The Melanoma data frame has data on 205 patients in Denmark with malignant melanoma.

Data Set	Class	Description
OME	data.frame	Experiments were performed on children on their ability to differentiate a signal in broadband noise. The noise was played from a pair of speakers, and a signal was added to just one channel; the subject had to turn his/her head to the channel with the added signal. The signal was either coherent (the amplitude of the noise was increased for a period) or incoherent (independent noise was added for the same period to form the same increase in power). The threshold used in the original analysis was the stimulus loudness needed to get 75% correct responses. Some of the children had suffered from otitis media with effusion (OME).
Pima.te	data.frame	A population of women who were at least 21 years old, of Pima Indian heritage, and living near Phoenix, Arizona, was tested for diabetes according to World Health Organization criteria. The data was collected by the National Institute of Diabetes and Digestive and Kidney Diseases. A total of 532 complete records was used, after dropping the (mainly missing) data on serum insulin.
Pima.tr	data.frame	A population of women who were at least 21 years old, of Pima Indian heritage, and living near Phoenix, Arizona, was tested for diabetes according to World Health Organization criteria. The data was collected by the National Institute of Diabetes and Digestive and Kidney Diseases. A total of 532 complete records was used, after dropping the (mainly missing) data on serum insulin.
Pima.tr2	data.frame	A population of women who were at least 21 years old, of Pima Indian heritage, and living near Phoenix, Arizona, was tested for diabetes according to World Health Organization criteria. The data was collected by the National Institute of Diabetes and Digestive and Kidney Diseases. A total of 532 complete records was used, after dropping the (mainly missing) data on serum insulin.
Rabbit	data.frame	Five rabbits were studied on two occasions, after treatment with saline (control) and after treatment with the 5-HT_3 antagonist MDL 72222. After each treatment, ascending doses of phenylbiguanide were injected intravenously at 10-minute intervals and the responses of mean blood pressure measured. The goal was to test whether the cardiogenic chemoreflex elicited by phenylbiguanide depends on the activation of 5-HT_3 receptors.
Rubber	data.frame	Data frame from accelerated testing of tire rubber.
SP500	numeric	Returns of the Standard & Poors 500 Index in the 1990s.
Sitka	data.frame	The Sitka data frame has 395 rows and 4 columns. It gives repeated measurements on the log-size of 79 Sitka spruce trees, 54 of which were grown in ozone-enriched chambers and 25 were controls. The size was measured five times in 1988, at roughly monthly intervals.
Sitka89	data.frame	The Sitka89 data frame has 632 rows and 4 columns. It gives repeated measurements on the log-size of 79 Sitka spruce trees, 54 of which were grown in ozone-enriched chambers and 25 were controls. The size was measured eight times in 1989, at roughly monthly intervals.
Skye	data.frame	The Skye data frame has 23 rows and 3 columns.
Traffic	data.frame	An experiment was performed in Sweden in 1961–1962 to assess the effect of a speed limit on the highway accident rate. The experiment was conducted on 92 days in each year, matched so that day j in 1962 was comparable to day j in 1961. On some days, the speed limit was in effect and enforced, while on other days there was no speed limit and cars tended to be driven faster. The speed limit days tended to be in contiguous blocks.

R Reference

Data Set	Class	Description
UScereal	data.frame	The UScereal data frame has 65 rows and 11 columns. The data comes from the 1993 American Statistical Association (ASA) Statistical Graphics Exposition and is taken from the mandatory Food and Drug Administration (FDA) food label. The data has been normalized here to a portion of one American cup.
UScrime	data.frame	Criminologists are interested in the effect of punishment regimes on crime rates. This has been studied using the aggregate data on 47 states of the United States for 1960 given in this data frame. The variables seem to have been rescaled to convenient numbers.
VA	data.frame	Veteran's Administration lung cancer trial from Kalbfleisch and Prentice.
abbey	numeric	A numeric vector of 31 determinations of nickel content (ppm) in a Canadian syenite rock.
accdeaths	ts	A regular time series giving the monthly totals of accidental deaths in the United States.
anorexia	data.frame	The anorexia data frame has 72 rows and 3 columns. Weight change data for young female anorexia patients.
bacteria	data.frame	Tests of the presence of the bacteria *H. influenzae* in children with otitis media in the Northern Territory of Australia.
beav1	data.frame	Reynolds describes a small part of a study of the long-term temperature dynamics of the beaver (*Castor canadensis*) in north-central Wisconsin. Body temperature was measured by telemetry every 10 minutes for four females, but data from a one period of less than a day for each of two animals is used here.
beav2	data.frame	Reynolds describes a small part of a study of the long-term temperature dynamics of the beaver (*Castor canadensis*) in north-central Wisconsin. Body temperature was measured by telemetry every 10 minutes for four females, but data from a period of less than a day for each of two animals is used here.
biopsy	data.frame	This breast cancer database was obtained from the University of Wisconsin Hospitals, Madison, from Dr. William H. Wolberg. He assessed biopsies of breast tumors for 699 patients up to July 15, 1992; each of nine attributes has been scored on a scale of 1 to 10, and the outcome is also known. There are 699 rows and 11 columns.
birthwt	data.frame	The birthwt data frame has 189 rows and 10 columns. The data was collected at Baystate Medical Center, Springfield, Massachusetts, during 1986.
cabbages	data.frame	The cabbages data set has 60 observations and 4 variables.
caith	data.frame	Data on the cross-classification of people in Caithness, Scotland, by eye and hair color. This region of the United Kingdom is particularly interesting as there is a mixture of people of Nordic, Celtic, and Anglo-Saxon origin.
cats	data.frame	The heart and body weights of samples of male and female cats used for digitalis experiments. The cats were all adult, over 2 kg in body weight.
cement	data.frame	Experiment on the heat evolved in the setting of each of 13 cements.
chem	numeric	A numeric vector of 24 determinations of copper in wholemeal flour, in parts per million.
coop	data.frame	Seven specimens were sent to six laboratories in three separate batches and each analyzed for analyte. Each analysis was duplicated.
cpus	data.frame	A relative performance measure and characteristics of 209 CPUs.

Data Set	Class	Description
crabs	data.frame	The crabs data frame has 200 rows and 8 columns, describing 5 morphological measurements on 50 crabs, each of two color forms and both sexes, of the species *Leptograpsus variegatus*, collected at Fremantle, Western Australia.
deaths	ts	A time series giving the monthly deaths from bronchitis, emphysema, and asthma in the United Kingdom, 1974–1979, for both sexes.
drivers	ts	A regular time series giving the monthly totals of car drivers in Great Britain killed or seriously injured from January 1969 to December 1984. Compulsory wearing of seat belts was introduced on January 31, 1983.
eagles	data.frame	Knight and Skagen collected data during a field study on the foraging behavior of wintering bald eagles in Washington State. The data concerned 160 attempts by one (pirating) bald eagle to steal a chum salmon from another (feeding) bald eagle.
epil	data.frame	Thall and Vail give a data set on 2-week seizure counts for 59 epileptics. The number of seizures was recorded for a baseline period of 8 weeks, and then patients were randomly assigned to a treatment group or a control group. Counts were then recorded for four successive 2-week periods. The subject's age is the only covariate.
farms	data.frame	The farms data frame has 20 rows and 4 columns. The rows are farms on the Dutch island of Terschelling, and the columns are factors describing the management of grassland.
fgl	data.frame	The fgl data frame has 214 rows and 10 columns. It was collected by B. German on fragments of glass collected in forensic work.
forbes	data.frame	A data frame with 17 observations on the boiling point of water and barometric pressure, in inches of mercury.
galaxies	numeric	A numeric vector of velocities, in kilometers/second, of 82 galaxies from 6 well-separated conic sections of an unfilled survey of the Corona Borealis region. Multimodality in such surveys is evidence for voids and superclusters in the far universe.
gehan	data.frame	A data frame from a trial of 42 leukemia patients. Some were treated with the drug 6-mercaptopurine, and the rest were controls. The trial was designed as matched pairs, both withdrawn from the trial when either came out of remission.
genotype	data.frame	Data from a foster feeding experiment with rat mothers and litters of four different genotypes: A, B, I and J. Rat litters were separated from their natural mothers at birth and given to foster mothers to rear.
geyser	data.frame	A version of the eruptions data from the Old Faithful geyser in Yellowstone National Park, Wyoming. This version comes from Azzalini and Bowman and is of continuous measurement from August 1 to August 15, 1985. Some nocturnal duration measurements were coded as 2, 3, or 4 minutes, having originally been described as "short," "medium," or "long."
gilgais	data.frame	This data set was collected on a line transect survey in gilgai territory in New South Wales, Australia. Gilgais are natural gentle depressions in otherwise flat land, and sometimes they seem to be regularly distributed. The data collection was stimulated by the question: are these patterns reflected in soil properties? At each of 365 sampling locations on a linear grid of 4 meters, spacing, samples were taken at depths 0–10 cm, 30–40 cm, and 80–90 cm below the surface. pH, electrical conductivity, and chloride content were measured on a 1:5 soil:water extract from each sample.
hills	data.frame	The record times in 1984 for 35 Scottish hill races.

Data Set	Class	Description
housing	data.frame	The housing data frame has 72 rows and 5 variables.
immer	data.frame	The immer data frame has 30 rows and 4 columns. Five varieties of barley were grown in six locations in 1931 and in 1932.
leuk	data.frame	A data frame of data from 33 leukemia patients.
mammals	data.frame	A data frame with average brain and body weights for 62 species of land mammals.
mcycle	data.frame	A data frame giving a series of measurements of head acceleration in a simulated motorcycle accident; used to test crash helmets.
menarche	data.frame	Proportions of female children at various ages during adolescence who have reached menarche.
michelson	data.frame	Measurements of the speed of light in air, made between June 5, and July 2, 1879. The data consists of 5 experiments, each consisting of 20 consecutive runs. The response is the speed of light, in kilometers/second, less 299,000. The currently accepted value, on this scale of measurement, is 734.5.
minn38	data.frame	Minnesota high school graduates of 1938 were classified according to four factors. The minn38 data frame has 168 rows and 5 columns.
motors	data.frame	The motors data frame has 40 rows and 3 columns. It describes an accelerated life test at each of four temperatures of 10 motorettes and has rather discrete times.
muscle	data.frame	The purpose of this experiment was to assess the influence of calcium in solution on the contraction of heart muscle in rats. The left auricle of 21 rat hearts was isolated, and on several occasions a constant-length strip of tissue was electrically stimulated and dipped into various concentrations of calcium chloride solution, after which the shortening of the strip was accurately measured as the response.
newcomb	numeric	A numeric vector giving the "Third Series" of measurements of the passage time of light recorded by Newcomb in 1882. The given values divided by 1,000 plus 24 give the time, in millionths of a second, for light to traverse a known distance. The "true" value is now considered to be 33.02.
nlschools	data.frame	Snijders and Bosker use as a running example a study of 2,287 eighth-grade pupils (aged about 11) in 132 classes in 131 schools in the Netherlands. Only the variables used in their examples are supplied.
npk	data.frame	A classical N, P, K (nitrogen, phosphate, potassium) factorial experiment on the growth of peas conducted on six blocks. Each half of a fractional factorial design confounding the NPK interaction was used on three of the plots.
npr1	data.frame	Data on the locations, porosity, and permeability (a measure of oil flow) on 104 oil wells in the U.S. Naval Petroleum Reserve No. 1 in California.
oats	data.frame	The yield of oats from a split-plot field trial using three varieties and four levels of manurial treatment. The experiment was laid out in six blocks of three main plots, each split into four subplots. The varieties were applied to the main plots and the manurial treatments to the subplots.
painters	data.frame	The subjective assessment, on an integer scale of 0 to 20, of 54 classical painters. The painters were assessed on four characteristics: composition, drawing, color, and expression. The data is due to the 18th-century art critic, de Piles.

Data Set	Class	Description
petrol	data.frame	The yield of a petroleum refining process with four covariates. The crude oil appears to come from only 10 distinct samples. This data was originally used by Prater to build an estimation equation for the yield of the refining process of crude oil to gasoline.
phones	list	A list object with the annual number of telephone calls in Belgium.
quine	data.frame	The quine data frame has 146 rows and 5 columns. Children from Walgett, New South Wales, Australia, were classified by culture, age, sex, and learner status, and the number of days absent from school in a particular school year was recorded.
road	data.frame	A data frame with the annual deaths in road accidents for half the U.S. states.
rotifer	data.frame	The data give the numbers of rotifers falling out of suspension for different fluid densities.
ships	data.frame	Data frame giving the number of damage incidents and aggregate months of service by ship type, year of construction, and period of operation.
shoes	list	A list of two vectors, giving the wear of shoes of materials A and B for one foot each of 10 boys.
shrimp	numeric	A numeric vector with 18 determinations by different laboratories of the amount (percentage of the declared total weight) of shrimp in shrimp cocktail.
shuttle	data.frame	The shuttle data frame has 256 rows and 7 columns. The first six columns are categorical variables giving example conditions; the seventh is the decision. The first 253 rows are the training set, the last 3 the test conditions.
snails	data.frame	Groups of 20 snails were held for periods of 1, 2, 3, or 4 weeks under carefully controlled conditions of temperature and relative humidity. There were two species of snail, A and B, and the experiment was designed as a 4-by-3-by-4-by-2 completely randomized design. At the end of the exposure time, the snails were tested to see if they had survived; the process itself is fatal for the animals. The object of the exercise was to model the probability of survival in terms of the stimulus variables and, in particular, to test for differences between species. The data are unusual in that, in most cases, fatalities during the experiment were fairly small.
steam	data.frame	Temperature and pressure in a saturated steam-driven experimental device.
stormer	data.frame	The stormer viscometer measures the viscosity of a fluid by measuring the time taken for an inner cylinder in the mechanism to perform a fixed number of revolutions in response to an actuating weight. The viscometer is calibrated by measuring the time taken with varying weights while the mechanism is suspended in fluids of accurately known viscosity. The data comes from such a calibration, and theoretical considerations suggest a nonlinear relationship among time, weight, and viscosity of the form $Time = (B1 * Viscosity)/(Weight - B2) + E$, where B1 and B2 are unknown parameters to be estimated, and E is error.
survey	data.frame	This data frame contains the responses of 237 Statistics I students at the University of Adelaide to a number of questions.
synth.te	data.frame	The synth.tr data frame has 250 rows and 3 columns. The synth.te data frame has 100 rows and 3 columns. It is intended that synth.tr be used from training and synth.te for testing.
synth.tr	data.frame	The synth.tr data frame has 250 rows and 3 columns. The synth.te data frame has 100 rows and 3 columns. It is intended that synth.tr be used from training and synth.te for testing.

Data Set	Class	Description
topo	data.frame	The `topo` data frame has 52 rows and 3 columns, of topographic heights within a 310-foot square.
waders	data.frame	The `waders` data frame has 15 rows and 19 columns. The entries are counts of waders in summer.
whiteside	data.frame	Mr. Derek Whiteside of the UK Building Research Station recorded the weekly gas consumption and average external temperature at his own house in southeast England for two heating seasons, one of 26 weeks before, and one of 30 weeks after cavity-wall insulation was installed. The object of the exercise was to assess the effect of the insulation on gas consumption.
wtloss	data.frame	This data frame gives the weight, in kilograms, of an obese patient at 52 time points over an 8-month period of a weight rehabilitation program.

methods

This package contains formally defined methods and classes for R objects, plus other programming tools.

Functions

Function	Description
@<-	Gets or sets information about the individual slots in an object.
MethodAddCoerce	Possibly modifies one or more methods to explicitly coerce this argument to `methodClass`, the class for which the method is explicitly defined.
Quote	These are utilities, currently in the `methods` package, that either provide some functionality needed by the package (e.g., element matching by name) or add compatibility with S-PLUS, or both.
S3Class, S3Class<-	`S3Class` extracts or replaces the S3-style class from an S4 class that was created from an S3 class through `setOldClass`.
S3Part, S3Part<-	The function `S3Part` extracts or replaces the S3 part of such an object.
addNextMethod	Generic function that finds the next method for the signature of the method definition method and caches that method in the method definition.
allNames	Returns the character vector of names (unlike names(), never returns NULL) for a method.
as, as<-	Manage the relations that allow coercing an object to a given class.
asMethodDefinition	Turns a function definition into an object of class `MethodDefinition`, corresponding to the given signature (by default, generates a default method with empty signature).
assignClassDef	Assigns the definition of the class to the specially named object.
assignMethodsMetaData	Utility to assign the metadata object recording the methods defined in a particular package.
balanceMethodsList	Called from setMethod to ensure that all nodes in the list have the same depth (i.e., the same number of levels of arguments).
body<-	Sets the body of a method.

Function	Description
cacheGenericsMetaData, cacheMetaData	Utilities for ensuring that the internal information about class and method definitions is up to date. Should normally be called automatically whenever needed (e.g., when a method or class definition changes or when a package is attached or detached).
cacheMethod	Stores the definition for this function and signature in the method metadata for the function.
callGeneric	The name and package of the current generic function is stored in the environment of the method definition object.
callNextMethod	A call to `callNextMethod` can only appear inside a method definition. It then results in a call to the first inherited method after the current method, with the arguments to the current method passed down to the next method. The value of that method call is the value of `callNextMethod`.
canCoerce	Tests if an object can be coerced to a given S4 class.
cbind2	Combines two matrix-like R objects by columns (`cbind2`) or rows (`rbind2`). These are (S4) generic functions with default methods.
checkSlotAssignment	Checks that the value provided is allowed for this slot, by consulting the definition of the class. Called from the C code that assigns slots.
classMetaName	A name for the object storing this class's definition.
classesToAM	Given a vector of class names or a list of class definitions, returns an adjacency matrix of the superclasses of these classes; i.e., a matrix with class names as the row and column names and with element [i, j] being 1 if the class in column j is a direct superclass of the class in row i, and 0 otherwise.
coerce, coerce<-	Manage the relations that allow coercing an object to a given class.
completeClassDefinition	Completes the definition of Class, relative to the class definitions visible from environment where. If doExtends is TRUE, completes the super- and subclass information.
completeExtends	Completes the extends information in the class definition, by following transitive chains.
completeSubclasses	
conformMethod	If the formal arguments, mnames, are not identical to the formal arguments to the function, fnames, conformMethod determines whether the signature and the two sets of arguments conform and returns the signature, possibly extended.
defaultDumpName	Default name to be used for dumping a method.
defaultPrototype	The prototype for a class that will have slots, is not a virtual class, and does not extend one of the basic classes. Both its class and its (R internal) type, `typeof()`, are "S4."
doPrimitiveMethod	Performs a primitive call to built-in function name the definition and call provided, and carried out in the specified environment.
dumpMethod	Dumps the method for this generic function and signature.
dumpMethods	Dumps all the methods for this generic.
elNamed, elNamed<-	Get or set the element of the vector corresponding to name.
existsFunction	Is there a function of this name? If generic is FALSE, generic functions are not counted.

Function	Description
existsMethod	Tests for the existence of a method corresponding to a given generic function and signature.
extends	Function to test inheritance relationships between an object and a class (is) or between two classes (extends) and to establish such relationships (setIs, an explicit alternative to the contains= argument to setClass).
findClass	Function to find and manipulate class definitions.
findFunction	Returns a list of either the positions on the search list or the current top-level environment on which a function object for name exists.
findMethod	Returns the package(s) in the search list (or in the packages specified by the where argument) that contain a method for this function and signature.
findMethodSignatures	Returns a character matrix whose rows are the class names from the signature of the corresponding methods; it operates either from a list returned by findMethods or by computing such a list itself, given the same arguments as findMethods.
findMethods	Returns a list of the method definitions currently existing for generic function f, limited to the methods defined in environment where if that argument is supplied and possibly limited to those including one or more of the specified classes in the method signature.
findUnique	Returns the list of environments (or equivalent) having an object named what, using environment where and its parent environments.
fixPre1.8	Beginning with R version 1.8.0, the class of an object contains the identification of the package in which the class is defined. The function fixPre1.8 fixes and reassigns objects missing that information (typically because they were loaded from a file saved with a previous version of R).
formalArgs	Returns the names of the formal arguments of this function.
functionBody, functionBody<-	These are utilities, currently in the methods package, that either provide some functionality needed by the package (e.g., element matching by name) or add compatibility with S-PLUS, or both.
generic.skeleton	Utility functions to support the definition and use of formal methods. Most of these functions will not normally be called directly by the user.
getAllSuperClasses	Gets the names of all the classes that this class definition extends.
getClass, getClassDef	Get the definition of a class.
getClasses	Function to find and manipulate class definitions.
getDataPart	Utility called to implement object@.Data.
getFromClassDef	Extracts one of the intrinsically defined class definition properties (".Properties", etc.). Strictly a utility function.
getFunction	These are utilities, currently in the methods package, that either provide some functionality needed by the package (e.g., element matching by name) or add compatibility with S-PLUS, or both.
getGeneric	Returns the definition of the function named f as a generic.

Function	Description
getGenerics	Returns the names of the generic functions that have methods defined on where; this argument can be an environment or an index into the search list.
getGroup	Returns the groups to which this generic belongs, searching from environment where (the global environment normally by default).
getGroupMembers	Returns all the members of the group generic function named group.
getMethod	Returns the method corresponding to a given generic function and signature.
getMethods	An older alternative to findMethods, returning information in the form of an object of class MethodsList, previously used for method dispatch.
getMethodsForDispatch	Support routine for computations on formal methods.
getMethodsMetaData	Utility to get the metadata object recording the methods defined in a particular package.
getPackageName	Returns the package associated with a particular environment or position on the search list, or the package containing a particular function.
getSlots	Returns a named character vector. The names are the names of the slots; the values are the classes of the corresponding slots.
getValidity	The validity of object related to its class definition is tested. If the object is valid, TRUE is returned; otherwise, either a vector of strings describing validity failures is returned or an error is generated (according to whether test is TRUE).
hasArg	Returns TRUE if name corresponds to an argument in the call, either a formal argument to the function or a component of . . ., and FALSE otherwise.
hasMethod	Tests for the existence of a method corresponding to a given generic function and signature.
hasMethods	Returns TRUE or FALSE according to whether there is a nonempty table of methods for function f in the environment or search position where (or anywhere on the search list if where is missing).
implicitGeneric	Returns the implicit generic version.
initialize	Given the name or the definition of a class, plus optionally data to be included in the object, initialize returns an object from that class.
is	Function to test inheritance relationships between an object and a class (is) or between two classes (extends) and to establish such relationships (setIs, an explicit alternative to the contains= argument to setClass).
isClass	Function to find and manipulate class definitions.
isClassDef	Is object a representation of a class?
isClassUnion	Tests if a class is a "ClassUnion".
isGeneric	Is there a function named f, and, if so, is it a generic?
isGroup	Manages collections of methods associated with a generic function, as well as providing information about the generic functions themselves.
isSealedClass, isSealedMethod	Check for either a method or a class that has been *sealed* when it was defined and therefore cannot be redefined.
isVirtualClass	Is the named class a virtual class?

Function	Description
isXS3Class	Old-style (S3) classes may be registered as S4 classes (by calling setOldClass), and many have been. These classes can then be contained in (i.e., superclasses of) regular S4 classes, allowing formal methods and slots to be added to the S3 behavior. The function S3Part extracts or replaces the S3 part of such an object. S3Class extracts or replaces the S3-style class. S3Class also applies to objects from an S4 class with S3methods=TRUE in the call to setClass.
listFromMethods	Support routine for computations on formal methods.
makeClassRepresentation	Constructs an object of class classRepresentation to describe a particular class. Mostly a utility function, but you can call it to create a class definition without assigning it, as setClass would do.
makeExtends	Converts the argument to a list defining the extension mechanism.
makeGeneric	Makes a generic function object corresponding to the given function name, optional definition, and optional default method.
makePrototypeFromClassDef	Makes the prototype implied by the class definition.
makeStandardGeneric	A utility function that makes a valid function calling standardGeneric for name f.
matchSignature	Matches the signature object (a partially or completely named subset of the signature arguments of the generic function object fun) and returns a vector of all the classes in the order specified by fun@signature.
metaNameUndo	As its name implies, this function undoes the name mangling used to produce metadata object names and returns an object of class ObjectsWithPackage.
method.skeleton	Writes a source file containing a call to setMethod to define a method for the generic function and signature supplied. By default, the method definition is in line in the call, but can be made an external (previously assigned) function.
methodSignatureMatrix	Returns a matrix with the contents of the specified slots as rows.
methodsPackageMetaName	A name-mangling device to hide metadata defining method and class information.
missingArg	Returns TRUE if the symbol supplied is missing from the call corresponding to the environment supplied (by default, environment of the call to missingArg).
new	Given the name or the definition of a class, plus optionally data to be included in the object, new returns an object from that class.
newBasic	The implementation of the function new for basic classes that don't have a formal definition.
newClassRepresentation	Various functions to support the definition and use of formal classes. Most of them are rarely suitable to be called directly. Others are somewhat experimental and/or partially implemented only. Do refer to setClass for normal code development.
newEmptyObject	Utility function to create an empty object into which slots can be set.
packageName	Returns the character-string name of the package (without the extraneous "package:" found in the search list).
packageSlot, packageSlot<-	Return or set the package name slot (currently an attribute, not a formal slot, but this may change someday).
possibleExtends	Finds the information that says whether one class extends another, directly or indirectly.
prohibitGeneric	Prevents your function from being made generic.

Function	Description
promptClass	Creates a help file for a class definition containing all relevant slot and method information for a class, with minimal markup for Rd processing; no QC facilities at present.
promptMethods	Generates a shell of documentation for the methods of a generic function.
prototype	In calls to setClass, this function constructs the prototype argument.
rbind2	Combines two matrix-like R objects by columns (cbind2) or rows (rbind2). These are (S4) generic functions with default methods.
reconcilePropertiesAndPrototype	Makes a list or a structure look like a prototype for the given class.
registerImplicitGenerics	Saves a set of implicit generic definitions in the cached table of the current session.
rematchDefinition	If the specified method in a call to setMethod specializes the argument list (by replacing ...), then rematchDefinition constructs the actual method stored.
removeClass	Function to find and manipulate class definitions.
removeGeneric	Removes all the methods for the generic function of this name and the function itself.
removeMethod	Creates and saves a formal method for a given function and list of classes.
removeMethods	Removes all the methods for the generic function of this name.
representation	In calls to setClass, this function constructs the representation argument.
requireMethods	Requires a subclass to implement methods for the generic functions, for this signature.
resetClass	Function to find and manipulate class definitions.
resetGeneric	Support routine for computations on formal methods.
sealClass	Function to find and manipulate class definitions.
selectMethod	Returns a method corresponding to a given generic function and signature.
selectSuperClasses	Returns superclasses of ClassDef, possibly only nonvirtual or direct or simple ones. This function is designed to be fast and, consequently, only works with the contains slot of the corresponding class definitions.
sessionData	Returns the index of the session data in the search list, attaching it if it is not attached.
setAs	Manages the relations that allow coercing an object to a given class.
setClass	Creates a class definition, specifying the representation (the slots) and/or the classes contained in this one (the superclasses), plus other optional details.
setClassUnion	A class may be defined as the *union* of other classes, i.e., as a virtual class defined as a superclass of several other classes. This function creates class unions.
setDataPart	Utility called to implement object@.Data. Calls to setDataPart are also used to merge the data part of a superclass prototype.
setGeneric	Creates a new generic function of the given name, i.e., a function that dispatches methods according to the classes of the arguments, from among the formal methods defined for this function.
setGenericImplicit	Turns a generic implicit.
setGroupGeneric	Creates a new generic function of the given name, i.e., a function that dispatches methods according to the classes of the arguments, from among the formal methods defined for this function.

Function	Description
setIs	Function to test inheritance relationships between an object and a class (is) or between two classes (extends) and to establish such relationships (setIs, an explicit alternative to the contains= argument to setClass).
setMethod	Creates and saves a formal method for a given function and list of classes.
setOldClass	Registers an old-style ("S3") class as a formally defined class. The Classes argument is the character vector used as the class attribute; in particular, if there is more than one string, old-style class inheritance is mimicked. Registering via setOldClass allows S3 classes to appear in method signatures, as a slot in an S4 class or as a superclass of an S4 class.
setPackageName	Used to establish a package name in an environment that would otherwise not have one. This allows you to create classes and/or methods in an arbitrary environment, but it is usually preferable to create packages by the standard R programming tools (package.skeleton, etc.).
setPrimitiveMethods	Utility functions to support the definition and use of formal methods. Most of these functions will not normally be called directly by the user.
setReplaceMethod	Manages collections of methods associated with a generic function, as well as providing information about the generic functions themselves.
setValidity	Sets the validity method of a class (but more normally, this method will be supplied as the validity argument to setClass).
show	Displays the object, by printing, plotting, or whatever suits its class. This function exists to be specialized by methods. The default method calls showDefault. Formal methods for show will usually be invoked for automatic printing (see the details).
showClass	Prints the information about a class definition.
showDefault	Utility used to enable show methods to be called by the automatic printing (via print.default).
showExtends	Prints the elements of the list of extensions; for printTo = FALSE, returns a list with components what and how; this is used, e.g., by promptClass().
showMethods	Shows a summary of the methods for one or more generic functions, possibly restricted to those involving specified classes.
sigToEnv	Turns the signature (a named vector of classes) into an environment with the classes assigned to the names.
signature	Returns a named list of classes to be matched to arguments of a generic function.
slot, slot<-	Return or set information about the individual slots in an object.
slotNames	Returns or sets information about the individual slots in an object.
slotsFromS3	Old-style (S3) classes may be registered as S4 classes (by calling setOldClass), and many have been. These classes can then be contained in (i.e., superclasses of) regular S4 classes, allowing formal methods and slots to be added to the S3 behavior. The function S3Part extracts or replaces the S3 part of such an object. S3Class extracts or replaces the S3-style class. S3Class also applies to objects from an S4 class with S3methods=TRUE in the call to setClass.
substituteDirect	Substitutes for the variables named in the second argument the corresponding objects; substituting into object.

Function	Description
standardGeneric	Dispatches a method from the current function call for the generic function.
substituteFunctionArgs	Utility function to support the definition and use of formal methods. Most of these functions will not normally be called directly by the user.
superClassDepth	`superClassDepth`, which is called from `getAllSuperClasses`, returns the same information, but as a list with components label and depth, the latter for the number of generations back each class is in the inheritance tree.
testInheritedMethods	A set of distinct inherited signatures is generated to test inheritance for all the methods of a specified generic function. If method selection is ambiguous for some of these, a summary of the ambiguities is attached to the returned object. This test should be performed by package authors *before* releasing a package.
testVirtual	Tests for a virtual class.
traceOff	The functions `traceOn` and `traceOff` have been replaced by extended versions of the functions `trace` and `untrace` and should not be used.
traceOn	The functions `traceOn` and `traceOff` have been replaced by extended versions of the functions `trace` and `untrace` and should not be used.
tryNew, trySilent	Tries to generate a new element from this class, but if the attempt fails (as, e.g., when the class is undefined or virtual) just returns `NULL`.
unRematchDefinition	Using knowledge of how rematchDefinition works, `unRematchDefinition` reverses the procedure; if given a function or method definition that does not correspond to this form, it just returns its argument.
validObject	The validity of an object related to its class definition is tested. If the object is valid, TRUE is returned; otherwise, either a vector of strings describing the validity failures is returned or an error is generated (according to whether `test` is TRUE).
validSlotNames	Returns names unless one of the names is reserved, in which case there is an error. (As of this writing, "class" is the only reserved slot name.)

mgcv

This package provides functions for generalized additive modeling and generalized additive mixed modeling. The term GAM is taken to include any GLM estimated by quadratically penalized (possibly quasi-) likelihood maximization. For more information on this package, see the help file.

nlme

This package provides functions for linear and nonlinear mixed-effects models. See the help file for more information.

nnet

This package provides functions for feed-forward neural networks and multinomial log-linear models.

Functions

Function	Description
class.ind	Generates a class indicator function from a given factor.
multinom	Fits multinomial log-linear models via neural networks.
nnet	Fits single-hidden-layer neural network, possibly with skip-layer connections.
nnetHess	Evaluates the Hessian (matrix of second derivatives) of the specified neural network. Normally called via argument Hess=TRUE to nnet or via vcov.multinom.
which.is.max	Finds the maximum position in a vector, breaking ties at random.

rpart

This package provides functions for recursive partitioning and regression trees.

Functions

Function	Description
meanvar	Creates a plot on the current graphics device of the deviance of the node divided by the number of observations at the node. Also returns the node number.
na.rpart	Handles missing values in an rpart object.
path.rpart	Returns a names list, where each element contains the splits on the path from the root to the selected nodes.
plotcp	Gives a visual representation of the cross-validation results in an rpart object.
post	Generates a PostScript presentation plot of an rpart object.
printcp	Displays the cp table for a fitted rpart object.
prune	Determines a nested sequence of subtrees of the supplied rpart object by recursively snipping off the least important splits, based on the complexity parameter (cp).
rpart	Fits an rpart model.
rpart.control	Various parameters that control aspects of the rpart fit.
rpconvert	Rpart objects changed (slightly) in their internal format in order to accommodate the changes for user-written split functions. This routine updates an old object to the new format.
rsq.rpart	Produces two plots. The first plots the r-square (apparent and apparent — from cross-validation) versus the number of splits. The second plots the relative error(cross-validation) +/− 1 − SE from cross-validation versus the number of splits.
snip.rpart	Creates a "snipped" rpart object, containing the nodes that remain after selected subtrees have been snipped off. The user can snip nodes using the toss argument or interactively by clicking the mouse button on specified nodes within the graphics window.
xpred.rpart	Gives the predicted values for an rpart fit, under cross-validation, for a set of complexity parameter values.

Data Sets

Data Set	Class	Description
car.test.frame	data.frame	The `car.test.frame` data frame has 60 rows and 8 columns, giving data on makes of cars taken from the April 1990 issue of *Consumer Reports*. This is part of a larger data set, some columns of which are given in `cu.summary`.
cu.summary	data.frame	The `cu.summary` data frame has 117 rows and 5 columns, giving data on makes of cars taken from the April 1990 issue of *Consumer Reports*.
kyphosis	data.frame	The `kyphosis` data frame has 81 rows and 4 columns, representing data on children who have had corrective spinal surgery.
solder	data.frame	The `solder` data frame has 720 rows and 6 columns, representing a balanced subset of a designed experiment varying 5 factors on the soldering of components on printed-circuit boards.

spatial

This package provides functions for Kriging and point pattern analysis.

Functions

Function	Description
Kaver	Forms the average of a series of (usually simulated) K functions.
Kenvl	Computes envelope (upper and lower limits) and average of simulations of K functions.
Kfn	Actually computes L = sqrt(K/pi).
Psim	Simulates binomial spatial point process.
SSI	Simulates SSI (sequential spatial inhibition) point process.
Strauss	Simulates Strauss spatial point process.
anova.trls	Computes analysis of variance tables for one or more fitted trend surface model objects; where `anova.trls` is called with multiple objects, it passes on the arguments to `anovalist.trls`.
anovalist.trls	Computes analysis of variance tables for one or more fitted trend surface model objects; where `anova.trls` is called with multiple objects, it passes on the arguments to `anovalist.trls`.
correlogram	Computes spatial correlograms of spatial data or residuals.
expcov	Spatial covariance function for use with `surf.gls`.
gaucov	Spatial covariance function for use with `surf.gls`.
plot.trls	Provides the basic quantities used in forming a variety of diagnostics for checking the quality of regression fits for trend surfaces calculated by `surf.ls`.
ppgetregion	Retrieves the rectangular domain (x1, xu) x (y1, yu) from the underlying C code.
ppinit	Reads a file in standard format and creates a point process object.

Function	Description
pplik	Pseudolikelihood estimation of a Strauss spatial point process.
ppregion	Sets the rectangular domain (xl, xu) x (yl, yu).
predict.trls	Predicted values based on trend surface model object.
prmat	Evaluates Kriging surface over a grid.
semat	Evaluates Kriging standard error of prediction over a grid.
sphercov	Spatial covariance function for use with surf.gls.
surf.gls	Fits a trend surface by generalized least squares.
surf.ls	Fits a trend surface by least squares.
trls.influence	Provides the basic quantities used in forming a variety of diagnostics for checking the quality of regression fits for trend surfaces calculated by surf.ls.
trmat	Evaluates trend surface over a grid.
variogram	Computes spatial (semi-)variogram of spatial data or residuals.

splines

This package provides functions for working with regression splines using the B-spline basis, bs, and the natural cubic spline basis, ns.

Functions

Function	Description
as.polySpline	Creates the piecewise polynomial representation of a spline object.
asVector	This is a generic function. Methods for this function coerce objects of given classes to vectors.
backSpline	Creates a monotone inverse of a monotone natural spline.
bs	Generates the B-spline basis matrix for a polynomial spline.
interpSpline	Creates an interpolation spline, either from x and y vectors or from a formula/ data.frame combination.
ns	Generates the B-spline basis matrix for a natural cubic spline.
periodicSpline	Creates a periodic interpolation spline, either from x and y vectors or from a formula/ data.frame combination.
polySpline	Creates the piecewise polynomial representation of a spline object.
spline.des	Evaluates the design matrix for the B-splines defined by knots at the values in x.
splineDesign	Evaluates the design matrix for the B-splines defined by knots at the values in x.
splineKnots	Returns the knot vector corresponding to a spline object.
splineOrder	Returns the order of a spline object.
xyVector	Creates an object to represent a set of x-y pairs.

stats

This package contains functions to perform a wide variety of statistical analyses.

Functions

Function	Description
AIC	Generic function for calculating the Akaike information criterion for one or several fitted model objects for which a log-likelihood value can be obtained, according to the formula $-2 * \text{log-likelihood} + k * n_{par}$, where n_{par} represents the number of parameters in the fitted model, and $k = 2$ for the usual AIC, or $k = \log(n)$ (n is the number of observations) for the so-called Bayesian information criterion (BIC) or Schwarz's Bayesian criterion (SBC).
ARMAacf	Computes the theoretical autocorrelation function or partial autocorrelation function for an autoregressive moving average (ARMA) process.
ARMAtoMA	Converts an ARMA process to an infinite moving average (MA) process.
Box.test	Computes the Box-Pierce or Ljung-Box test statistic for examining the null hypothesis of independence in a given time series. These are sometimes known as "portmanteau" tests.
C	Sets the `"contrasts"` attribute for the factor.
D	Computes derivatives of simple expressions, symbolically.
Gamma	Family object for Gamma distributions (used by functions such as `glm`).
HoltWinters	Computes Holt-Winters filtering of a given time series. Unknown parameters are determined by minimizing the squared prediction error.
IQR	Computes the interquartile range of the x values.
KalmanForecast, KalmanLike, KalmanRun, KalmanSmooth	Use Kalman filtering to find the (Gaussian) log-likelihood, or for forecasting or smoothing.
NLSstAsymptotic	Fits the asymptotic regression model, in the form `b0 + b1*(1-exp(-exp(lrc) * x)`, to the xy data. This can be used as a building block in determining starting estimates for more complicated models.
NLSstClosestX	Uses inverse linear interpolation to approximate the x value at which the function represented by xy is equal to yval.
NLSstLfAsymptote	Provides an initial guess at the horizontal asymptote on the left side (i.e., small values of x) of the graph of y versus x from the xy object. Primarily used within `initial` functions for self-starting nonlinear regression models.
NLSstRtAsymptote	Provides an initial guess at the horizontal asymptote on the right side (i.e., large values of x) of the graph of y versus x from the xy object. Primarily used within `initial` functions for self-starting nonlinear regression models.
PP.test	Computes the Phillips-Perron test for the null hypothesis that x has a unit root against a stationary alternative.
SSD	Function to compute the matrix of residual sums of squares and products, or the estimated variance matrix for multivariate linear models.

Function	Description
SSasymp	This selfStart model evaluates the asymptotic regression function and its gradient. It has an initial attribute that will evaluate initial estimates of the parameters Asym, R0, and lrc for a given set of data.
SSasympOff	This selfStart model evaluates an alternative parametrization of the asymptotic regression function and the gradient with respect to those parameters. It has an initial attribute that creates initial estimates of the parameters Asym, lrc, and c0.
SSasympOrig	This selfStart model evaluates the asymptotic regression function through the origin and its gradient. It has an initial attribute that will evaluate initial estimates of the parameters Asym and lrc for a given set of data.
SSbiexp	This selfStart model evaluates the biexponential model function and its gradient. It has an initial attribute that creates initial estimates of the parameters A1, lrc1, A2, and lrc2.
SSfol	This selfStart model evaluates the first-order compartment function and its gradient. It has an initial attribute that creates initial estimates of the parameters lKe, lKa, and lCl.
SSfpl	This selfStart model evaluates the four-parameter logistic function and its gradient. It has an initial attribute that will evaluate initial estimates of the parameters A, B, xmid, and scal for a given set of data.
SSgompertz	This selfStart model evaluates the Gompertz growth model and its gradient. It has an initial attribute that creates initial estimates of the parameters Asym, b2, and b3.
SSlogis	This selfStart model evaluates the logistic function and its gradient. It has an initial attribute that creates initial estimates of the parameters Asym, xmid, and scal.
SSmicmen	This selfStart model evaluates the Michaelis-Menten model and its gradient. It has an initial attribute that will evaluate initial estimates of the parameters Vm and K.
SSweibull	This selfStart model evaluates the Weibull model for growth curve data and its gradient. It has an initial attribute that will evaluate initial estimates of the parameters Asym, Drop, lrc, and pwr for a given set of data.
StructTS	Fits a structural model for a time series by maximum likelihood.
TukeyHSD	Creates a set of confidence intervals on the differences between the means of the levels of a factor with the specified family-wise probability of coverage. The intervals are based on the Studentized range statistic, Tukey's honest significant difference method. There is a plot method.
TukeyHSD.aov	Creates a set of confidence intervals on the differences between the means of the levels of a factor with the specified family-wise probability of coverage. The intervals are based on the Studentized range statistic, Tukey's honest significant difference method. There is a plot method.
acf	The function acf computes (and by default plots) estimates of the autocovariance or autocorrelation function. The function pacf is the function used for partial autocorrelations. The function ccf computes the cross-correlation or cross-covariance of two univariate series.
acf2AR	Computes an AR process exactly fitting an autocorrelation function.

Function	Description
add.scope	add.scope and drop.scope compute those terms that can be individually added to or dropped from a model while respecting the hierarchy of terms.
add1	Computes all the single terms in the scope argument that can be added to or dropped from the model, fits those models, and computes a table of the changes in fit.
addmargins	For a given table, one can specify which of the classifying factors to expand by one or more levels to hold margins to be calculated. One may, for example, form sums and means over the first dimension and medians over the second. The resulting table will then have two extra levels for the first dimension and one extra level for the second. The default is to sum over all margins in the table. Other possibilities may give results that depend on the order in which the margins are computed. This is flagged in the printed output from the function.
aggregate	Splits the data into subsets, computes summary statistics for each, and returns the result in a convenient form.
alias	Finds aliases (linearly dependent terms) in a linear model specified by a formula.
anova	Computes analysis of variance (or deviance) tables for one or more fitted model objects.
anova.lmlist	Computes an analysis of variance table for one or more linear model fits.
ansari.test	Performs the Ansari-Bradley two-sample test for a difference in scale parameters.
aov	Fits an analysis of variance model by a call to lm for each stratum.
approx	Returns a list of points that linearly interpolate given data points, or a function performing the linear (or constant) interpolation.
approxfun	Returns a list of points that linearly interpolate given data points, or a function performing the linear (or constant) interpolation.
ar	Fits an autoregressive time series model to the data, by default selecting the complexity by AIC.
arima	Fits an ARIMA model to a univariate time series.
arima.sim	Simulates from an ARIMA model.
arima0	Fits an ARIMA model to a univariate time series and forecasts from the fitted model.
as.dendrogram	Coerces an object to class "dendrogram" (which provides general functions for handling treelike structures).
as.dist	Coerces to a dist object (a matrix returned by the dist function).
as.formula	The generic function formula and its specific methods provide a way of extracting formulas that have been included in other objects. as.formula is almost identical, additionally preserving attributes when object already inherits from "formula".
as.hclust	Converts objects from other hierarchical clustering functions to class "hclust".
as.stepfun	Given the vectors $(x[1],..., x[n])$ and $(y[0],y[1],..., y[n])$ (one value more!), stepfun(x, y,...) returns an interpolating "step" function, say fn. That is, $fn(t) = c\_i\{[i]\}$ (constant) for t in $(x[i], x[i+1])$ and at the abscissa values, if (by default) right = FALSE, $fn(x\_i) = y\_i\{fn(x[i]) = y[i]\}$ and for right = TRUE, $fn(x[i]) = y[i-1]$, for $i=1,...,n$.
as.ts	Coerces an object to a ts object.

Function	Description
asOneSidedFormula	Names, expressions, numeric values, and character strings are converted to one-sided formulas. If `object` is a formula, it must be one sided, in which case it is returned unaltered.
ave	Subsets of `x[ ]` are averaged, where each subset consists of those observations with the same factor levels.
bandwidth.kernel	Returns the equivalent bandwidth for a `tskernel` object.
bartlett.test	Performs Bartlett's test of the null that the variances in each of the groups (samples) are the same.
binom.test	Performs an exact test of a simple null hypothesis about the probability of success in a Bernoulli experiment.
binomial	Family function for binomial distributions (used by functions such as `glm`).
biplot	Plots a biplot on the current graphics device.
bw.SJ, bw.bcv, bw.nrd, bw.nrd0, bw.ucv	Bandwidth selectors for Gaussian kernels in `density`.
cancor	Computes the canonical correlations between two data matrices.
case.names	Simple utility returning (nonmissing) case names and (noneliminated) variable names.
ccf	The function `acf` computes (and by default plots) estimates of the autocovariance or autocorrelation function. The function `pacf` is the function used for partial autocorrelations. The function `ccf` computes the cross-correlation or cross-covariance of two univariate series.
chisq.test	Performs chi-squared contingency table tests and goodness-of-fit tests.
clearNames	Sets the `names` attribute of `object` to `NULL` and returns the object.
cmdscale	Classical multidimensional scaling of a data matrix. Also known as *principal coordinates analysis*.
coef, coefficients	`coef` is a generic function that extracts model coefficients from objects returned by modeling functions. `coefficients` is an alias for it.
complete.cases	Returns a logical vector indicating which cases are complete, i.e., have no missing values.
confint	Computes confidence intervals for one or more parameters in a fitted model.
constrOptim	Minimizes a function subject to linear inequality constraints using an adaptive barrier algorithm.
contr.SAS, contr.helmert, contr.poly, contr.sum, contr.treatment	Return a matrix of contrasts.
contrasts, contrasts<-	Set and view the contrasts associated with a factor.
convolve	Uses the fast Fourier transform to compute the several kinds of convolutions of two sequences.
cooks.distance	Computes "Cook's distance" on a model object.
cophenetic	Computes the cophenetic distances for a hierarchical clustering.

Function	Description
cor	Computes the correlation of two vectors, or the columns of two matrices.
cor.test	Tests for association between paired samples, using one of Pearson's product-moment correlation coefficient, Kendall's tau, or Spearman's rho.
cov	Computes the covariance of two vectors, or the columns of two matrices.
cov.wt	Returns a list containing estimates of the weighted covariance matrix and the mean of the data and optionally of the (weighted) correlation matrix.
cov2cor	var, cov, and cor compute the variance of x and the covariance or correlation of x and y if these are vectors. If x and y are matrices, then the covariances (or correlations) between the columns of x and the columns of y are computed. cov2cor scales a covariance matrix into the corresponding correlation matrix *efficiently*.
covratio	Returns the covariance ratio (for regression diagnostics) on a model object.
cpgram	Plots a cumulative periodogram.
cutree	Cuts a tree, e.g., resulting from hclust, into several groups by specifying either the desired number(s) of groups or the cut height(s).
cycle	time creates the vector of times at which a time series was sampled. cycle gives the positions in the cycle of each observation. frequency returns the number of samples per unit time, and deltat gives the time interval between observations (see ts).
dbeta	Density function for the Beta distribution.
dbinom	Density function for the binomial distribution.
dcauchy	Density function for the Cauchy distribution.
dchisq	Density function for the chi-squared distribution.
decompose	Decomposes a time series into seasonal, trend, and irregular components using moving averages. Deals with additive or multiplicative seasonal components.
delete.response	Returns a terms object for the same model but with no response variable.
deltat	time creates the vector of times at which a time series was sampled. cycle gives the positions in the cycle of each observation. frequency returns the number of samples per unit time, and deltat gives the time interval between observations (see ts).
dendrapply	Applies function FUN to each node of a dendrogram recursively. When y <- dendrapply(x, fn), then y is a dendrogram of the same graph structure as x and for each node, y.node[j] <- FUN(x.node[j], ...) (where y.node[j] is an (invalid!) notation for the jth node of y).
density	The (S3) generic function density computes kernel density estimates. Its default method does so with the given kernel and bandwidth for univariate observations.
density.default	The (S3) generic function density computes kernel density estimates. Its default method does so with the given kernel and bandwidth for univariate observations.
deriv, deriv3	Compute derivatives of simple expressions, symbolically.
deviance	Returns the deviance of a fitted model object.
dexp	Density function for the exponential distribution.

R Reference

Function	Description
df	Density, distribution function, quantile function, and random generation for the *F*-distribution with df1 and df2 degrees of freedom (and optional noncentrality parameter ncp).
df.kernel	The "tskernel" class is designed to represent discrete symmetric normalized smoothing kernels. These kernels can be used to smooth vectors, matrices, or time series objects. There are print, plot, and [methods for these kernel objects.
df.residual	Returns the residual degrees of freedom extracted from a fitted model object.
dfbeta	Returns dfbeta for a model object (for regression diagnostics).
dfbetas	Returns dfbetas for a model object (for regression diagnostics).
dffits	Returns dffits for a model object (for regression diagnostics).
dgamma	Density function for the Gamma distribution.
dgeom	Density, distribution function, quantile function, and random generation for the geometric distribution with parameter prob.
dhyper	Density function for the hypergeometric distribution.
diff.ts	Methods for objects of class "ts", typically the result of ts.
diffinv	Computes the inverse function of the lagged differences function diff.
dist	Computes and returns the distance matrix computed by using the specified distance measure to compute the distances between the rows of a data matrix.
dlnorm	Density function for the log-normal distribution.
dlogis	Density function for the logistic distribution.
dmultinom	Generates multinomially distributed random number vectors and computes multinomial probabilities.
dnbinom	Density function for the negative binomial distribution.
dnorm	Density function for the normal distribution.
dpois	Density function for the Poisson distribution.
drop.scope	add.scope and drop.scope compute those terms that can be individually added to or dropped from a model while respecting the hierarchy of terms.
drop.terms	delete.response returns a terms object for the same model, but with no response variable. drop.terms removes variables from the righthand side of the model. There is also a "[.terms" method to perform the same function (with keep.response=TRUE). reformulate creates a formula from a character vector.
drop1	Computes all the single terms in the scope argument that can be added to or dropped from the model, fits those models, and computes a table of the changes in fit.
dsignrank	Density, distribution function, quantile function, and random generation for the distribution of the Wilcoxon signed rank statistic obtained from a sample with size n.
dt	Density, distribution function, quantile function, and random generation for the *t*-distribution with df degrees of freedom (and optional noncentrality parameter ncp).

Function	Description
dummy.coef	Extracts coefficients in terms of the original levels of the coefficients rather than the coded variables.
dunif	Density function for the uniform distribution.
dweibull	Density function for the Weibull distribution.
dwilcox	Density function for the distribution of the Wilcoxon rank sum statistic.
ecdf	Computes or plots an empirical cumulative distribution function.
eff.aovlist	Computes the efficiencies of fixed-effects terms in an analysis of variance model with multiple strata.
effects	Returns (orthogonal) effects from a fitted model, usually a linear model. This is a generic function, but currently only has a method for objects inheriting from classes "lm" and "glm".
embed	Embeds the time series x into a low-dimensional Euclidean space.
end	Extracts and encodes the times the first and last observations were taken. Provided only for compatibility with S version 2.
estVar	Function to compute matrix of residual sums of squares and products, or the estimated variance matrix for multivariate linear models.
expand.model.frame	Evaluates new variables as if they had been part of the formula of the specified model. This ensures that the same na.action and subset arguments are applied and allows, for example, x to be recovered for a model using sin(x) as a predictor.
extractAIC	Computes the (generalized) Akaike information criterion for a fitted parametric model.
factanal	Performs maximum likelihood factor analysis on a covariance matrix or data matrix.
family	Family objects provide a convenient way to specify the details of the models used by functions such as glm. See the documentation for glm for the details on how such model fitting takes place.
fft	Performs the fast Fourier transform of an array.
filter	Applies linear filtering to a univariate time series or to each series separately of a multivariate time series.
fisher.test	Performs Fisher's exact test for testing the null of independence of rows and columns in a contingency table with fixed marginals.
fitted, fitted.values	fitted is a generic function that extracts fitted values from objects returned by modeling functions. fitted.values is an alias for it.
fivenum	Returns Tukey's five-number summary (minimum, lower-hinge, median, upper-hinge, maximum) for the input data.
fligner.test	Performs a Fligner-Killeen (median) test of the null that the variances in each of the groups (samples) are the same.
formula	The generic function formula and its specific methods provide a way of extracting formulas that have been included in other objects.
frequency	Returns the number of samples per unit time from a ts object.
friedman.test	Performs a Friedman rank sum test with unreplicated blocked data.
ftable	Creates "flat" contingency tables.

Function	Description
gaussian	Family object for Gaussian functions (used by functions such as glm).
getInitial	Evaluates initial parameter estimates for a nonlinear regression model.
get_all_vars	Returns a data.frame containing the variables used in formula plus those specified. Unlike model.frame.default, it returns the input variables and not those resulting from function calls in formula.
glm	Used to fit generalized linear models, specified by giving a symbolic description of the linear predictor and a description of the error distribution.
glm.control	Auxiliary function as user interface for glm fitting. Typically only used when calling glm or glm.fit.
glm.fit	glm is used to fit generalized linear models, specified by giving a symbolic description of the linear predictor and a description of the error distribution.
hasTsp	tsp returns the tsp attribute (or NULL). It is included for compatibility with S version 2. tsp<- sets the tsp attribute. hasTsp ensures x has a tsp attribute, by adding one if needed.
hat, hatvalues, hatvalues.lm	Return the hat matrix for a model object (for regression diagnostics).
hclust	Hierarchical cluster analysis on a set of dissimilarities and methods for analyzing it.
heatmap	Plots a heat map object (an image with an accompanying dendrogram).
influence	Provides the basic quantities that are used in forming a wide variety of diagnostics for checking the quality of regression fits.
influence.measures	Produces a class "infl" object tabular display showing the DFBETAS for each model variable, DFFITS, covariance ratios, Cook's distances, and the diagonal elements of the hat matrix.
integrate	Adaptive quadrature of functions of one variable over a finite or infinite interval.
interaction.plot	Plots the mean (or other summary) of the response for two-way combinations of factors, thereby illustrating possible interactions.
inverse.gaussian	Family object for inverse Gaussian distributions (used by functions such as glm).
is.empty.model	R model notation allows models with no intercept and no predictors. These require special handling internally. is.empty.model() checks whether an object describes an empty model.
is.leaf	Class "dendrogram" provides general functions for handling treelike structures. It is intended as a replacement for similar functions in hierarchical clustering and classification/regression trees, such that all of these can use the same engine for plotting or cutting trees. The code is still in the testing stage, and the API may change in the future.
is.mts	Tells whether an object is of class mts.
is.stepfun	Tells whether an object is a function of class stepfun.
is.ts	Tells whether an object is of class ts.
is.tskernel	Tells whether an object is of class tskernel.
isoreg	Computes the isotonic (monotonically increasing nonparametric) least squares regression that is piecewise constant.
kernapply	Computes the convolution between an input sequence and a specific kernel.
kernel	Constructs a general kernel or named specific kernels (returns a "tskernel" object).

Function	Description
kmeans	Performs *k*-means clustering on a data matrix.
knots	Extracts the knots from a step function (returned by `stepfun`).
kruskal.test	Performs a Kruskal-Wallis rank sum test.
ks.test	Performs one- or two-sample Kolmogorov-Smirnov tests.
ksmooth	The Nadaraya-Watson kernel regression estimate.
lag	Computes a lagged version of a time series, shifting the time base back by a given number of observations.
lag.plot	Plots time series against lagged versions of themselves. Helps visualizing "autodependence" even when autocorrelations vanish.
line	Fits a line robustly.
lines.ts	Plotting method for objects inheriting from class `"ts"`.
lm	`lm` is used to fit linear models. It can be used to carry out regression, single-stratum analysis of variance, and analysis of covariance (although `aov` may provide a more convenient interface for these).
lm.fit	Basic computing engines called by `lm` and used to fit linear models. These should usually *not* be used directly unless by experienced users.
lm.influence	Provides the basic quantities that are used in forming a wide variety of diagnostics for checking the quality of regression fits.
lm.wfit	Basic computing engines called by `lm` and used to fit linear models. These should usually *not* be used directly unless by experienced users.
loadings	Extracts or prints loadings in factor analysis (or principal components analysis).
loess	Fits a polynomial surface determined by one or more numerical predictors, using local fitting.
loess.control	Sets control parameters for `loess` fits.
loess.smooth	Plots and adds a smooth curve computed by `loess` to a scatter plot.
logLik	Extracts the log-likelihood value from an object (usually a model).
loglin	Used to fit log-linear models to multidimensional contingency tables by iterative proportional fitting.
lowess	Performs the computations for locally weighted scatter plot smoothing (LOWESS), smoother which uses locally weighted polynomial regression.
ls.diag	Computes basic statistics, including standard errors, t-, and p-values, for the regression coefficients.
ls.print	Computes basic statistics, including standard errors, t-, and p-values, for the regression coefficients and prints them if `print.it` is TRUE.
lsfit	Finds the least squares estimate of β in the model $Y = X\beta + \varepsilon$.
mad	Computes the median absolute deviation, i.e., the (lo-/hi-) median of the absolute deviations from the median and (by default) adjusts by a factor for asymptotically normal consistency.
mahalanobis	Returns the squared Mahalanobis distance of all rows in x and the vector $\mu =$ center with respect to $\Sigma = $ cov. This is (for vector x) defined as $D^2 = (x - \mu)' \Sigma^{-1} (x - \mu)$.

Function	Description
make.link	This function is used with the `family` functions in `glm()`. Given the name of a link, it returns a link function, an inverse link function, the derivative $d\mu / d\eta$ and a function for domain checking.
makeARIMA	Uses Kalman filtering to find the (Gaussian) log-likelihood, or for forecasting or smoothing.
makepredictcall	Utility to help `model.frame.default` create the right matrices when predicting from models with terms like `poly` or `ns`.
manova	A class for the multivariate analysis of variance.
mantelhaen.test	Performs a Cochran-Mantel-Haenszel chi-squared test of the null that two nominal variables are conditionally independent in each stratum, assuming that there is no three-way interaction.
mauchly.test	Tests whether a Wishart-distributed covariance matrix (or transformation thereof) is proportional to a given matrix.
mcnemar.test	Performs McNemar's chi-squared test for symmetry of rows and columns in a two-dimensional contingency table.
median	Computes the sample median.
median.default	Computes the sample median.
medpolish	Fits an additive model using Tukey's median polish procedure.
model.extract	Returns the response, offset, subset, weights, or other special components of a model frame passed as optional arguments to `model.frame`.
model.frame	`model.frame` (a generic function) and its methods return a `data.frame` with the variables needed to use `formula` and any `...` arguments.
model.matrix	Creates a design matrix.
model.offset	Returns the offset of a model frame.
model.response	Returns the response of a model frame.
model.tables	Computes summary tables for model fits, especially complex aov fits.
model.weights	Returns the weights of a model frame.
monthplot	Plots seasonal (or other) subseries from a time series.
mood.test	Performs Mood's two-sample test for a difference in scale parameters.
mvfft	Performs the fast Fourier transform of an array.
na.action	Extracts information on the NA action used to create an object.
na.contiguous	Finds the longest consecutive stretch of nonmissing values in a time series object. (In the event of a tie, the first such stretch.)
na.exclude	`na.exclude` returns the object with incomplete cases removed and with the `na.action` attribute set to "exclude". (Usually used as an `na.action` argument for a modeling function.)
na.fail	Returns the object if it does not contain any missing values and signals an error otherwise. (Usually used as an `na.action` argument for a modeling function.)
na.omit	Returns the object with incomplete cases removed. (Usually used as an `na.action` argument for a modeling function.)

Function	Description
na.pass	Returns an object unchanged. (Usually used as an na.action argument for a modeling function.)
napredict	Uses missing value information to adjust residuals and predictions.
naprint	Uses missing value information to report the effects of an na.action.
naresid	Uses missing value information to adjust residuals and predictions.
nextn	Returns the smallest integer, greater than or equal to n, that can be obtained as a product of powers of the values contained in factors.
nlm	Carries out a minimization of the function f using a Newton-type algorithm.
nlminb	Unconstrained and constrained optimization using PORT routines.
nls	Determines the nonlinear (weighted) least squares estimates of the parameters of a nonlinear model.
nls.control	Allows the user to set some characteristics of the nonlinear least squares algorithm.
numericDeriv	Numerically evaluates the gradient of an expression.
offset	An offset is a term to be added to a linear predictor, such as in a generalized linear model, with known coefficient 1 rather than an estimated coefficient.
oneway.test	Tests whether two or more samples from normal distributions have the same means. The variances are not necessarily assumed to be equal.
optim	General-purpose optimization based on Nelder-Mead, quasi-Newton, and conjugate-gradient algorithms. It includes an option for box-constrained optimization and simulated annealing.
optimise, optimize	The function optimize searches the interval from lower to upper for a minimum or maximum of the function f with respect to its first argument. optimise is an alias for optimize.
order.dendrogram	Returns the order (index) or the "label" attribute for the leaves in a dendrogram. These indices can then be used to access the appropriate components of any additional data.
p.adjust	Given a set of p-values, returns p-values adjusted using one of several methods.
pacf	Computes partial autocorrelations.
pairwise.prop.test	Calculates pairwise comparisons between pairs of proportions with correction for multiple testing.
pairwise.t.test	Calculates pairwise comparisons between group levels with corrections for multiple testing.
pairwise.table	Creates a table of p-values for pairwise comparisons with corrections for multiple testing.
pairwise.wilcox.test	Calculates pairwise comparisons between group levels with corrections for multiple testing.
pbeta	Distribution function for the Beta distribution.
pbinom	Distribution function for the binomial distribution.
pbirthday	Computes the probability of a coincidence for a generalized birthday paradox problem.
pcauchy	Distribution function for the Cauchy distribution.

Function	Description
pchisq	Distribution function for the chi-squared distribution.
pexp	Distribution function for the exponential distribution.
pf	Distribution function for the F-distribution.
pgamma	Distribution function for the Gamma distribution.
pgeom	Distribution function for the geometric distribution.
phyper	Distribution function for the hypergeometric distribution.
plclust	Hierarchical cluster analysis on a set of dissimilarities and methods for analyzing it.
plnorm	Distribution function for the log-normal distribution.
plogis	Distribution function for the logistic distribution.
plot.TukeyHSD	Creates a set of confidence intervals on the differences between the means of the levels of a factor with the specified family-wise probability of coverage. The intervals are based on the Studentized range statistic, Tukey's honest significant difference method. There is a `plot` method.
plot.density	The `plot` method for density objects.
plot.ecdf	Computes or plots an empirical cumulative distribution function.
plot.lm	Plots diagnostics for an lm object.
plot.mlm	Plots diagnostics for an mlm object.
plot.spec, plot.spec.coherency, plot.spec.phase	Plotting methods for objects of class `"spec"`. For multivariate time series, they plot the marginal spectra of the series or pairs plots of the coherency and phase of the cross-spectra.
plot.stepfun	Method of the generic `plot` for `stepfun` objects and utility for plotting piecewise-constant functions.
plot.ts	Plotting method for objects inheriting from class `"ts"`.
pnbinom	Distribution function for the negative binomial distribution.
pnorm	Distribution function for the normal distribution.
poisson	Family objects for Poisson distributions (used by functions such as `glm`).
poisson.test	Performs an exact test of a simple null hypothesis about the rate parameter in a Poisson distribution or for the ratio between two rate parameters.
poly, polym	Return or evaluate orthogonal polynomials of degree 1 to `degree` over the specified set of points x. These are all orthogonal to the constant polynomial of degree 0. Alternatively, evaluate raw polynomials.
power	Creates a link object based on the link function $\eta = \mu^\lambda$.
power.anova.test	Computes power of test or determines parameters to obtain target power.
power.prop.test	Computes power of test or determines parameters to obtain target power.
power.t.test	Computes power of test or determines parameters to obtain target power.
ppoints	Generates the sequence of probability points `(1:m - a)/(m + (1-a)-a)`, where m is either n, if `length(n)==1`, or `length(n)`.
ppois	Distribution function for the Poisson distribution.
ppr	Fits a projection pursuit regression model.

Function	Description
prcomp	Performs a principal components analysis on the given data matrix and returns the results as an object of class `prcomp`.
predict	Generic function for predictions from the results of various model-fitting functions.
preplot	Computes an object to be used for plots relating to the given model object.
princomp	Performs a principal components analysis on the given numeric data matrix and returns the results as an object of class `princomp`.
printCoefmat	Utility function to be used in higher-level `print` methods, such as `print.summary.lm`, `print.summary.glm`, and `print.anova`. The goal is to provide a flexible interface with smart defaults such that often only x needs to be specified.
profile	Investigates the behavior of an objective function near the solution.
proj	Returns a matrix or list of matrices giving the projections of the data onto the terms of a linear model. It is most frequently used for `aov` models.
promax	These functions "rotate" loading matrices in factor analysis.
prop.test	Used for testing the null that the proportions (probabilities of success) in several groups are the same or that they equal certain given values.
prop.trend.test	Performs a chi-squared test for trend in proportions, i.e., a test asymptotically optimal for local alternatives where the log odds vary in proportion with `score`. By default, `score` is chosen as the group numbers.
psignrank	Distribution function for the distribution of the Wilcoxon signed rank statistic.
pt	Distribution function for the t-distribution.
ptukey	Distribution function for the Studentized range.
punif	These functions provide information about the uniform distribution on the interval from `min` to `max`. `dunif` gives the density, `punif` gives the distribution function, `qunif` gives the quantile function, and `runif` generates random deviates.
pweibull	Distribution function for the Weibull distribution.
pwilcox	Distribution function for the distribution of the Wilcoxon rank sum statistic.
qbeta	Quantile function for the Beta distribution.
qbinom	Quantile function for the binomial distribution.
qbirthday	Computes the number of observations needed to have a specified probability of coincidence for a generalized birthday paradox problem.
qcauchy	Quantile function for the Cauchy distribution.
qchisq	Quantile function for the chi-squared distribution.
qexp	Quantile function for the exponential distribution.
qf	Quantile function for the F-distribution.
qgamma	Quantile function for the Gamma distribution.
qgeom	Quantile function for the geometric distribution.
qhyper	Quantile function for the hypergeometric distribution.
qlnorm	Quantile function for the log-normal distribution.
qlogis	Quantile function for the logistic distribution.
qnbinom	Quantile function for the negative binomial distribution.

Function	Description
qnorm	Quantile function for the normal distribution.
qpois	Quantile function for the Poisson distribution.
qqline	Adds a line to a normal Q-Q plot (usually generated by qqnorm or qqplot) that passes through the first and third quartiles.
qqnorm	Generic function the default method of which produces a normal Q-Q plot of the values in y.
qqplot	Produces a Q-Q plot of two data sets.
qsignrank	Density, distribution function, quantile function, and random generation for the distribution of the Wilcoxon signed rank statistic obtained from a sample with size n.
qt	Quantile function for the t-distribution.
qtukey	Function of the distribution of the Studentized range, R/s, where R is the range of a standard normal sample and $df*s^2$ is independently distributed as chi-squared with df degrees of freedom; see pchisq.
quade.test	Performs a Quade test with unreplicated blocked data.
quantile	The generic function quantile produces sample quantiles corresponding to the given probabilities. The smallest observation corresponds to a probability of 0 and the largest to a probability of 1.
quantile.default	The generic function quantile produces sample quantiles corresponding to the given probabilities. The smallest observation corresponds to a probability of 0 and the largest to a probability of 1.
quasi	Family object for the quasi distribution (used by functions such as glm).
quasibinomial	Family object for the quasibinomial distribution (used by functions such as glm).
quasipoisson	Family object for the quasi-Poisson distribution (used by functions such as glm).
qunif	Quantile function for the uniform distribution.
qweibull	Quantile function for the Weibull distribution.
qwilcox	Quantile function for the Wilcoxon rank sum statistic.
r2dtable	Generates random two-way tables with given marginals using Patefield's algorithm.
rbeta	Random number generation for the Beta distribution.
rbinom	Random number generation for the binomial distribution.
rcauchy	Random number generation for the Cauchy distribution.
rchisq	Random number generation for the chi-squared distribution.
read.ftable	Reads, writes, and coerces "flat" contingency tables.
rect.hclust	Draws rectangles around the branches of a dendrogram, highlighting the corresponding clusters. First, the dendrogram is cut at a certain level, and then a rectangle is drawn around selected branches.
reformulate	Creates a formula from a character vector.
relevel	The levels of a factor are reordered so that the level specified by ref is first, and the others are moved down. This is useful for contr.treatment contrasts, which take the first level as the reference.

Function	Description
reorder	`reorder` is a generic function. Its `"factor"` method reorders the levels of a factor depending on values of a second variable, usually numeric. The `"character"` method is a convenient alias.
replications	Returns a vector or a list of the number of replicates for each term in the formula.
reshape	Reshapes a data frame between "wide" format with repeated measurements in separate columns of the same record and "long" format with the repeated measurements in separate records.
resid	Generic function that extracts model residuals from objects returned by modeling functions. The abbreviated form `resid` is an alias for `residuals`.
residuals	Generic function that extracts model residuals from objects returned by modeling functions.
rexp	Random generation for the exponential distribution.
rf	Random generation for the F-distribution.
rgamma	Random generation for the Gamma distribution.
rgeom	Random generation for the geometric distribution.
rhyper	Random generation for the hypergeometric distribution.
rlnorm	Random generation for the log-normal distribution.
rlogis	Random generation for the logistic distribution.
rmultinom	Generates multinomially distributed random number vectors and computes multinomial probabilities.
rnbinom	Random generation for the negative binomial distribution.
rnorm	Random generation for the normal distribution.
rpois	Random generation for the Poisson distribution.
rsignrank	Random generation for the distribution of the Wilcoxon signed rank statistic.
rstandard	Returns the standardized residuals from a model object.
rstudent	Returns the Studentized residuals from a model object.
rt	Random generation for the t-distribution.
runif	Generates random numbers from the uniform distribution.
runmed	Computes running medians of odd span. This is the "most robust" scatter plot smoothing possible. For efficiency (and historical reasons), you can use one of two different algorithms giving identical results.
rweibull	Random generation for the Weibull distribution.
rwilcox	Random generation for the distribution of the Wilcoxon rank sum statistic.
scatter.smooth	Plots and adds a smooth curve computed by `loess` to a scatter plot.
screeplot	Plots the variances against the number of the principal component. This is also the `plot` method for classes `"princomp"` and `"prcomp"`.
sd	Computes the standard deviation of the values in `x`.
se.contrast	Returns the standard errors for one or more contrasts in an `aov` object.
selfStart	Constructs self-starting nonlinear models.

Function	Description
setNames	This is a convenience function that sets the names on an object and returns the object. It is most useful at the end of a function definition where one is creating the object to be returned and would prefer not to store it under a name just so the names can be assigned.
shapiro.test	Performs the Shapiro-Wilk test of normality.
simulate	Simulates one or more responses from the distribution corresponding to a fitted model object.
smooth	Tukey's smoothers, *3RS3R*, *3RSS*, *3R*, etc.
smooth.spline	Fits a cubic smoothing spline to the supplied data.
smoothEnds	Smooths end points of a vector y using subsequently smaller medians and Tukey's end point rule at the very end.
sortedXyData	This is a constructor function for the class of `sortedXyData` objects. These objects are mostly used in the `initial` function for a self-starting nonlinear regression model, which will be of the `selfStart` class.
spec.ar	Fits an AR model to x (or uses the existing fit) and computes (and by default plots) the spectral density of the fitted model.
spec.pgram	Calculates the periodogram using a fast Fourier transform and optionally smooths the result with a series of modified Daniell smoothers (moving averages giving half weight to the end values).
spec.taper	Applies a cosine-bell taper to a time series.
spectrum	Estimates the spectral density of a time series.
spline	Performs cubic spline interpolation of given data points, returning either a list of points obtained by the interpolation or a *function* performing the interpolation. Returns a list containing components x and y, which give the ordinates where interpolation took place and the interpolated values.
splinefun	Performs cubic spline interpolation of given data points, returning either a list of points obtained by the interpolation or a *function* performing the interpolation. Returns a function with formal arguments x and `deriv`, the latter defaulting to 0.
splinefunH	Performs Hermite spline interpolation of given data points, returning either a list of points obtained by the interpolation or a *function* performing the interpolation.
start	Extracts and encodes the times the first and last observations were taken. Provided only for compatibility with S version 2.
stat.anova	Utility function, used in `lm` and `glm` methods for `anova(..., test != NULL)` and should not be used by the average user.
step	Selects a formula-based model by AIC.
stepfun	Returns an interpolating step function from two sets of vectors.
stl	Decomposes a time series into seasonal, trend, and irregular components using `loess`.
supsmu	Smooths the (x, y) values by Friedman's supersmoother.
symnum	Symbolically encodes a given numeric or logical vector or array. Particularly useful for visualization of structured matrices, e.g., correlation, sparse, or logical ones.
t.test	Performs one- and two-sample *t*-tests on vectors of data.

Function	Description
termplot	Plots regression terms against their predictors, optionally with standard errors and partial residuals added.
terms	Generic function that can be used to extract *terms* objects from various kinds of R data objects.
time	Creates the vector of times at which a time series was sampled.
toeplitz	Forms a symmetric Toeplitz matrix given its first row.
ts	Used to create time series objects.
ts.intersect	Binds time series that have a common frequency. `ts.intersect` is restricted to the time covered by all the series.
ts.plot	Plots several time series on a common plot. Unlike `plot.ts`, the series can have different time bases, but they should have the same frequency.
ts.union	Binds time series that have a common frequency. `ts.union` pads with NAs to the total time coverage.
tsSmooth	Performs fixed-interval smoothing on a univariate time series via a state-space model.
tsdiag	Generic function to plot time series diagnostics.
tsp, tsp<-	`tsp` returns the `tsp` attribute (or NULL). It is included for compatibility with S version 2. `tsp<-` sets the `tsp` attribute.
uniroot	Searches the interval from `lower` to `upper` for a root (i.e., 0) of the function `f` with respect to its first argument.
update	Updates and (by default) refits a model. It does this by extracting the call stored in the object, updating the call and (by default) evaluating that call.
var	Computes the variance of a vector.
var.test	Performs an *F*-test to compare the variances of two samples from normal populations.
variable.names	Simple utility returning (nonmissing) case names and (noneliminated) variable names.
varimax	These functions "rotate" loading matrices in factor analysis.
vcov	Returns the variance-covariance matrix of the main parameters of a fitted model object.
weighted.mean	Computes a weighted mean of a numeric vector.
weighted.residuals	Computes weighted residuals from a linear model fit.
weights	All these functions are `methods` for class `"lm"` objects.
wilcox.test	Performs one- and two-sample Wilcoxon tests on vectors of data; the latter is also known as the Mann-Whitney test.
window, window<-	`window` is a generic function that extracts the subset of the object x observed between the times `start` and end. If a frequency is specified, the series is then resampled at the new frequency.
write.ftable	Reads, writes, and coerces "flat" contingency tables.
xtabs	Creates a contingency table from cross-classifying factors, usually contained in a data frame, using a formula interface.

Data Set

Data Set	Class	Description
p.adjust.methods	character	Allowed methods for `p.adjust`.

stats4

This package contains statistical functions using S4 methods and classes.

Functions

Function	Description
AIC	Calculates the Akaike information criterion for one or several fitted model objects for which a log-likelihood value can be obtained.
BIC	Calculates the Bayesian information criterion (BIC), also known as Schwarz's Bayesian criterion (SBC), for one or several fitted model objects for which a log-likelihood value can be obtained, according to the formula $-2 * \text{log-likelihood} + n_{par} * \log(n_{obs})$, where n_{par} represents the number of parameters and n_{obs} the number of observations in the fitted model.
coef	Extracts model coefficients from objects returned by modeling functions.
confint	Computes confidence intervals for one or more parameters in a fitted model.
logLik	Extracts the log-likelihood from a model object.
mle	Estimates parameters by the method of maximum likelihood.
plot	Generic function for plotting an R object.
profile	Investigates behavior of objective function near the solution represented by fitted.
summary	Generic function used to produce result summaries of the results of various model-fitting functions.
update	Updates and (by default) refits a model.
vcov	Returns the variance-covariance matrix of the main parameters of a fitted model object.

survival

This package contains functions for survival analysis.

Functions

Function	Description
Surv	Creates a survival object, usually used as a response variable in a model formula.
aareg	Returns an object of class `"aareg"` that represents an Aalen model.
attrassign	The `"assign"` attribute on model matrices describes which columns come from which terms in the model formula.

Function	Description
basehaz	Computes the baseline survival curve for a Cox model.
cch	Returns estimates and standard errors from relative risk regression fit to data from case-cohort studies, cohort data, and Borgan II, a generalization of the Lin-Ying estimator.
clogit	Estimates a logistic regression model by maximizing the conditional likelihood.
cluster	This is a special function used in the context of survival models. It identifies correlated groups of observations and is used on the righthand side of a formula.
cox.zph	Tests the proportional hazards assumption for a Cox regression model fit (`coxph`).
coxph	Fits a Cox proportional hazards regression model.
coxph.control	Used to set various numeric parameters controlling a Cox model fit. Typically, it would only be used in a call to `coxph`.
coxph.detail	Returns the individual contributions to the first and second derivative matrix, at each unique event time.
coxph.fit	Internal survival function.
dsurvreg	Density, cumulative probability, and quantiles for the set of distributions supported by the `survreg` function.
format.Surv	Creates a survival object, usually used as a response variable in a model formula.
frailty	Adds a simple random-effects term to a Cox or survreg model.
is.Surv	Tests for a survival object.
is.na.Surv	Tests for NA values in a survival object.
is.na.ratetable	Matches variable names in data to those in a rate table for `survexp`.
is.ratetable	Verifies not only the `class` attribute but also the structure of the object.
labels.survreg	Finds a suitable set of labels from a survival object for use in printing or plotting, for example.
pspline	Specifies a penalized spline basis for the predictor.
psurvreg	Density, cumulative probability, and quantiles for the set of distributions supported by the `survreg` function.
pyears	Computes the person-years of follow-up time contributed by a cohort of subjects, stratified into subgroups.
qsurvreg	Density, cumulative probability, and quantiles for the set of distributions supported by the `survreg` function.
ratetable	Matches variable names in data to those in a rate table for `survexp`.
ridge	Specifies a ridge regression term when used in a coxph or survreg model formula.
strata	This is a special function used in the context of the Cox survival model. It identifies stratification variables when they appear on the righthand side of a formula.
survConcordance	Computes the concordance between a right-censored survival time and a single continuous covariate.
survSplit	Given a survival data set and a set of specified cut times, splits each record into multiple subrecords at each cut time. The new data set will be in "counting process" format, with a start time, stop time, and event status for each record.

Function	Description
survdiff	Tests if there is a difference between two or more survival curves using the G^ρ family of tests, or for a single curve against a known alternative.
survexp	Returns either the expected survival of a cohort of subjects or the individual expected survival for each subject.
survfit	Computes an estimate of a survival curve for censored data using either the Kaplan-Meier or the Fleming-Harrington method or computes the predicted survivor function.
survobrien	O'Brien's test for association of a single variable with survival.
survreg	Fits a parametric survival regression model. These are location-scale models for an arbitrary transform of the time variable; the most common cases use a log transformation, leading to accelerated failure time models.
survreg.control	Checks and packages the fitting options for `survreg`.
survreg.fit	Internal survival function.
survregDtest	This routine is called by `survreg` to verify that a distribution object is valid.
tcut	Attaches categories for person-year calculations to a variable without losing the underlying continuous representation.
untangle.specials	Given a `terms` structure and a desired special name, this returns an index appropriate for subscripting the `terms` structure and another appropriate for the data frame.

Data Sets

Data Set	Class	Description
aml	data.frame	Survival in patients with acute myelogenous leukemia. The question at the time was whether the standard course of chemotherapy should be extended ("maintenance") for additional cycles.
bladder	data.frame	Data on recurrences of bladder cancer, used by many people to demonstrate methodology for recurrent event modeling. Bladder1 is the full data set from the study. This data set contains only the 85 subjects with nonzero follow-up who were assigned to either thiotepa or placebo.
bladder1	data.frame	Data on recurrences of bladder cancer, used by many people to demonstrate methodology for recurrent event modeling. Bladder1 is the full data set from the study. It contains all three treatment arms and all recurrences for 118 subjects; the maximum observed number of recurrences is 9.
bladder2	data.frame	Data on recurrences of bladder cancer, used by many people to demonstrate methodology for recurrent event modeling. Bladder2 uses the same subset of subjects as bladder, but formatted in the (start, stop] or Anderson-Gill style.
cancer	data.frame	Survival in patients with advanced lung cancer from the North Central Cancer Treatment Group. Performance scores rate how well the patient can perform normal daily activities.
cgd	data.frame	Data is from a placebo controlled trial of gamma interferon in chronic granulotomous disease (CGD).
colon	data.frame	Data from one of the first successful trials of adjuvant chemotherapy for colon cancer.

Data Set	Class	Description
heart, jasa, jasa1	data.frame	Survival of patients on the waiting list for the Stanford heart transplant program.
kidney	data.frame	Data on the recurrence times to infection, at the point of insertion of the catheter, for kidney patients using portable dialysis equipment.
leukemia	data.frame	Survival in patients with acute myelogenous leukemia. The question at the time was whether the standard course of chemotherapy should be extended ("maintenance") for additional cycles.
lung	data.frame	Survival in patients with advanced lung cancer from the North Central Cancer Treatment Group. Performance scores rate how well the patient can perform normal daily activities.
mgus, mgus1, mgus2	data.frame	Natural history of 241 subjects with monoclonal gammapathy of undetermined significance (MGUS).
nwtco	data.frame	Missing data/measurement error example. Tumor histology predicts survival, but prediction is stronger with central lab histology than with the local institution determination.
ovarian	data.frame	Survival in a randomized trial comparing two treatments for ovarian cancer.
pbc	data.frame	This data is from the Mayo Clinic trial in primary biliary cirrhosis (PBC) of the liver conducted between 1974 and 1984.
pbcseq	data.frame	This data is a continuation of the PBC data set and contains the follow-up laboratory data for each study patient.
rats	data.frame	Forty-eight rats were injected with a carcinogen and then randomized to either drug or placebo. The number of tumors ranged from 0 to 13; all rats were censored at 6 months after randomization.
stanford2	data.frame	This contains the Stanford heart transplant data in a different format. The main data set is in heart.
survexp.mn	ratetable	Census data sets for the expected-survival and person-year functions.
survexp.mnwhite	ratetable	Census data sets for the expected-survival and person-year functions.
survexp.us	ratetable	Census data sets for the expected-survival and person-year functions.
survexp.usr	ratetable	Census data sets for the expected-survival and person-year functions.
survreg.distributions	list	List of distributions for accelerated failure models. These are location-scale families for some transformation of time.
tobin	data.frame	Economists fit a parametric censored data model called the *tobit*. The data come from Tobin's original paper.
veteran	data.frame	Randomized trial of two treatment regimens for lung cancer. This is a standard survival analysis data set.

tcltk

The package contains interface and language bindings to Tcl/Tk GUI elements. Please see the online help for more details.

tools

This package provides tools for developing packages.

Functions

Function	Description
Rd2HTML	This (experimental) function converts from an R help page to an HTML document.
Rd2ex	This (experimental) function converts from an R help page to the format used by example.
Rd2latex	This (experimental) function converts from an R help page to a LaTeX document.
Rd2txt	This (experimental) function converts from an R help page to a text document.
Rd_db	Builds a simple database of all R documentation (Rd) sources in a package, as a list of character vectors with the lines of the Rd files in the package.
Rdiff	Given two R output files, computes differences, ignoring headers, footers, and some encoding differences.
Rdindex	Prints a two-column index table with names and titles from given R documentation files to a given output file or connection. The titles are nicely formatted between two column positions (typically 25 and 72, respectively).
buildVignettes	Runs Sweave and texi2dvi on all vignettes of a package.
checkDocFiles	Checks, for all Rd files in a package, whether all arguments shown in the usage sections of the Rd file are documented in its arguments section.
checkDocStyle	Investigates how (S3) methods are shown in the usages of the Rd files in a package.
checkFF	Performs checks on calls to compiled code from R code.
checkMD5sums	Checks the files against a file "MD5".
checkNEWS	Reads R's NEWS file or a similarly formatted one. This is an experimental feature, new in R 2.4.0, and may change in several ways.
checkRd	These experimental functions take the output of the parse_Rd function and check it or produce a help page from it. Their interfaces (and existence!) are subject to change.
checkReplaceFuns	Checks whether replacement functions or S3/S4 replacement methods in the package R code have their final argument named value.
checkS3methods	Checks whether all S3 methods defined in the package R code have all arguments of the corresponding generic, with positional arguments of the generics in the same positions for the method.
checkTnF	Checks the specified R package or code file for occurrences of T or F and gathers the expressions containing these.
checkVignettes	Checks all Sweave files of a package by running Sweave and/or Stangle on them.
codoc	Compares names and optionally also corresponding positions and default values of the arguments of functions.

Function	Description
codocClasses	Finds inconsistencies between actual and documented "structure" of R objects in a package. codoc compares names and optionally also corresponding positions and default values of the arguments of functions. codocClasses and codocData compare slot names of S4 classes and variable names of data sets, respectively.
codocData	Compares slot names of S4 classes.
delimMatch	Matches delimited substrings in a character vector, with proper nesting.
dependsOnPkgs	Finds "reverse" dependencies of packages, i.e., those packages that depend on this one and (optionally) so on recursively.
encoded_text_to_latex	Translates non-ASCII characters in text to LaTeX escape sequences.
file_path_as_absolute	Turns a possibly relative file path absolute, performing tilde expansion, if necessary.
file_path_sans_ext	Returns the file paths without extension.
findHTMLlinks	Finds HTML links in an R help file.
getDepList	Given a dependency matrix, creates a DependsList object for that package, which will include the dependencies for that matrix, which ones are installed, which unresolved dependencies were found online, which unresolved dependencies were not found online, and any R dependencies.
installFoundDepends	Takes the Found element of a pkgDependsList object and attempts to install all of the listed packages from the specified repositories.
list_files_with_exts	Returns the paths or names of the files in directory dir with extensions matching one of the elements of exts.
list_files_with_type	Returns the paths of the files in dir of the given "type," as determined by the extensions recognized by R.
md5sum	Computes the 32-byte MD5 checksums of one or more files.
package.dependencies	Parses and checks the dependencies of a package against the currently installed version of R (and other packages).
parse_Rd	Reads an Rd file and parses it, for processing by other functions. It is *experimental*.
pkgDepends	Convenience function that wraps getDepList and takes as input a package name.
pkgVignettes	Runs Sweave and texi2dvi on all vignettes of a package.
read.00Index	Reads item/description information from 00Index-style files.
readNEWS	Read R's NEWS file or a similarly formatted one. This is an experimental feature, new in R 2.4.0, and may change in several ways.
showNonASCII	Prints elements of a character vector that contain non-ASCII bytes, printing such bytes as an escape like <fc>.
testInstalledBasic	Allows an installed package to be tested by running the basic tests.
testInstalledPackage	Allows an installed package to be tested.
testInstalledPackages	Allows all base and recommended packages to be tested.
texi2dvi	Runs latex and bibtex until all cross-references are resolved and creates either a device independent (DVI) or a PDF file.

Function	Description
undoc	Finds the objects in a package that are undocumented, in the sense that they are visible to the user (or data objects or S4 classes provided by the package), but no documentation entry exists.
vignetteDepends	Given a vignette name, creates a DependsList object that reports information about the packages the vignette depends on.
write_PACKAGES	Generates *PACKAGES* and PACKAGES.gz files for a repository of source or Mac/Windows binary packages.
xgettext, xgettext2pot, xngettext	For each file in the *R* directory (including system-specific subdirectories) of a package, extract the unique arguments passed to stop, warning, message, gettext, and gettextf, or to ngettext.

Data Sets

Data Set	Class	Description
Adobe_glyphs	data.frame	A data frame that gives Adobe glyph names for Unicode points.
charset_to_Unicode	hexmode	A matrix of Unicode points with columns for the common 8-bit encodings.

utils

This package contains a variety of utility functions for R, including package management, file reading and writing, and editing.

Functions

Function	Description
?	Documentation on a topic.
RShowDoc	Utility function to find and display R documentation.
RSiteSearch	Searches for keywords or phrases in the R-help mailing list archives, help pages, vignettes, or task views, using the search engine at *http://search.r-project.org*, and displays the results in a web browser.
Rprof	Enables or disables profiling of the execution of R expressions.
Rprofmem	Enables or disables reporting of memory allocation in R.
Rtangle	A driver for Stangle that extracts R code chunks.
RtangleSetup	A driver for Stangle that extracts R code chunks.
RtangleWritedoc	These functions are handy for writing Sweave drivers and currently not documented. Look at the source code of the Sweave Latex driver (in this package) or the HTML driver (in the R2HTML package from CRAN) to see how they can be used.
RweaveChunkPrefix	These functions are handy for writing Sweave drivers and currently not documented. Look at the source code of the Sweave Latex driver (in this package) or the HTML driver (in the R2HTML package from CRAN) to see how they can be used.

Function	Description
RweaveEvalWithOpt	These functions are handy for writing Sweave drivers and currently not documented. Look at the source code of the Sweave Latex driver (in this package) or the HTML driver (in the R2HTML package from CRAN) to see how they can be used.
RweaveLatex	A driver for Sweave that translates R code chunks in LaTeX files.
RweaveLatexFinish	These functions are handy for writing Sweave drivers and currently not documented. Look at the source code of the Sweave Latex driver (in this package) or the HTML driver (in the R2HTML package from CRAN) to see how they can be used.
RweaveLatexOptions	These functions are handy for writing Sweave drivers and currently not documented. Look at the source code of the Sweave Latex driver (in this package) or the HTML driver (in the R2HTML package from CRAN) to see how they can be used.
RweaveLatexSetup	A driver for Sweave that translates R code chunks in LaTeX files.
RweaveLatexWritedoc	These functions are handy for writing Sweave drivers and currently not documented. Look at the source code of the Sweave Latex driver (in this package) or the HTML driver (in the R2HTML package from CRAN) to see how they can be used.
RweaveTryStop	These functions are handy for writing Sweave drivers and currently not documented. Look at the source code of the Sweave Latex driver (in this package) or the HTML driver (in the R2HTML package from CRAN) to see how they can be used.
Stangle	A frontend to Sweave using a simple driver by default, which discards the documentation and concatenates all code chunks the current S engine understands.
Sweave	Sweave provides a flexible framework for mixing text and S code for automatic report generation. The basic idea is to replace the S code with its output, such that the final document only contains the text and the output of the statistical analysis.
SweaveSyntConv	This function converts the syntax of files in Sweave format to another Sweave syntax definition.
URLdecode	Function to decode characters in URLs.
URLencode	Function to encode characters in URLs.
View	Invokes a spreadsheet-style data viewer on a matrix-like R object.
alarm	Gives an audible or visual signal to the user.
apropos	apropos() returns a character vector giving the names of all objects in the search list matching a specified value.
argsAnywhere	Returns the arguments for all functions with a name matching its argument, whether visible on the search path, registered as an S3 method, or in a namespace but not exported.
as.person	A class and utility method for holding information about persons such as name and email address.
as.personList	A class and utility method for holding information about persons such as name and email address.
as.relistable	relist() is an S3 generic function with a few methods in order to allow easy inversion of unlist(obj) when that is used with an object of (S3) class "relistable".
as.roman	Manipulates integers as roman numerals.
assignInNamespace	Utility function to access and replace the nonexported functions in a namespace, for use in developing packages with namespaces.

Function	Description
available.packages	Used to automatically compare the version numbers of installed packages with the newest available version on the repositories and update outdated packages on the fly.
browseEnv	Opens a browser with list of objects currently in the `sys.frame()` environment.
browseURL	Loads a given URL into a web browser.
browseVignettes	Lists available vignettes in an HTML browser with links to PDF, LaTeX/noweb source, and (tangled) R code (if available).
bug.report	Invokes an editor to write a bug report and optionally mail it to the automated r-bugs repository at *r-bugs@r-project.org*. Some standard information on the current version and configuration of R are included automatically.
capture.output	Evaluates its arguments with the output being returned as a character string or sent to a file. Related to `sink` in the same way that `with` is related to `attach`.
checkCRAN	Functions helping to maintain CRAN, some of which may also be useful to administrators of other repository networks.
chooseCRANmirror	Interacts with the user to choose a CRAN mirror.
citEntry	Creates "citation" objects, which are modeled after BibTeX entries.
citFooter	Creates a footer in a CITATION file.
citHeader	Creates a header in a CITATION file.
citation	Shows how to cite R and R packages in publications.
close.socket	Closes the socket and frees the space in the file descriptor table. The port may not be freed immediately.
combn	Generates all combinations of the elements of x taken m at a time. If x is a positive integer, returns all combinations of the elements of `seq(x)` taken m at a time. If argument FUN is not NULL, applies a function given by the argument to each point. If simplify is FALSE, returns a list; otherwise, returns an `array`, typically a `matrix`. `...` are passed unchanged to the FUN function, if specified.
compareVersion	Compares two package version numbers to see which is later.
contrib.url	Used to automatically compare the version numbers of installed packages with the newest available version on the repositories and update outdated packages on the fly.
count.fields	Counts the number of fields, as separated by sep, in each of the lines of file read.
data	Loads specified data sets or lists the available data sets.
data.entry, dataentry, de, de.ncols, de.restore, de.setup	Spreadsheet-like editors for entering or editing data.
debugger	Function to dump the evaluation environments (frames) and to examine dumped frames.
demo	User-friendly interface for running some demonstration R scripts. `demo()` gives the list of available topics.
download.file	Used to download a file from the Internet.
download.packages	Used to automatically compare the version numbers of installed packages with the newest available version on the repositories and update outdated packages on the fly.

Function	Description
dump.frames	Function to dump the evaluation environments (frames) and to examine dumped frames.
edit	Invokes an editor on an R object.
emacs	Invokes the text editor emacs on an R object.
example	Runs all the R code from the *Examples* part of R's online help.
file.edit	Edits one or more files in a text editor.
file_test	Utility for shell-style file tests.
find	Returns a character vector giving the names of all objects in the search list matching a given value.
fix	Invokes edit on x and assigns the new (edited) version of x in the user's workspace.
fixInNamespace	Utility function to access and replace the nonexported functions in a namespace, for use in developing packages with namespaces.
flush.console	On the Mac OS X and Windows GUIs, ensures that the display of output in the console is current, even if output buffering is on. (This does nothing except on console-based versions of R.)
formatOL, formatUL	Format unordered (itemize) and ordered (enumerate) lists.
getAnywhere	Locates and returns all objects with a name matching its argument, whether visible on the search path, registered as an S3 method, or in a namespace but not exported.
getCRANmirrors	Interacts with the user to choose a CRAN mirror.
getFromNamespace	Utility function to access and replace the nonexported functions in a namespace, for use in developing packages with namespaces.
getS3method	Gets a method for an S3 generic, possibly from a namespace.
getTxtProgressBar	Text progress bar in the R console.
glob2rx	Changes *wildcard* (aka *globbing*) patterns into the corresponding regular expressions (regexp).
head	Returns the first or last parts of a vector, matrix, table, data frame, or function. Since head() and tail() are generic functions, they may also have been extended to other classes.
help	The primary interface to R's help system.
help.request	Prompts users to check they have done all that is expected of them before sending a post to the R-help mailing list, provides a template for the post with session information included, and optionally sends the email (on Unix systems).
help.search	Allows for searching the help system for documentation matching a given character string in the (file) name, alias, title, concept, or keyword entries (or any combination thereof), using either fuzzy matching or regular expression matching. Names and titles of the matched help entries are displayed nicely formatted.
help.start	Starts the hypertext (currently HTML) version of R's online documentation.
history	Loads or saves or displays the commands history.
index.search	Used to search the indexes for help files, possibly under aliases.

Function	Description
install.packages	Used to automatically compare version numbers of installed packages with the newest available version on the repositories and update outdated packages on the fly.
installed.packages	Finds (or retrieves) details of all packages installed in the specified libraries.
is.relistable	relist() is an S3 generic function with a few methods in order to allow easy inversion of unlist(obj) when that is used with an object of (S3) class "relistable".
limitedLabels	Allows the user to browse directly on any of the currently active function calls and is suitable as an error option. The expression options(error=recover) will make this the error option.
loadhistory	Loads or saves or displays the commands history.
localeToCharset	Aims to find a suitable coding for the locale named, by default the current locale, and if it is a UTF-8 locale, a suitable single-byte encoding.
ls.str, lsf.str	ls.str and lsf.str are variations of ls applying str() to each matched name.
make.packages.html	Updates HTML documentation files.
make.socket	With server = FALSE, attempts to open a client socket to the specified port and host. With server = TRUE, listens on the specified port for a connection and then returns a server socket. It is a good idea to use on.exit to ensure that a socket is closed, as you only get 64 of them.
makeRweaveLatexCodeRunner	These functions are handy for writing Sweave drivers and currently not documented. Look at the source code of the Sweave Latex driver (in this package) or the HTML driver (in the R2HTML package from CRAN) to see how they can be used.
memory.limit	Gets or sets the memory limit on Microsoft Windows platforms.
memory.size	Checks the current memory usage on Microsoft Windows platforms.
menu	Presents the user with a menu of choices labeled from 1 to the number of choices. To exit without choosing an item, select 0.
methods	Lists all available methods for an S3 generic function or all methods for a class.
mirror2html	Functions helping to maintain CRAN, some of which may also be useful to administrators of other repository networks.
modifyList	Modifies a possibly nested list recursively by changing a subset of elements at each level to match a second list.
new.packages	Used to automatically compare the version numbers of installed packages with the newest available version on the repositories and update outdated packages on the fly.
normalizePath	Converts file paths to canonical form for the platform, to display them in a user-understandable form.
nsl	Interface to gethostbyname.
object.size	Provides an estimate of the memory that is being used to store an R object.
old.packages	Used to automatically compare the version numbers of installed packages with the newest available version on the repositories and update outdated packages on the fly.

Function	Description
package.skeleton	Automates some of the setup for a new source package. It creates directories; saves functions, data, and R code files to appropriate places; and creates skeleton help files and a *Read-and-delete-me* file describing further steps in packaging.
packageDescription	Parses and returns the *DESCRIPTION* file of a package.
packageStatus	Summarizes information about installed packages and packages available at various repositories, and automatically upgrades outdated packages.
page	Displays a representation of the object named by x in a pager via `file.show`.
person	Creates a "person" object.
personList	Creates a "personList" object.
pico	Invokes a text editor on an R object.
prompt	Facilitates the construction of files documenting R objects.
promptData	Generates a shell of documentation for a data set.
promptPackage	Generates a shell of documentation for an installed or source package.
read.DIF	Reads a file in Data Interchange Format (DIF) and creates a data frame from it. DIF is a format for data matrices such as single spreadsheets.
read.csv, read.csv2, read.delim, read.delim2	Read a file in table format and create a data frame from it, with cases corresponding to lines and variables to fields in the file.
read.fortran	Reads fixed-format data files using FORTRAN-style format specifications.
read.fwf	Reads a table of fixed-width-formatted data into a `data.frame`.
read.socket	`read.socket` reads a string from the specified socket; `write.socket` writes to the specified socket. There is very little error checking done by either.
read.table	Reads a file in table format and creates a data frame from it, with cases corresponding to lines and variables to fields in the file.
readCitationFile	The *CITATION* file of R packages contains an annotated list of references that should be used for citing the packages.
recover	Allows the user to browse directly on any of the currently active function calls and is suitable as an error option. The expression `options(error=recover)` will make this the error option.
relist	`relist()` is an S3 generic function with a few methods in order to allow easy inversion of `unlist(obj)` when that is used with an object of (S3) class `"relistable"`.
remove.packages	Removes installed packages/bundles and updates index information as necessary.
rtags	Provides etags-like indexing capabilities for R code, using R's own parser.
savehistory	Loads or saves or displays the commands history.
select.list	Selects item(s) from a character vector.
sessionInfo	Prints version information about R and attached or loaded packages.
setRepositories	Interacts with the user to choose the package repositories to be used.
setTxtProgressBar	Text progress bar in the R console.
stack	Stacking vectors concatenates multiple vectors into a single vector along with a factor indicating where each observation originated; unstacking reverses this.

Function	Description
str	Compactly displays the internal structure of an R object; the idea is to give reasonable output for *any* R object.
strOptions	`strOptions()` is a convenience function for setting `options(str = .)`.
summaryRprof	Summarizes the output of the `Rprof` function to show the amount of time used by different R functions.
tail	Returns the first or last parts of a vector, matrix, table, data frame, or function. Since `head()` and `tail()` are generic functions, they may also have been extended to other classes.
timestamp	Loads or saves or displays the commands history.
toBibtex	Converts R objects to character vectors with BibTeX markup.
toLatex	Converts R objects to character vectors with LaTeX markup.
txtProgressBar	Text progress bar in the R console.
type.convert	Converts a character vector to logical, integer, numeric, complex, or factor, as appropriate.
unstack	Stacking vectors concatenates multiple vectors into a single vector along with a factor indicating where each observation originated; unstacking reverses this.
unzip	Extracts files from or lists a zip archive.
update.packageStatus	Summarizes information about installed packages and packages available at various repositories and automatically upgrades outdated packages.
update.packages	Used to automatically compare the version numbers of installed packages with the newest available version on the repositories and update outdated packages on the fly.
upgrade	Summarizes information about installed packages and packages available at various repositories and automatically upgrades outdated packages.
url.show	Extension of `file.show` to display text files from a remote server.
vi	Invokes a text editor on an R object.
vignette	Views a specified vignette or lists the available ones.
write.csv, write.csv2	Convenience wrappers to `write.table` for producing CSV files from an R object.
write.socket	`read.socket` reads a string from the specified socket; `write.socket` writes to the specified socket. There is very little error checking done by either.
write.table	Prints its required argument x (after converting it to a data frame if it is not one nor a matrix) to a file or connection.
wsbrowser	The `browseEnv` function opens a browser with list of objects currently in the `sys.frame()` environment.
xedit	Invokes the xedit editor on an R object.
xemacs	Invokes the xemacs editor on an R object.
zip.file.extract	Extracts the file named `file` from the zip archive, if possible, and writes it in a temporary location.

Bibliography

[Adler2006] Joseph Adler, *Baseball Hacks: Tips and Tools for Analyzing and Winning with Statistics*, O'Reilly Media, 9780596009427

[Albert2009] Jim Albert, *Bayesian Computation with R*, Springer, 9780387922973

[Bivand2008] Roger Bivand, *Applied Spatial Data Analysis with R*, Springer, 9780387781709

[Chambers2008] John Chambers, *Software for Data Analysis: Programming with R*, Springer, 9780387759357

[Chatfield2003] Chris Chatfield, *The Analysis of Time Series*, Sixth Edition, Chapman and Hall/CRC, 9781584883173

[Cleveland1993] William Cleveland, *Visualizing Data*, Hobart Press, 9780963488404

[Dalgaard2008] Peter Dalgaard, *Introductory Statistics with R*, Second Edition, Springer, 9780387790534

[Dobson2008] Anette Dobson and Adrian Barnett, *An Introduction to Generalized Linear Models*, Third Edition, Chapman and Hall/CRC, 9781584889502

[Drake1967] Alvin Drake, *Fundamentals of Applied Probability Theory*, McGraw-Hill, 070178151

[Ewens2005] Warren Ewens and Gregory Grant, *Statistical Methods in Bioinformatics: An Introduction*, Second Edition, Springer, 9780387400822

[Foulkes2009] Andrea Foulkes, *Applied Statistical Genetics with R: For Population-Based Association Studies*, Springer, 9780387895536

[Gelman2003] Andrew Gelman, John Carlin, Hal Stern, and Donald Rubin, *Bayesian Data Analysis*, Second Edition, Chapman and Hall/CRC, 9781584883883

[Gentleman2005] Robert Gentleman, Vincent Carey, Wolfgang Huber, Rafael Irizarry, and Sandrine Dudoit, *Bioinformatics and Computational Biology Solutions Using R and Bioconductor*, Springer, 9780387251462

[Greene2007] William Greene, *Econometric Analysis*, Sixth Edition, Prentice Hall, 9780135132456

[Hahne2008] Florian Hahne, Wolfgang Huber, Robert Gentleman, and Seth Falcon, *Bioconductor Case Studies*, Springer, 9780387772394

[Hastie2009] Trevor Hastie, Robert Tibshirani, and Jerome Friedman, *The Elements of Statistical Learning: Data Mining, Inference, and Prediction*, Second Edition, Springer, 9780387848570

[RADMIN2009] The R Development Core Team, *R Installation and Administration* Version 2.9.2 (2009-08-24), 3900051097, *http://cran.r-project.org/doc/manuals/R-admin.html*

[RDDATA2009] The R Development Core Team, *R Data Import/Export* Version 2.9.2 (2009-08-24), 3900051100, *http://cran.r-project.org/doc/manuals/R-data.html*

[REXT2009] The R Development Core Team, *Writing R Extensions* Version 2.9.2 (2009-08-24), 3900051119, *http://cran.r-project.org/doc/manuals/R-exts.html*

[RInternals2009] The R Development Core Team, *R Internals* Version 2.9.2 (2009-08-24), 3900051143, *http://cran.r-project.org/doc/manuals/R-ints.html*

[RIntro2009] W. Venables and D. Smith, The R Development Core Team, *An Introduction to R* Version 2.9.2 (2009-08-24), 3900051127, *http://cran.r-project.org/doc/manuals/R-intro.html*

[RLang2009] The R Development Core Team, *R Language Definition* Version 2.9.2 (2009-08-24), 3900051135, *http://cran.r-project.org/doc/manuals/R-lang.html*

[Sarkar2008] Deepayan Sarkar, *Lattice: Multivariate Data Visualization with R*, Springer, 9780387759685

[Venables2002] W. Venables and B. Ripley, *Modern Applied Statistics with S*, Fourth Edition, Springer, 9780387954578

Index

We'd like to hear your suggestions for improving our indexes. Send email to *index@oreilly.com*.

redefining, 64
summarizing, 194–205
Fundamentals of Queueing Theory, 136

G

gam(), 430
gamboost(), 415
Gamma function family, 394
garbage collection and gc(), 137, 138
Gauss-Markov Theorem, 384
gaussian function family, 394
generalized additive models
    gam(), 430
    gamboost(), 415
generalized linear models (see GLMs
        (generalized linear models))
generic functions, 25, 54, 113, 121, 128,
        132
GEO accession numbers, 471
    getGEO(), 474
get(), 98
getAnywhere(), 133
getMethod(), 129
getMKLthreads(), 140
getOption(), 36
gets (assignment) operator (<-), 20
getS3method(), 133
getSlots(), 127
getSYMBOL(), 480
glmboost(), 415
GLMs (generalized linear models), 392
    function families, 394
    glm(), 393, 394, 419, 436
    glmboost(), 415
global environment, 99
globalenv(), 99
GNU General Public Licence, xvi
Gnu readline library, 13
Google
    R Style guide, 78
    searching for packages using, 40
graphical parameter by name, 250–254
graphical user interfaces (GUIs), 7–10, 38
graphics devices, 243
graphics, lattice, 536–541
    arguments to specify data, 264
    customizing, 308–318
    density plots, 281–282

functions compared to standard
        graphics, 266
high-level plotting functions, 268–307
histograms, 278
lattice(), 264
lattice.getOption(), 276
overview, 31, 264
printing, 26
residual and fit-spread plots, 306
strip plots, 282
trellis.device(), 274
trellis.par.get(), 311
trellis.skeleton, 309
trivariate trellis plots, 301–307
univariate quantile-quantile plots,
        284–293
univariate trellis plots, 269
graphics, standard, 28–32
    annotation of, 245
    categorical data and variables, 224–
        229
    conditional density plot, 224
    customization of, 244–254
    customizing, 244–254
    distributions, 237–240
    formats supported, 243
    functions, 254, 318
    list of parameters, 250–254
    mosaicplot(), 227–229, 254
    package, 211, 532–535
    package for, 211
    pie charts and pie(), 223, 254
    plot(), 239, 254, 335, 381
    plot.default(), 212
    plot.earth(), 422
    plot.rpart(), 410
    plot.trellis(), 315
    plotcp(), 411
    plotmo(), 422
    princomp(), 331
    setting parameters, 244
    spineplot(), 229, 254
    text properties for, 248
    time series, 218
grDevices, 528–532
grid package, 535
grouping variables, 191, 263, 265
gzfile(), 92

L

labelling plot points, 215
labelling plots, 251
lambda values, 391
language object, 57
language object type, 81
lapply(), 188
lars(), 390
Lasso, 390
layout
    layout(), 248
LazyData and LazyLoad, 44
LDAs (linear discriminant analysis) and
        lda(), 440
least angle regression and lars(), 390
least median squares (LMS), 386
least trimmed squared (LTS), 386
least-squares regression, 383, 384
legends
    adding to bar charts, 220
    legend(), 258
length attribute, 83
Leopard, 5
levelplot(), 301
levels, 85
levels attribute, 93
lexical versus dynamic scope, 100
library(), 36, 38
licensing terms, xvi
limits of precision, 62
line properties, 249, 251
line segments and segments(), 258
line wrapping, 12
linear classification models, 435–445
linear discriminant analysis (LDAs), 440
linear models, 373–382
linear predictor, 392
linear regression, 26, 373–382
lines(), 255
Linux
    command-line interface, 7, 13
    data editor, 150
    launching R from, 8
    loading/installing packages in, 38
    R installation on, 3, 5
LISP, 61
lists, 73

applying function to, 188
as arguments, 82
converting objects to, 57, 109
defined, 23, 83
object type, 80
lm(), 26, 375, 382, 387, 395, 419
lm.ridge(), 390
lmRob(), 387
LMS, 386
load(), 151
loading objects, 151
loadings(), 330
local(), 102
locator(), 213
locfit(), 405
locpoly(), 404
loess(), 403
log linear models, 444
    loglin(), 444
    loglm(), 445
logical object type, 80
logistic regression, 435–440
logit(), 436
loglin(), 444
loglm(), 445
long data, 204
lookup tables, 136
looping extensions, 71
loops, 70–72
lqs(), 386, 387
lssvm(), 432
LTS, 386

M

Mac OS X
    Apple development tools, 165
    building R on, 144
    command line interface, 13
    data editor, 148
    graphics devices on, 243
    history() command error in, 13
    installing SQLiteodbc driver, 165
    loading/installing packages on, 38, 41
    R GUI, 7
    R installation, 5
    running R from terminal, 8
machine learning algorithms, 405–417
    for classification, 445–452

for data mining, 453–462
make.groups(), 191, 374
mantelhaen.test(), 367
mapply(), 189
maptree library, 447
margins
    add text to, 261
    in charts, 245
market basket analysis, 453–458
MARS (multivariate adaptive regression
        splines), 418
MASS packages, 389, 541–544
math functions, built-in, 135
matlines(), 255
matplot(), 216, 254
matrices and matrix(), 22, 84, 186
    and data frames, 200
    perspective, 262
    variance-covariance, 380
max(), 323
McNemar's chi-squared test and
        mcnemar.test(), 367
MDA (mixture discriminant analysis) and
        mda(), 444
mean(), 323
means, comparing, 344–348, 357–360
meanSdPlot(), 473
memory, 137–139
    mem.limits(), 138
    memory.limit(), 138
    memory.profile(), 137
    memory.size(), 138
merge(), 184
message(), 105
methods, 25, 67, 118, 128–130, 550
mgcv package, 431, 557
MGED Society, 488
MIAME, 488
min(), 323
Minimum Information About a
        Microarray Experiment
        (MIAME), 488
mixture discriminant analysis (MDA) and
        mda(), 444
model
    refining, 382
model.tables(), 351
modelling functions, 101

models, 26, 28
    getting information, 376
    specifying, 376
    viewing, 376
modulo operator (%%), 493
modulus operator (%%), 64
mona(), 462
monitoring memory, 137–139
Mood's two-sample test and mood.test(),
        360
mosaicplot(), 227–229, 254
mtext(), 261
multicollinearity, 389
multidimensional data, 74, 84
multinomial log-linear models and
        multinom(), 438
multiple charts, plotting, 245
multiple data types, 23
multiple inheritance, 119
multiplication operator (*), 64
multivariate adaptive regression splines
        (MARS), 418
mvr(), 392
MySQL Query Browser, 156

N

NA values, 53, 75, 83, 323
na.omit(), 419
named components, 23
names attribute, 93
naming
    conventions for objects, 78
    list elements, 83
    of arguments, 113
    probability distribution type
            arguments, 338
    variables, 88
NaN values, 54, 62
National Center for Biotechnology
        Information (NCBI), 470
NATURAL JOIN, 184
NCBI, 470
ncv.test(), 385
negation (not) operator (!), 493
negation operator (-), 65
negative integers, 74
Nelder, John, 392
network server, running Rserve as, 15

polymorphism, 118
polynomial and poly(), 376
polynomial surfaces, 403
popping off stack, 100
POSIXct, 91
POSIXlt, 91
pound sign (#) comment, 19
power.anova.test(), 372
power.prop.test(), 371
power.t.test(), 370
PowerPC
    R installation, 5
ppr() (projection pursuit regression), 427
prcomp(), 328
preallocating memory, 137
precedence of operators, 66
precompiled binaries, 6
predicting values and predict(), 378, 466
predictors, 373
PRIM, 414
prim(), 414
primitive object types, 79–82
principal components analysis, 328–331
    prc(), 391
    princomp(), 328, 329, 331
printing
    factors compared to character vectors,
        85
    for time series, 90
    from R console, 56
    loop results not printed, 71
    print(), 26, 71
    print.ts(), 90
    printcp(), 447
probability distribution, 335
proj(), 353
projection pursuit regression and ppr(),
    427
promise object, 96, 98
promise object type, 81
prompt (command prompt), 12
prompt (documentation) functions, 44
properties of functions, 111
proportion test
    power.prop.test(), 371
    prop.test(), 360, 371
proportional odds linear regression and
    polr(), 440

prune(), 411
public package repositories, 35
pushing onto stack, 100

Q

qc() (quality control for affymetrix data),
    472
quadratic discriminant analysis (QDA)
    and qda(), 441
quantile plot
    qnorm(), 336
    quantile(), 324
quantile-quantile (Q-Q) plot
    qq(), 302
    qqmath(), 284
    qqnorm(), 240, 255
    qqplot(), 240, 255
quasi function family, 394
quasibinomial function family, 394
quasipoisson function family, 394
quote(), 57, 96

R

R
    basic operations, 17
        (see also packages)
    building from source, 4, 141–144
    code style standards, 77
    combining with other applications, 15
    console, 11–13, 41, 55, 147
    data editor, 151
    downloading, 3
    for 64-bit Windows, 5
    graphical user interfaces (GUIs), 7–10,
        148
    interactive mode, 12
    interpreter, 11, 55
    language objects, 79, 81, 130
    reserved words in, 64
    running in Excel, 14
    seeing how it works, 57–59
    strengths and weaknesses, xvi
    upgrading, 5
    versions, 3
R Foundation, 39
R Tools, 141–143
R-Forge packages, 40

variances of two populations, comparing, 349
vectors, 82–83
  and arrays, 22
  applying functions to, 188
  assembling, 83
  basic, 79
  built-in types, 130
  defined, 17
  expand length of, 83
  numeric, 61
  primitive object types in, 80
Venables, W. N., 413
vignette(), 34, 490
virtual classes, 126
Vista install issues, 4
Visualizing Data (Cleveland), 536
vsn(), 472
vsn2(), 472
vsnrma(), 472

W

warning(), 105
warnings (see errors and warnings)
weakref object type, 82
web applications, incorporating R into, 15
web, browsing for packages on, 40
websites (see URLs)
Wedderburn, Robert, 392
Welch Two Sample t-test, 346
while loops, 70
whiskers, 240, 296
wide data, 204
Wilcoxon test and wilcox.test(), 357
Windows, Microsoft
  building R in, 141–143
  command-line editing, 13
  data editor, 148
  graphics devices on, 243
  installing R on, 4
  loading/installing packages on, 38
  location of install files, 8
  R GUI in, 7
  SQLiteODBC, 167
  Vista install issues, 4
wire-frame plots and wireframe(), 306
with() and within(), 103

write. functions
  chart of, 161
  write.csv() and write.csv2(), 161
  write.table(), 161
Writing R Extensions manual, 45

X

X Windows (see Linux)
xpred.rpart(), 405
xtabs(), 199
xyplot(), 264, 293, 374

Y

Yates' continuity correction, 366
Yum (Yellow Dog Updater, Modified), 5

Z

zero, 89

About the Author

Joseph Adler has many years of experience in data mining and data analysis at various companies, including DoubleClick, American Express, and VeriSign. He graduated from MIT with an Sc.B. and M.Eng. in computer science and electrical engineering. He is the inventor on several patents for computer security and cryptography, and is the author of *Baseball Hacks* (O'Reilly).

Colophon

The animal on the cover of *R in a Nutshell* is a harpy eagle (*Harpia harpyja*). Black feathers line the top half of the bird, while white feathers mostly make up the balance, although the underside of its wings may be striped black-and-white. Unlike other species of birds, male and female harpy eagles appear virtually identical.

These eagles—the most powerful, carnivorous raptors in the Americas—typically inhabit tropical rain forests. They prey upon animals that live in trees: sloths, monkeys, opossums, and even other birds, such as macaws.

The eagle is named after the harpies of ancient Greek mythology, female wind spirits who were said to be human from the chest to their ankles and eagle from the neck up. Mythological harpies tormented people as they carried them to the underworld with their clawed feet; perhaps similarly, harpy eagles' talons violently pierce and subdue their prey before the eagles carry them back to their nests.

Harpy eagles also inspire modern-day life: the eagle is the national bird of Panama and is pictured on the country's coat of arms. The bird also inspired the design of Fawkes the Phoenix in the Harry Potter film series.

The cover image is from *Cassell's Natural History*. The cover font is Adobe ITC Garamond. The text font is Linotype Birka; the heading font is Adobe Myriad Condensed; and the code font is LucasFont's TheSansMonoCondensed.

Get even more for your money.

Join the O'Reilly Community, and register the O'Reilly books you own. It's free, and you'll get:

- 40% upgrade offer on O'Reilly books
- Membership discounts on books and events
- Free lifetime updates to electronic formats of books
- Multiple ebook formats, DRM FREE
- Participation in the O'Reilly community
- Newsletters
- Account management
- 100% Satisfaction Guarantee

Signing up is easy:

1. **Go to: oreilly.com/go/register**
2. **Create an O'Reilly login.**
3. **Provide your address.**
4. **Register your books.**

Note: English-language books only

To order books online:

oreilly.com/order_new

For questions about products or an order:

orders@oreilly.com

To sign up to get topic-specific email announcements and/or news about upcoming books, conferences, special offers, and new technologies:

elists@oreilly.com

For technical questions about book content:

booktech@oreilly.com

To submit new book proposals to our editors:

proposals@oreilly.com

Many O'Reilly books are available in PDF and several ebook formats. For more information:

oreilly.com/ebooks

O'REILLY®

Spreading the knowledge of innovators   www.oreilly.com

Buy this book and get access to the online edition for 45 days—for free!

R in a Nutshell

By Joseph Adler
December 2009, $49.99
ISBN 9780596801700

With Safari Books Online, you can:

Access the contents of thousands of technology and business books

- Quickly search over 7000 books and certification guides
- Download whole books or chapters in PDF format, at no extra cost, to print or read on the go
- Copy and paste code
- Save up to 35% on O'Reilly print books
- **New!** Access mobile-friendly books directly from cell phones and mobile devices

Stay up-to-date on emerging topics before the books are published

- Get on-demand access to evolving manuscripts.
- Interact directly with authors of upcoming books

Explore thousands of hours of video on technology and design topics

- Learn from expert video tutorials
- Watch and replay recorded conference sessions

To try out Safari and the online edition of this book FREE for 45 days,
go to **www.oreilly.com/go/safarienabled** and enter the coupon code SALXAZG.
To see the complete Safari Library, visit safari.oreilly.com.

Spreading the knowledge of innovators   safari.oreilly.com